FOURTH EDITION

Introduction to Random Signals and Applied Kalman Filtering

WITH MATLAB EXERCISES

FOURTH EDITION

Introduction to Random Signals and Applied Kalman Filtering

WITH MATLAB EXERCISES

Robert Grover Brown
Professor Emeritus
Iowa State University

Patrick Y. C. Hwang
Rockwell Collins, Inc.

John Wiley & Sons, Inc.

VP and Publisher	Don Fowley
Associate Publisher	Dan Sayre
Editorial Assistant	Charlotte Cerf
Senior Marketing Manager	Christopher Ruel
Senior Production Manager	Janis Soo
Senior Production Editor	Joyce Poh
Designer	James O'Shea

This book was set in 10/13 Photina Regular by Thomson Digital. Printed and bound by Courier Westford. The cover was printed by Courier Westford.

This book is printed on acid free paper.

Library of Congress Cataloging-in-Publication Data

Brown, Robert Grover.
  Introduction to random signals and applied Kalman filtering : with MATLAB exercises / Robert Grover Brown, Patrick Y.C. Hwang.—4th ed.
    p. cm.
  Includes index.
  ISBN 978-0-470-60969-9 (hardback)
  1. Signal processing—Data processing. 2. Random noise theory. 3. Kalman filtering—Data processing. 4. MATLAB. I. Hwang, Patrick Y. C. II. Title.
  TK5102.9.B75 2012
  621.382'2—dc23   2011042847

Printed in the United States of America
10 9 8 7 6 5 4 3 2 1

Preface To The Fourth Edition

We are happy to report that Kalman filtering is alive and well after 50 years of existence. New applications keep appearing regularly which broadens the interest throughout engineering. This, of course, has all been made possible by the fantastic advances in computer technology over the past few decades. If it were not for that, the neat recursive solution for the Wiener filter problem that R. E. Kalman introduced in 1960 would probably only be remembered as an interesting academic curiosity, in engineering literature at least. Enough of the history; clearly, Kalman filtering is here to stay. It is eminently practical, and it has withstood the test of time.

This text is a revision of the Third Edition of *Introduction to Random Signals and Applied Kalman Filtering with MATLAB Exercises*. Kalman filtering has now reached a stage of maturity where a variety of extensions and variations on the basic theory have been introduced since the Third Edition was published in 1997. We have included some of these developments, especially those that deal with nonlinear systems. The extended Kalman filter is included because it is used widely and it is still the preferred solution to many integrated navigation systems today, just as it was a few decades ago.

Our intent is still to keep the book at the introductory level with emphasis on applications. We feel that this material is suitable for undergraduate as well as beginning graduate courses. The necessary prerequisite material for this book will be found in a number of books on linear systems analysis or linear control systems. We think that we have improved the organization of the text material in this edition by making a distinct separation of the random signals background material (Part 1) and the main subject of applied Kalman filtering (Part 2). Students who have already had a course in random processes and response of linear systems to random inputs may want to move directly to Part 2 and simply use Part 1 as reference material.

Part 1 (Chapters 1 through 3) includes the essential notions of probability, an introduction to random signals and response to linear systems, state-space modeling, and Monte Carlo simulation. We have found from teaching both university-credit and continuing-education courses that the main impediment to learning about Kalman filtering is not the mathematics. Rather, it is the background material in random processes that usually causes the difficulty. Thus, the background material in Part 1 is most important.

Part 2 (Chapters 4 through 9) contains the main theme of the book, namely applied Kalman filtering. This part begins with the basic filter derivation using the minimum-mean-square-error approach. This is followed by various embellishments on the basic theory such as the information filter, suboptimal analysis, conditional density viewpoint, Bayesian estimation, relationship to least squares and other estimators, smoothing, and other methods of coping with nonlinearities. Chapter 8 is a completely new chapter that expands on the complementary filter material in the Third Edition. Chapter 9 deals entirely with applications, mainly those in the engineering field of positioning, navigation, and timing (PNT).

The authors wish to thank all of our former students and colleagues for their many helpful comments and suggestions over the years. Special thanks goes to the late Larry Levy. His generous counsel and advice on many aspects of Kalman filtering will certainly be missed.

Robert Grover Brown
Patrick Y. C. Hwang

Brief Contents

Contents

2 Mathematical Description of Random Signals 57

3 Linear Systems Response, State-Space Modeling, and Monte Carlo Simulation 105

PART 2 KALMAN FILTERING AND APPLICATIONS 139

4 Discrete Kalman Filter Basics 141

PART 1

Random Signals Background

1

Probability and Random Variables: A Review

1.1 RANDOM SIGNALS

Nearly everyone has some notion of random or noiselike signals. One has only to tune a radio away from a station, turn up the volume, and the result is static, or noise. If one were to look at a recording of such a signal, it would appear to wander on aimlessly with no apparent order in its amplitude pattern, as shown in Fig. 1.1. Signals of this type cannot be described with explicit mathematical functions such as sine waves, step functions, and the like. Their description must be put in probabilistic terms. Early investigators recognized that random signals could be described loosely in terms of their spectral content, but a rigorous mathematical description of such signals was not formulated until the 1940s, most notably with the work of Wiener and Rice (1, 2).

Probability plays a key role in the description of noiselike or random signals. It is especially important in Kalman filtering, which is also sometimes referred to as statistical filtering. This is a bit of a misnomer though, because Kalman filtering is based on probabilistic descriptors of the signals and noise, and these are assumed to be known at the beginning of the filtering problem. Recall that in probability we assume that we have some a priori knowledge about the likelihood of certain elemental random outcomes. Then we wish to predict the theoretical relative frequency of occurrence of combinations of these outcomes. In statistics we do just the reverse. We make observations of random outcomes. Then, based on these observations, we seek mathematical models that faithfully represent what we have observed. In Kalman filtering it may well be that some statistical methods were used in arriving at the necessary probabilistic descriptors. But, nevertheless, once the descriptors are determined (or assumed) it is all probability the rest of the way.

Our treatment of probability must necessarily be brief and directed toward the specific needs of subsequent chapters. We make no apology for this, because many fine books have been written on probability in a rigorous sense. Our main objective here is the study of random signal analysis and Kalman filtering, and we wish to move on to this as quickly as possible. First, though, we must at least review the bare essentials of probability with special emphasis on random variables.

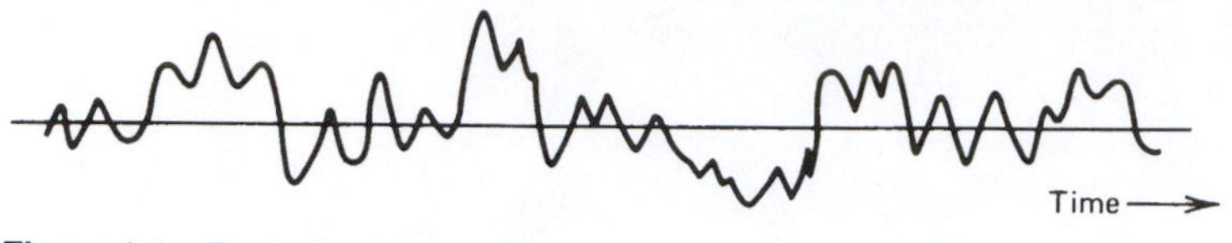

Figure 1.1 Typical noise waveform.

1.2 INTUITIVE NOTION OF PROBABILITY

Most engineering and science students have had some acquaintance with the intuitive concepts of probability. Typically, with the intuitive approach we first consider all possible outcomes of a chance experiment as being equally likely, and then the probability of a particular combination of outcomes, say, event A, is defined as

$$P(A) = \frac{\text{Possible outcomes favoring event } A}{\text{Total possible outcomes}} \tag{1.2.1}$$

where we read $P(A)$ as "probability of event A." This concept is then expanded to include the relative-frequency-of-occurrence or statistical viewpoint of probability. With the relative-frequency concept, we imagine a large number of trials of some chance experiment and then define probability as the relative frequency of occurrence of the event in question. Considerations such as what is meant by "large" and the existence of limits are normally avoided in elementary treatments. This is for good reason. The idea of limit in a probabilistic sense is subtle.

Although the older intuitive notions of probability have limitations, they still play an important role in probability theory. The ratio-of-possible-events concept is a useful problem-solving tool in many instances. The relative-frequency concept is especially helpful in visualizing the statistical significance of the results of probability calculations. That is, it provides the necessary tie between the theory and the physical situation. An example that illustrates the usefulness of these intuitive notions of probability should now prove useful.

EXAMPLE 1.1

In straight poker, each player is dealt 5 cards face down from a deck of 52 playing cards. We pose two questions:

(a) What is the probability of being dealt four of a kind, that is, four aces, four kings, and so forth?
(b) What is the probability of being dealt a straight flush, that is, a continuous sequence of five cards in any suit?

Solution to Question (a) This problem is relatively complicated if you think in terms of the sequence of chance events that can take place when the cards are dealt one at a time. Yet the problem is relatively easy when viewed in terms of the ratio of favorable to total number of outcomes. These are easily counted in this case. There are only 48 possible hands containing 4 aces; another 48 containing 4 kings; etc. Thus, there are $13 \cdot 48$ possible four-of-a-kind hands. The total number of possible poker hands of any kind is obtained from the combination formula for "52 things taken 5 at a time" (3). This is given by the binomial coefficient

$$\binom{52}{5} = \frac{52!}{5!(52-5)!} = \frac{52 \cdot 51 \cdot 50 \cdot 49 \cdot 48}{5 \cdot 4 \cdot 3 \cdot 2 \cdot 1} = 2,598,960 \tag{1.2.2}$$

Therefore, the probability of being dealt four of a kind is

$$P(\text{four of a kind}) = \frac{13 \cdot 48}{2,598,960} = \frac{624}{2,598,960} \approx .00024 \tag{1.2.3}$$

Solution to Question (b) Again, the direct itemization of favorable events is the simplest approach. The possible sequences in each of four suits are: AKQJ10, KQJ109, . . . , 5432A. (*Note:* We allow the ace to be counted either high or low.) Thus, there are 10 possible straight flushes in each suit (including the royal flush of the suit) giving a total of 40 possible straight flushes. The probability of a straight flush is, then,

$$P(\text{Straight flush}) = \frac{40}{2,598,960} \approx .000015 \tag{1.2.4}$$

We note in passing that in poker a straight flush wins over four of a kind; and, rightly so, since it is the rarer of the two hands.

■

1.3 AXIOMATIC PROBABILITY

It should be apparent that the intuitive concepts of probability have their limitations. The ratio-of-outcomes approach requires the equal-likelihood assumption for all outcomes. This may fit many situations, but often we wish to consider "unfair" chance situations as well as "fair" ones. Also, there are situations where all possible outcomes simply cannot be enumerated. The relative-frequency approach is intuitive by its very nature. Intuition should never be ignored; but, on the other hand, it can lead one astray in complex situations. For these reasons, the axiomatic formulation of probability theory is now almost universally favored among both applied and theoretical scholars in this area. As we would expect, axiomatic probability is compatible with the older, more heuristic probability theory.

Axiomatic probability begins with the concept of a *sample space*. We first imagine a conceptual chance experiment. The sample space is the set of all possible *outcomes* of this experiment. The individual outcomes are called *elements* or *points* in the sample space, We denote the sample space as S and its set of elements as $\{s_1, s_2, s_3, \ldots\}$. The number of points in the sample space may be finite, countably infinite, or simply infinite, depending on the experiment under consideration.

It should be noted that elements of a sample space must always be *mutually exclusive* or *disjoint*. On a given trial, the occurrence of one excludes the occurrence of another. There is no overlap of points in a sample space.

In axiomatic probability, the term event has special meaning and should not be used interchangeably with outcome. An *event* is a special subset of the sample space S. We usually wish to consider various events defined on a sample space, and they will be denoted with uppercase letters such as A, B, C, . . . , or perhaps A_1, A_2, . . . , etc. Also, we will have occasion to consider the set of operations of union, intersection, and complement of our defined events. Thus, we must be careful in our definition of events to make the set sufficiently complete such that these set operations also yield properly defined events. In discrete problems, this can always be done by defining the set of events under consideration to be all possible subsets of the sample space S. We will tacitly assume that the null set is a subset of every set, and that every set is a subset of itself.

One other comment about events is in order before proceeding to the basic axioms of probability. The event A is said to occur if *any* point in A occurs.

The three axioms of probability may now be stated. Let S be the sample space and A be any event defined on the sample space. The first two axioms are

$$\textit{Axiom 1:}\quad P(A) \geq 0 \tag{1.3.1}$$

$$\textit{Axiom 2:}\quad P(S) = 1 \tag{1.3.2}$$

Now, let $A_1, A_2, A_3, \ \ldots$ be mutually exclusive (disjoint) events defined on S. The sequence may be finite or countably infinite. The third axiom is then

$$\begin{aligned} \textit{Axiom 3:}\quad & P(A_1 \cup A_2 \cup A_3 \cup \ldots) \\ & = P(A_1) + P(A_2) + P(A_3) + \cdots \end{aligned} \tag{1.3.3}$$

Axiom 1 simply says that the probability of an event cannot be negative. This certainly conforms to the relative-frequency-of-occurrence concept of probability. Axiom 2 says that the event S, which includes all possible outcomes, must have a probability of unity. It is sometimes called the certain event. The first two axioms are obviously necessary if axiomatic probability is to be compatible with the older relative-frequency probability theory. The third axiom is not quite so obvious, perhaps, and it simply must be assumed. In words, it says that when we have nonoverlapping (disjoint) events, the probability of the union of these events is the sum of the probabilities of the individual events. If this were not so, one could easily think of counterexamples that would not be compatible with the relative-frequency concept. This would be most undesirable.

We now recapitulate. There are three essential ingredients in the formal approach to probability. First, a sample space must be defined that includes all possible outcomes of our conceptual experiment. We have some discretion in what we call outcomes, but caution is in order here. The outcomes must be disjoint and all-inclusive such that $P(S) = 1$. Second, we must carefully define a set of events on the sample space, and the set must be closed such that me operations of union, intersection, and complement also yield events in the set. Finally, we must assign probabilities to all events in accordance with the basic axioms of probability. In physical problems, this assignment is chosen to be compatible with what we feel to be reasonable in terms of relative frequency of occurrence of the events. If the sample space S contains a finite number of elements, the probability assignment is usually made directly on the elements of S. They are, of course, elementary events

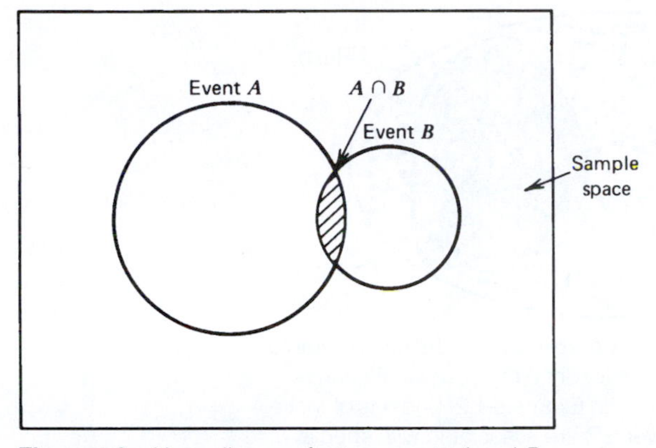

Figure 1.2 Venn diagram for two events A and B.

themselves. This, along with Axiom 3, then indirectly assigns a probability to all other events defined on the sample space. However, if the sample space consists of an infinite "smear" of points, the probability assignment must be made on events and not on points in the sample space.

Once we have specified the sample space, the set of events, and the probabilities associated with the events, we have what is known as a *probability space.* This provides the theoretical structure for the formal solution of a wide variety of probability problems.

In addition to the set operations of union and complementation, the operation of intersection is also useful in probability theory. The intersection of two events A and B is the event containing points that are common to both A and B. This is illustrated in Fig. 1.2 with what is sometimes called a Venn diagram. The points lying within the heavy contour comprise the union of A and B, denoted as $A \cup B$ or "A or B." The points within the shaded region are the event "A intersection B," which is denoted $A \cap B$, or sometimes just "A and B,"* The following relationship should be apparent just from the geometry of the Venn diagram:

$$P(A \cup B) = P(A) + P(B) - P(A \cap B) \tag{1.3.6}$$

The subtractive term in Eq. (1.3.6) is present because the probabilities in the overlapping region have been counted twice in the summation of $P(A)$ and $P(B)$.

The probability $P(A \cap B)$ is known as the *joint* probability of A and B and will be discussed further in Section 1.5. We digress for the moment, though, to look at an example.

EXAMPLE 1.2

Return to Example 1.1(a) where we are dealt 5 cards from a deck of 52 cards. Clearly the sample space is quite large in this example; it contains 2,598,960 outcomes! Yet, this is quite legitimate with the associated probability of each outcome being 1/2,598,960. Now if we were to group all the possible hands into just two groups, the first group containing all the possible 4-of-a-kind hands, and the second group containing all the hands that do not have 4-of-a-kind, our event

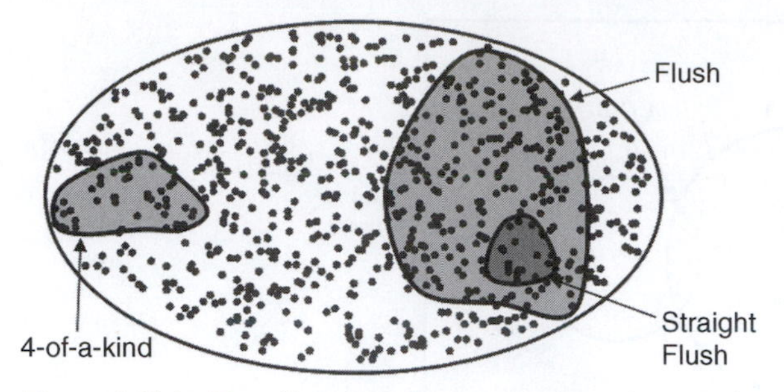

Figure 1.2(*a*) The ellipsoidal set contains the entire sample space of elemental outcomes (dots) that represent each 5-card poker hand described in Example 1.2. Subsets of these outcomes (known as "events") are associated with special card combinations such as "4-of-a-kind," "flush," or "straight flush" hands. Such event sets may be mutually disjointed, overlapping or a full subset of one another.

space would be reduced to just two events. The first would have a probability of 624/2,598,960 and the second would have a probability of 2,598,336/2,598,960. This would be a legitimate event space that would satisfy all the basic requirements just enumerated above.

On the other hand, suppose we form events as follows:

(a) All possible 4-of-a-kind hands;
(b) All possible straight flush hands;
(c) All possible flush hands;
(d) All hands not containing any of the hands specified in (a), (b), or (c).

Certainly we could calculate the associated probabilities for each of the four groupings using the ratio-of-events rule. But, would this be a legitimate events space?

The answer is no, because the defined events are not disjointed. There is overlap between Event (b) and Event (c). See Fig. 1.2*a*. Also, the sum of the event probabilities would not be unity.

■

1.4 RANDOM VARIABLES

In the study of noiselike signals, we are nearly always dealing with physical quantities such as voltage, force, distance, and so forth, which can be measured in physical units. In these cases, the chance occurrences are related to real numbers, not just objects, like playing cards, dots on the dice, and the like. This brings us to the notion of a *random variable*. Let us say that we have a conceptual experiment for which we have defined a suitable events space and a probability assignment for the set of events. A random variable is simply a function that maps every point in the events space on to points on the real line. The probability assignments in the events space transfer over to the corresponding points on the real line. This will now be illustrated with an example.

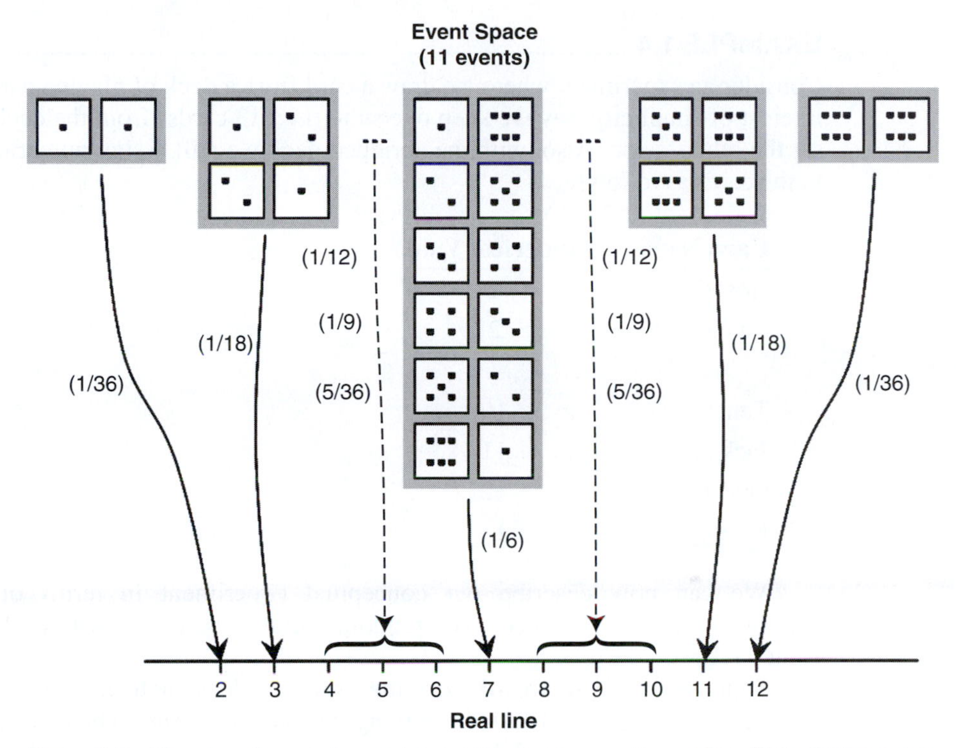

Figure 1.3 Mapping of events in event-space into numbers on the real line. (Probabilities being transfered are shown in parentheses.)

EXAMPLE 1.3

There are many games of chance that involve the throw of two dice, and it is the sum of dots on the throw that is of special interest. Monopoly and casino craps are notable examples. We will use this as our first example of a random variable. The setting for this is shown in Fig. 1.3. In this example we begin with 36 possible elemental outcomes that could occur with the throw of two dice. We will assume that each outcome has a probability of 1/36. We then group the elemental outcomes according to the sum of the dots on the top of the dice. This results in just 11 items in the new events space. Finally, these are then mapped into the integer numbers 2 through 12 as shown in the referenced figure. Note that the probabilities assigned in the events space are transferred directly over to the corresponding random variables. The events are disjoint, and their probabilities sum to unity, so we have a legitimate set of random variables.

1.5 JOINT AND CONDITIONAL PROBABILITY, BAYES RULE, AND INDEPENDENCE

In many applications we have need to consider two or more chance situations where there may be probabilistic connections among the events. A variety of these connections that are most commonly encountered will be covered in this section. We will illustrate these with an example.

EXAMPLE 1.4

Consider an experiment where we draw a card from a deck of playing cards. In the interest of simplicity, say we keep one suite (i.e., 13 cards) from the deck, and we set the others aside. Also, with the abridged deck we will assign numerical values to the cards as follows:

Card Name	Numerical Value
Ace	1
Two	2
⋮	⋮
Ten	10
Jack	11
Queen	12
King	13

We can now describe our conceptual experiment in terms of random variables rather than objects. The random variable space consists of 13 integer values 1, 2, . . . , 13.

Suppose we draw a card from the abridged deck, note its value, and then replace the card. (This is called sampling with replacement.) Then we shuffle the deck and draw another card. We will denote the first card value as random variable X and the second card value as Y. We can now keep repeating this pairwise experiment with replacement on and on as long as we like, conceptually, at least, and collect a large amount of statistical data. We now have two joint random variables and their probabilities to keep track of. This calls for an array of probabilities as shown in Table 1.1.

Clearly, in this experiment with replacement the result of the first draw does not affect the second draw in any way, so X and Y are said to be independent. (This will be formalized presently.) The entries in the array of Table 1.1 represent the joint occurrence of X_i and Y_j for all i and j ranging from 1 to 13. (We will subsequently omit the subscripts i and j to eliminate the clutter.) We can now think of the pairwise X and Y as defining a new sample space with the probabilities as indicated in Table 1.1. They are the probabilities of the joint occurrence of X and Y, and they will be denoted simply as p_{XY}. The entries in the

Table 1.1 Joint Probabilities for Two Draws with Replacement

X 1st Draw / Y 2nd Draw	1	2	3	. . .	13	Marginal Probability
1	1/169	1/169	1/169	. . .	1/169	1/13
2	1/169	1/169	1/169	. . .	1/169	1/13
3	. . .	. . . . .	. . .			. . .
. . .						
13	1/169	1/169	1/169	. . .	1/169	1/13
Marginal Probability	1/13	1/13	1/13		1/13	Sum = 1

array are all nonnegative and sum to unity, so the conditions for a legitimate sample space that were stated in Section 1.3 are satisfied.

In addition to the joint probabilities in Table 1.1, the sum of the entries in the respective rows and columns are also of interest. They are sometimes called the marginal or *unconditional probabilities*. Thinking in terms of statistics, they represent the relative frequency of occurrence of a particular value of one of the random variables irrespective of the other member of the pair. For example, consider the marginal value shown adjacent to the first row in the array. This is the probability of $Y=1$, irrespective of the value of its X partner. This is, of course, just 1/13 in this example. We will denote the unconditional probabilities of X and Y as p_X and p_Y, as the case may be. Also, we will usually omit the word "unconditional" because it is redundant. The single subscript on p tacitly indicates there is no conditioning.

Finally, there are two other distributions that are also of interest when considering joint random variables. These are the conditional probabilities of X given Y and Y given X, and they will be denoted as $p_{X|Y}$ and $p_{Y|X}$. Note here that we consider the "given" variable as being fixed, and the probability distribution is on the "first" variable. One might think at first glance that the conditional distributions come directly from the rows and columns of the p_{XY} array such as what is shown in Table 1.2. However, even though the respective rows and columns do provide the correct relative frequency-of-occurrences, their sums do not equal unity, so they cannot be legitimate probability distributions.

This brings us to Bayes Rule (or Theorem).

Bayes Rule:

$$p_{X|Y} = \frac{p_{Y|X}p_X}{p_Y} \tag{1.5.1}$$

Conditional and Joint Relationships:

$$p_{X|Y} = \frac{p_{XY}}{p_Y} \tag{1.5.2}$$

or

$$p_{Y|X} = \frac{p_{XY}}{p_X} \tag{1.5.3}$$

Note that the probabilities p_X and p_Y are the "normalizing" factors in the denominators of Eqs. (1.5.2) and (1.5.3) that take us from the joint row and column distributions to the respective conditional distributions. It is also worth noting that the joint distribution of X and Y contains all the necessary information to get the conditional distributions either way, so to speak.

Finally, we come to the term independence as it is used in random variable theory. Two discrete random variables X and Y are said to be independent if

$$p_{XY} = p_X p_Y$$

That is, in words, X and Y are independent if and only if their joint probability distribution is equal to the product of their respective p_X and p_Y distributions. It is tacitly implied that this must be true for all permissible values of X and Y. Before moving on to continuous random variables, we will look at an example where we do not have independence.

EXAMPLE 1.5.

Card Drawing Without Replacement Let us return to the card-drawing example (Example 1.4) where the abridged card deck is limited to just one suit (i.e., 13 cards), and the cards are given numerical values of 1 through 13. This time, though, consider that the two successive draws are made without replacement. Clearly, the result of the second draw is not independent of the first draw. The initial card from the first draw is missing from the deck of the second draw, so the probability of drawing that same card is zero. Thus, the joint probability table is modified as shown in Table 1.2.

This array of joint probabilities is quite different from the corresponding array in Table 1.1 because of the zeros along the major diagonal. The off-diagonal terms also differ slightly from those in Table 1.1, because their row and column sums must be unity. Note, though, that all the marginal probabilities are 1/13 just as before in the experiment with replacement. This makes sense when viewed from a statistical viewpoint. Imagine a large number of pairs being drawn without replacement. The statistics of pairwise combinations will differ radically between the with-replacement and without-replacement experiments. But, in the larger sense, irrespective of which draw is which, there is no special preference given to any of the numerical realizations (i.e., 1 through 13). So, you would expect all of the marginal probabilities to be the same, namely 1/13.

Before leaving the card-drawing experiment, it is worth mentioning that we do not have to rely on the intuitive notion of independence when we know the joint probability distribution in all its detail. Look first at the array in Table 1.1. Clearly, all the entries (which are p_{XY}) are equal to the respective products of the marginal probabilities (which are p_X and p_Y). Thus, we have formal verification of the independence of X and Y for the "with-replacement" case. Next look at the entries in the array of Table 1.2. Clearly, here the entries in the array are not equal to the products of the respective marginal probabilities, so we do not have independence for the "without-replacement" experiment. This is consistent with our intuitive notion of independence.

Table 1.2 Joint Probabilities for Two Draws Without Replacement

X 1st Draw / Y 2nd Draw	1	2	3	...	13	Marginal Probability
1	0	1/156	1/156	...	1/156	1/13
2	1/156	0	1/156	...	1/156	1/13
3	1/156	1/156	0			...
...	...	...	...			
13	1/156	1/156	1/156	...	0	1/13
Marginal Probability	1/13	1/13	1/13		1/13	Sum = 1

1.6 CONTINUOUS RANDOM VARIABLES AND PROBABILITY DENSITY FUNCTION

The emphasis in the preceding paragraphs has been on discrete random variables. We will now extend these ideas to continuous random variables. We will begin the discussion with a simple spin-the-pointer example.

EXAMPLE 1.5A

In many games, the player spins a pointer that is mounted on a circular card of some sort and is free to spin about its center. This is depicted in Fig. 1.3*a* and the circular card is intentionally shown without any markings along its edge. Suppose we define the outcome of an experiment as the location on the periphery of the card at which the pointer stops. The sample space then consists of an infinite number of points along a circle. For analysis purposes, we might wish to identify each point in the sample space in terms of an angular coordinate measured in radians. The functional mapping that maps all points on a circle to corresponding points on the real line between 0 and 2π would then define an appropriate random variable.

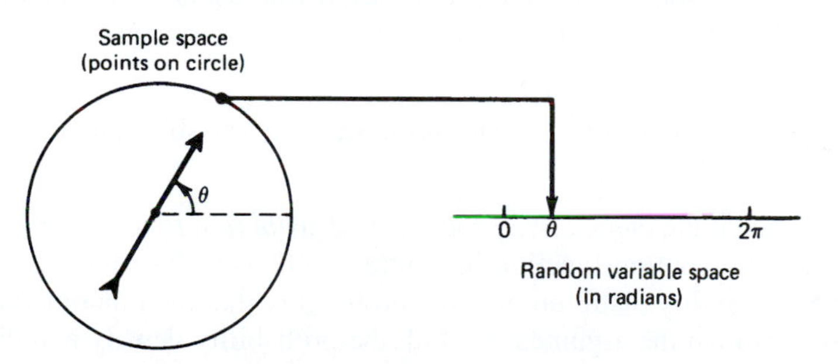

Figure 1.3(*a*) Mapping for spin-the-pointer example.

In the spin-the-pointer example just presented, the random variable sample space is seen to be a continuous smear of points on the real line between 0 and 2π. In this chance situation the probability of achieving any particular point along the real line is zero. But yet we know that the probability of achieving a point within a nonzero interval between 0 and 2π is also nonzero. So, we should be able to describe this chance situation in terms of a probability density rather than just probability. This is analogous to the long thin rod problem in engineering mechanics where we describe the mass of the rod in terms of mass per unit length rather than just mass. The total mass is then the integral of the mass density. So be it with probability. In the spin-the-pointer example the probability that the pointer will lie between θ_1 and θ_2 (where $\theta_2 > \theta_1$) is

$$P(\theta_1 < X < \theta_2) = \int_{\theta_1}^{\theta_2} f_X(\theta) d\theta \tag{1.6.1}$$

where X is the random variable, f_x is the probability density function, θ is a dummy variable of integration, $\theta_2 > \theta_1$, and both θ_1 and θ_2 are assumed to lie within the larger $[0, 2\pi)$ interval.

In the example at hand, we would usually assume that the pointer would be equally likely to stop anywhere in the $[0, 2\pi]$ interval; so in this case $f_X(\theta)$ would be a constant, namely

$$f_X(\theta) = \frac{1}{2\pi}, \quad 0 \le \theta < 2\pi \tag{1.6.2}$$

In this way,

$$\int_0^{2\pi} f_X(\theta) d\theta = 1 \tag{1.6.3}$$

which is correct for the certain event.

It works out that the integral of the probability density function is also a useful descriptor of a continuous random variable. It is usually defined with the integral's lower limit being set at the random variable's smallest possible realization, and the upper limit is set at some dummy variable, say x, where x is within the range of the random variable space. The integral is then

$$F_X(x) = \int_0^x f_X(\theta) d\theta, \quad x \text{ within the random variable space} \tag{1.6.4}$$

and $F_X(x)$ is called the *cumulative probability distribution* function (or sometimes, for short, just distribution function in contrast to density function). In words, the cumulative probability distribution is the probability that the random variable X is equal to or less than the argument x. Both the probability density and distribution functions for the spin-the-pointer example are shown in Fig. 1.4.

In summary, the probability density and distribution functions are alternative ways of describing the relative distribution of the random variable. Both functions are useful in random-variable analysis, and you should always keep in mind the derivative/integral relationship between the two. As a matter of notation, we will normally use an uppercase symbol for the distribution function and the corresponding lowercase symbol for the density function. The subscript in each case indicates the random variable being considered. The argument of the function is a dummy variable and may be almost anything.

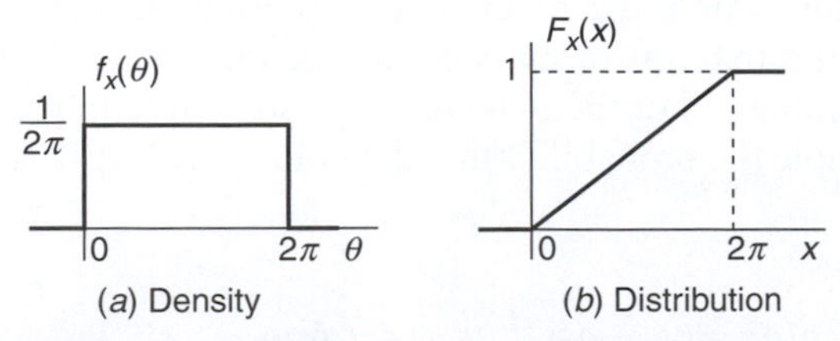

Figure 1.4 Probability density and cumulative distribution functions for the spin-the-pointer example.

1.7 EXPECTATION, AVERAGES, AND CHARACTERISTIC FUNCTION

The idea of averaging is so commonplace that it may not seem worthy of elaboration. Yet there are subtleties, especially as averaging relates to probability. Thus we need to formalize the notion of average.

Perhaps the first thing to note is that we always average over numbers and not "things." There is no such thing as the average of apples and oranges. When we compute a student's average grades, we do not average over A, B, C, and so on; instead, we average over numerical equivalents that have been arbitrarily assigned to each grade. Also, the quantities being averaged may or may not be governed by chance. In either case, random or deterministic, the average is just the sum of the numbers divided by the number of quantities being averaged. In the random case, the *sample average* or *sample mean* of a random variable X is defined as

$$\bar{X} = \frac{X_1 + X_2 + \cdots + X_N}{N} \tag{1.7.1}$$

where the bar over X indicates average, and $X_1, X_2, \ldots,$ are sample realizations obtained from repeated trials of the chance situation under consideration. We will use the terms *average* and *mean* interchangeably, and the adjective *sample* serves as a reminder that we are averaging over a finite number of trials as in Eq. (1.7.1).

In the study of random variables we also like to consider the conceptual average that would occur for an infinite number of trials. This idea is basic to the relative-frequency concept of probability. This hypothetical average is called *expected value* and is aptly named; it simply refers to what one would "expect" in the typical statistical situation. Beginning with discrete probability, imagine a random variable whose n possible realizations are $x_1, x_2, \ldots, x_n$. The corresponding probabilities are $p_1, p_2, \ldots, p_n$. If we have N trials, where N is large, we would expect approximately $p_1N\ x_1$'s, $p_2N\ x_2$'s, etc. Thus, the sample average would be

$$\bar{X}_{\text{sample}} \approx \frac{(p_1N)x_1 + (p_2N)x_2 + \cdots + (p_nN)x_n}{N} \tag{1.7.2}$$

This suggests the following definition for expected value for the discrete probability case:

$$\text{Expected value of } X = E(X) = \sum_{i=1}^{n} p_i x_i \tag{1.7.3}$$

where n is the number of allowable values of the random variable X.

Similarly, for a continuous random variable X, we have

$$\text{Expected value of } X = E(X) = \int_{-\infty}^{\infty} x f_X(x)dx \tag{1.7.4}$$

It should be mentioned that Eqs. (1.7.3) and (1.7.4) are definitions, and the arguments leading up to these definitions were presented to give a sensible rationale for the definitions, and not as a proof. We can use these same arguments for defining the expectation of a function of X, as well as for X. Thus, we have the following:

Discrete case:

$$E(g(X)) = \sum_{i}^{n} p_i g(x_i) \tag{1.7.5}$$

Continuous case:

$$E(g(X)) = \int_{-\infty}^{\infty} g(x) f_X(x) dx \tag{1.7.6}$$

As an example of the use of Eq. (1.7.6), let the function $g(X)$ be X^k. Equation (1.7.6) [or its discrete counterpart Eq. (1.7.5)] then provides an expression for the kth *moment* of X, that is,

$$E(X^k) = \int_{-\infty}^{\infty} x^k f_X(x) dx \tag{1.7.7}$$

The second moment of X is of special interest, and it is given by

$$E(X^2) = \int_{-\infty}^{\infty} x^2 f_X(x) dx \tag{1.7.8}$$

The first moment is, of course, just the expectation of X, which is also known as the *mean* or *average value* of X. Note that when the term *sample* is omitted, we tacitly assume that we are referring to the hypothetical infinite-sample average.

We also have occasion to look at the second moment of X "about the mean." This quantity is called the *variance* of X and is defined as

$$\text{Variance of } X = E[(X - E(X))^2] \tag{1.7.9}$$

In a qualitative sense, the variance of X is a measure of the dispersion of X about its mean. Of course, if the mean is zero, the variance is identical to the second moment.

The expression for variance given by Eq. (1.7.9) can be reduced to a more convenient computational form by expanding the quantity within the brackets and then noting that the expectation of the sum is the sum of the expectations. This leads to

$$\begin{aligned} \text{Var}\, X &= E[X^2 - 2X \cdot E(X) + (E(X))^2] \\ &= E(X^2) - (E(X))^2 \end{aligned} \tag{1.7.10}$$

The square root of the variance is also of interest, and it has been given the name *standard deviation*, that is,

$$\text{Standard deviation of } X = \sqrt{\text{Variance of } X} \tag{1.7.11}$$

EXAMPLE 1.6

Let X be uniformly distributed in the interval $(0, 2\pi)$. This leads to the probability density function (see Section 1.6).

$$f_X(x) = \begin{cases} \dfrac{1}{2\pi}, & 0 \le x < 2\pi \\ 0, & \text{elsewhere} \end{cases}$$

Find the mean, variance, and standard deviation of X.

The mean is just the expectation of X and is given by Eq. (1.7.4).

$$\begin{aligned} \text{Mean of } X = E(X) &= \int_0^{2\pi} x \cdot \frac{1}{2\pi} dx \\ &= \frac{1}{2\pi} \cdot \frac{x^2}{2}\bigg|_0^{2\pi} = \pi \end{aligned} \tag{1.7.12}$$

Now that we have computed the mean, we are in a position to find the variance from Eq. (1.7.10).

$$\begin{aligned} \text{Var}\, X &= \int_0^{2\pi} x^2 \frac{1}{2\pi} dx - \pi^2 \\ &= \frac{4}{3}\pi^2 - \pi^2 \\ &= \frac{1}{3}\pi^2 \end{aligned} \tag{1.7.13}$$

The standard deviation is now just the square root of the variance:

$$\begin{aligned} \text{Standard deviation of } X &= \sqrt{\text{Var}\, X} \\ &= \sqrt{\frac{1}{3}\pi^2} = \frac{1}{\sqrt{3}}\pi \end{aligned} \tag{1.7.14}$$

■

The *characteristic function* associated with the random variable X is defined as

$$\psi_X(\omega) = \int_{-\infty}^{\infty} f_X(x) e^{j\omega x} dx \tag{1.7.15}$$

It can be seen that $\psi_X(\omega)$ is just the Fourier transform of the probability density function with a reversal of sign on ω. Thus, the theorems (and tables) of Fourier transform theory can be used to advantage in evaluating characteristic functions and their inverses.

The characteristic function is especially useful in evaluating the moments of X. This can be demonstrated as follows. The moments of X may be written as

$$E(X) = \int_{-\infty}^{\infty} x f_X(x) dx \tag{1.7.16}$$

$$\begin{aligned} E(X^2) &= \int_{-\infty}^{\infty} x^k f_X(x) dx \\ &\vdots \\ &\text{etc.} \end{aligned} \tag{1.7.17}$$

Now consider the derivatives of $\psi_x(\omega)$ evaluated at $\omega = 0$.

$$\left[\frac{d\psi_X}{d\omega}\right]_{\omega=0} = \left[\int_{-\infty}^{\infty} jx f_X(x) e^{j\omega x} dx\right]_{\omega=0} = \int_{-\infty}^{\infty} jx f_X(x) dx \tag{1.7.18}$$

$$\begin{aligned} \left[\frac{d^2\psi_X}{d\omega^2}\right]_{\omega=0} &= \left[\int_{-\infty}^{\infty} (jx)^2 f_X(x) e^{j\omega x} dx\right]_{\omega=0} = \int_{-\infty}^{\infty} j^2 x^2 f_X(x) dx \\ &\vdots \\ &\text{etc.} \end{aligned} \tag{1.7.19}$$

It can be seen that

$$E(X) = \frac{1}{j}\left[\frac{d\psi_X}{d\omega}\right]_{\omega=0} \tag{1.7.20}$$

$$\begin{aligned} E(X^2) &= \frac{1}{j^2}\left[\frac{d^2\psi_X}{d\omega^2}\right]_{\omega=0} \\ &\vdots \\ &\text{etc.} \end{aligned} \tag{1.7.21}$$

Thus, with the help of a table of Fourier transforms, you can often evaluate the moments without performing the integrations indicated in their definitions.

1.8 NORMAL OR GAUSSIAN RANDOM VARIABLES

The random variable X is called *normal* or *Gaussian* if its probability density function is

$$f_X(x) = \frac{1}{\sqrt{2\pi}\sigma} \exp\left[-\frac{1}{2\sigma^2}(x - m_X)^2\right] \tag{1.8.1}$$

Note that this density function contains two parameters m_X and σ^2. These are the random variable's mean and variance. That is, for the f_X specified by Eq. (1.8.1),

$$\int_{-\infty}^{\infty} x f_X(x) dx = m_X \tag{1.8.2}$$

and

$$\int_{-\infty}^{\infty} (x - m_X)^2 f_X(x) dx = \sigma^2 \tag{1.8.3}$$

Note that the normal density function is completely specified by assigning numerical values to the mean and variance. Thus, a shorthand notation has come into common usage to designate a normal random variable. When we write

$$X \sim \mathcal{N}(m_X, \sigma^2) \tag{1.8.4}$$

we mean X is normal with its mean given by the first argument in parentheses and its variance by the second argument. Also, as a matter of terminology, the terms normal and Gaussian are used interchangeably in describing normal random variables, and we will make no distinction between the two.

The normal density and distribution functions are sketched in Figs. 1.5*a* and 1.5*b*. Note that the density function is symmetric and peaks at its mean. Qualitatively, then, the mean is seen to be the most likely value, with values on either side of the mean gradually becoming less and less likely as the distance from the mean becomes larger. Since many natural random phenomena seem to exhibit this

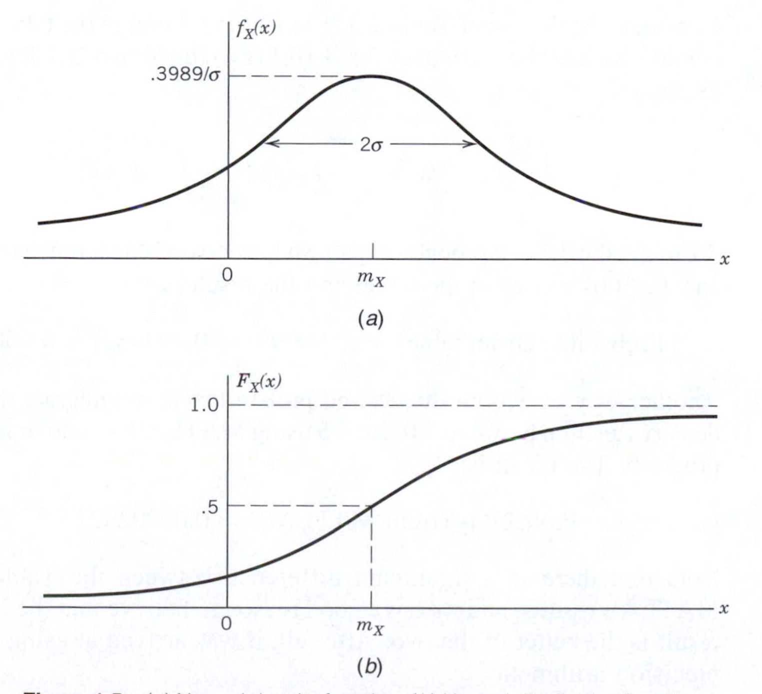

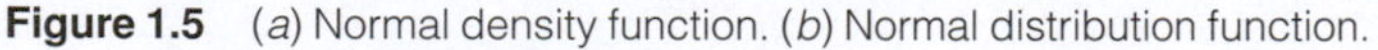

Figure 1.5 (*a*) Normal density function. (*b*) Normal distribution function.

central-tendency property, at least approximately, the normal distribution is encountered frequently in applied probability. Recall that the variance is a measure of dispersion about the mean. Thus, small σ corresponds to a sharp-peaked density curve, whereas large σ will yield a curve with a flat peak.

The normal distribution function is, of course, the integral of the density function:

$$F_X(x) = \int_{-\infty}^{\infty} f_X(u)du = \int_{-\infty}^{\infty} \frac{1}{\sqrt{2\pi}\sigma} \exp\left[-\frac{1}{2\sigma^2}(u - m_X)^2\right] du \qquad (1.8.5)$$

Unfortunately, this integral cannot be represented in closed form in terms of elementary functions. Thus, its value must be obtained from tables or by numerical integration. A brief tabulation for zero mean and unity variance is given in Table 1.4. A quick glance at the table will show that the distribution function is very close to unity for values of the argument greater than 4.0 (i.e., 4σ). In our table, which was taken from Feller (5), the tabulation quits at about this point. In some applications, though, the difference between $F_X(x)$ and unity [i.e., the area under the "tail" of $f_X(x)$] is very much of interest, even though it is quite small. Tables such as the one given here are not of much use in such cases, because the range of x is limited and the resolution is poor. Fortunately, it is relatively easy to integrate the normal density function numerically using software such as MATLAB's quad. An example will illustrate this.

EXAMPLE 1.7

Let X be a normal random variable with zero mean and unity variance, and say we want to find the probability that $4.0 < x < 4.5$.

(a) Consider first the use of Table 1.3. The $F_X(x)$ column in the table gives the cumulative distribution values for $\mathcal{N}(0,1)$ as a function of x. Clearly, in this example:

$$\int_{4.0}^{4.5} f_X(x)dx = \int_{-\infty}^{4.5} f_X(x)dx - \int_{-\infty}^{4.0} f_X(x)dx$$

So, to get the desired probability, all we have to do is difference the $F_X(4.5)$ and $F_X(4.0)$ entries in the table, and the result is:

$$\text{Probability (from table)} = (0.999997 - 0.999968) = 0.000029$$

(b) Another way of getting the desired probability is to integrate the normal density function between 4.0 and 4.5 using MATLAB's quad or integration program. The result is:

$$\text{Probability (from MATLAB)} \approx 0.00002827$$

Note that there is a significant difference between the Table 1.3 and MATLAB results, and we have good reason to believe that the MATLAB result is the better of the two. After all, it was arrived at using very high precision arithmetic.

■

Table 1.3 The Normal Density and Distribution Functions for Zero Mean and Unity Variance

$$f_X(x) = \frac{1}{\sqrt{2\pi}} e^{-x^2/2}, \quad F_X(x) = \int_{-\infty}^{x} \frac{1}{\sqrt{2\pi}} e^{-u^2/2} du$$

x	$f_X(x)$	$F_X(x)$	x	$f_X(x)$	$F_X(x)$
.0	.398 942	.500 000	2.3	.028 327	.989 276
.1	.396 952	.539 828	2.4	.022 395	.991 802
.2	.391 043	.579 260	2.5	.017 528	.993 790
.3	.381 388	.617 911	2.6	.013 583	.995 339
.4	.368 270	.655 422	2.7	.010 421	.996 533
.5	.352 065	.691 462	2.8	.007 915	.997 445
.6	.333 225	.725 747	2.9	.005 953	.998 134
.7	.312 254	.758 036	3.0	.004 432	.998 650
.8	.289 692	.788 145	3.1	.003 267	.999 032
.9	.266 085	.815 940	3.2	.002 384	.999 313
1.0	.241 971	.841 345	3.3	.001 723	.999 517
1.1	.217 852	.864 334	3.4	.001 232	.999 663
1.2	.194 186	.884 930	3.5	.000 873	.999 767
1.3	.171 369	.903 200	3.6	.000 612	.999 841
1.4	.149 727	.919 243	3.7	.000 425	.999 892
1.5	.129 581	.933 193	3.8	.000 292	.999 928
1.6	.110 921	.945 201	3.9	.000 199	.999 952
1.7	.094 049	.955 435	4.0	.000 134	.999 968
1.8	.078 950	.964 070	4.1	.000 089	.999 979
1.9	.065 616	.971 283	4.2	.000 059	.999 987
2.0	.053 991	.977 250	4.3	.000 039	.999 991
2.1	.043 984	.982 136	4.4	.000 025	.999 995
2.2	.035 475	.986 097	4.5	.000 016	.999 997

In spite of the previous remarks, tables of probabilities, both normal and otherwise, can be useful for quick and rough calculations. A word of caution is in order, though, relative to normal distribution tables. They come in a variety of forms. For example, some tables give the one-sided area under the normal density curve from 0 to X, rather than from $-\infty$ to X as we have done in Table 1.3. Other tables do something similar by tabulating a function known as the error function, which is normalized differently than the usual distribution function. Thus a word of warning is in order. Be wary of using unfamiliar tables or computer library software!

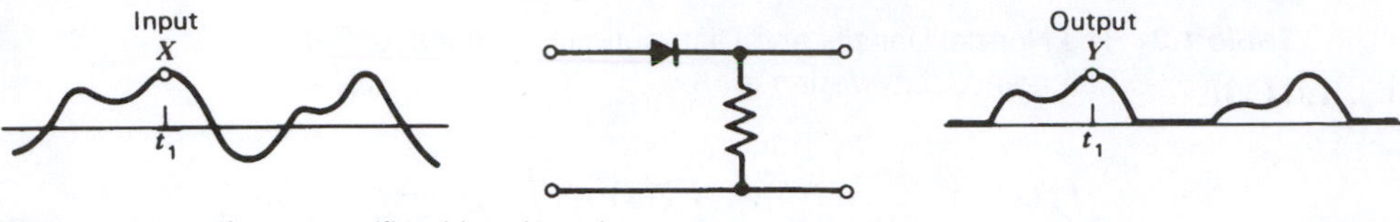

Figure 1.6 Half-wave rectifier driven by noise.

1.9 IMPULSIVE PROBABILITY DENSITY FUNCTIONS

In the case of the normal distribution and many others, the probability associated with the random variable X is smoothly distributed over the real line from $-\infty$ to ∞. The corresponding probability density function is then continuous, and the probability that any particular value of X, say, x_0, is realized is zero. This situation is common in physical problems, but we also have occasion to consider cases where the random variable has a mixture of discrete and smooth distribution. Rectification or any sort of hard limiting of noise leads to this situation, and an example will illustrate how this affects the probability density and distribution functions.

Consider a simple half-wave rectifier driven by noise as shown in Fig. 1.6. For our purpose here, it will suffice to assume that the amplitude of the noise is normally distributed with zero mean. That is, if we were to sample the input at any particular time t_1, the resultant sample is a random variable, say, X, whose distribution is $\mathcal{N}(0, \sigma_x^2)$. The corresponding output sample is, of course, a different random variable; it will be denoted as Y.

Because of our assumption of normality, X will have probability density and distribution functions as shown in Fig. 1.5, but with $m_X = 0$. The sample space in this case may be thought of as all points on the real line, and the function defining the random variable X is just a one-to-one mapping of the sample space into X. Not so with Y though; all positive points in the sample space map one-to-one, but the negative points all map into zero because of the diode! This means that a total probability of $\frac{1}{2}$ in the sample space gets squeezed into a single point, zero, in the Y space. The effect of this on the density and distribution functions for Y is shown in Figs. 1.7*a* and 1.7*b*. Note that in order to have the area under the density function be $\frac{1}{2}$ at $y = 0$, we must have a Dirac delta (impulse) function at the origin. This, in turn, gives rise to a jump or discontinuity in the corresponding distribution function. It should be mentioned that at the point of discontinuity, the value of the distribution function is $\frac{1}{2}$. That is, the distribution function is continuous from the right and not from the left. This is due to the "*equal to* or less than . . . " statement in the definition of the probability distribution function (see Section 1.6).

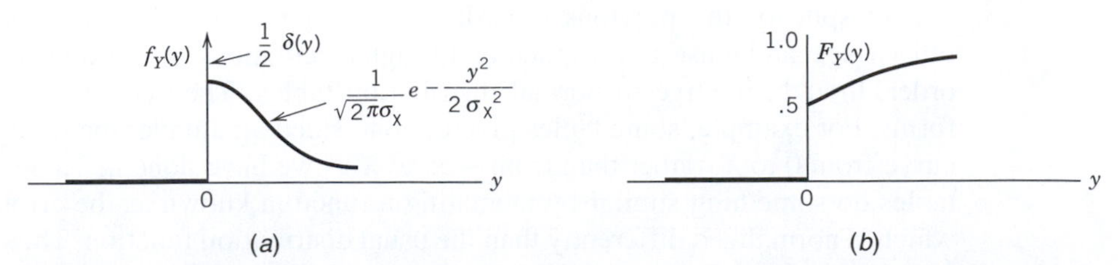

Figure 1.7 Output density and distribution functions for diode example. (*a*) Probability density function for Y. (*b*) Probability distribution function for Y.

1.10 JOINT CONTINUOUS RANDOM VARIABLES

In the subsequent chapters we will frequently have occasion to deal with joint continuous random variables. These are also referred to as multivariate random variables with the case of two joint random variables, called the bivariate case, being encountered frequently. We will begin with the bivariate case and then extend the discussion to higher-order cases later.

Recall from Sec. 1.6 that in the single-variate case we have the probability interpretation of probability density:

$$P(x_0 \leq X \leq x_0 + dx) = f_X(x_0)dx \tag{1.10.1}$$

where x_0 is within the random variable space. The corresponding equation in the bivariate case is then

$$P(\{x_0 \leq X \leq x_0 + dx\} \; and \; \{y_0 \leq Y \leq y_0 + dy\}) = f_{XY}\{x_0, y_0)dxdy \tag{1.10.2}$$

It then follows directly from integral calculus that the probability that X and Y both lie within the region shown as R in Fig. 1.8 is:

$$P(X \; and \; Y \; lie \; within \; R) = \iint_R f_{XY}(x, y)dxdy \tag{1.10.3}$$

It should now be apparent that we can also define a joint cumulative distribution function for the bivariate case as:

$$F_{XY}(x_0, y_0) = P(\{X \leq x_0\} \; and \; \{Y \leq y_0\}) = \int_{-\infty}^{y_0} \int_{-\infty}^{x_0} f_{XY}(x, y)dxdy \tag{1.10.4}$$

The integration region for this double integral is the open lower-left quadrant bounded on the East by x_0 and on the North by y_0. It should also be apparent that we

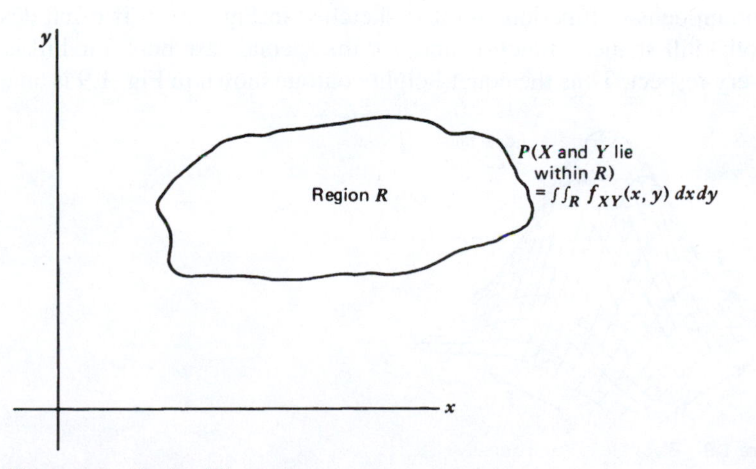

Figure 1.8 Region *R* in *xy* plane.

have a differential/integral relationship between the density and cumulative distribution functions that is similar to what we have in the single-variate case. The only difference is that we have a double integral for the integration and the second partial derivative for the derivative part of the analogy.

The bivariate normal random variable arises frequently in applied situations, so it deserves special attention. Example 1.8 will illustrate a case where the two random variables have zero means, equal variances, and are independent.

EXAMPLE 1.8

Consider a dart-throwing game in which the target is a conventional *xy* coordinate system. The player aims each throw at the origin according to his or her best ability. Since there are many vagaries affecting each throw, we can expect a scatter in the hit pattern. Also, after some practice, the scatter should be unbiased, left-to-right and up-and-down. Let the coordinate of a hit be a bivariate random variable (X, Y). In this example we would not expect the x coordinate to affect y in any way; therefore statistical independence of X and Y is a reasonable assumption. Also, because of the central tendency in X and Y, the assumption of normal distribution would appear to be reasonable. Thus we assume the following probability densities:

$$f_X(x) = \frac{1}{\sqrt{2\pi}\sigma} e^{-x^2/2\sigma^2} \tag{1.10.5}$$

$$f_Y(y) = \frac{1}{\sqrt{2\pi}\sigma} e^{-y^2/2\sigma^2} \tag{1.10.6}$$

The joint density is then given by

$$f_{XY}(x, y) = \frac{1}{\sqrt{2\pi}\sigma} e^{-x^2/2\sigma^2} \cdot \frac{1}{\sqrt{2\pi}\sigma} e^{-y^2/2\sigma^2} = \frac{1}{2\pi\sigma^2} e^{-(x^2+y^2)/2\sigma^2} \tag{1.10.7}$$

Equation (1.10.7) is the special case of a bivariate normal density function in which X and Y are independent, unbiased, and have equal variances. This is an important density function and it is sketched in Fig. 1.9. It is often described as a smooth "hill-shaped" function and, for the special case here, the hill is symmetric in every respect. Thus the equal-height contour shown in Fig. 1.9 is an exact circle.

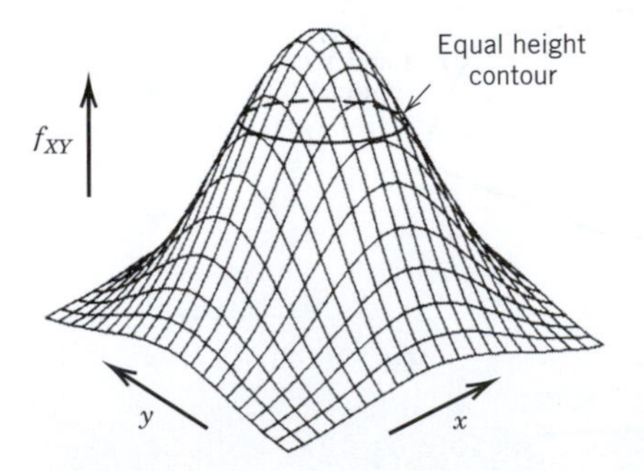

Figure 1.9 Bivariate normal density function.

It is worthy of mention that only the *joint* probability density function gives the complete description of the probabilistic relationship between the X and Y random variables. The other densities (i.e., conditional and marginal) contain specialized information which may be important in certain applications, but each does not, when considered alone, "tell the whole story". Only the joint density function does that.

Bayes Rule, independence, and the equations for marginal density carry over directly from discrete probability to the continuous case. They are given by the following.

Bayes Rule (Continuous Case):

$$f_{X|Y} = \frac{f_{Y|X} f_X}{f_Y} \tag{1.10.8}$$

Conditional Densities (Continuous Case):

$$f_{X|Y} = \frac{f_{XY}}{f_Y} \tag{1.10.9}$$

Also,

$$f_{Y|X} = \frac{f_{XY}}{f_X} \tag{1.10.10}$$

Independence:
Random variables X and Y are independent if and only if

$$f_{XY} = f_X f_Y \tag{1.10.11}$$

Marginal probability density:

$$f_Y = \int_{-\infty}^{\infty} f_{XY}(x, y) dx \tag{1.10.12}$$

and

$$f_X = \int_{-\infty}^{\infty} f_{XY}(x, y) dy \tag{1.10.13}$$

The indicated integrations are analogous to summing out a row or column (as the case may be) in discrete probability (see Sec. 1.5). One should always remember that to get marginal probability from joint probability, you must always "sum out," not simply substitute a fixed value for the random variable being held fixed. Example 1.9 will illustrate this.

EXAMPLE 1.9

Return to the dart throwing example (Example 1.7). The end result for f_{XY} was

$$f_{XY}(x, y) = \frac{1}{2\pi\sigma^2} e^{-(x^2+y^2)/2\sigma^2} \tag{1.10.14}$$

Now, say we want to get f_X from f_{XY}. To do this we need to integrate out on y. This yields

$$\begin{aligned} f_X &= \int_{-\infty}^{\infty} \frac{1}{2\pi\sigma^2} e^{-(x^2+y^2)/2\sigma^2} dy \\ &= \frac{1}{\sqrt{2\pi}\sigma} e^{-x^2/2\sigma^2} \int_{-\infty}^{\infty} \frac{1}{2\pi\sigma^2} e^{-y^2/2\sigma^2} dy \\ &= \frac{1}{\sqrt{2\pi}\sigma} e^{-x^2/2\sigma^2} \end{aligned} \tag{1.10.15}$$

which is the correct f_X for this example where X and Y are independent.

The extension of the bivariate case to multivariate situation is fairly obvious. Consider the trivariate case, for example. Here the probability density is a function of three variables instead of two. To get the probability that the 3-tuple random variable lies within a given region, we must integrate over a volume rather than an area. Also, the Bayes Rule for the trivariate case is:

$$f_{XY|Z} = \frac{f_{Z|XY} f_{XY}}{f_Z} \tag{1.10.16}$$

Or, if the variable to be held fixed is both Y and Z, we would have

$$f_{X|YZ} = \frac{f_{YZ|X} f_X}{f_{YZ}} \tag{1.10.17}$$

Also, the marginal probability densities are obtained by double integration. For example,

$$f_X = \int_{-\infty}^{\infty} \int_{-\infty}^{\infty} f_{XYZ}(x, y, z) dy dz$$

and so forth.

1.11 CORRELATION, COVARIANCE, AND ORTHOGONALITY

The expectation of the product of two random variables X and Y is of special interest. In general, it is given by

$$E(XY) = \int_{-\infty}^{\infty} \int_{-\infty}^{\infty} xy f_{XY}(x, y) dx\, dy \tag{1.11.1}$$

There is a special simplification of Eq. (1.11.1) that occurs when X and Y are independent. In this case, f_{XY} may be factored (see Eq. 1.10.10). Equation (1.11.1) then reduces to

$$E(XY) = \int_{-\infty}^{\infty} x f_X(x) dx \int_{-\infty}^{\infty} y f_Y(y) dy = E(X)E(Y) \tag{1.11.2}$$

If X and Y possess the property of Eq. (1.11.2), that is, the expectation of the product is the product of the individual expectations, they are said to be *uncorrelated*. Obviously, if X and Y are independent, they are also uncorrelated. However, the converse is not true, except in a few special cases (see Section 1.15).

As a matter of terminology, if

$$E(XY) = 0 \tag{1.11.3}$$

X and Y are said to be *orthogonal*.

The *covariance* of X and Y is also of special interest, and it is defined as

$$\text{Cov of } X \text{ and } Y = E(X - m_X)(Y - m_Y)] \tag{1.11.4}$$

With the definition of Eq. (1.11.4) we can now define the *correlation coefficient* for two random variables as

$$\begin{aligned} \text{Correlation coefficient} = \rho &= \frac{\text{Cov of } X \text{ and } Y}{\sqrt{\text{Var } X}\sqrt{\text{Var } Y}} \\ &= \frac{E[(X - m_X)(Y - m_Y)]}{\sqrt{\text{Var } X}\sqrt{\text{Var } Y}} \end{aligned} \tag{1.11.5}$$

The correlation coefficient is a normalized measure of the degree of correlation between two random variables, and the normalization is such that ρ always lies within the range $-1 \leq \rho \leq 1$. This will be demonstrated (not proved) by looking at three special cases:

1. $Y = X$ (maximum positive correlation):
   When $Y = X$, Eq. (1.11.5) becomes

$$\rho = \frac{E[(X - m_X)(X - m_X)]}{\sqrt{E(X - m_X)^2}\sqrt{E(X - m_X)^2}} = 1$$

2. $Y = -X$ (maximum negative correlation):
   When $Y = -X$, Eq. (1.11.5) becomes

$$\rho = \frac{E[(X - m_X)(-X + m_X)]}{\sqrt{E(X - m_X)^2}\sqrt{E(-X + m_X)^2}} = -1$$

3. X and Y uncorrelated, that is, $E(XY) = E(X)E(Y)$:
   Expanding the numerator of Eq. (1.11.5) yields

$$\begin{aligned} &E(XY - m_X Y - m_Y X + m_X m_Y) \\ &\quad = E(XY) - m_X E(Y) - m_Y E(X) + m_X m_Y \\ &\quad = m_X m_Y - m_X m_Y - m_Y m_X + m_X m_Y = 0 \end{aligned} \tag{1.11.5}$$

Thus, $\rho = 0$.

We have now examined the extremes of positive and negative correlation and zero correlation; there can be all shades of gray in between. [For further details, see Papoulis (7).]

1.12 SUM OF INDEPENDENT RANDOM VARIABLES AND TENDENCY TOWARD NORMAL DISTRIBUTION

Since we frequently need to consider additive combinations of independent random variables, this will now be examined in some detail. Let X and Y be independent random variables with probability density functions $f_X(x)$ and $f_Y(y)$. Define another random variable Z as the sum of X and Y:

$$Z = X + Y \tag{1.12.1}$$

Given the density functions of X and Y, we wish to find the corresponding density of Z.

Let z be a particular realization of the random variable Z, and think of z as being fixed. Now consider all possible realizations of X and Y that yield z. Clearly, they satisfy the equation

$$x + y = z \tag{1.12.2}$$

and the locus of points in the x, y plane is just a straight line, as shown in Fig. 1.10.

Next, consider an incremental perturbation of z to $z + dz$, and again consider the locus of realizations of X and Y that will yield $z + dz$. This locus is also a straight line, and it is shown as the upper line in Fig. 1.10. It should be apparent that all x and y lying within the differential strip between the two lines map into points between z and $z + dz$ in the z space. Therefore,

$$\begin{aligned} P(z \le Z \le z + dz) &= P(x \text{ and } y \text{ lie in differential strip}) \\ &= \iint\limits_{\substack{\text{Diff.}\\ \text{strip}}} f_X(x) f_Y(y)\, dx\, dy \end{aligned} \tag{1.12.3}$$

But within the differential strip, y is constrained to x according to

$$y = z - x \tag{1.12.4}$$

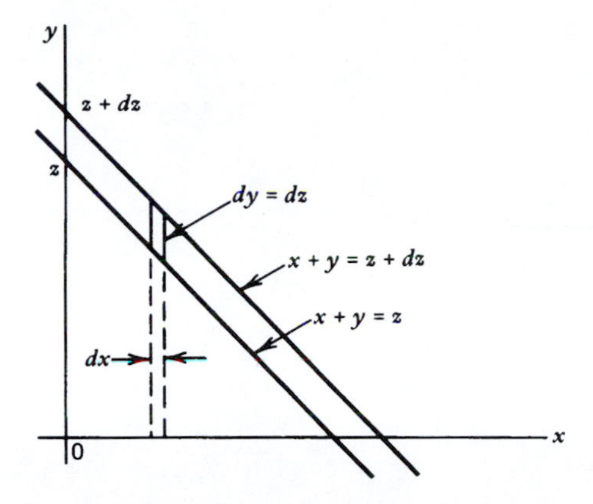

Figure 1.10 Differential strip used for deriving $f_Z(Z)$.

Also, since the strip is only of differential width, the double integral of Eq. (1.12.3) reduces to a single integral. Choosing x as the variable of integration and noting that $dy = dz$ lead to

$$P(z \leq Z \leq z + dz) = \left[\int_{-\infty}^{\infty} f_X(x) f_Y(z - x) dx \right] dz \tag{1.12.5}$$

It is now apparent from Eq. (1.12.5) that the quantity within the brackets is the desired probability density function for Z. Thus,

$$f_Z(z) = \int_{-\infty}^{\infty} f_X(x) f_Y(z - x) dx \tag{1.12.6}$$

It is of interest to note that the integral on the right side of Eq. (1.12.6) is a convolution integral. Thus, from Fourier transform theory, we can then write

$$\mathfrak{F}[f_Z] = \mathfrak{F}[f_X] \cdot \mathfrak{F}[f_Y] \tag{1.12.7}$$

where $\mathfrak{F}[\cdot]$ denotes "Fourier transform of $[\cdot]$." We now have two ways of evaluating the density of Z: (1) We can evaluate the convolution integral directly, or (2) we can first transform f_X and f_Y, then form the product of the transforms, and finally invert the product to get f_Z. Examples that illustrate each of these methods follow.

EXAMPLE 1.11

Let X and Y be independent random variables with identical rectangular density functions as shown in Fig. 1.11*a*. We wish to find the density function for their sum, which we will call Z.

Note first that the density shown in Fig. 1.11*a* has even symmetry. Thus $f_Y(z - x) = f_Y(x - z)$. The convolution integral expression of Eq. (1.12.6) is then the integral of a rectangular pulse multiplied by a similar pulse shifted to the right

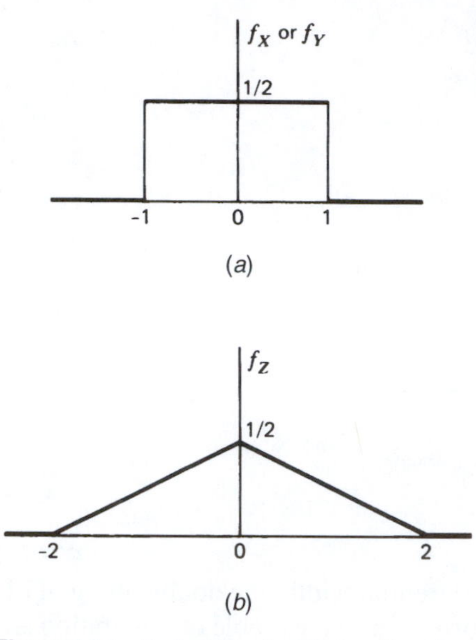

Figure 1.11 Probability density functions for X and Y and their sum. (a) Density function for both X and Y. (b) Density function for Z, where $Z = X + Y$.

amount z. When $z > 2$ or $z < 2$, there is no overlap in the pulses so their product is zero. When $-2 \leq z \leq 0$, there is a nontrivial overlap that increases linearly beginning at $z = -2$ and reaching a maximum at $z = 0$. The convolution integral then increases accordingly as shown in Fig. 1.11b. A similar argument may be used to show that $f_Z(z)$ decreases linearly in the interval where $0 \leq z \leq 2$. This leads to the triangular density function of Fig. 1.11b.

We can go one step further now and look at the density corresponding to the sum of three random variables. Let W be defined as

$$W = X + Y + V \tag{1.12.8}$$

where X, Y, and V are mutually independent and have identical rectangular densities as shown in Fig. 1.11a. We have already worked out the density for $X + Y$, so the density of W is the convolution of the two functions shown in Figs. 1.11a and 1.11b. We will leave the details of this as an exercise, and the result is shown in Fig. 1.12. Each of the segments labeled 1, 2, and 3 is an arc of a parabola. Notice the smooth central tendency. With a little imagination one can see a similarity between this and a zero-mean normal density curve. If we were to go another step and convolve a rectangular density with that of Fig. 1.12, we would get the density for the sum of four independent random variables. The resulting function would consist of connected segments of cubic functions extending from -4 to $+4$. Its appearance, though not shown, would resemble the normal curves even more than that of Fig. 1.12. And on and on—each additional convolution results in a curve that resembles the normal curve more closely than the preceding one.

This simple example is intended to demonstrate (not prove) that a superposition of independent random variables always tends toward normality, regardless of

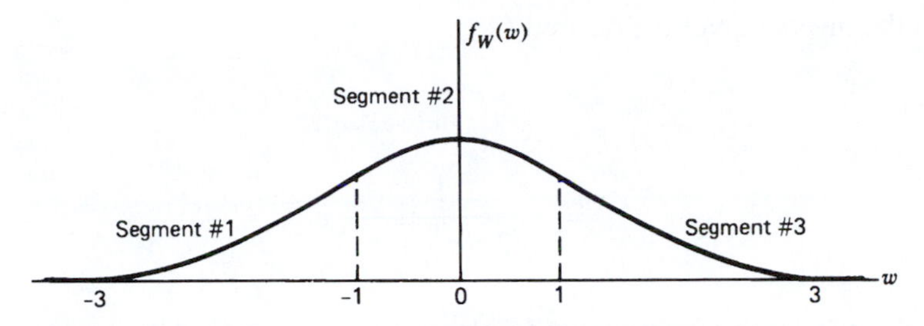

Figure 1.12 Probability density for the sum of three independent random variables with identical rectangular density functions.

the distribution of the individual random variables contributing to the sum. This is known as the *central limit theorem* of statistics. It is a most remarkable theorem, and its validity is subject to only modest restrictions (3). In engineering applications the noise we must deal with is frequently due to a superposition of many small contributions. When this is so, we have good reason to make the assumption of normality. The central limit theorem says to do just that. Thus we have here one of the reasons for our seemingly exaggerated interest in normal random variables—they are a common occurrence in nature.

EXAMPLE 1.11

Let X and Y be independent normal random variables with zero means and variances σ_X^2 and σ_Y^2. We wish to find the probability density function for the sum of X and Y, which will again be denoted as Z. For variety, we illustrate the Fourier transform approach. The explicit expressions for f_X and f_Y are

$$f_X(t) = \frac{1}{\sqrt{2\pi}\sigma_X} e^{-t^2/2\sigma_X^2} \qquad (1.12.9)$$

$$f_Y(t) = \frac{1}{\sqrt{2\pi}\sigma_Y} e^{-t^2 2\sigma_Y^2} \qquad (1.12.10)$$

Note that we have used t as the dummy argument of the functions. It is of no consequence because it is integrated out in the transformation to the ω domain. Using Fourier transform tables, we find the transforms of f_X and f_Y to be

$$\mathfrak{F}[f_X] = e^{-\sigma_X^2\omega^2/2} \qquad (1.12.11)$$

$$\mathfrak{F}[f_Y] = e^{-\sigma_Y^2\omega^2/2} \qquad (1.12.12)$$

Forming their product yields

$$\mathfrak{F}[f_X]\mathfrak{F}[f_Y] = e^{-(\sigma_X^2+\sigma_Y^2)\omega^2/2} \qquad (1.12.13)$$

Then the inverse gives the desired f_Z:

$$
\begin{aligned}
f_Z(z) &= \mathfrak{F}^{-1}[e^{-(\sigma_X^2+\sigma_Y^2)\omega^2/2}] \\
&= \frac{1}{\sqrt{2\pi(\sigma_X^2+\sigma_Y^2)}} e^{-z^2/2(\sigma_X^2+\sigma_Y^2)}
\end{aligned} \tag{1.12.14}
$$

Note that the density function for Z is also normal in form, and its variance is given by

$$
\sigma_Z^2 = \sigma_X^2 + \sigma_Y^2 \tag{1.12.15}
$$

The summation of any number of random variables can always be thought of as a sequence of summing operations on two variables; therefore, it should be clear that summing any number of independent normal random variables leads to a normal random variable. This rather remarkable result can be generalized further to include the case of dependent normal random variables, which we will discuss later.

■

1.13 TRANSFORMATION OF RANDOM VARIABLES

A mathematical transformation that takes one set of variables (say, inputs) into another set (say, outputs) is a common situation in systems analysis. Let us begin with a simple single-input, single-output situation where the input–output relationship is governed by the algebraic equation

$$
y = g(x) \tag{1.13.1}
$$

Here we are interested in random inputs, so think of x as a realization of the input random variable X, and y as the corresponding realization of the output Y. Assume we know the probability density function for X, and would like to find the corresponding density for Y. It is tempting to simply replace x in $f_X(x)$ with its equivalent in terms of y and pass it off at that. However, it is not quite that simple, as will be seen presently.

First, let us assume that the transformation $g(x)$ is one-to-one for all permissible x. By this we mean that the functional relationship given by Eq. (1.13.1) can be reversed, and x can be written uniquely as a function of y. Let the "reverse" relationship be

$$
x = h(y) \tag{1.13.2}
$$

The probabilities that X and Y lie within corresponding differential regions must be equal. That is,

$$
P(X \text{ is between } x \text{ and } x + dx) = P(Y \text{ is between } y \text{ and } y + dy) \tag{1.13.3}
$$

or

$$\int_x^{x+dx} f_X(u)du = \begin{cases} \int_y^{y+dy} f_Y(u)du, & \text{for } dy \text{ positive} \\ -\int_y^{y+dy} f_Y(u)du, & \text{for } dy \text{ negative} \end{cases} \tag{1.13.4}$$

One of the subtleties of this problem should now be apparent from Eq. (1.13.4). If positive dx yields negative dy (i.e., a negative derivative), the integral of f_Y must be taken from $y + dy$ to y in order to yield a positive probability. This is the equivalent of interchanging the limits and reversing the sign as shown in Eq. (1.13.4).

The differential equivalent of Eq. (1.13.4) is

$$f_X(x)dx = f_Y(y)|dy| \tag{1.13.5}$$

where we have tacitly assumed dx to be positive. Also, x is constrained to be $h(y)$. Thus, we have

$$f_Y(y) = \left|\frac{dx}{dy}\right| f_X(h(y)) \tag{1.13.6}$$

or, equivalently,

$$f_Y(y) = |h'(y)| f_X(h(y)) \tag{1.13.7}$$

where $h'(y)$ indicates the derivative of h with respect to y. A simple example where the transformation is linear will now be presented.

EXAMPLE 1.12

Find the appropriate output density functions for the case where the input X is $\mathcal{N}(0, \sigma_X^2)$ and the transformation is

$$y = Kx \quad (K \text{ is a given constant}) \tag{1.13.8}$$

We begin with the scale-factor transformation indicated by Eq. (1.13.8). We first solve for x in terms of y and then form the derivative. Thus,

$$x = \frac{1}{K} y \tag{1.13.9}$$

$$\left|\frac{dx}{dy}\right| = \left|\frac{1}{K}\right| \tag{1.13.10}$$

We can now obtain the equation for f_Y from Eq. (1.13.6). The result is

$$f_Y(y) = \frac{1}{|K|} \cdot \frac{1}{\sqrt{2\pi}\sigma_X} \exp\left[-\frac{\left(\frac{y}{K}\right)^2}{2\sigma_X^2}\right] \tag{1.13.11}$$

Or rewriting Eq. (1.13.11) in standard normal form yields

$$f_Y(y) = \frac{1}{\sqrt{2\pi(K\sigma_X)^2}} \exp\left[-\frac{y^2}{2(K\sigma_X)^2}\right] \tag{1.13.12}$$

It can now be seen that transforming a zero-mean normal random variable with a simple scale factor yields another normal random variable with a corresponding scale change in its standard deviation. It is important to note that normality is preserved in a linear transformation.

■

Our next example will illustrate a nonlinear transformation of bivariate random variable.

EXAMPLE 1.13

In the dart-throwing example (Example 1.8), the hit location was described in terms of rectangular coordinates. This led to the joint probability density function

$$f_{XY}(x, y) = \frac{1}{2\pi\sigma^2} e^{-(x^2+y^2)/2\sigma^2} \tag{1.13.13}$$

This is a special case of the bivariate normal density function where the two variates are independent and have zero means and equal variances. Suppose we wish to find the corresponding density in terms of polar coordinates r and θ. Formally, then, we wish to define new random variables R and Θ such that pairwise realizations (x, y) transform to (r, θ) in accordance with

$$\begin{aligned} r &= \sqrt{x^2 + y^2}, \quad r \geq 0 \\ \theta &= \tan^{-1}\frac{y}{x}, \quad 0 \leq \theta < 2\pi \end{aligned} \tag{1.13.14}$$

Or, equivalently,

$$\begin{aligned} x &= r\cos\theta \\ y &= r\sin\theta \end{aligned} \tag{1.13.15}$$

We wish to find $f_{R\Theta}(r, \theta)$ and the unconditional density functions $f_R(r)$ and $f_\Theta(\theta)$.

The probability that a hit will lie within an area bounded by a closed contour C is given by

$$P(\text{Hit lies within } C) = \iint\limits_{\substack{\text{Area}\\ \text{enclosed}\\ \text{by } C}} f_{XY}(x, y)\, dx\, dy \tag{1.13.16}$$

We know from multivariable calculus that the double integral in Eq. (1.13.16) can also be evaluated in the r, θ coordinate frame as

$$\iint\limits_{\substack{\text{Region}\\ \text{encosed}\\ \text{by } C}} f_{XY}(x,y)\,dx\,dy = \iint\limits_{\substack{\text{Region}\\ \text{encosed}\\ \text{by } C'}} f_{XY}(x(r,\theta),y(r,\theta))\left|J\left(\frac{x,y}{r,\theta}\right)\right| dr\,d\theta \tag{1.13.17}$$

where C' is the contour in the r, θ plane corresponding to C in the x, y plane. That is, points within C map into points within C'. (Note that it is immaterial as to how we draw the "picture" in the r, θ coordinate frame. For example, if we choose to think of r and θ as just another set of Cartesian coordinates for plotting purposes, C' becomes a distortion of C.) The J quantity in Eq. (1.13.17) is the Jacobian of the transformation, defined as

$$J\left(\frac{x,y}{r,\theta}\right) = \text{Det}\begin{bmatrix} \dfrac{\partial x}{\partial r} & \dfrac{\partial y}{\partial r} \\ \dfrac{\partial x}{\partial \theta} & \dfrac{\partial y}{\partial \theta} \end{bmatrix} \tag{1.13.18}$$

The vertical bars around J in Eq. (1.13.17) indicate absolute magnitude. We can argue now that if Eq. (1.13.17) is true for regions in general, it must also be true for differential regions. Let the differential region in the r, q domain be bounded by r and $r + dr$ in one direction and by θ and $\theta + d\theta$ in the other. If it helps, think of plotting r and θ as Cartesian coordinates. The differential region in the r, θ domain is then rectangular, and the corresponding one in the x, y domain is a curvilinear differential rectangle (see Fig. 1.13). Now, by the very definition of joint density, the quantity multiplying $dr\,d\theta$ in Eq. (1.13.17) is seen to be the desired density function. That is,

$$f_{R\Theta}(r,\theta) = f_{XY}[x(r,\theta),y(r,\theta)]\left|J\left(\frac{x,y}{r,\theta}\right)\right| \tag{1.13.19}$$

In the transformation of this example,

$$J\left(\frac{x,y}{r,\theta}\right) = \text{Det}\begin{bmatrix} \cos\theta & \sin\theta \\ -r\sin\theta & r\cos\theta \end{bmatrix} = r \tag{1.13.20}$$

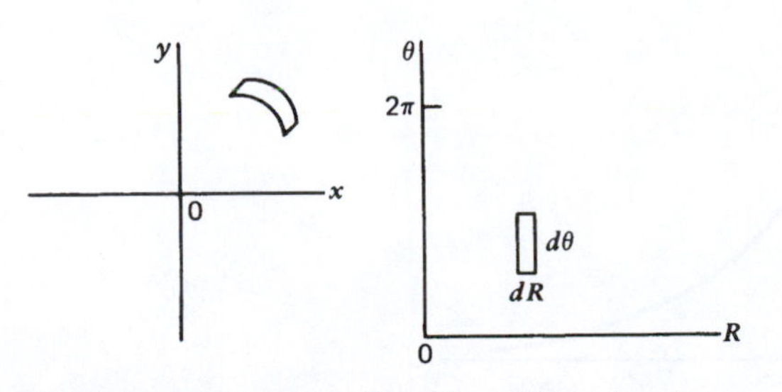

Figure 1.13 Corresponding differential regions for transformation of Example 1.13.

Since the radial coordinate r is always positive, $|J| = r$. We can now substitute Eqs. (1.13.13) and (1.13.20) into (1.13.19) and obtain

$$\begin{aligned} f_{R\Theta}(r,\theta) &= r\frac{1}{2\pi\sigma^2}\exp\left[-\frac{(r\cos\theta)^2 + (r\sin\theta)^2}{2\sigma^2}\right] \\ &= \frac{r}{2\pi\sigma^2}e^{-r^2/2\sigma^2} \end{aligned} \tag{1.13.21}$$

Note that the density function of this example has no explicit dependence on θ. In other words, all angles between 0 and 2π are equally likely, which is what we would expect in the target-throwing experiment.

We get the unconditional density functions by integrating $f_{R\Theta}$ with respect to the appropriate variables. That is,

$$\begin{aligned} f_R(r) &= \int_0^{2\pi} f_{R\Theta}(r,\theta)\,d\theta = \frac{r}{2\pi\sigma^2}e^{-r^2/2\sigma^2}\int_0^{2\pi} d\theta \\ &= \frac{r}{\sigma^2}e^{-r^2/2\sigma^2} \end{aligned} \tag{1.13.22}$$

and

$$f_\Theta(\theta) = \int_0^\infty f_{R\Theta}(r,\theta)\,dr = \begin{cases} \dfrac{1}{2\pi}, & 0 \le \theta < 2\pi \\ 0, & \text{otherwise} \end{cases} \tag{1.13.23}$$

The single-variate density function given by Eq. (1.13.22) is called the *Rayleigh* density function. It is of considerable importance in applied probability, and it is sketched in Fig. 1.14. It is easily verified that the mode (peak value) of the Rayleigh density is equal to standard deviation of the x and y normal random variables from which it was derived. Thus, we see that similar independent, zero-mean normal densities in the x, y domain correspond to Rayleigh and uniform densities in the r, θ domain.

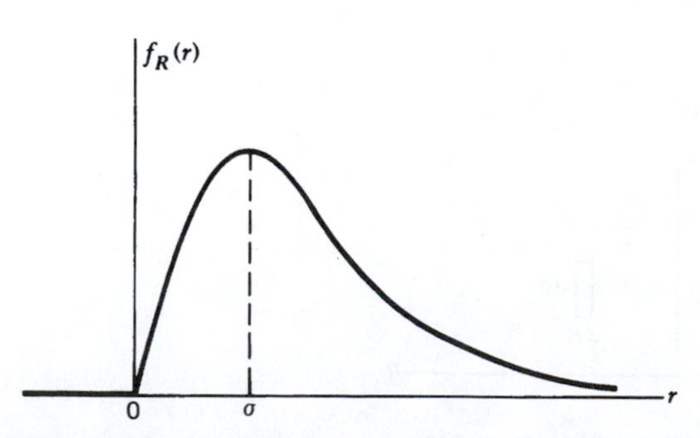

Figure 1.14 Rayleigh density function.

1.14
MULTIVARIATE NORMAL DENSITY FUNCTION

In Sections 1.8 and 1.10 examples of the single and bivariate normal density functions were presented. We work so much with normal random variables that we need to elaborate further and develop a general form for the n-dimensional normal density function. One can write out the explicit equations for the single and bivariate cases without an undue amount of "clutter" in the equations. However, beyond this, matrix notation is a virtual necessity. Otherwise, the explicit expressions are completely unwieldy.

Consider a set of n random variables $X_1, X_2, \ldots, X_n$ (also called variates). We define a vector random variable $\mathbf{X}$ as*

$$\mathbf{X} = \begin{bmatrix} X_1 \\ X_2 \\ \vdots \\ X_n \end{bmatrix} \tag{1.14.1}$$

In general, the components of $\mathbf{X}$ may be correlated and have nonzero means. We denote the respective means as $m_1, m_2, \ldots, m_n$, and thus, we define a mean vector $\mathbf{m}$ as

$$\mathbf{m} = \begin{bmatrix} m_1 \\ m_2 \\ \vdots \\ m_n \end{bmatrix} \tag{1.14.2}$$

Also, if $x_1, x_2, \ldots, x_n$ is a set of realizations of $X_1, X_2, \ldots, X_n$, we can define a vector realization of $\mathbf{X}$ as

$$\mathbf{x} = \begin{bmatrix} x_1 \\ x_2 \\ \vdots \\ x_n \end{bmatrix} \tag{1.14.3}$$

We next define a matrix that describes the variances and correlation structure of the n variates. The *covariance matrix* for $\mathbf{X}$ is defined as

$$\mathbf{C} = \begin{bmatrix} E[(X_1 - m_1)^2] & E[(X_1 - m_1)(X_2 - m_2)] & \cdots & \\ E[(X_2 - m_2)(X_1 - m_1)] & \ddots & & \\ \vdots & & & E[(X_n - m_n)^2] \end{bmatrix} \tag{1.14.4}$$

*Note that uppercase $\mathbf{X}$ denotes a column vector in this section. This is a departure from the usual notation of matrix theory, but it is necessitated by a desire to be consistent with the previous uppercase notation for random variables. The reader will simply have to remember this minor deviation in matrix notation. It appears in this section and also occasionally in Chapter 2.

The terms along the major diagonal of $\mathbf{C}$ are seen to be the variances of the variates, and the off-diagonal terms are the covariances.

The random variables $X_1, X_2, \ldots, X_n$ are said to be *jointly normal* or *jointly Gaussian* if their joint probability density function is given by

$$f_X(\mathbf{x}) = \frac{1}{(2\pi)^{n/2}|\mathbf{C}|^{1/2}} \exp\left\{ -\frac{1}{2}\left[(\mathbf{x}-\mathbf{m})^T \mathbf{C}^{-1}(\mathbf{x}-\mathbf{m})\right] \right\} \tag{1.14.5}$$

where $\mathbf{x}$, $\mathbf{m}$, and $\mathbf{C}$ are defined by Eqs. (1.14.2) to (1.14.4) and $|\mathbf{C}|$ is the determinant of $\mathbf{C}$. "Super -1" and "super T" denote matrix inverse and transpose, respectively. Note that the defining function for f_X is scalar and is a function of $x, x_2, \ldots, x_n$ when written out explicitly. We have shortened the indicated functional dependence to $\mathbf{x}$ just for compactness in notation. Also note that $\mathbf{C}^{-1}$ must exist in order for f_X to be properly defined by Eq. (1.14.5). Thus, $\mathbf{C}$ must be nonsingular. More will be said of this later.

Clearly, Eq. (1.14.5) reduces to the standard normal form for the single variate case. For the bivariate case, we may write out f_X explicitly in terms of x_1 and x_2 without undue difficulty. Proceeding to do this, we have

$$\mathbf{X} = \begin{bmatrix} X_1 \\ X_2 \end{bmatrix}, \quad \mathbf{x} = \begin{bmatrix} x_1 \\ x_2 \end{bmatrix}, \quad \mathbf{m} = \begin{bmatrix} m_1 \\ m_2 \end{bmatrix} \tag{1.14.6}$$

and

$$\mathbf{C} = \begin{bmatrix} E[(X_1-m_1)^2] & E[(X_1-m_1)(X_2-m_2)] \\ E[(X_1-m_1)(X_2-m_2)] & E[(X_2-m_2)^2] \end{bmatrix} = \begin{bmatrix} \sigma_1^2 & \rho\sigma_1\sigma_2 \\ \rho\sigma_1\sigma_2 & \sigma_2^2 \end{bmatrix} \tag{1.14.7}$$

The second form for $\mathbf{C}$ in Eq. (1.14.7) follows directly from the definitions of variance and correlation coefficient. The determinant of $\mathbf{C}$ and its inverse are given by

$$|\mathbf{C}| = \begin{vmatrix} \sigma_1^2 & \rho\sigma_1\sigma_2 \\ \rho\sigma_1\sigma_2 & \sigma_2^2 \end{vmatrix} = (1-\rho^2)\sigma_1^2\sigma_2^2 \tag{1.14.8}$$

$$\mathbf{C}^{-1} = \begin{bmatrix} \dfrac{\sigma_2^2}{|\mathbf{C}|} & -\dfrac{\rho\sigma_1\sigma_2}{|\mathbf{C}|} \\ -\dfrac{\rho\sigma_1\sigma_2}{|\mathbf{C}|} & \dfrac{\sigma_1^2}{|\mathbf{C}|} \end{bmatrix} = \begin{bmatrix} \dfrac{1}{(1-\rho^2)\sigma_1^2} & \dfrac{-\rho}{(1-\rho^2)\sigma_1\sigma_2} \\ \dfrac{-\rho}{(1-\rho^2)\sigma_1\sigma_2} & \dfrac{1}{(1-\rho^2)\sigma_2^2} \end{bmatrix} \tag{1.14.9}$$

Substituting Eqs. (1.14.8) and (1.14.9) into Eq. (1.14.5) then yields the desired density function in terms of x_1 and x_2.

$$f_{x_1x_2}(x_1, x_2) = \frac{1}{2\pi\sigma_1\sigma_2\sqrt{1-\rho^2}} \exp\left\{ -\frac{1}{2(1-\rho^2)}\left[\frac{(x_1-m_1)^2}{\sigma_1^2} - \frac{2\rho(x_1-m_1)(x_2-m_2)}{\sigma_1\sigma_2} + \frac{(x_2-m_2)^2}{\sigma_2^2}\right]\right\} \tag{1.14.10}$$

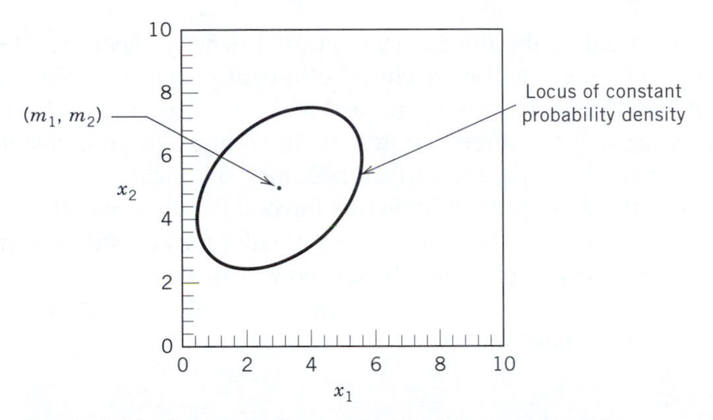

Figure 1.15 Contour projection of a bivariate normal density function to the (X_1, X_2) plane.

It should be clear in Eq. (1.14.10) that $f_{x_1x_2}(x_1, x_2)$ is intended to mean the same as $f_X(\mathbf{x})$ in vector notation.

As mentioned previously, the third- and higher-order densities are very cumbersome to write out explicitly; therefore, we will examine the bivariate density in some detail in order to gain insight into the general multivariate normal density function. The bivariate normal density function is a smooth hill-shaped surface over the x_1, x_2 plane. This was sketched previously in Fig. 1.9 for the special case where $\sigma_1 = \sigma_2$ and $\rho = 0$. In the more general case, a constant probability density contour projects into the x_1, x_2 plane as an ellipse with its center at (m_1, m_2) as shown in Fig. 1.15. The orientation of the ellipse in Fig. 1.15 corresponds to a positive correlation coefficient. Points on the ellipse may be thought of as equally likely combinations of x_1 and x_2. If $\rho = 0$, we have the case where X_1 and X_2 are uncorrelated, and the ellipses have their semimajor and semiminor axes parallel to the x_1 and x_2 axes. If we specialize further and let $\sigma_1 = \sigma_2$ (and $\rho = 0$), the ellipses degenerate to circles. In the other extreme, as $|\rho|$ approaches unity, the ellipses become more and more eccentric.

The uncorrelated case where $\rho = 0$ is of special interest, and in this case $f_{x_1x_2}$ reduces to the form given in Eq. (1.14.11).

For uncorrelated X_1 and X_2

$$\begin{aligned} f_{X_1X_2}(x_1, x_2) &= \frac{1}{2\pi\sigma_1\sigma_2}\exp\left\{-\frac{1}{2}\left[\frac{(x_1 - m_1)^2}{\sigma_1^2} + \frac{(x_2 - m_2)^2}{\sigma_2^2}\right]\right\} \\ &= \frac{1}{\sqrt{2\pi}\sigma_1}e^{-(x_1-m_1)^2/2\sigma_1^2} \cdot \frac{1}{\sqrt{2\pi}\sigma_2}e^{-(x_2-m_2)^2/2\sigma_2^2} \end{aligned} \tag{1.14.11}$$

The joint density function is seen to factor into the product of $f_{X_1}(x_1)$ and $f_{X_2}(x_2)$. Thus, *two normal random variables that are uncorrelated are also statistically independent*. It is easily verified from Eq. (1.14.5) that this is also true for any number of uncorrelated normal random variables. This is exceptional, because in general zero correlation does not imply statistical independence. It does, however, in the Gaussian case.

Now try to visualize the three-variate normal density function. The locus of constant $f_{X_1X_2X_3}(x_1, x_2, x_3)$ will be a closed elliptically shaped surface with three axes of symmetry. These axes will be aligned with the x_1, x_2, x_3 axes for the case of zero correlation among the three variates. If, in addition to zero correlation, the variates have equal variances, the surface becomes spherical.

If we try to extend the geometric picture beyond the three-variate case, we run out of Euclidean dimensions. However, conceptually, we can still envision equal-likelihood surfaces in hyperspace, but there is no way of sketching a picture of such surfaces. Some general properties of multivariate normal random variables will be explored further in the next section.

1.15 LINEAR TRANSFORMATION AND GENERAL PROPERTIES OF NORMAL RANDOM VARIABLES

The general linear transformation of one set of normal random variables to another is of special interest in noise analysis. This will now be examined in detail.

We have just seen that the density function for jointly normal random variables $X_1, X_2, \ldots, X_n$ can be written in matrix form as

$$f_X(\mathbf{x}) = \frac{1}{(2\pi)^{n/2}|\mathbf{C}_X|^{1/2}} \exp\left\{-\frac{1}{2}\left[(\mathbf{x} - \mathbf{m}_X)^T \mathbf{C}_X^{-1}(\mathbf{x} - \mathbf{m}_X)\right]\right\} \tag{1.15.1}$$

We have added the subscript X to $\mathbf{m}$ and $\mathbf{C}$ to indicate that these are the mean and covariance matrices associated with the $\mathbf{X}$ random variable. We now define a new set of random variables $Y_1, Y_2, \ldots, Y_n$ that are linearly related to $X_1, X_2, \ldots, X_n$ via the equation

$$\mathbf{y} = \mathbf{A}\mathbf{x} + \mathbf{b} \tag{1.15.2}$$

where lowercase $\mathbf{x}$ and $\mathbf{y}$ indicate realizations of $\mathbf{X}$ and $\mathbf{Y}$, $\mathbf{b}$ is a constant vector, and $\mathbf{A}$ is a square matrix that will be assumed to be nonsingular (i.e., invertible). We wish to find the density function for $\mathbf{Y}$, and we can use the methods of Section 1.13 to do so. In particular, the transformation is one-to-one; therefore, a generalized version of Eq. (1.13.32) may be used.

$$f_Y(\mathbf{y}) = f_X(\mathbf{x}(\mathbf{y}))\left|J\left(\frac{\mathbf{x}}{\mathbf{y}}\right)\right| \tag{1.15.3}$$

We must first solve Eq. (1.15.2) for $\mathbf{x}$ in terms of $\mathbf{y}$. The result is

$$\mathbf{x} = \mathbf{A}^{-1}\mathbf{y} - \mathbf{A}^{-1}\mathbf{b} \tag{1.15.4}$$

Let the individual terms of $\mathbf{A}^{-1}$ be denoted as

$$\mathbf{A}^{-1} = \begin{bmatrix} d_{11} & d_{12} & \cdots & d_{1n} \\ d_{21} & d_{22} & \cdots & \\ \cdot & \cdot & \cdots & \\ d_{n1} & \cdot & \cdots & d_{nn} \end{bmatrix} \tag{1.15.5}$$

The scalar equations represented by Eq. (1.15.4) are then

$$\begin{aligned} x_1 &= (d_{11}y_1 + d_{12}y_2 + \cdots) - (d_{11}b_1 + d_{12}b_2 + \cdots) \\ x_2 &= (d_{21}y_1 + d_{22}y_2 + \cdots) - (d_{21}b_1 + d_{22}b_2 + \cdots) \\ x_3 &= \cdots \text{etc.} \end{aligned} \tag{1.15.6}$$

The Jacobian of the transformation is then

$$\begin{aligned} J\left(\frac{\mathbf{x}}{\mathbf{y}}\right) &= J\left(\frac{x_1, x_2, \ldots}{y_1, y_2, \ldots}\right) = \text{Det}\begin{bmatrix} \dfrac{\partial x_1}{\partial y_1} & \dfrac{\partial x_2}{\partial y_1} & \cdots \\ \dfrac{\partial x_1}{\partial y_2} & \dfrac{\partial x_2}{\partial y_2} & \cdots \\ \cdots\cdots & \cdots\cdots & \cdots \end{bmatrix} \\ &= \text{Det}\begin{bmatrix} d_{11} & d_{21} & \cdots \\ d_{12} & d_{22} & \cdots \\ \cdots & \cdots & \cdots \end{bmatrix} = \text{Det}(\mathbf{A}^{-1})^T = \text{Det}(\mathbf{A}^{-1}) \end{aligned} \tag{1.15.7}$$

We can now substitute Eqs. (1.15.4) and (1.15.7) into Eq. (1.15.3). The result is

$$\begin{aligned} f_Y(\mathbf{y}) = &\frac{|\text{Det}(\mathbf{A}^{-1})|}{(2\pi)^{n/2}|\mathbf{C}_X|^{1/2}} \\ &\times \exp\left\{-\frac{1}{2}\left[(\mathbf{A}^{-1}\mathbf{y} - \mathbf{A}^{-1}\mathbf{b} - \mathbf{m}_X)^T \mathbf{C}_X^{-1} (\mathbf{A}^{-1}\mathbf{y} - \mathbf{A}^{-1}\mathbf{b} - \mathbf{m}_X)\right]\right\} \end{aligned} \tag{1.15.8}$$

We find the mean of $\mathbf{Y}$ by taking the expectation of both sides of the linear transformation

$$\mathbf{Y} = \mathbf{AX} + \mathbf{b}$$

Thus,

$$\mathbf{m}_Y = \mathbf{A}\mathbf{m}_X + \mathbf{b} \tag{1.15.9}$$

The exponent in Eq. (1.15.8) may now be written as

$$
\begin{aligned}
&-\frac{1}{2}[(\mathbf{A}^{-1}\mathbf{y}-\mathbf{A}^{-1}\mathbf{b}-\mathbf{A}^{-1}\mathbf{A}\mathbf{m}_X)^T\mathbf{C}_X^{-1}(\mathbf{A}^{-1}\mathbf{y}-\mathbf{A}^{-1}\mathbf{b}-\mathbf{A}^{-1}\mathbf{A}\mathbf{m}_X)] \\
&\quad=-\frac{1}{2}[(\mathbf{y}-\mathbf{m}_Y)^T(\mathbf{A}^{-1})^T\mathbf{C}_X^{-1}\mathbf{A}^{-1}(\mathbf{y}-\mathbf{m}_Y)] \\
&\quad=-\frac{1}{2}\left[(\mathbf{y}-\mathbf{m}_Y)^T(\mathbf{A}\mathbf{C}_X\mathbf{A}^T)^{-1}(\mathbf{y}-\mathbf{m}_Y)\right]
\end{aligned}
\tag{1.15.10}
$$

The last step in Eq. (1.15.10) is accomplished by using the reversal rule for the inverse of triple products and noting that the order of the transpose and inverse operations may be interchanged. Also note that

$$
|\mathrm{Det}(\mathbf{A}^{-1})|=\frac{1}{|\mathrm{Det}\,\mathbf{A}|}=\frac{1}{|\mathrm{Det}\,\mathbf{A}|^{1/2}\cdot|\mathrm{Det}\,\mathbf{A}^T|^{1/2}} \tag{1.15.11}
$$

Substitution of the forms given in Eqs. (1.15.10) and (1.15.11) into Eq. (1.15.8) yields for f_Y

$$
\begin{aligned}
f_Y(\mathbf{y}) &= \frac{1}{(2\pi)^{n/2}|\mathbf{A}\mathbf{C}_X\mathbf{A}^T|^{1/2}} \\
&\quad\times\exp\left\{-\frac{1}{2}\left[(\mathbf{y}-\mathbf{m}_Y)^T(\mathbf{A}\mathbf{C}_X\mathbf{A}^T)^{-1}(\mathbf{y}-\mathbf{m}_Y)\right]\right\}
\end{aligned}
\tag{1.15.12}
$$

It is apparent now that f_Y is also normal in form with the mean and covariance matrix given by

$$
\mathbf{m}_Y=\mathbf{A}\mathbf{m}_X+\mathbf{b} \tag{1.15.13}
$$

and

$$
\mathbf{C}_Y=\mathbf{A}\mathbf{C}_X\mathbf{A}^T \tag{1.15.14}
$$

Thus, we see that *normality is preserved in a linear transformation.* All that is changed is the mean and the covariance matrix; the *form* of the density function remains unchanged.

There are, of course, an infinite number of linear transformations one can make on a set of normal random variables. Any transformation, say, **S**, that produces a new covariance matrix $\mathbf{S}\mathbf{C}_X\mathbf{S}^T$ that is diagonal is of special interest. Such a transformation will yield a new set of normal random variables that are uncorrelated, and thus they are also statistically independent. In a given problem, we may not choose to actually make this change of variables, but it is important just to know that the variables can be decoupled and under what circumstances this can be done. It works out that a diagonalizing transformation will always exist if $\mathbf{C}_X$ is positive definite (8).* In the case of a covariance matrix, this implies that all the correlation coefficients are less than unity in magnitude. This will be demonstrated for the bivariate case, and the extension to higher-order cases is fairly obvious.

A symmetric matrix $\mathbf{C}$ is said to be positive definite if the scalar $\mathbf{x}^T\mathbf{C}\mathbf{x}$ is positive for all nontrivial $\mathbf{x}$, that is, $\mathbf{x} \neq 0$. Writing out $\mathbf{x}^T\mathbf{C}\mathbf{x}$ explicitly for the 2×2 case yields

$$[x_1 x_2]\begin{bmatrix} c_{11} & c_{12} \\ c_{12} & c_{22} \end{bmatrix}\begin{bmatrix} x_1 \\ x_2 \end{bmatrix} = c_{11}x_1^2 + 2c_{12}x_1x_2 + c_{22}x_2^2 \tag{1.15.15}$$

But if $\mathbf{C}$ is a covariance matrix,

$$c_{11} = \sigma_1^2, \quad c_{12} = \rho\sigma_1\sigma_2, \quad c_{22} = \sigma_2^2 \tag{1.15.16}$$

Therefore, $\mathbf{x}^T\mathbf{C}\mathbf{x}$ is

$$\mathbf{x}^T\mathbf{C}\mathbf{x} = (\sigma_1 x_1)^2 + 2\rho(\sigma_1 x_1)(\sigma_2 x_2) + (\sigma_2 x_2)^2 \tag{1.15.17}$$

Equation (1.15.17) now has a simple geometric interpretation. Assume $|\rho| < 1$; ρ can be related to the negative cosine of some angle θ, where $0 < \theta < \pi$. Equation (1.15.17) will then be recognized as the equation for the square of the "opposite side" of a general triangle; and this, of course, must be positive. Thus, a 2×2 covariance matrix is positive definite, provided $|\rho| < 1$.

It is appropriate now to summarize some of the important properties of multivariate normal random variables:

1. The probability density function describing a vector random variable $\mathbf{X}$ is completely defined by specifying the mean and covariance matrix of $\mathbf{X}$.
2. The covariance matrix of $\mathbf{X}$ is positive definite. The magnitudes of all correlation coefficients are less than unity.
3. If normal random variables are uncorrelated, they are also statistically independent.
4. A linear transformation of normal random variables leads to another set of normal random variables. A decoupling (decorrelating) transformation will always exist if the original covariance matrix is positive definite.
5. If the joint density function for n random variables is normal in form, all marginal and conditional densities associated with the n variates will also be normal in form.

1.16 LIMITS, CONVERGENCE, AND UNBIASED ESTIMATORS

No discussion of probability could be complete without at least some mention of limits and convergence. To put this in perspective, we first review the usual

* MATLAB's Cholesky factorization function is helpful in determining a diagonalizing transformation. This is discussed in detail in Section 3.10 on Monte Carlo simulation.

deterministic concept of convergence. As an example, recall that the Maclaurin series for e^x is

$$e^x = 1 + x + \frac{x^2}{2!} + \frac{x^3}{3!} + \cdots \tag{1.16.1}$$

This series converges uniformly to e^x for all real x in any finite interval. By convergence we mean that if a given accuracy figure is specified, we can find an appropriate number of terms such that the specified accuracy is met by a truncated version of the series. In particular, note that once we have determined how many terms are needed in the truncated series, this same number is good for all x within the interval, and there is nothing "chancy" about it. In contrast, we will see presently that such "100 percent sure" statements cannot be made in probabilistic situations. A look at the sample mean of n random variables will serve to illustrate this.

Let $X_1, X_2, \ldots, X_n$ be independent random variables with identical probability density functions $f_X(x)$. In terms of an experiment, these may be thought of as ordered samples of the random variable X. Next, consider a sequence of random variables defined as follows:

$$\begin{aligned}
Y_1 &= X_1 \\
Y_2 &= \frac{X_1 + X_2}{2} \\
Y_3 &= \frac{X_1 + X_2 + X_3}{3} \\
&\vdots \\
Y_n &= \frac{X_1 + X_2 + \cdots X_n}{n}
\end{aligned} \tag{1.16.2}$$

The random variable Y_n is, of course, just the sample mean of the random variable X. We certainly expect Y_n to get closer to $E(X)$ as n becomes large. But closer in what sense? This is the crucial question. It should be clear that any particular experiment could produce an "unusual" event in which the sample mean would differ from $E(X)$ considerably. On the other hand, quite by chance, a similar experiment might yield a sample mean that was quite close to $E(X)$. Thus, in this probabilistic situation, we cannot expect to find a fixed number of samples n that will meet a specified accuracy figure for all experiments. No matter how large we make n, there is always some nonzero probability that the very unusual thing will happen, and a particular experiment will yield a sample mean that is outside the specified accuracy. Thus, we can only hope for convergence in some sort of average sense and not in an absolute (100 percent sure) sense.

Let us now be more specific in this example, and let X (and thus $X_1, X_2, \ldots, X_n$) be normal with mean m_X and variance σ_X^2. From Section 1.15 we also know that the sample mean Y_n is a normal random variable. Since a normal random variable is characterized by its mean and variance, we now examine these parameters for Y_n.

The expectation of a sum of elements is the sum of the expectations of the elements. Thus,

$$\begin{aligned} E(Y_n) &= E\left(\frac{X_1 + X_2 + \cdots X_n}{n}\right) \\ &= \frac{1}{n}[E(X_1) + E(X_2) + \cdots] \\ &= \frac{1}{n}[nE(X)] = m_X \end{aligned} \tag{1.16.3}$$

The sample mean is, of course, an estimate of the true mean of X, and we see from Eq. (1.16.3) that it at least yields $E(X)$ "on the average." Estimators that have this property are said to be unbiased. That is, an estimator is said to be *unbiased* if

$$E(\text{Estimate of } X) = E(X) \tag{1.16.4}$$

Consider next the variance of Y_n. Using Eq. (1.16.3) and recalling that the expectation of the sum is the sum of the expectations, we obtain

$$\begin{aligned} \text{Var}\, Y_n &= E[Y_n - E(Y_n)]^2 \\ &= E(Y_n^2 - 2Y_n m_X + m_X^2) \\ &= E(Y_n^2) - m_X^2 \end{aligned} \tag{1.16.5}$$

The sample mean Y_n may now be replaced with $(1/n)(X_1 + X_2 \cdots + X_n)$; and, after squaring and some algebraic simplification, Eq. (1.16.5) reduces to

$$\begin{aligned} \text{Var}\, Y_n &= \frac{1}{n}\text{Var}\, X \\ &= \frac{\sigma_X^2}{n} \end{aligned} \tag{1.16.6}$$

Thus, we see that the variance of the sample mean decreases with increasing n and eventually goes to zero as $n \to \infty$.

The probability density functions associated with the sample mean are shown in Fig. 1.16 for three values of n. It should be clear from the figure that convergence of some sort takes place as $n \to \infty$. However, no matter how large we make n, there will still remain a nonzero probability that Y_n will fall outside some specified accuracy interval. Thus, we have convergence in only a statistical sense and not in an absolute deterministic sense.

There are a number of types of statistical convergence that have been defined and are in common usage (9, 10). We look briefly at two of these. Consider a sequence of random variables $Y_1, Y_2, \ldots, Y_n$. The sequence Y_n is said to *converge in the mean* (or mean square) to Y if

$$\lim_{n\to\infty} E[(Y_n - Y)^2] = 0 \tag{1.16.7}$$

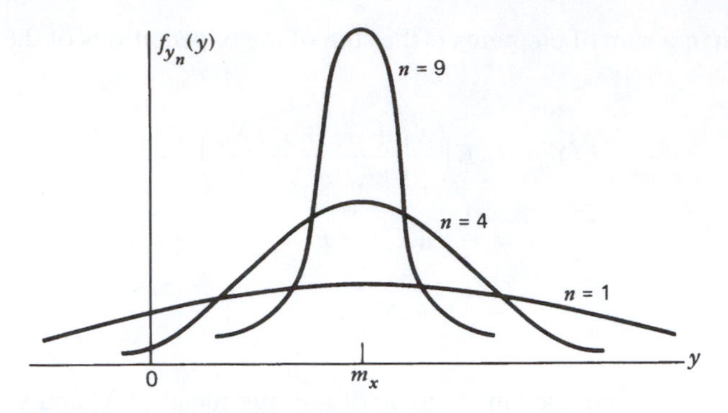

Figure 1.16 Probability density functions illustrating convergence of the sample mean.

Convergence in the mean is sometimes abbreviated as

$$\text{l.i.m. } Y_n = Y \tag{1.16.8}$$

where l.i.m. denotes "limit in the mean."

The sequence Y_n *converges in probability* to Y if

$$\lim_{n \to \infty} P(|Y_n - Y| \geq \varepsilon) = 0 \tag{1.16.9}$$

where ε is an arbitrarily small positive number.

It should be clear from Eqs. (1.16.3) and (1.16.6) that the sample mean converges "in the mean" to m_X. It also converges in probability because the area under the "tails" of the probability density outside a specified interval about m_X goes to zero as $n \to \infty$. Roughly speaking, convergence in the mean indicates that the dispersion (variance) about the limiting value shrinks to zero in the limit. Similarly, convergence in probability means than an arbitrarily small accuracy criterion is met with a probability of one as $n \to \infty$. Davenport and Root (9) point out that convergence in the mean is a more severe requirement than convergence in probability. Thus, if a sequence converges in the mean, we are also assured that it will converge in probability. The converse is not true though, because convergence in probability is a "looser" sort of criterion than convergence in the mean.

1.17
A NOTE ON STATISTICAL ESTIMATORS

Books on statistics usually give a whole host of types of estimators, each having its own descriptive title. For example, a *consistent* estimator is one that continually gets better and better with more and more observations; estimation efficiency is a measure of the accuracy of the estimate relative to what we could ever expect to achieve with a given set of observations; and so forth. We will not go on further. Most of these descriptors pertain to estimating parameters such as mean and/or variance of the random variable under consideration. These are important parameters, and then accurate determination from observations is important in real-life applications. However, estimating parameters is a statistical problem, not a filtering problem. In the filtering problem we usually assume that the key statistical parameters have already been determined, and the remaining problem is that of estimating the random process itself as it evolves with time. This is the central theme of Kalman filtering, which is the primary subject of this book, so we will defer on the statistics part of the overall problem.

However, all the foregoing comments being said, certain statistical terms carry over to filtering, and we need to give these terms precise mathematical meaning:

(a) Linear estimate. A linear estimate of a random variable is one that is formed as a linear function of the measurements, both past and present. We think of the random variable and its estimate as evolving with time; i.e., in general, they are not constants. Prior information about the process being estimated may also be included in the estimate, but if so, it must be accounted for linearly.

(b) Unbiased estimate. This estimate is one where

$$E[\hat{x}] = x, \ (x \text{ is the random variable, and } \hat{x} \text{ is its estimate}) \qquad (1.17.1)$$

If the above statement is satisfied, then the expectation of the error is zero, i.e.,

$$E[x - \hat{x}] = E[e] = 0 \qquad (1.17.2)$$

(c) Minimum-mean-square-error estimate. This estimate is formed such that $E\left[(x - \hat{x})^2\right]$ is made as small as possible. Simply stated, no further mathematical manipulation will do any better.

(d) Minimum variance estimate. Here, the variance of the estimate is made as small as possible. Clearly, if x itself has zero mean, then the minimum-mean-square-error estimate, or Item (c), is the same as the minimum variance estimate.

(e) Consistent estimate. When associated with the filtering of a dynamic random process, a consistent estimate is defined as one where the estimation error is zero mean with the covariance matching that calculated by the filter.

PROBLEMS

1.1 In straight poker, five cards are dealt to each player from a deck of ordinary playing cards. What is the probability that a player will be dealt a flush (i.e., five cards all of one suit)?

1.2 In the game of blackjack, the player is initially dealt two cards from a deck of ordinary playing cards. Without going into all the details of the game, it will suffice to say here that the best possible hand one could receive on the initial deal is a combination of an ace of any suit and any face card or 10. What is the probability that the player will be dealt this combination?

1.3 During most of 2009 the U.S. Congress was debating wide-ranging changes in the health care system in the United States. During this period a prominent midwestern newspaper conducted a poll in an attempt to assess the public's feelings about the proposed changes. The results were published in the form of the following table of percentages:

	Democratic	**Republican**	**Independent**
Strongly in favor	36	4	15
Just in favor	10	6	7
Don't much care	30	27	32
Strongly oppose	20	51	36
Just oppose	4	11	8
Not sure	2	1	2

Even allowing for some roundoff, note that the sum of all the percentages in the table is not 100 percent. Thus, this cannot be a table of joint probabilities.

(a) If not joint probabilities, what do the numerical values represent in terms of probabilities (if anything)?

(b) If we have two sets of random outcomes, not necessarily disjoint, and we have the table of joint probabilities, then we can always get all of the conditional and unconditional probabilities from the joint table. This is to say that the joint table tells the "whole story." Can the table of joint probabilities be obtained from the table given in this problem? Explain your answer.

1.4 Imagine a simple dice game where three dice are rolled simultaneously. Just as in craps, in this game we are only interested in the sum of the dots on any given roll.

(a) Describe, in words, the sample space for this probabilistic scenario. (You do not need to calculate all the associated probabilities—just a few at the low end of the realizable sums will suffice.)

(b) What is the probability of rolling a 3?

(c) What is the probability of rolling a 4?

(d) What is the probability of rolling a 3 or 4?

1.5 Roulette is a popular table game in casinos throughout the world. In this game a ball is spun around a wheel rotating in the opposite direction. As the wheel and ball slow down, the ball drops into a pocket on the wheel at random. In United States casinos the pockets are numbered 1 through 36 plus two more that are numbered 0 and 00. The probability of the ball falling into any particular pocket is then 1/38. The player can make a variety of bets. The simplest is to bet that the ball falls into one particular pocket, and the payout on this bet is 35 to 1. That is, in the event of a win the casino returns the player's wager plus 35 times the wager.

(a) Compute the player's average return when betting on a single number. Express the return as a percentage.

(b) Also compute the average percentage "take" for the casino.

(c) The layout of the numbers on the table is such that it is easy to bet on selected groups of four numbers. The payout on this bet is 8 to 1. What is the player's average return on this 4-number group bet? (This is also known as a "corner" bet.)

1.6 Contract bridge is played with an ordinary deck of 52 playing cards. There are four players with players on opposite sides of the table being partners. One player acts as dealer and deals each player 13 cards. A bidding sequence then takes place, and this establishes the trump suit and names the player who is to attempt to take a certain number of tricks. This player is called the declarer. The play begins with the player to the declarer's left leading a card. The declarer's partner's hand is then laid out on the table face up for everyone to see. This then enables the declarer to see a total of 27 cards as the play begins. Knowledge of these cards will, of course, affect his or her strategy in the subsequent play.

Suppose the declarer sees 11 of the 13 trump cards as the play begins. We will assume that the opening lead was not a trump, which leaves 2 trumps outstanding in the opponents' hands. The disposition of these is, of course, unknown to the declarer. There are, however, a limited number of possibilities:

(a) Both trumps lie with the opponent to the left and none to the right.
(b) Both trumps are to the right and none to the left.
(c) The two trumps are split, one in each opponent's hand.

Compute the probabilities for each of the (a), (b), and (c) possibilities.

(*Hint:* Rather than look at all possible combinations, look at numbers of combinations for 25 specific cards held by the opponents just after the opening lead. Two of these will, of course, be specific trump cards. The resulting probability will be the same regardless of the particular choice of specific cards.)

1.7 In the game of craps the casinos have ways of making money other than betting against the player rolling the dice. The other participants standing around the table may also place side bets with the casino while waiting their turn with the dice, one such side bet is to bet that the next roll will be 2 (i.e.,"snake eyes"). This is a one roll bet and the payout is 30 to 1. That is, the casino returns the player's bet plus 30 times the amount bet with a win.

(a) Compute the average percentage return for the player making this bet.
(b) Also compute the average percentage casino "take" for this bet. would you say this is a good bet relative to throwing the dice?

1.8 Cribbage is an excellent two-player card game. It is played with a 52-card deck of ordinary player cards. The initial dealer is determined by a cut of the cards, and the dealer has a slight advantage over the nondealer because of the order in which points are counted. Six cards are dealt to each player and, after reviewing their cards, they each contribute two cards to the crib. The non-dealer then cuts the remaining deck for the dealer who turns a card. This card is called the starter, if it turns out to be a jack, the dealer immediately scores 2 points. Thus, the jack is an important card in cribbage.

(a) What is the unconditional probability that the starter card will be a jack?
(b) Now consider a special case where both the dealer and non-dealer have looked at their respective six cards, and the dealer observes that there is no jack in his hand. At this point the deck has not yet been cut for the starter card. From the dealer's viewpoint, what is the probability that the starter card will be a jack?
(c) Now consider another special case where the dealer, in reviewing his initial six cards, notes that one of them is a jack. Again, from his vantage point,

what is the probability that the starter card will be a jack? [Note in both parts (a) and (b) we are considering conditional rather than unconditional probability].

1.9 Assume equal likelihood for the birth of boys and girls. What is the probability that a four-child family chosen at random will have two boys and two girls, irrespective of the order of birth?

[*Note:* The answer is not $\frac{1}{2}$ as might be suspected at first glance.)

1.10 Consider a sequence of random binary digits, zeros and ones. Each digit may be thought of as an independent sample from a sample space containing two elements, each having a probability of $\frac{1}{2}$. For a six-digit sequence, what is the probability of having:

(a) Exactly 3 zeros and 3 ones arranged in any order?
(b) Exactly 4 zeros and 2 ones arranged in any order?
(c) Exactly 5 zeros and 1 one arranged in any order?
(d) Exactly 6 zeros?

1.11 A certain binary message is n bits in length. If the probability of making an error in the transmission of a single bit is p, and if the error probability does not depend on the outcome of any previous transmissions, show that the probability of occurrence of exactly k bit errors in a message is

$$P(k\text{ errors}) = \binom{n}{k} p^k (1-p)^{n-k} \tag{P1.10}$$

The quantity $\binom{n}{k}$ denotes the number of combinations of n things taken k at a time. (This is a generalization of Problems 1.9 and 1.10.)

1.12 Video poker has become a popular game in casinos in the United States (4). The player plays against the machine in much the same way as with slot machines, except that the machine displays cards on a video screen instead of the familiar bars, etc., on spinning wheels. When a coin is put into the machine, it immediately displays five cards on the screen. After this initial five-card deal, the player is allowed to discard one to five cards at his or her discretion and obtain replacement cards (i.e., this is the "draw"). The object, of course, is to try to improve the poker hand with the draw.

(a) Suppose the player is dealt the 3, 7, 8, 10 of hearts and the queen of spades on the initial deal. The player then decides to keep the four hearts and discard the queen of spades in hopes of getting another heart on the draw, and thus obtain a flush (five cards, all of the same suit). The typical video poker machine pays out five coins for a flush. Assume that this is the payout and that the machine is statistically fair. What is the expected (i.e., average) return for this draw situation? (Note that an average return of 1.0 is the break-even return.)
(b) Some of the Las Vegas casinos advertise 100 percent (or "full pay") video poker machines. These machines have pay tables that are such that the machines will return slightly greater than 100 percent under the right conditions. How can this be? The answer is that most of the players do not play a perfect game in terms of their choices on the draw. Suppose, hypothetically, that only 10 percent of the players make perfect choices on

the draw and they achieve a 100.2 percent; and the other 90 percent of the players only achieve a 98 percent return. What would be the casino percentage return under these circumstances?

1.13 The random variable X may take on all values between 0 and 2, with all values within this range being equally likely.

(a) Sketch the probability density function for X.
(b) Sketch the cumulative probability distribution function for X.
(c) Calculate $E(X)$, $E(X^2)$, and Var X.

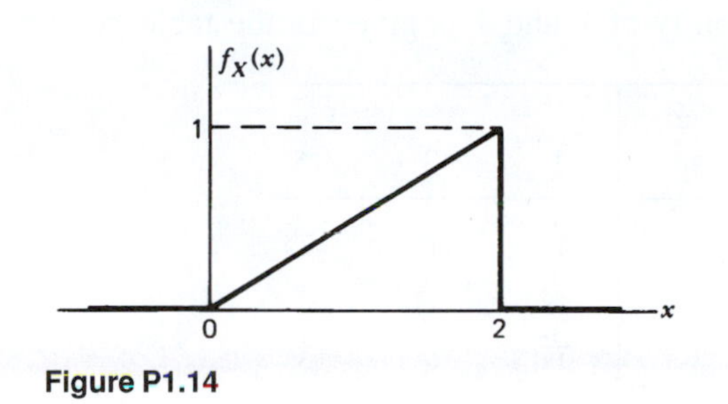

Figure P1.14

1.14 A random variable X has a probability density function as shown.

(a) Sketch the cumulative distribution function for X.
(b) What is the variance of X?

1.15 A random variable X whose probability density function is given by

$$f_X(x) = \begin{cases} \alpha e^{-\alpha x}, & x \geq 0 \\ 0, & x < 0 \end{cases}$$

is said to have an exponential probability density function. This density function is sometimes used to describe the failure of equipment components (13). That is, the probability that a particular component will fail within time T is

$$P(\text{failure}) = \int_0^T f_X(x)\,dx \tag{P1.15}$$

Note that α is a parameter that may be adjusted to fit the situation at hand. Find α for an electronic component whose average lifetime is 10,000 hours. ("Average" is used synonymously with "expectation" here.)

1.16 Consider a sack containing several identical coins whose sides are labeled +1 and −1. A certain number of coins are withdrawn and tossed simultaneously. The algebraic sum of the numbers resulting from the toss is a discrete random variable. Sketch the probability density function associated with the random variable for the following situations:

(a) One coin is tossed.
(b) Two coins are tossed.
(c) Five coins are tossed.
(d) Ten coins are tossed.

The density functions in this case consist of impulses. In the sketches, represent impulses with "arrows" whose lengths are proportional to the magnitudes of the impulses. (This is the discrete analog of Example 1.11. Note the tendency toward central distribution.)

1.17 Let the sum of the dots resulting from the throw of two dice be a discrete random variable X. The probabilities associated with their permissible values 2, 3, . . . , 12 are easily found by itemizing all possible results of the throw (or from Fig. 1.3). Find $E(X)$ and Var X.

1.18 Discrete random variables X and Y may each take on integer values 1, 3, and 5, and the joint probability of X and Y is given in the table below.

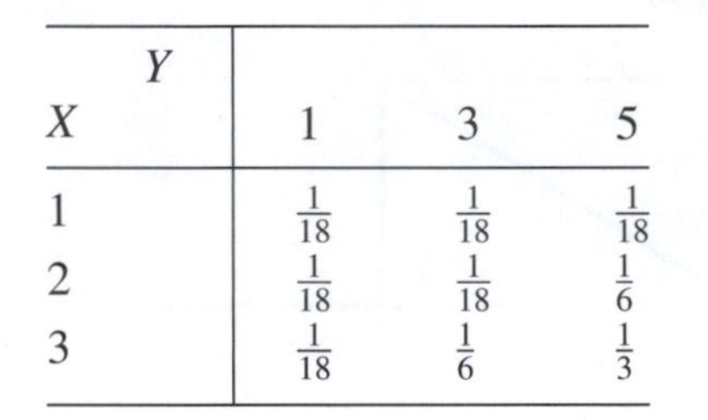

X \ Y	1	3	5
1	$\frac{1}{18}$	$\frac{1}{18}$	$\frac{1}{18}$
2	$\frac{1}{18}$	$\frac{1}{18}$	$\frac{1}{6}$
3	$\frac{1}{18}$	$\frac{1}{6}$	$\frac{1}{3}$

(a) Are random variables X and Y independent?
(b) Find the unconditional probability $P(Y=5)$.
(c) What is the conditional probability $P(Y = 5|X = 3)$?

1.19 The diagram shown as Fig. P1.19 gives the error characteristics of a hypothetical binary transmission system. The numbers shown next to the arrows are the conditional probabilities of Y given X. The unconditional probabilities for X are shown to the left of the figure. Find:

(a) The conditional probabilities $P(X = 0|Y = 1)$ and $P(X = 0|Y = 0)$.
(b) The unconditional probabilities $P(Y=0)$ and $P(Y=1)$.
(c) The joint probability array for $P(X, Y)$.

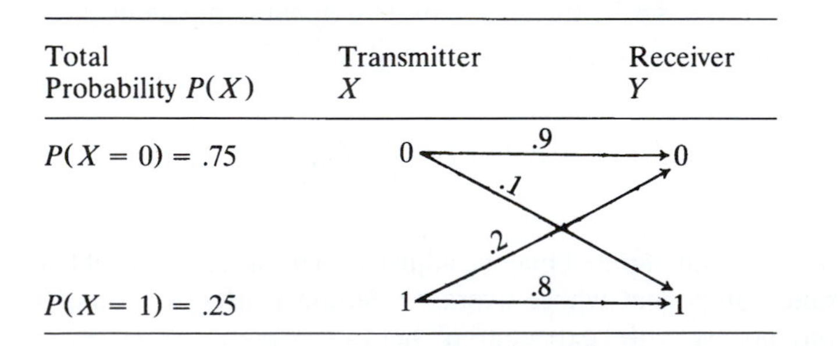

Figure P1.19

1.20 The Rayleigh probability density function is defined as

$$f_R(r) = \frac{r}{\sigma^2} e^{-r^2/2\sigma^2} \tag{P1.20}$$

where σ^2 is a parameter of the distribution (see Example 1.13).

(a) Find the mean and variance of a Rayleigh distributed random variable R.
(b) Find the mode of R (i.e., the most likely value of R).

1.21 The target shooting example of Section 1.13 led to the Rayleigh density function specified by Eq. (1.13.22).

(a) Show that the probability that a hit will lie within a specified distance R_0 from the origin is given by

$$P(\text{Hit lies within } R_0) = 1 - e^{-R_0^2/2\sigma^2} \quad \text{(P1.21)}$$

(b) The value of R_0 in Eq. (P1.21) that yields a probability of .5 is known as the circular probable error (or circular error probable, CEP). Find the CEP in terms of σ.

(c) Navigation engineers also frequently use a 95 percent accuracy figure in horizontal positioning applications. (This is in contrast to CEP.) The same circular symmetry assumptions used in parts (a) and (b) apply here in computing the 95 percent radius. To be specific, find the R_{95} which is such that 95 percent of the horizontal error lies within a circle of radius R_{95}. Just as in part (b) express R_{95} in terms of σ.

1.22 Consider a random variable X with an exponential probability density function

$$f_X(x) = \begin{cases} e^{-x}, & x \geq 0 \\ 0, & x < 0 \end{cases}$$

Find:

(a) $P(X \geq 2)$.
(b) $P(1 \leq X \leq 2)$.
(c) $E(X)$, $E(X^2)$, and Var X.

1.23 Random variables X and Y have a joint probability density function defined as follows:

$$f_{XY}(x, y) = \begin{cases} .25, & -1 \leq x \leq 1 \quad \text{and} \quad -1 \leq y \leq 1 \\ 0, & \text{otherwise} \end{cases}$$

Are random variables X and Y statistically independent?

[*Hint:* Integrate with respect to appropriate dummy variables to obtain $f_X(x)$ and $f_Y(y)$. Then check to see if the product of f_X and f_Y is equal to f_{XY}.]

1.24 Random variables X and Y have a joint probability density function

$$f_{XY}(x, y) = \begin{cases} e^{-(x+y)}, & x \geq 0 \quad \text{and} \quad y \geq 0 \\ 0, & \text{otherwise} \end{cases}$$

Find:

(a) $P(X \leq \frac{1}{2})$
(b) $P(X + Y) \leq 1]$.
(c) $P[(X \text{ or } Y) \geq 1]$.
(d) $P[(X \text{ and } Y) \geq 1]$.

1.25 Are the random variables of Problem 1.24 statistically independent?

1.26 Random variables X and Y are statistically independent and their respective probability density functions are

$$f_X(x) = \tfrac{1}{2}e^{-|x|}$$
$$f_Y(y) = e^{-2|y|}$$

Find the probability density function associated with $X + Y$.

(*Hint:* Fourier transforms are helpful here.)

1.27 Random variable X has a probability density function

$$f_X(x) = \begin{cases} \frac{1}{2}, & -1 \le x \le 1 \\ 0, & \text{otherwise} \end{cases}$$

Random variable Y is related to X through the equation

$$y = x^3 + 1$$

What is the probability density function for Y?

1.28 The vector Gaussian random variable

$$\mathbf{X} = \begin{bmatrix} X_1 \\ X_2 \end{bmatrix}$$

is completely described by its mean and covariance matrix. In this example, they are

$$\mathbf{m}_X = \begin{bmatrix} 1 \\ 2 \end{bmatrix}$$
$$\mathbf{C}_X = \begin{bmatrix} 4 & 1 \\ 1 & 1 \end{bmatrix}$$

Now consider another vector random variable $\mathbf{Y}$ that is related to $\mathbf{X}$ by the equation

$$\mathbf{y} = \mathbf{A}\mathbf{x} + \mathbf{b}$$

where

$$\mathbf{A} = \begin{bmatrix} 2 & 1 \\ 1 & -1 \end{bmatrix}, \quad \mathbf{b} = \begin{bmatrix} 1 \\ 1 \end{bmatrix}$$

Find the mean and covariance matrix for $\mathbf{Y}$.

1.29 A pair of random variables, X and Y, have a joint probability density function

$$f_{XY}(x, y) = \begin{cases} 1, & 0 \le y \le 2x \quad \text{and} \quad 0 \le x \le 1 \\ 0, & \text{elsewhere} \end{cases}$$

Find:

$$E(X|Y = .5)$$

[*Hint:* Find $f_{X|Y}(x)$ for $y = .5$, and then integrate $xf_{X|Y}(x)$ to find $E(X|Y = .5)$.]

1.30 Two continuous random variables X and Y have a joint probability density function that is uniform inside the unit circle and zero outside, that is,

$$f_{XY}(x, y) = \begin{cases} 1/\pi, & (x^2 + y^2) \leq 1 \\ 0, & (x^2 + y^2) > 1 \end{cases}$$

(a) Find the unconditional probability density function for the random variable Y and sketch the probability density as a function of Y.

(b) Are the random variables X and Y statistically independent?

1.31 A normal random variable X is described by $\mathcal{N}(0, 4)$. Similarly, Y is normal and is described by $\mathcal{N}(1, 9)$. X and Y are independent. Another random variable Z is defined by the additive combination

$$Z = X + 2Y$$

Write the explicit expression for the probability density function for Z.

1.32 Consider a random variable that is defined to be the sum of the squares of n independent normal random variables, all of which are $\mathcal{N}(0, 1)$. The parameter n is any positive integer. Such a sum-of-the-squares random variable is called a chi-square random variable with n degrees of freedom. The probability density function associated with a chi-square random variable X is

$$f_X(x) = \begin{cases} \dfrac{x^{(n/2)-1} e^{-x/2}}{2^{n/2}\Gamma\left(\dfrac{n}{2}\right)}, & x > 0 \\ 0, & x \leq 0 \end{cases}$$

where Γ indicates the gamma function (10). It is not difficult to show that the mean and variance of X are given by

$$E(X) = n$$
$$\operatorname{Var} X = 2n$$

[This is easily derived by noting that the defining integral expressions for the first and second moments of X are in the exact form of a single-sided Laplace transform with $s = \frac{1}{2}$. See Appendix A for a table of Laplace transforms and note that $n! = \Gamma(n + 1)$.]

(a) Make rough sketches of the chi-square probability density for $n = 1$, 2, and 4. (You will find MATLAB useful here.)

(b) Note that the chi-square random variable for $n > 1$ is the sum of independent random variables that are radically non-Gaussian (e.g., look at the sketch of f_X for $n = 1$). According to the central limit theorem, there should be a tendency toward normality as we sum more and more such random variables. Using MATLAB, create an *m*-file for the chi-square density function for $n = 16$. Plot this function along with the normal density function for a $\mathcal{N}(16, 32)$ random variable (same mean and sigma as the chi-square random variable). This is intended to demonstrate the tendency toward normality, even when the sum contains only 16 terms.

REFERENCES CITED IN CHAPTER 1

1. N. Wiener, *Extrapolation, Interpolation, and Smoothing of Stationary Time Series*, Cambridge, MA: MIT Press and New York: Wiley, 1949.
2. S. O. Rice, "Mathematical Analysis of Noise," *Bell System Tech. J.*, 23, 282–332 (1994); 24, 46–256 (1945).
3. A. M. Mood, F. A. Graybill, and D. C. Boes, *Introduction to the Theory of Statistics*, 3rd edition, New York: McGraw-Hill, 1974.
4. S. Bourie, *American Casino Guide – 2005 Edition*, Dania, Florida, Casino Vacations, 2005.
5. W. Feller, *An Introduction to Probability Theory and Its Applications*, Vol. 1, 2nd edition, New York: Wiley, 1957.
6. P. Beckmann, *Probability in Communication Engineering*, New York: Harcourt, Brace, and World, 1967.
7. A. Papoulis, *Probability, Random Variables, and Stochastic Processes*, 2nd edition, New York: McGraw-Hill, 1984.
8. G. H. Golub and C. F. Van Loan, *Matrix Computations*, 2nd edition, Baltimore, MD: The Johns Hopkins University Press, 1989.
9. W. B. Davenport, Jr. and W. L. Root, *An Introduction to the Theory of Random Signals and Noise*, New York: McGraw-Hill, 1958.
10. K. S. Shanmugan and A. M. Breipohl, *Random Signals: Detection, Estimation, and Data Analysis*, New York: Wiley, 1988.
11. Anthony Hayter, *Probability and Statistics for Engineers and Scientists*, 3rd Edition, Duxbury Press, 2006.
12. Alfredo H-S Ang, Wilson H. Tang, *Probability Concepts in Engineering: Emphasis on Applications to Civil and Environmental Engineering*, 2nd Edition, New York: Wiley, 2006.
13. William W. Hines, Douglas C. Montgomery, David M. Goldsman, Connie M. Borror, *Probability and Statistics in Engineering,* 4th Edition, New York: Wiley, 2003.
14. William DeCoursey, *Statistics and Probability for Engineering Applications*, Newnes, 2003
15. Frederick Mosteller, *Fifty Challenging Problems in Probability with Solutions*, Dover, 1987.

2

Mathematical Description of Random Signals

The concept of frequency spectrum is familiar from elementary physics, and so it might seem appropriate to begin our discussion of noiselike signals with their spectral description. This approach, while intuitively appealing, leads to all sorts of difficulties. The only really careful way to describe noise is to begin with a probabilistic description and then proceed to derive the associated spectral characteristics from the probabilistic model. We now proceed toward this end.

2.1 CONCEPT OF A RANDOM PROCESS

We should begin by distinguishing between deterministic and random signals. Usually, the signals being considered here will represent some physical quantity such as voltage, current, distance, temperature, and so forth. Thus, they are real variables. Also, time will usually be the independent variable, although this does not necessarily need to be the case. A signal is said to be *deterministic* if it is exactly predictable for the time span of interest. Examples would be

(a) $x(t) = 10 \sin 2\pi t$ (sine wave)

(b) $x(t) = \begin{cases} 1, & t \geq 0 \\ 0, & t < 0 \end{cases}$ (unit step)

(c) $x(t) = \begin{cases} 1 - e^{-t}, & t \geq 0 \\ 0, & t < 0 \end{cases}$ (exponential response)

Notice that there is nothing "chancy" about any of these signals. They are described by functions in the usual mathematical sense; that is, specify a numerical value of t and the corresponding value of x is determined. We are usually able to write the functional relationship between x and t explicitly. However, this is not really necessary. All that is needed is to know conceptually that a functional relationship exists.

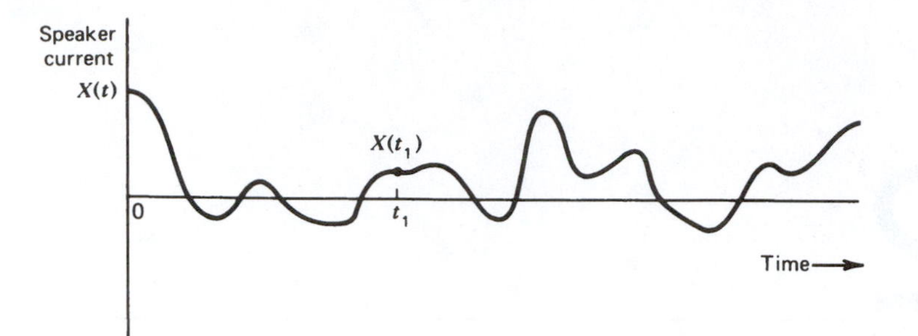

Figure 2.1 Typical radio noise signal.

In contrast with a deterministic signal, a *random signal* always has some element of chance associated with it. Thus, it is not predictable in a deterministic sense. Examples of random signals are:

(d) $X(t) = 10\ \sin(2\pi t + \theta)$, where θ is a random variable uniformly distributed between 0 and 2π.

(e) $X(t) = A\ \sin(2\pi t + \theta)$, where θ and A are independent random variables with known distributions.

(f) $X(t) =$ A noiselike signal with no particular deterministic structure—one that just wanders on aimlessly ad infinitum.

Since all of these signals have some element of chance associated with them, they are random signals. Signals such as (d), (e), and (f) are formally known as *random* or *stochastic processes,* and we will use the terms *random* and *stochastic* interchangeably throughout the remainder of the book.*

Let us now consider the description of signal (f) in more detail. It might be the common audible radio noise that was mentioned in Chapter 1. If we looked at an analog recording of the radio speaker current, it might appear as shown in Fig. 2.1. We might expect such a signal to have some kind of spectral description, because the signal is audible to the human ear. Yet the precise mathematical description of such a signal is remarkably elusive, and it eluded investigators prior to the 1940s (3, 4).

Imagine sampling the noise shown in Fig. 2.1 at a particular point in time, say, t_1. The numerical value obtained would be governed largely by chance, which suggests it might be considered to be a random variable. However, with random variables we must be able to visualize a conceptual statistical experiment in which samples of the random variable are obtained under identical chance circumstances. It would not be proper in this case to sample X by taking successive time samples of the same signal, because, if they were taken very close together, there would be a close statistical connection among nearby samples. Therefore, the conceptual experiment in this case must consist of many "identical" radios, all playing simultaneously, all

*We need to recognize a notational problem here. Denoting the random process as $X(t)$ implies that there is a functional relationship between X and t. This, of course, is not the case because $X(t)$ is governed by chance. For this reason, some authors (1, 2) prefer to use a subscript notation, that is, X_t rather than $X(t)$, to denote a random time signal. X_t then "looks" like a random variable with time as a parameter, which is precisely what it is. This notation, however, is not without its own problems. Suffice it to say, in most engineering literature, time random processes are denoted with "parentheses t" rather than "subscript t." We will do likewise, and the reader will simply have to remember that $X(t)$ does not mean function in the usual mathematical sense when $X(t)$ is a random process.

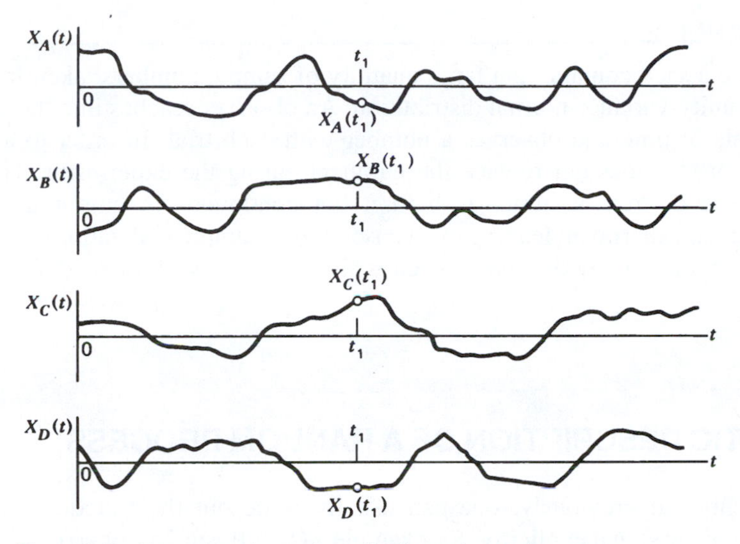

Figure 2.2 Ensemble of sample realizations of a random process.

being tuned away from regular stations in different portions of the broadcast band, and all having their volumes turned up to the same sound level. This then leads to the notion of an ensemble of similar noiselike signals as shown in Fig. 2.2.

It can be seen then that a random process is a set of random variables that unfold with time in accordance with some conceptual chance experiment. Each of the noiselike time signals so generated is called a *sample realization* of the process. Samples of the individual signals at a particular time t_1 would then be sample realizations of the random variable $X(t_1)$. Four of these are illustrated in Fig. 2.2 as $X_A(t_1)$, $X_B(t_1)$, $X_C(t_1)$, and $X_D(t_1)$. If we were to sample at a different time, say, t_2, we would obtain samples of a different random variable $X(t_2)$, and so forth. Thus, in this example, an infinite set of random variables is generated by the random process $X(t)$.

The radio experiment just described is an example of a *continuous-time random process* in that time evolves in a continuous manner. In this example, the probability density function describing the amplitude variation also happens to be continuous. However, random processes may also be discrete in either time or amplitude, as will be seen in the following two examples.

EXAMPLE 2.1

Consider a card player with a deck of standard playing cards numbered from 1 (ace) through 13 (king). The deck is shuffled and the player picks a card at random and observes its number. The card is then replaced, the deck reshuffled, and another card observed. This process is then repeated at unit intervals of time and continued on ad infinitum. The random process so generated would be discrete in both time and "amplitude," provided we say that the observed number applies only at the precise instant of time it is observed.

The preceding description would, of course, generate only one sample realization of the process. In order to obtain an ensemble of sample signals, we need to imagine an ensemble of card players, each having similar decks of cards and each generating a different (but statistically similar) sample realization of the process. ■

EXAMPLE 2.2

Imagine a sack containing a large quantity of sample numbers taken from a zero-mean, unity-variance normal distribution. An observer reaches into the sack at unit intervals of time and observes a number with each trial. In order to avoid exact repetition, he does not replace the numbers during the experiment. This process would be discrete in time, as before, but continuous in amplitude. Also, the conceptual experiment leading to an ensemble of sample realizations of the process would involve many observers, each with a separate sack of random numbers.

2.2 PROBABILISTIC DESCRIPTION OF A RANDOM PROCESS

As mentioned previously, one can usually write out the functional form for a deterministic signal explicitly; for example, $s(t) = 10 \sin 2\pi t$, or $s(t) = t^2$, and so on. No such deterministic description is possible for random signals because the numerical value of the signal at any particular time is governed by chance. Thus, we should expect our description of noiselike signals to be somewhat vaguer than that for deterministic signals. One way to specify a random process is to describe in detail the conceptual chance experiment giving rise to the process. Examples 2.1 and 2.2 illustrated this way of describing a random process. The following two examples will illustrate this further.

EXAMPLE 2.3

Consider a time signal (e.g., a voltage) that is generated according to the following rules: (a) The waveform is generated with a sample-and-hold arrangement where the "hold" interval is 1 sec; (b) the successive amplitudes are independent samples taken from a set of random numbers with uniform distribution from -1 to $+1$; and (c) the first switching time after $t = 0$ is a random variable with uniform distribution from 0 to 1. (This is equivalent to saying the time origin is chosen at random.) A typical sample realization of this process is shown in Fig. 2.3. Note that the process mean is zero and its mean-square value works out to be one-third. [This is obtained from item (b) of the description and integrating $\int_{-1}^{1} x^2 p(x)dx$.]

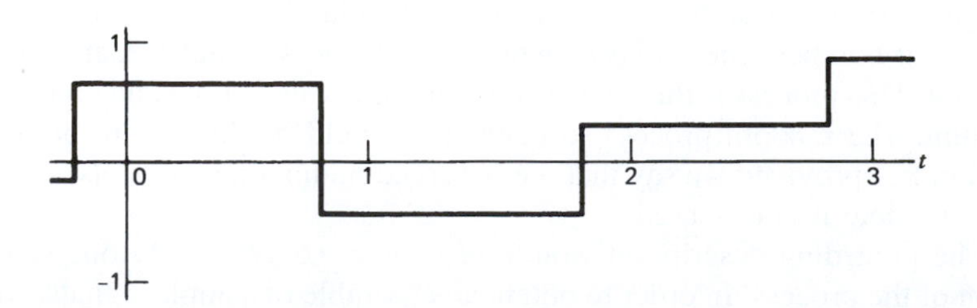

Figure 2.3 Sample signal for Example 2.3.

EXAMPLE 2.4

Consider another time function generated with a sample-and-hold arrangement with these properties: (a) The "hold" interval is 0.2 sec, (b) the successive amplitudes are independent samples obtained from a zero-mean normal distribution with a variance of one-third, and (c) the switching points occur at multiples of .2 units of time; that is, the time origin is not chosen at random in this case. A sketch of a typical waveform for this process is shown in Fig. 2.4.

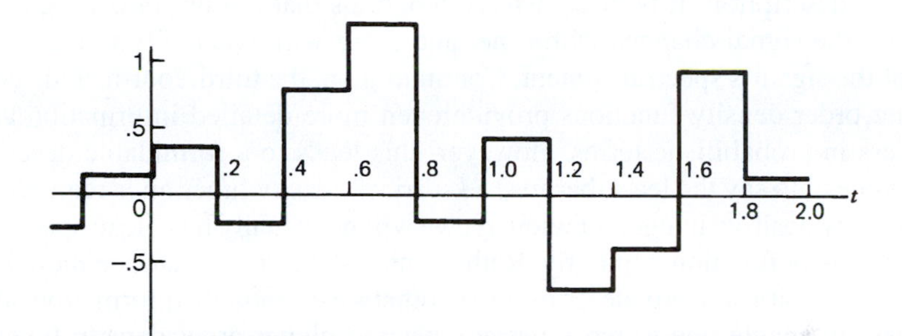

Figure 2.4 Typical waveform for Example 2.4.

Now, from Examples 2.3 and 2.4 it should be apparent that if we simply say, "Noiselike waveform with zero mean and mean-square value of one-third," we really are not being very definite. Both processes of Examples 2.3 and 2.4 would satisfy these criteria, but yet they are quite different. Obviously, more information than just mean and variance is needed to completely describe a random process. We will now explore the "description" problem in more detail.

A more typical "noiselike" signal is shown in Fig. 2.5. The times indicated, $t_1, t_2, \ldots, t_k$, have been arranged in ascending order, and the corresponding sample values $X_1, X_2, \ldots, X_k$ are, of course, random variables. Note that we have abbreviated the notation and have let $X(t_1) = X_1$, $X(t_2) = X_2, \ldots,$ and so on. Obviously, the first-order probability density functions $f_{X_1}(x), f_{X_2}(x), \ldots, f_{X_k}(x)$, are important in describing the process because they tell us something about the process amplitude distribution. In Example 2.3, $f_{X_1}(x), f_{X_2}(x), \ldots, f_{X_k}(x)$, are all identical density functions and are given by [using $f_{X_1}(x)$ as an example]

$$f_{X_1}(x) \begin{cases} \dfrac{1}{2}, & -1 \leq x \leq 1 \\ 0, & |x| > 1 \end{cases}$$

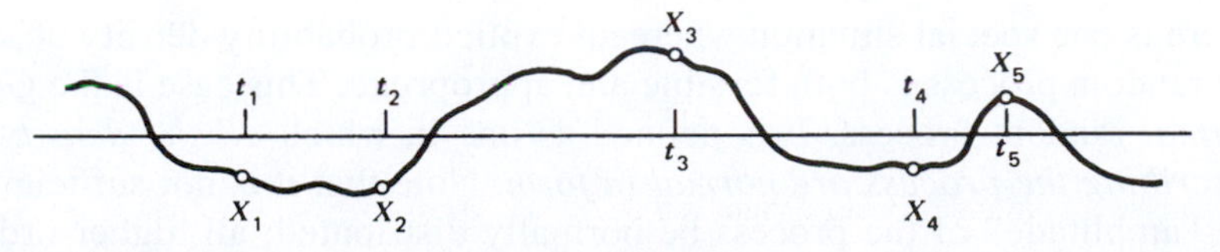

Figure 2.5 Sample signal of a typical noise process.

The density functions are not always identical for $X_1, X_2, \ldots, X_k$; they just happened to be in this simple example. In Example 2.4, the density functions describing the amplitude distribution of the $X_1, X_2, \ldots, X_k$ random variables are again all the same, but in this case they are normal in form with a variance of one-third. Note that the first-order densities tell us something about the relative distribution of the process amplitude as well as its mean and mean-square value.

It should be clear that the joint densities relating any pair of random variables, for example, $f_{X_1X_2}(x_1, x_2)$, $f_{X_1X_3}(x_1, x_3)$, and so forth, are also important in our process description. It is these density functions that tell us something about how rapidly the signal changes with time, and these will eventually tell us something about the signal's spectral content. Continuing on, the third, fourth, and subsequent higher-order density functions provide even more detailed information about the process in probabilistic terms. However, this leads to a formidable description of the process, to say the least, because a k-variate density function is required where k can be any positive integer. Obviously, we will not usually be able to specify this kth order density function explicitly. Rather, this usually must be done more subtly by providing, with a word description or otherwise, enough information about the process to enable one to write out any desired higher-order density function; but the actual "writing it out" is usually not done.

Recall from probability theory that two random variables X and Y are said to be statistically independent if their joint density function can be written in product form

$$f_{XY}(x, y) = f_X(x)f_Y(y) \tag{2.2.1}$$

Similarly, random processes $X(t)$ and $Y(t)$ are statistically independent if the joint density for any combination of random variables of the two processes can be written in product form, that is, $X(t)$ and $Y(t)$ are independent if

$$f_{X_1X_2\ldots Y_1Y_2\ldots} = f_{X_1X_2\ldots}f_{Y_1Y_2\ldots} \tag{2.2.2}$$

In Eq. (2.2.2) we are using the shortened notation $X_1 = X(t_1)$, $X_2 = X(t_2), \ldots,$ and $Y_1 = Y_1(t_1')$, $Y_2 = Y_2(t_2'), \ldots,$ where the sample times do not have to be the same for the two processes.

In summary, the test for completeness of the process description is this: Is enough information given to enable one, conceptually at least, to write out the kth order probability density function for any k? If so, the description is as complete as can be expected; if not, it is incomplete to some extent, and radically different processes may fit the same incomplete description.

2.3 GAUSSIAN RANDOM PROCESS

There is one special situation where an explicit probability density description of the random process is both feasible and appropriate. This case is the *Gaussian* or *normal* random process. It is defined as one in which *all the density functions describing the process are normal in form*. Note that it is not sufficient that just the "amplitude" of the process be normally distributed; all higher-order density functions must also be normal! As an example, the process defined in Example 2.4

has a normal first-order density function, but closer scrutiny will reveal that its second-order density function is not normal in form. Thus, the process is not a Gaussian process.

The multivariate normal density function was discussed in Section 1.14. It was pointed out there that matrix notation makes it possible to write out all k-variate density functions in the same compact matrix form, regardless of the size of k. All we have to do is specify the vector random-variable mean and covariance matrix, and the density function is specified. In the case of a Gaussian random process the "variates" are the random variables $X(t_1), X(t_2), \dots, X(t_k)$, where the points in time may be chosen arbitrarily. Thus, enough information must be supplied to specify the mean and covariance matrix regardless of the choice of $t_1, t_2, \dots, t_k$. Examples showing how to do this will be deferred for the moment, because it is expedient first to introduce the basic ideas of stationarity and correlation functions.

2.4 STATIONARITY, ERGODICITY, AND CLASSIFICATION OF PROCESSES

A random process is said to be *time stationary* or simply *stationary* if the density functions describing the process are invariant under a translation of time. That is, if we consider a set of random variables $X_1 = X(t_1), X_2 = X(t_2), \dots, X_k = X(t_k)$, and also a translated set $X_1' = X(t_1 + \tau)$, $X_2' = X(t_1 + \tau)$, $X_k' = X(t_k + \tau)$, the density functions $f_{X_1}, f_{X_1X_2}, \dots, f_{X_1X_2\dots X_k}$ describing the first set would be identical in form to those describing the translated set. Note that this applies to all the higher-order density functions. The adjective *strict* is also used occasionally with this type of stationarity to distinguish it from wide-sense stationarity, which is a less restrictive form of stationarity. This will be discussed later in Section 2.5 on correlation functions.

A random process is said to be *ergodic* if time averaging is equivalent to ensemble averaging. In a qualitative sense this implies that a single sample time signal of the process contains all possible statistical variations of the process. Thus, no additional information is to be gained by observing an ensemble of sample signals over the information obtained from a one-sample signal, for example, one long data recording. An example will illustrate this concept.

EXAMPLE 2.5

Consider a somewhat trivial process defined to be a constant with time, the constant being a random variable with zero-mean normal distribution. An ensemble of sample realizations for this process is shown in Fig. 2.6. A common physical situation where this kind of process model would be appropriate is random instrument bias. In many applications, some small residual random bias will remain in spite of all attempts to eliminate it, and the bias will be different for each instrument in the batch. In Fig. 2.6 we see that time samples collected from a single sample signal, say the first one, will all have the same value a_0. The average of these is, of course, just a_0. On the other hand, if we were to collect samples in an ensemble

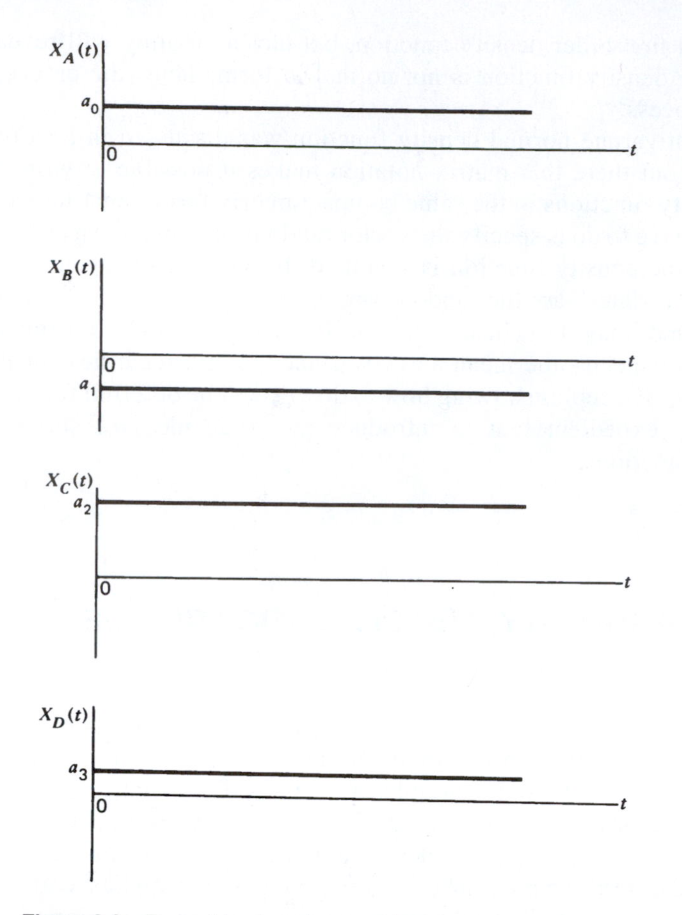

Figure 2.6 Ensemble of random constants.

sense, the values $a_0, a_1, a_2, \ldots, a_n$, would be obtained. These would have a normal distribution with zero mean. Obviously, time and ensemble sampling do not lead to the same result in this case, so the process is not ergodic. It is, however, a stationary process because the "statistics" of the process do not change with time.

In the case of physical noise processes, one can rarely justify strict stationarity or ergodicity in a formal sense. Thus, we often lean on heuristic knowledge of the processes involved and simply make assumptions accordingly.

Random processes are sometimes classified according to two categories, *deterministic* and *nondeterministic*. As might be expected, a deterministic random process resembles a deterministic nonrandom signal in that it has some special deterministic structure. Specifically, if the process description is such that knowledge of a sample signal's past enables exact prediction of its future, it is classified as a deterministic random process. Examples are:

1. $X(t) = a$; a is normal, $\mathcal{N}(m, \sigma^2)$.
2. $X(t) = A \sin \omega t$; A is Rayleigh distributed, and ω is a known constant.
3. $X(t) = A \sin(\omega t + \theta)$; A and θ are independent, and Rayleigh and uniformly distributed, respectively.

In each case, if one were to specify a particular sample signal prior to some time, say, t_1, the sample realizations for that particular signal would be indirectly specified, and the signal's future values would be exactly predictable.

Random processes that are not deterministic are classified as nondeterministic. These processes have no special functional structure that enables their exact prediction by specification of certain key parameters or their past history. Typical "noise" is a good example of a nondeterministic random process. It wanders on aimlessly, as determined by chance, and has no particular deterministic structure.

2.5 AUTOCORRELATION FUNCTION

The autocorrelation function for a random process $X(t)$ is defined as*

$$R_X(t_1, t_2) = E[X(t_1)X(t_2)] \tag{2.5.1}$$

where t_1 and t_2 are arbitrary sampling times. Clearly, it tells how well the process is correlated with itself at two different times. If the process is stationary, its probability density functions are invariant with time, and the autocorrelation function depends only on the time difference $t_2 - t_1$. Thus, R_X reduces to a function of just the time difference variable τ, that is,

$$R_X(\tau) = E[X(t)X(t+\tau)] \quad \text{(stationary case)} \tag{2.5.2}$$

where t_1 is now denoted as just t and t_2 is $(t+\tau)$. Stationarity assures us that the expectation is not dependent on t.

Note that the autocorrelation function is the ensemble average (i.e., expectation) of the product of $X(t_1)$ and $X(t_2)$; therefore, it can formally be written as

$$R_X(t_1, t_2) = E[X_1 X_2] = \int_{-\infty}^{\infty}\int_{-\infty}^{\infty} x_1 x_2 f_{X_1X_2}(x_1, x_2)\,dx_1\,dx_2 \tag{2.5.3}$$

where we are using the shortened notation $X_1 = X(t_1)$ and $X_2 = X(t_2)$. However, Eq. (2.5.3) is often not the simplest way of determining R_X because the joint density function $f_{X_1X_2}(x_1, x_2)$ must be known explicitly in order to evaluate the integral. If the ergodic hypothesis applies, it is often easier to compute R_X as a time average rather than an ensemble average. An example will illustrate this.

* In describing the correlation properties of random processes, some authors prefer to work with the *autocovariance function* rather than the autocorrelation function as defined by Eq. (2.5.1). The autocovariance function is defined as

$$\text{Autocovariance function} = E\{[X(t_1) - m_x(t_1)][X(t_2) - m_x(t_2)]\}$$

The two functions are obviously related. In one case the mean is included in the product (autocorrelation), and in the other the mean is subtracted out (autocovariance). That is the essential difference. The two functions are, of course, identical for zero-mean processes. The autocorrelation function is probably the more common of the two in engineering literature, so it will be used throughout this text.

EXAMPLE 2.6

Consider the same process defined in Example 2.3. A typical sample signal for this process is shown in Fig. 2.7 along with the same signal shifted in time an amount τ. Now, the process under consideration in this case is ergodic, so we should be able to interchange time and ensemble averages. Thus, the autocorrelation function can be written as

$$\begin{aligned} R_X(\tau) &= \text{time average of } X_A(t) \cdot X_A(t+\tau) \\ &= \lim_{T\to\infty} \frac{1}{T} \int_0^T X_A(t) \cdot X_A(t+\tau)dt \end{aligned} \tag{2.5.4}$$

It is obvious that when $\tau = 0$, the integral of Eq. (2.5.4) is just the mean square value of $X_A(t)$, which is $\frac{1}{3}$ in this case. On the other hand, when τ is unity or larger, there is no overlap of the correlated portions of $X_A(t)$ and $X_A(t+\tau)$, and thus the average of the product is zero. Now, as the shift τ is reduced from 1 to 0, the overlap of correlated portions increases linearly until the maximum overlap occurs at $\tau = 0$. This then leads to the autocorrelation function shown in Fig. 2.8. Note that for stationary ergodic processes, the direction of time shift τ is immaterial, and hence the autocorrelation function is symmetric about the origin. Also, note that we arrived at $R_X(\tau)$ without formally finding the joint density function $f_{X_1X_2}(x_1, x_2)$.

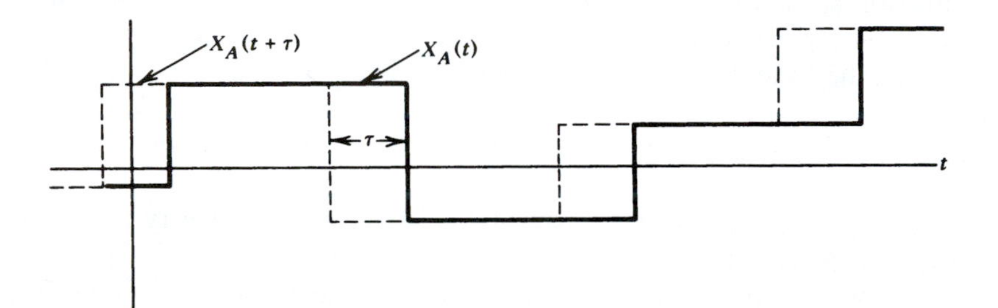

Figure 2.7 Random waveform for Example 2.6.

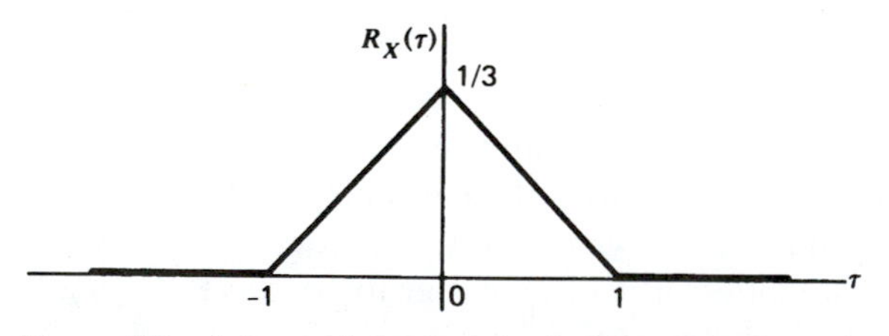

Figure 2.8 Autocorrelation function for Example 2.6.

Sometimes, the random process under consideration is not ergodic, and it is necessary to distinguish between the usual autocorrelation function (ensemble average) and the time-average version. Thus, we define the *time autocorrelation function* as

$$\Re_{X_A}(\tau) = \lim_{T\to\infty} \frac{1}{T} \int_0^T X_A(t)X_A(t+\tau)dt \tag{2.5.4a}$$

where $X_A(t)$ denotes a sample realization of the $X(t)$ process. There is the tacit assumption that the limit indicated in Eq. (2.5.4a) exists. Also note that script $\mathfrak{R}$ rather than italic R is used as a reminder that this is a time average rather than an ensemble average.

EXAMPLE 2.7

To illustrate the difference between the usual autocorrelation function and the time autocorrelation function, consider the deterministic random process

$$X(t) = A \sin \omega t \tag{2.5.5}$$

where A is a normal random variable with zero mean and variance σ^2, and ω is a known constant. Suppose we obtain a single sample of A and its numerical value is A_1. The corresponding sample of $X(t)$ would then be

$$X_A(t) = A_1 \sin \omega t \tag{2.5.6}$$

According to Eq. (2.5.4a), its time autocorrelation function would then be

$$\begin{aligned} \mathfrak{R}_{X_A}(\tau) &= \lim_{T\to\infty} \frac{1}{T} \int_0^T A_1 \sin \omega t \cdot A_1 \sin \omega(t+\tau) dt \\ &= \frac{A_1^2}{2} \cos \omega\tau \end{aligned} \tag{2.5.7}$$

On the other hand, the usual autocorrelation function is calculated as an ensemble average, that is, from Eq. (2.5.1). In this case, it is

$$\begin{aligned} R_X(t_1, t_2) &= E[X(t_1)X(t_2)] \\ &= E[A \sin \omega t_1 \cdot A \sin \omega t_2] \\ &= \sigma^2 \sin \omega t_1 \sin \omega t_2 \end{aligned} \tag{2.5.8}$$

Note that this expression is quite different from that obtained for $\mathfrak{R}_{X_A}(\tau)$. Clearly, time averaging does not yield the same result as ensemble averaging, so the process is not ergodic. Furthermore, the autocorrelation function given by Eq. (2.5.8) does not reduce to simply a function of $t_2 - t_1$. Therefore, the process is not stationary. ■

General Properties of Autocorrelation Functions

There are some general properties that are common to all autocorrelation functions for stationary processes. These will now be enumerated with a brief comment about each:

1. $R_X(0)$ is the mean-square value of the process $X(t)$. This is self-evident from Eq. (2.5.2).
2. $R_X(\tau)$ is an even function of τ. This results from the stationarity assumption. [In the nonstationary case there is symmetry with respect to the two

arguments t_1 and t_2. In Eq. (2.5.1) it certainly makes no difference in which order we multiply $X(t_1)$ and $X(t_2)$. Thus, $R_X(t_1, t_2) = R_X(t_2, t_1)$.]

3. $|R_X(\tau)| \leq R_X(0)$ for all τ. We have assumed $X(t)$ is stationary and thus the mean-square values of $X(t)$ and $X(t+\tau)$ must be the same. Also the magnitude of the correlation coefficient relating two random variables is never greater than unity. Thus, $R_X(\tau)$ can never be greater in magnitude than $R_X(0)$.
4. If $X(t)$ contains a periodic component, $R_X(t)$ will also contain a periodic component with the same period. This can be verified by writing $X(t)$ as the sum of the nonperiodic and periodic components and then applying the definition given by Eq. (2.5.2). It is of interest to note that if the process is ergodic as well as stationary and if the periodic component is sinusoidal, then $R_X(\tau)$ will contain no information about the phase of the sinusoidal component. The harmonic component always appears in the autocorrelation function as a cosine function, irrespective of its phase.
5. If $X(t)$ does not contain any periodic components, $R_X(\tau)$ tends to zero as $\tau \to \infty$. This is just a mathematical way of saying that $X(t+\tau)$ becomes completely uncorrelated with $X(t)$ for large τ if there are no hidden periodicities in the process. Note that a constant is a special case of a periodic function. Thus, $R_X(\infty) = 0$ implies zero mean for the process.
6. The Fourier transform of $R_X(\tau)$ is real, symmetric, and nonnegative. The real, symmetric property follows directly from the even property of $R_X(\tau)$. The nonnegative property is not obvious at this point. It will be justified later in Section 2.7, which deals with the spectral density function for the process.

It was mentioned previously that strict stationarity is a severe requirement, because it requires that all the higher-order probability density functions be invariant under a time translation. This is often difficult to verify. Thus, a less demanding form of stationarity is often used, or assumed. A random process is said to be *covariance stationary* or *wide-sense stationary* if $E[X(t_1)]$ is independent of t_1 and $E[X(t_1)X(t_2)]$ is dependent only on the time difference $t_2 - t_1$. Obviously, if the second-order density $f_{X_1 X_2}(x_1, x_2)$ is independent of the time origin, the process is covariance stationary.

Further examples of autocorrelation functions will be given as this chapter progresses. We will see that the autocorrelation function is an important descriptor of a random process and one that is relatively easy to obtain because it depends on only the second-order probability density for the process.

2.6 CROSSCORRELATION FUNCTION

The crosscorrelation function between the processes $X(t)$ and $Y(t)$ is defined as

$$R_{XY}(x_{t_1}, t_2) = E[X(t_1)Y(t_2)] \tag{2.6.1}$$

Again, if the processes are stationary, only the time *difference* between sample points is relevant, so the crosscorrelation function reduces to

$$R_{XY}(\tau) = E[X(t)Y(t+\tau)] \quad \text{(stationary case)} \tag{2.6.2}$$

Just as the autocorrelation function tells us something about how a process is correlated with itself, the crosscorrelation function provides information about the mutual correlation between the two processes.

Notice that it is important to order the subscripts properly in writing $R_{XY}(\tau)$. A skew-symmetric relation exists for stationary processes as follows. By definition,

$$R_{XY}(\tau) = E[X(t)Y(t+\tau)] \tag{2.6.3}$$

$$R_{YX}(\tau) = E[Y(t)X(t+\tau)] \tag{2.6.4}$$

The expectation in Eq. (2.6.4) is invariant under a translation of $-\tau$. Thus, $R_{YX}(\tau)$ is also given by

$$R_{YX}(\tau) = E[Y(t-\tau)X(t)] \tag{2.6.5}$$

Now, comparing Eqs. (2.6.3) and (2.6.5), we see that

$$R_{YX}(\tau) = R_{YX}(-\tau) \tag{2.6.6}$$

Thus, interchanging the order of the subscripts of the crosscorrelation function has the effect of changing the sign of the argument.

EXAMPLE 2.8

Let $X(t)$ be the same random process of Example 2.6 and illustrated in Fig. 2.7. Let $Y(t)$ be the same signal as $X(t)$, but delayed one-half unit of time. The crosscorrelation $R_{XY}(\tau)$ would then be shown in Fig. 2.9. Note that $R_{XY}(\tau)$ is not an even function of τ, nor does its maximum occur at $\tau = 0$. Thus, the crosscorrelation function lacks the symmetry possessed by the autocorrelation function.

■

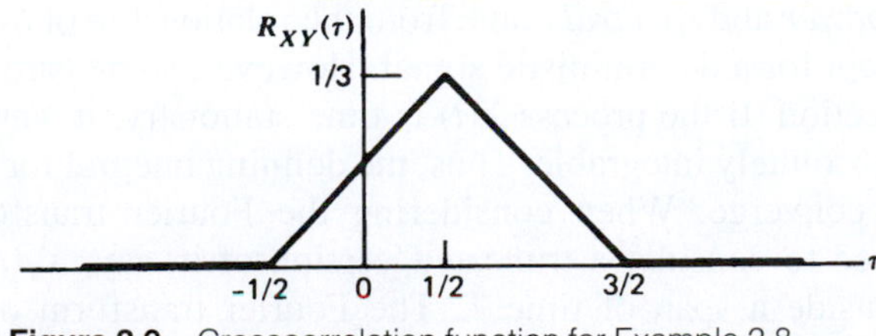

Figure 2.9 Crosscorrelation function for Example 2.8.

We frequently need to consider additive combinations of random processes. For example, let the process $Z(t)$ be the sum of stationary processes $X(t)$ and $Y(t)$:

$$Z(t) = X(t) + Y(t) \tag{2.6.7}$$

The autocorrelation function of the summed process is then

$$\begin{aligned} R_Z(\tau) &= E\{[X(t) + Y(t)][X(t+\tau) + Y(t+\tau)]\} \\ &= E[X(t)X(t+\tau) + E[Y(t)X(t+\tau)] + E[X(t)Y(t+\tau) + E[Y(t)Y(t+\tau)] \\ &= R_X(\tau) + R_{YX}(\tau) + R_{XY}(\tau) + R_Y(\tau) \end{aligned} \tag{2.6.8}$$

Now, if X and Y are zero-mean uncorrelated processes, the middle terms of Eq. (2.6.8) are zero, and we have

$$R_Z(\tau) = R_X(\tau) + R_X(\tau) \quad \text{(for zero crosscorrelation)} \tag{2.6.9}$$

This can obviously be extended to the sum of more than two processes. Equation (2.6.9) is a much-used relationship, and it should always be remembered that it applies only when the processes being summed have zero crosscorrelation.

2.7 POWER SPECTRAL DENSITY FUNCTION

It was mentioned in Section 2.6 that the autocorrelation function is an important descriptor of a random process. Qualitatively, if the autocorrelation function decreases rapidly with τ, the process changes rapidly with time; conversely, a slowly changing process will have an autocorrelation function that decreases slowly with τ. Thus, we would suspect that this important descriptor contains information about the frequency content of the process; and this is in fact the case. For stationary processes, there is an important relation known as the *Wiener–Khinchine relation:*

$$S_X(j\omega) = \mathfrak{F}[R_X(\tau)] = \int_{-\infty}^{\infty} R_X(\tau) e^{-j\omega\tau} d\tau \tag{2.7.1}$$

where $\mathfrak{F}[\cdot]$ indicates Fourier transform and ω has the usual meaning of (2π) (frequency in hertz). S_X is called the *power spectral density function* or simply the *spectral density function* of the process.

The adjectives *power* and *spectral* come from the relationship of $S_X(j\omega)$ to the usual spectrum concept for a deterministic signal. However, some care is required in making this connection. If the process $X(t)$ is time stationary, it wanders on ad infinitum and is not absolutely integrable. Thus, the defining integral for the Fourier transform does not converge. When considering the Fourier transform of the process, we are forced to consider a truncated version of it, say, $X_T(t)$, which is truncated to zero outside a span of time T. The Fourier transform of a sample realization of the truncated process will then exist.

Let $\mathfrak{F}\{X_T\}$ denote the Fourier transform of $X_T(t)$, where it is understood that for any given ensemble of samples of $X_T(t)$ there will be corresponding ensemble of $\mathfrak{F}\{X_T(t)\}$. That is, $\mathfrak{F}\{X_T(t)\}$ has stochastic attributes just as does $X_T(t)$. Now look at the following expectation:

$$E\left[\frac{1}{T}|\mathfrak{F}\{X_T(t)\}|^2\right]$$

For any particular sample realization of $X_T(t)$, the quantity inside the brackets is known as the *periodogram* for that particular signal. It will now be shown that averaging over an ensemble of periodograms for large T yields the power spectral density function.

The expectation of the periodogram of a signal spanning the time interval $[0, T]$ can be manipulated as follows:

$$\begin{aligned}
&E\left[\frac{1}{T}|\mathfrak{F}\{X_T(t)\}|^2\right] \\
&E\left[\frac{1}{T}\int_0^T X(t)e^{-j\omega t}dt \int_0^T X(s)e^{j\omega t}ds\right] \\
&\frac{1}{T}\int_0^T \int_0^T E[X(t)X(s)e^{-j\omega(t-s)}dt\,ds
\end{aligned} \tag{2.7.2}$$

Note that we were able to drop the subscript T on $X(t)$ because of the restricted range of integration. If we now assume $X(t)$ is stationary, $E[X(t)X(s)]$ becomes $R_X(t-s)$ and Eq. (2.7.2) becomes

$$E\left[\frac{1}{T}|\mathfrak{F}\{X_T(t)\}|^2\right] = \frac{1}{T}\int_0^T \int_0^T R_X(t-s)e^{-j\omega(t-s)}dt\,ds \tag{2.7.3}$$

The appearance of $t-s$ in two places in Eq. (2.7.3) suggests a change of variables. Let

$$\tau = t - s \tag{2.7.4}$$

Equation (2.7.3) then becomes

$$\frac{1}{T}\int_0^T \int_0^T R_X(t-s)e^{-j\omega(t-s)}dt\,ds = -\frac{1}{T}\int_0^T \int_0^T R_X(t-s)e^{-j\omega\tau}d\tau\,dt \tag{2.7.5}$$

The new region of integration in the τt plane is shown in Fig. 2.10.

Next we interchange the order of integration and integrate over the two triangular regions separately. This leads to

$$\begin{aligned}
&E\left[\frac{1}{T}|\mathfrak{F}\{X_T(t)\}|^2\right] \\
&= \frac{1}{T}\int_{-T}^T \int_0^{\tau+T} R_X(\tau)e^{-j\omega\tau}dt\,d\tau + \frac{1}{T}\int_0^T \int_\tau^T R_X(\tau)e^{-j\omega\tau}dt\,d\tau
\end{aligned} \tag{2.7.6}$$

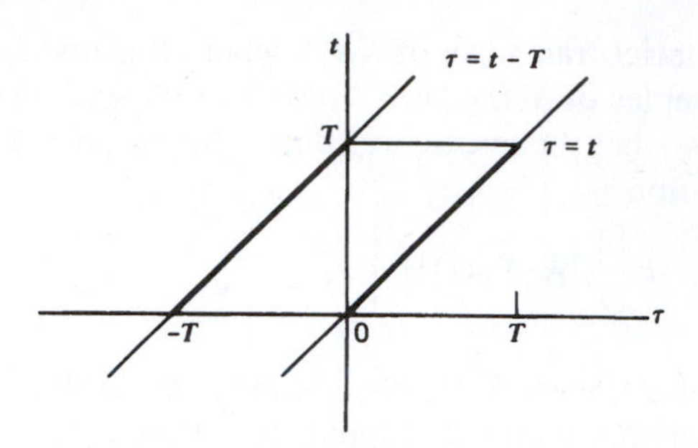

Figure 2.10 Region of integration in the τt plane.

We now integrate with respect to t with the result

$$E\left[\frac{1}{T}|\mathfrak{F}\{X_T(t)\}|^2\right] = \frac{1}{T}\int_{-T}^{0}(\tau+T)R_X(\tau)e^{-j\omega\tau}d\tau + \frac{1}{T}\int_{\tau}^{T}(T-\tau)R_X(\tau)e^{-j\omega\tau}\,d\tau \tag{2.7.7}$$

Finally, Eq. (2.7.7) may be written in more compact form as

$$E\left[\frac{1}{T}|\mathfrak{F}\{X_T(t)\}|^2\right] = \int_{-T}^{T}\left(1-\frac{|\tau|}{T}\right)R_X(\tau)e^{-j\omega\tau}\,d\tau \tag{2.7.8}$$

The factor $1-|\tau|/T$ that multiplies $R_X(\tau)$ may be thought of as a triangular weighting factor that approaches unity as T becomes large; at least this is true if $R_X(\tau)$ approaches zero as τ becomes large, which it will do if $X(t)$ contains no periodic components. Thus, as T becomes large, we have the following relationship:

$$E\left[\frac{1}{T}|\mathfrak{F}\{X_T(t)\}|^2\right] \Rightarrow \int_{-\infty}^{\infty} R_X(\tau)e^{-j\omega\tau}\,d\tau \quad \text{as } T\to\infty \tag{2.7.9}$$

Or, in other words,

$$\text{Average periodogram for large } T \Rightarrow \text{power spectral density} \tag{2.7.10}$$

Note especially the "for large T" qualification in Eq. (2.7.10). (This is pursued further in Section 2.13.)

Equation (2.7.9) is a most important relationship, because it is this that ties the spectral function $S_X(j\omega)$ to "spectrum" as thought of in the usual deterministic sense. Remember that the spectral density function, as formally defined by Eq. (2.7.1), is a probabilistic concept. On the other hand, the periodogram is a spectral concept in the usual sense of being related to the Fourier transform of a time signal. The relationship given by Eq. (2.7.9) then provides the tie between the probabilistic and spectral descriptions of the process, and it is this equation that suggests the name for $S_X(j\omega)$, power *spectral* density function. More will be said of this in Section 2.13, which deals with the determination of the spectral function from experimental data.

Because of the spectral attributes of the autocorrelation function $R_X(\tau)$, its Fourier transform $S_X(j\omega)$ always works out to be a real, nonnegative, symmetric function of ω. This should be apparent from the left side of Eq. (2.7.9), and will be illustrated in Example 2.9.

EXAMPLE 2.9

Consider a random process $X(t)$ whose autocorrelation function is given by

$$R_X(\tau) = \sigma^2 e^{-\beta|\tau|} \tag{2.7.11}$$

where σ^2 and β are known constants. The spectral density function for the $X(t)$ process is

$$S_X(j\omega) = \mathfrak{F}[R_X(\tau)] = \frac{\sigma^2}{j\omega + \beta} + \frac{\sigma^2}{-j\omega + \beta} = \frac{2\sigma^2\beta}{\omega^2 + \beta^2} \tag{2.7.12}$$

Both R_X and S_X are sketched in Fig. 2.11.

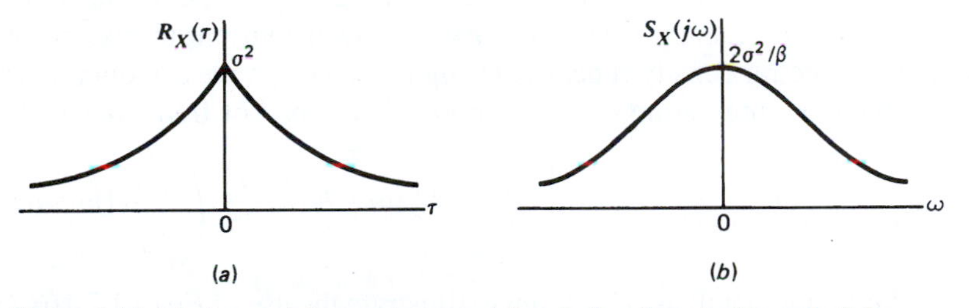

Figure 2.11 Autocorrelation and spectral density functions for Example 2.9. (*a*) Autocorrelation function. (*b*) Spectral function.

Occasionally, it is convenient to write the spectral density function in terms of the complex frequency variable s rather than ω. This is done by simply replacing $j\omega$ with s; or, equivalently, replacing ω^2 with $-s^2$. For Example 2.9, the spectral density function in terms of s is then

$$S_X(s) = \left.\frac{2\sigma^2\beta}{\omega^2 + \beta^2}\right|_{\omega^2 = -s^2} = \frac{2\sigma^2\beta}{-s^2 + \beta^2} \tag{2.7.13}$$

It should be clear now why we chose to include the "j" with ω in $S_X(j\omega)$, even though $S_X(j\omega)$ always works out to be a real function of ω. By writing the argument of $S_X(j\omega)$ as $j\omega$, rather than just ω, we can use the same symbol for spectral function in either the complex or real frequency domain. That is,

$$S_X(s) = S_X(j\omega) \tag{2.7.14}$$

is correct notation in the usual mathematical sense.

From Fourier transform theory, we know that the inverse transform of the spectral function should yield the autocorrelation function, that is,

$$\mathfrak{F}^{-1}[S_X(j\omega)] = \frac{1}{2\pi}\int_{-\infty}^{\infty} S_X(j\omega)e^{j\omega\tau}\,d\omega = R_X(\tau) \tag{2.7.15}$$

If we let $\tau = 0$ in Eq. (2.7.15), we get

$$R_X(0) = E\left[X^2(t)\right] = \frac{1}{2\pi}\int_{-\infty}^{\infty} S_X(j\omega)\,d\omega \tag{2.7.16}$$

Equation (2.7.16) provides a convenient means of computing the mean square value of a stationary process, given its spectral function. As mentioned before, it is sometimes convenient to use the complex frequency variable s rather than $j\omega$. If this is done, Eq. (2.7.16) becomes

$$E\left[X^2(t)\right] = \frac{1}{2\pi j}\int_{-j\infty}^{j\infty} S_X(s)\,ds \tag{2.7.17}$$

Equation (2.7.16) suggests that we can consider the signal power as being distributed in frequency in accordance with $S_X(j\omega)$, thus, the terms *power* and *density* in power spectral density function. Using this concept, we can obtain the power in a finite band by integrating over the appropriate range of frequencies, that is,

$$\begin{bmatrix} \text{“Power” in} \\ \text{range } \omega_1 \le \omega \le \omega_2 \end{bmatrix} = \frac{1}{2\pi}\int_{-\omega_2}^{-\omega_1} S_X(j\omega)\,d\omega + \frac{1}{2\pi}\int_{\omega_1}^{\omega_2} S_X(j\omega)\,d\omega \tag{2.7.18}$$

An example will now be given to illustrate the use of Eqs. (2.7.16) and (2.7.17).

EXAMPLE 2.10

Consider the spectral function of Example 2.9:

$$S_X(j\omega) = \frac{2\sigma^2\beta}{\omega^2 + \beta^2} \tag{2.7.19}$$

Application of Eq. (2.7.16) should yield the mean square value σ^2. This can be verified using conventional integral tables.

$$E(X^2) = \frac{1}{2\pi}\int_{-\infty}^{\infty} \frac{2\sigma^2\beta}{\omega^2 + \beta^2}\,d\omega = \frac{\sigma^2\beta}{\pi}\left[\frac{1}{\beta}\tan^{-1}\frac{\omega}{\beta}\right]_{-\infty}^{\infty} = \sigma^2 \tag{2.7.20}$$

Or equivalently, in terms of s:

$$E(X^2) = \frac{1}{2\pi j}\int_{-j\infty}^{j\infty} \frac{2\sigma^2\beta}{s^2 + \beta^2}\,ds = \sigma^2 \tag{2.7.21}$$

More will be said about evaluating integrals of the type in Eq. (2.7.21) later, in Chapter 3.

■

In summary, we see that the autocorrelation function and power spectral density function are Fourier transform pairs. Thus, both contain the same basic information about the process, but in different forms. Since we can easily transform back and forth between the time and frequency domains, the manner in which the information is presented is purely a matter of convenience for the problem at hand.

Before leaving the subject of power spectral density, it is worthy of mention that when two processes x and y are uncorrelated, the spectral density of the sum is given by

$$S_{x+y}(j\omega) = S_x(j\omega) + S_y(j\omega)$$

This follows directly from the corresponding autocorrelation equation, Eq. (2.6.9).

2.8 WHITE NOISE

White noise is defined to be a stationary random process having a constant spectral density function. The term "white" is a carryover from optics, where white light is light containing all visible frequencies. Denoting the white noise spectral amplitude as A, we then have

$$S_{wn}(j\omega) = A \tag{2.8.1}$$

The corresponding autocorrelation function for white noise is then

$$R_{wn}(\tau) = A\delta(\tau) \tag{2.8.2}$$

These functions are sketched in Fig. 2.12.

In analysis, one frequently makes simplifying assumptions in order to make the problem mathematically tractable. White noise is a good example of this. However, by assuming the spectral amplitude of white noise to be constant for all frequencies (for the sake of mathematical simplicity), we find ourselves in the awkward situation of having defined a process with infinite variance. Qualitatively, white noise is sometimes characterized as noise that is jumping around infinitely far,

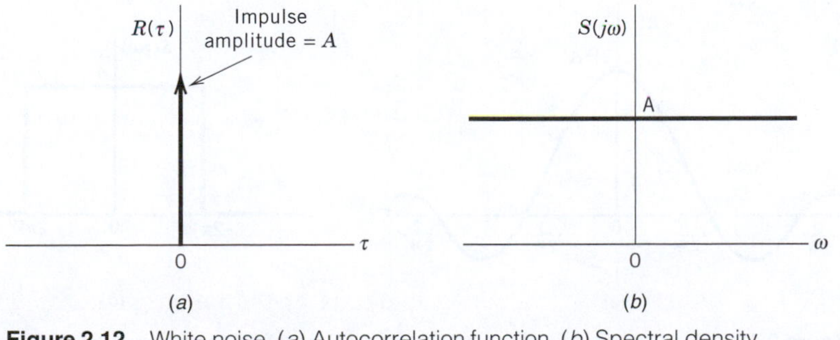

Figure 2.12 White noise. (*a*) Autocorrelation function. (*b*) Spectral density function.

infinitely fast! This is obviously physical nonsense but it is a useful abstraction. The saving feature is that all physical systems are bandlimited to some extent, and a bandlimited system driven by white noise yields a process that has finite variance; that is, the end result makes sense. We will elaborate on this further in Chapter 3.

Bandlimited white noise is a random process whose spectral amplitude is constant over a finite range of frequencies, and zero outside that range. If the bandwidth includes the origin (sometimes called baseband), we then have

$$S_{bwn}(j\omega) = \begin{cases} A, & |\omega| \le 2\pi W \\ 0, & |\omega| > 2\pi W \end{cases} \tag{2.8.3}$$

where W is the physical bandwidth in hertz. The corresponding autocorrelation function is

$$R_{bwn}(\tau) = 2WA \frac{\sin(2\pi W\tau)}{2\pi W\tau} \tag{2.8.4}$$

Both the autocorrelation and spectral density functions for baseband bandlimited white noise are sketched in Fig. 2.13. It is of interest to note that the autocorrelation function for baseband bandlimited white noise is zero for $\tau = 1/2W, 2/2W, 3/2W$, etc. From this we see that if the process is sampled at a rate of $2W$ samples/second (sometimes called the Nyquist rate), the resulting set of random variables are uncorrelated. Since this usually simplifies the analysis, the white bandlimited assumption is frequently made in bandlimited situations.

The frequency band for bandlimited white noise is sometimes offset from the origin and centered about some center frequency W_0. It is easily verified that the autocorrelation/spectral-function pair for this situation is

$$S(j\omega) = \begin{cases} A, & 2\pi W_1 \le |\omega| \le 2\pi W_2 \\ 0, & |\omega| < 2\pi W_1 \text{ and } |\omega| > 2\pi W_2 \end{cases} \tag{2.8.5}$$

$$\begin{aligned} R(\tau) &= A\left[2W_2 \frac{\sin 2\pi W_2\tau}{2\pi W_2\tau} - 2W_1 \frac{\sin 2\pi W_1\tau}{2\pi W_1\tau}\right] \\ &= A2\,\Delta W \frac{\sin \pi\Delta W\tau}{\pi\Delta W\tau} \cos 2\pi W_o\tau \end{aligned} \tag{2.8.6}$$

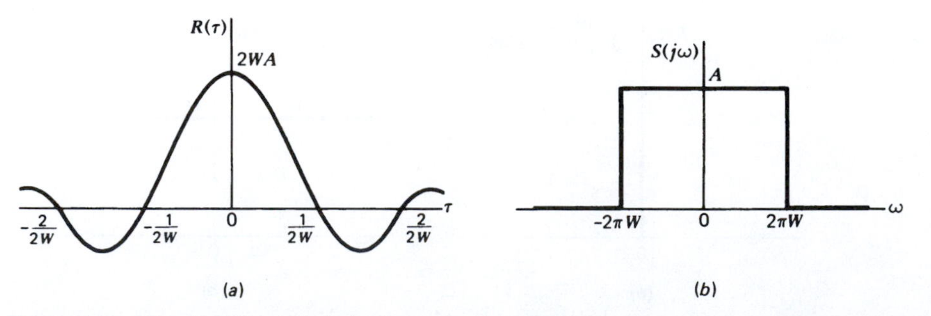

Figure 2.13 Baseband bandlimited white noise. (*a*) Autocorrelation function. (*b*) Spectral density function.

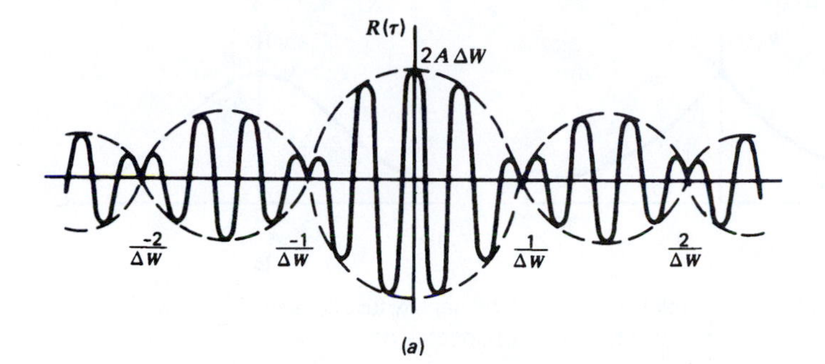

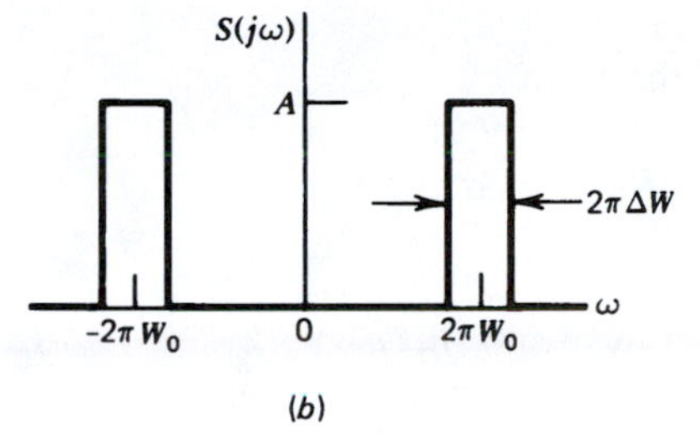

Figure 2.14 Bandlimited white noise with center frequency W_0. (*a*) Autocorrelation function. (*b*) Spectral density.

where

$$\Delta W = W_2 - W_1 \text{ Hz}$$

$$W_0 = \frac{W_1 + W_2}{2} \text{Hz}$$

These functions are sketched in Fig. 2.14.

It is worth noting the bandlimited white noise has a finite mean-square value, and thus it is physically plausible, whereas pure white noise is not. However, the mathematical forms for the autocorrelation and spectral functions in the bandlimited case are more complicated than for pure white noise.

Before leaving the subject of white noise, it is worth mentioning that the analogous discrete-time process is referred to as a white sequence. A *white sequence* is defined simply as a sequence of zero-mean, uncorrelated random variables. That is, all members of the sequence have zero means and are mutually uncorrelated with all other members of the sequence. If the random variables are also normal, then the sequence is a *Gaussian* white sequence.

2.9 GAUSS–MARKOV PROCESSES

A stationary Gaussian process $X(t)$ that has an exponential autocorrelation is called a *Gauss–Markov* process. The autocorrelation and spectral functions for this

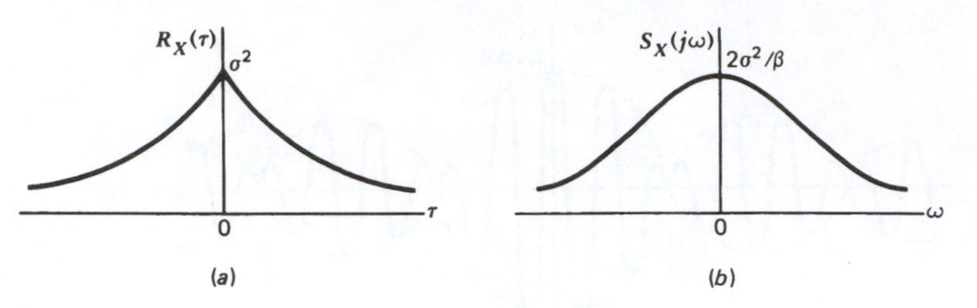

Figure 2.15 Autocorrelation and spectral density functions for Gauss–Markov process. (*a*) Autocorrelation function. (*b*) Spectral density.

process are then of the form

$$R_X(\tau) = \sigma^2 e^{-\beta|\tau|} \tag{2.9.1}$$

$$S_X(j\omega) = \frac{2\sigma^2\beta}{\omega^2 + \beta^2} \left[\text{or } S_X(s) = \frac{2\sigma^2\beta}{-s^2 + \beta^2}\right] \tag{2.9.2}$$

These functions are sketched in Fig. 2.15. The mean-square value and time constant for the process are given by the σ^2 and $1/\beta$ parameters, respectively. The process is nondeterministic, so a typical sample time function would show no deterministic structure and would look like typical "noise." The exponential autocorrelation function indicates that sample values of the process gradually become less and less correlated as the time separation between samples increases. The autocorrelation function approaches zero as $\tau \to \infty$, and thus the mean value of the process must be zero. The reference to Markov in the name of this process is not obvious at this point, but it will be after the discussion on optimal prediction in Chapter 4.

The Gauss–Markov process is an important process in applied work because (1) it seems to fit a large number of physical processes with reasonable accuracy, and (2) it has a relatively simple mathematical description. As in the case of all stationary Gaussian processes, *specification of the process autocorrelation function completely defines the process*. This means that any desired higher-order probability density function for the process may be written out explicitly, given the autocorrelation function. An example will illustrate this.

EXAMPLE 2.11

Let us say that a Gauss–Markov process *X*(*t*) has autocorrelation function

$$R_X(\tau) = 100e^{-2|\tau|} \tag{2.9.3}$$

We wish to write out the third-order probability density function

$$f_{X_1X_2X_3}(x_1, x_2, x_3) \quad \text{where } X_1 = X(0),\ X_2 = X(.5),\ \text{and } X_3 = X(1)$$

First we note that the process mean is zero. The covariance matrix in this case is a 3×3 matrix and is obtained as follows:

$$C_X = \begin{bmatrix} E(X_1^2) & E(X_1X_2) & E(X_1X_3) \\ E(X_2X_1) & E(X_2^2) & E(X_2X_3) \\ E(X_3X_1) & E(X_3X_2) & E(X_3^2) \end{bmatrix} = \begin{bmatrix} R_X(0) & R_X(.5) & R_X(1) \\ R_X(.5) & R_X(0) & R_X(.5) \\ R_X(1) & R_X(.5) & R_X(0) \end{bmatrix}$$

$$= \begin{bmatrix} 100 & 100e^{-1} & 100e^{-2} \\ 100e^{-1} & 100 & 100e^{-1} \\ 100e^{-2} & 100e^{-1} & 100 \end{bmatrix} \tag{2.9.4}$$

Now that $\mathbf{C}_X$ has been written out explicitly, we can use the general normal form given by Eq. (1.14.5). The desired density function is then

$$f_X(\mathbf{x}) = \frac{1}{(2\pi)^{3/2}|\mathbf{C}_X|^{1/2}} e^{-\frac{1}{2}[\mathbf{x}^T \mathbf{C}_x^{-1} \mathbf{x}]} \tag{2.9.5}$$

where $\mathbf{x}$ is the 3-tuple

$$\mathbf{x} = \begin{bmatrix} x_1 \\ x_2 \\ x_3 \end{bmatrix} \tag{2.9.6}$$

and $\mathbf{C}_x$ is given by Eq. (2.9.4).

■

The simple scalar Gauss–Markov process whose autocorrelation function is exponential is sometimes referred to as a first-order Gauss–Markov process. This is because the discrete-time version of the process is described by a first-order difference equation of the form

$$X(t_{k+1}) = e^{-\beta \Delta t} X(t_k) + W(t_k)$$

where $W(t_k)$ is an uncorrelated zero-mean Gaussian sequence. Discrete-time Gaussian processes that satisfy higher-order difference equations are also often referred to as Gauss–Markov processes of the appropriate order. Such processes are best described in vector form, and this is discussed in detail in Section 3.8. For now we will simply illustrate the benefit of higher-order models with a second-order process example.

EXAMPLE 2.12 SECOND-ORDER GAUSS–MARKOV PROCESS

Suppose we have a dynamic situation where the position and velocity are to be modeled as random processes, and the usual exact derivative relationship between x and $\dot{x}$ applies. We will assume that both processes are stationary and have finite variances. In this case we dare not model the position as a simple first-order Markov process, because the corresponding velocity would have infinite variance. This is obvious from their respective power spectral densities, and the methods for computing mean square value discussed in Chapter 3, Sections 3.2 and 3.3. The spectral densities are

$$S_{pos}(j\omega) = \frac{2\sigma^2\beta}{\omega^2 + \beta^2}, \text{ and } S_{vel}(j\omega) = \frac{2\sigma^2\beta\omega^2}{\omega^2 + \beta^2}$$

The integral of $S_{pos}(j\omega)$ over all ω will exist, but the integral of $S_{vel}(j\omega)$ will not. Thus, we need to look further for a model where both position and velocity have finite variance.

So, consider the following tentative power spectral density for x:

$$S_x(j\omega) = \frac{2\sqrt{2}\omega_0^3\sigma^2}{\omega^4 + \omega_0^4} \tag{2.9.7}$$

where σ^2 is the position variance parameter and ω_0 is the natural frequency parameter (in rad/s). This power spectral density leads to an especially simple autocorrelation function of the form:

$$R_x(\tau) = \begin{cases} \sigma^2 e^{-a\tau}(\cos a\tau + \sin a\tau), & \tau \geq 0 \\ \sigma^2 e^{a\tau}[\cos(-a\tau) + \sin(-a\tau)], & \tau < 0 \end{cases} \tag{2.9.8}$$

where

$$a = \frac{\omega_0}{\sqrt{2}} \tag{2.9.9}$$

Qualitatively, this will be recognized as a damped oscillatory function with its maximum at the origin. The σ^2 and ω_0 parameters can be chosen to roughly match the σ^2 and β parameters of a corresponding first-order Markov process. Note that the tentative $S_x(j\omega)$ is fourth-order in ω in the denominator and zero-order in the numerator, so both the position and velocity variances will be finite for this model.

In order to check the reasonableness of the second-order Gauss–Markov model, it was simulated using MATLAB with $\sigma^2 = 1\ \text{m}^2$ and $\omega_0 = 0.1$ rad/s. For comparison, a first-order Markov process was also simulated with its σ^2 and β set at $1\ \text{m}^2$ and 0.1 rad/s respectively. The results for a 100-second time span are shown in Figs. 2.16(*a*) and (*b*). Both processes were assigned the same variance, but different seeds were used in the Monte Carlo simulations, so there is no way to make an exact comparison. The most conspicuous difference in the two sample realizations shown in the figure lies in the high frequency components. Clearly, the high frequency content is much smaller in the second-order model than in the first-order model. This was expected because of the ω^4 the denominator of $S_x(j\omega)$ for the second-order model, versus ω^2 in the denominator of the first-order $S_x(j\omega)$.

■

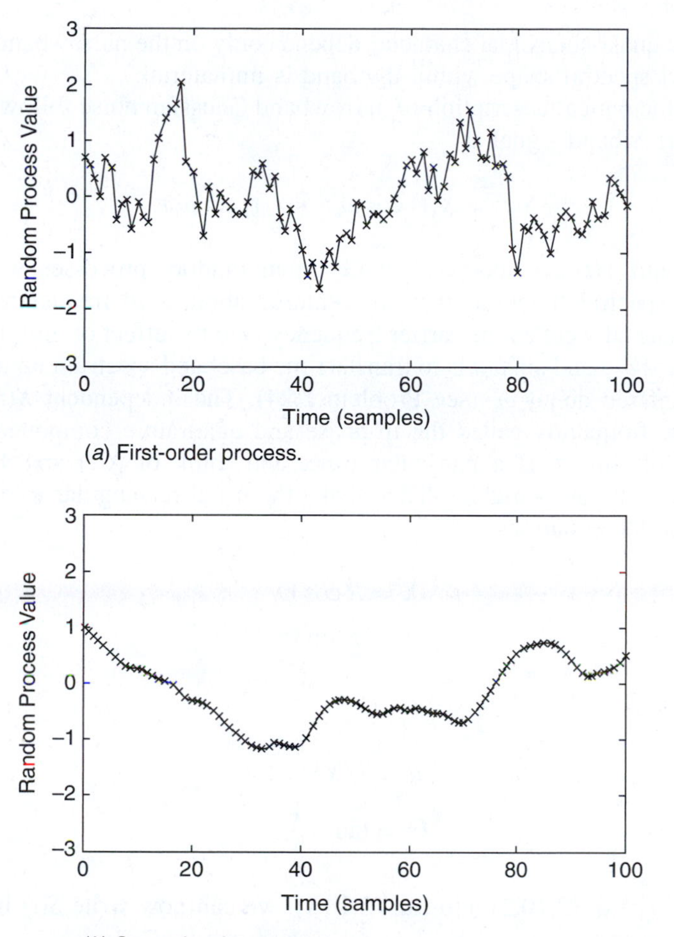

(a) First-order process.

(b) Second-order process.

Figure 2.16 First- and second-order Gauss–Markov processes are compared.

2.10 NARROWBAND GAUSSIAN PROCESS

In both control and communication systems, we frequently encounter situations where a very narrowband system is excited by wideband Gaussian noise. A high-Q tuned circuit and/or a lightly damped mass–spring arrangement are examples of narrowband systems. The resulting output is a noise process with essentially all its spectral content concentrated in a narrow frequency range. If one were to observe the output of such a system, the time function would appear to be nearly sinusoidal, especially if just a few cycles of the output signal were observed. However, if one were to carefully examine a long record of the signal, it would be seen that the quasi-sinusoid is slowly varying in both amplitude and phase. Such a signal is called narrowband noise and, if it is the result of passing wideband Gaussian noise through a linear narrowband system, then it is also Gaussian. We are assured of this because any linear operation on a set of normal variates results in another set of normal variates. The quasi-sinusoidal character depends only on the narrowband property, and the exact spectral shape within the band is immaterial.

The mathematical description of narrowband Gaussian noise follows. We first write the narrowband signal as

$$S(t) = X(t)\cos\omega_c t - Y(t)\sin\omega_c t \tag{2.10.1}$$

where $X(t)$ and $Y(t)$ are independent Gaussian random processes with similar narrowband spectral functions that are centered about zero frequency. The frequency ω_c is usually called the carrier frequency, and the effect of multiplying $X(t)$ and $Y(t)$ by $\cos\omega_c t$ and $\sin\omega_c t$ is to translate the baseband spectrum up to a similar spectrum centered about ω_c (see Problem 2.24). The independent $X(t)$ and $Y(t)$ processes are frequently called the in-phase and quadrature components of $S(t)$. Now, think of time t as a particular time, and think of $X(t)$ and $Y(t)$ as the corresponding random variables. Then make the usual rectangular to polar transformation via the equations

$$\begin{aligned} X &= R\cos\Theta \\ Y &= R\sin\Theta \end{aligned} \tag{2.10.2}$$

or, equivalently,

$$\begin{aligned} R &= \sqrt{X^2 + Y^2} \\ \Theta &= \tan^{-1}\frac{Y}{X} \end{aligned} \tag{2.10.3}$$

By substituting Eq. (2.10.2) into Eq. (2.10.1), we can now write $S(t)$ in the form

$$\begin{aligned} S(t) &= R(t)\cos\Theta(t)\cos\omega_c t - R(t)\sin\Theta(t)\sin\omega_c t \\ &= R(t)\cos\left[\omega_c t + \Theta(t)\right] \end{aligned} \tag{2.10.4}$$

It is from Eq. (2.10.4) that we get the physical interpretation of "slowly varying envelope (amplitude) and phase."

Before we proceed, a word or two about the probability densities for X, Y, R, and Θ is in order. If X and Y are independent normal random variables with the same variance σ^2, their individual and joint densities are

$$f_X(x) = \frac{1}{\sqrt{2\pi}\sigma} e^{-x^2/2\sigma^2} \tag{2.10.5}$$

$$f_Y(y) = \frac{1}{\sqrt{2\pi}\sigma} e^{-y^2/2\sigma^2} \tag{2.10.6}$$

and

$$f_{XY}(x, y) = \frac{1}{2\pi\sigma^2} e^{-(x^2+y^2)/2\sigma^2} \tag{2.10.7}$$

The corresponding densities for R and Θ are Rayleigh and uniform (see Example 1.13). The mathematical forms are

$$f_R(r) = \frac{r}{\sigma^2} e^{-r^2/2\sigma^2}, \quad r \geq 0 \quad \text{(Rayleigh)} \tag{2.10.8}$$

$$f_\Theta(\theta) = \begin{cases} \dfrac{1}{2\pi}, & 0 \leq \theta < 2\pi \\ 0, & \text{otherwise} \end{cases} \quad \text{(uniform)} \tag{2.10.9}$$

Also, the joint density function for R and Θ is

$$f_{R\Theta}(r, \theta) = \frac{r}{2\pi\sigma^2} e^{-r^2/2\sigma^2}, \quad r \geq 0 \text{ and } 0 \leq \theta < 2\pi \tag{2.10.10}$$

It is of interest to note here that if we consider simultaneous time samples of envelope and phase, the resulting random variables are statistically independent. However, the *processes* $R(t)$ and $\Theta(t)$ are not statistically independent (5). This is due to the fact that the joint probability density associated with adjacent samples cannot be written in product form, that is,

$$f_{R_1R_2\Theta_1\Theta_2}(r_1, r_2, \theta_1, \theta_2) \neq f_{R_1R_2}(r_1, r_2) f_{\Theta_1\Theta_2}(\theta_1, \theta_2) \tag{2.10.11}$$

We have assumed that $S(t)$ is a Gaussian process, and from Eq. (2.10.1) we see that

$$\text{Var}\, S = \tfrac{1}{2}(\text{Var}\, X) + \tfrac{1}{2}(\text{Var}\, Y) = \sigma^2 \tag{2.10.12}$$

Thus,

$$f_S(s) = \frac{1}{\sqrt{2\pi}\sigma} e^{-s^2/2\sigma^2} \tag{2.10.13}$$

The higher-order density functions for S will, of course, depend on the specific shape of the spectral density for the process.

2.11 WIENER OR BROWNIAN-MOTION PROCESS

Suppose we start at the origin and take n steps forward or backward at random, with equal likelihood of stepping in either direction. We pose two questions: After taking n steps, (1) what is the average distance traveled, and (2) what is the variance of the distance? This is the classical random-walk problem of statistics. The averages considered here must be taken in an ensemble sense; for example, think of running simultaneous experiments and then averaging the results for a given number of steps. It should be apparent that the average distance traveled is zero, provided we say "forward" is positive and "backward" is negative. However, the square of the distance is always positive (or zero), so its average for a large number of trials will not be zero. It is shown in elementary statistics that the variance after n unit steps is just n, or the standard deviation is $\sqrt{n}$ (see Problem 2.13). Note that this increases without bound as n increases, and thus the process is nonstationary.

The continuous analog of random-walk is the output of an integrator driven with white noise. This is shown in block-diagram form in Fig. 2.17*a*. Here we consider the input switch as closing at $t = 0$ and the initial integrator output as being zero. An ensemble of typical output time signals is shown in Fig. 2.17*b*. The system response $X(t)$ is given by

$$X(t) = \int_0^t F(u)\, du \tag{2.11.1}$$

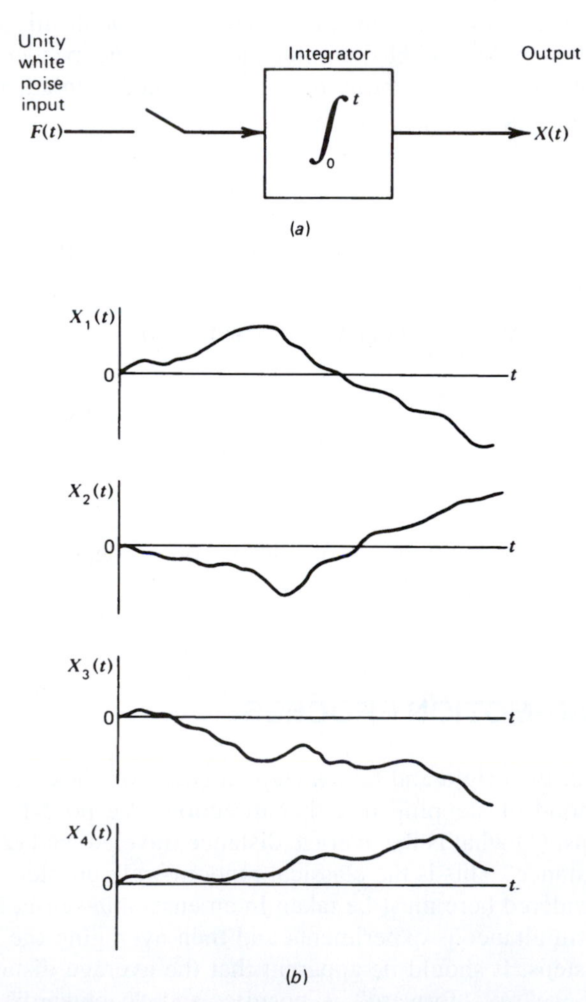

Figure 2.17 Continuous analog of random walk. (*a*) Block diagram. (*b*) Ensemble of output signals.

Clearly, the average of the output is

$$E[X(t)] = E\left[\int_0^t F(u)\,du\right] = \int_0^t E[F(u)]\,dy = 0 \tag{2.11.2}$$

Also, the mean-square-value (variance) is

$$E[X^2(t)] = E\left[\int_0^t F(u)\,du \int_0^t F(\upsilon)\,d\upsilon\right] = \int_0^t \int_0^t E[F(u)F(\upsilon)]\,du\,d\upsilon \tag{2.11.3}$$

But $E[F(u)F(\upsilon)]$ is just the autocorrelation function $R_F(u-\upsilon)$, which in this case is a Dirac delta function. Thus,

$$E[X^2(t)] = \int_0^t \int_0^t \delta(u-\upsilon)\,du\,d\upsilon = \int_0^t d\upsilon = t \tag{2.11.14}$$

So, $E[X^2(t)]$ increases linearly with time and the rms value increases in accordance with $\sqrt{t}$ (for unity white noise input). (Problem 2.26 provides a demonstration of this.)

Now, add the further requirement that the input be *Gaussian* white noise. The output will then be a Gaussian process because integration is a linear operation on the input. The resulting continuous random-walk process is known as the *Wiener* or *Brownian-motion* process. The process is nonstationary, it is Gaussian, and its mean, mean-square value, and autocorrelation function are given by

$$E[X(t)] = 0 \tag{2.11.5}$$

$$E[X^2(t)] = t \tag{2.11.6}$$

$$\begin{aligned} R_X(t_1, t_2) = E[X(t_1)X(t_2)] &= E\left[\int_0^{t_1} F(u)\,du \cdot \int_0^{t_2} F(\upsilon)\,d\upsilon\right] \\ &= \int_0^{t_2}\int_0^{t_1} E[F(u)F(\upsilon)]\,du\,d\upsilon = \int_0^{t_2}\int_0^{t_1} \delta(u-\upsilon)\,du\,d\upsilon \end{aligned}$$

Evaluation of the double integral yields

$$R_X(t_1, t_2) = \begin{cases} t_2, & t_1 \geq t_2 \\ t_1, & t_1 < t_2 \end{cases} \tag{2.11.7}$$

Since the process is nonstationary, the autocorrelation function is a general function of the two arguments t_1 and t_2. With a little imagination, Eq. (2.11.7) can be seen to describe two faces of a pyramid with the sloping ridge of the pyramid running along the line $t_1 = t_2$.

It was mentioned before that there are difficulties in defining directly what is meant by Gaussian white noise. This is because of the "infinite variance" problem. The Wiener process is well behaved, though. Thus, we can reverse the argument given here and begin by arbitrarily defining it as a Gaussian process with an autocorrelation function given by Eq. (2.11.7). This completely specifies the process. We can now describe Gaussian white noise in terms of its integral. That is, Gaussian white noise is that hypothetical process which, when integrated, yields a Wiener process.

2.12 PSEUDORANDOM SIGNALS

As the name implies, pseudorandom signals have the appearance of being random, but are not truly random. In order for a signal to be truly random, there must be some uncertainty about it that is governed by chance. Pseudorandom signals do not have this "chance" property. For example, consider a hypothetical situation where we have a very long record of a particular realization of a Gauss–Markov process. After the fact, nothing remains to chance insofar as the observer is concerned. Next, imagine folding the record back on itself into a single loop. Then imagine playing the loop back over and over again. Clearly, the resulting signal is periodic and completely known to the original observer. Yet, to a second observer it would appear to be pure random noise, especially if the looped-back splice were done carefully so as to not have an abrupt discontinuity at the splice.

It also should be clear that if the looped signal is periodic, then its spectrum would be line-type rather than continuous. The line spacing would be quite small if the original record length was large; but, nevertheless, the true spectrum would consist of discrete lines rather than a "smear" as in the case of true random noise. Line-type spectrum is characteristic of all pseudorandom signals.

Pseudorandom noise of the type just described would not be of much value in today's digital world. Yet, computer-generated pseudorandom noise has proved quite useful in a variety of practical applications, and two simplified examples will illustrate this.

EXAMPLE 2.13

Binary sequences generated by shift registers with feedback have periodic properties and have found extensive application in ranging and communication systems (6, 7, 8, 9). We will use the simple 5-bit shift register shown in Fig. 2.18 to demonstrate how a pseudorandom binary signal can be generated. In this system the bits are shifted to the right with each clock pulse, and the input on the left is determined by the feedback arrangement. For the initial condition shown, it can be verified that the output sequence is

1111100011011101010000100101100 1111100.........

31 bits same 31 bits etc.

Note that the sequence repeats itself after 31 bits. This periodic property is characteristic of a shift register with feedback. The maximum length of the sequence (before repetition) is given by $(2^n - 1)$, where n is the register length (9). The 5-bit

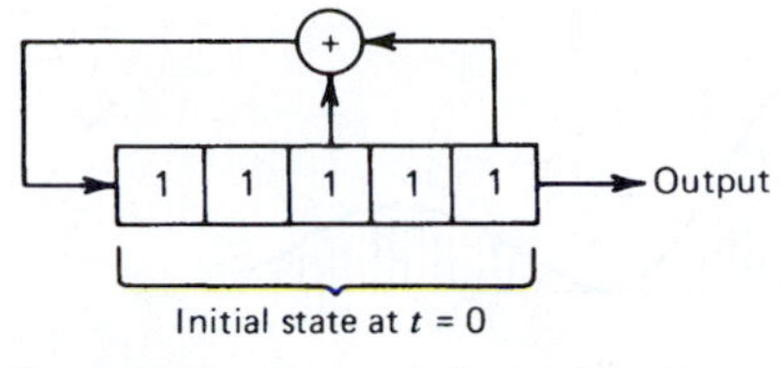

Figure 2.18 Binary shift register with feedback.

example used here is then a maximum-length sequence. These sequences are especially interesting because of their pseudorandom appearance. Note that there are nearly the same number of zeros and ones in the sequence (16 ones and 15 zeros), and that they appear to occur more or less at random. If we consider a longer shift register, say, one with 10 bits, its maximum-length sequence would be 1023 bits. It would have 512 ones and 511 zeros; again, these would appear to be distributed at random. A casual look at any interior string of bits would not reveal anything systematic. Yet the string of bits so generated is entirely deterministic. Once the feedback arrangement and initial condition are specified, the output is determined forever after. Thus, the sequence is pseudorandom, not random.

When converted to voltage waveforms, maximum-length sequences also have special autocorrelation properties. Returning to the 31-bit example, let binary one be 1 V and binary zero be -1 V, and let the voltage level be held constant during the clock-pulse interval. The resulting time waveform is shown in Fig. 2.19, and its time autocorrelation function is shown in Fig. 2.20. Note that the unique distribution of zeros and ones for this sequence is such that the autocorrelation function is a small constant value after a shift of one unit of time (i.e., one bit). This is typical of all

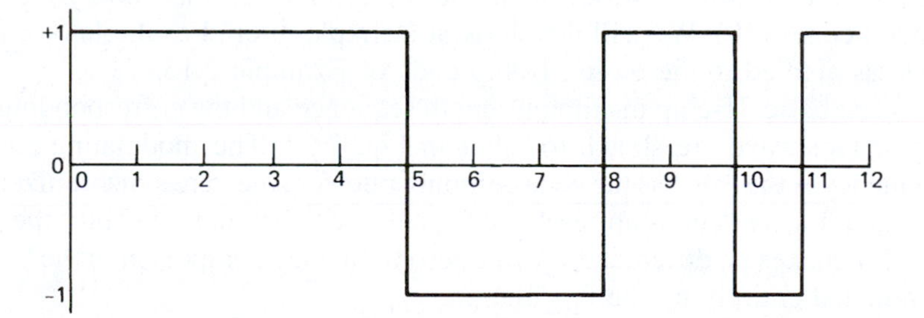

Figure 2.19 Pseudorandom binary waveform.

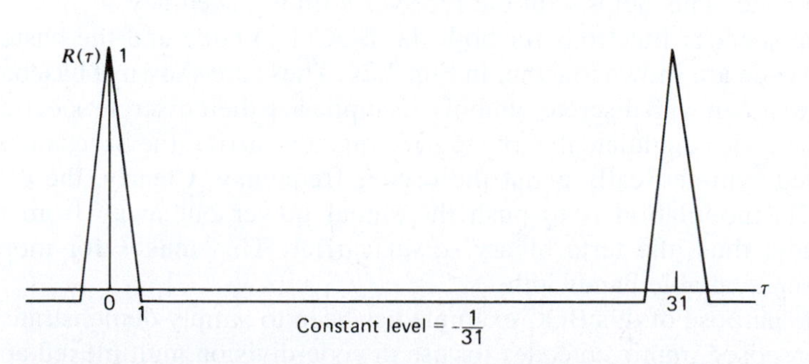

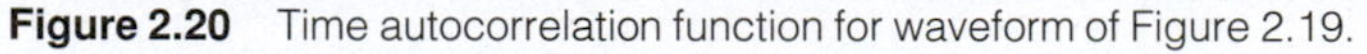

Figure 2.20 Time autocorrelation function for waveform of Figure 2.19.

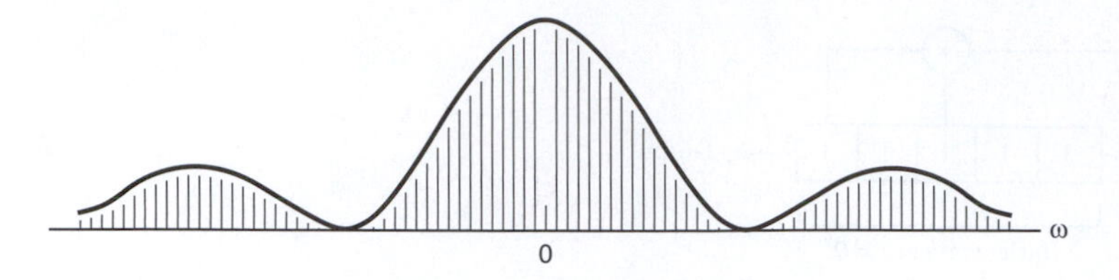

Figure 2.21 Spectral density for pseudorandom binary waveform.

maximum-length sequences. When the sequence length is long, the correlation after a shift of one unit is quite small. This has obvious advantages in correlation detection schemes, and such schemes have been used extensively in electronic ranging applications (7, 8).

The spectral density function for the waveform of Fig. 2.19 is shown in Fig. 2.21. As with all pseudorandom signals, the spectrum is line-type rather than continuous (6).

EXAMPLE 2.14 BOC (1.1) CODE

In code-division multiplexed systems such as GPS, the various codes' bandwidths and correlation structures are important considerations. There is usually no shortage of possible codes; and, as one might expect, some are better than others. One possible way of improving on a basic set of codes is to modulate each code in the set with a periodic square wave. Such codes are referred to as Binary Offset Carrier (BOC) codes (10). We will now look at a simple tutorial example of a BOC (1,1) code as applied to the basic 31-chip code of Example 2.13.

The basic 31-chip maximum-length sequence and the corresponding modulating square wave are shown together in Fig. 2.22. The modulating square wave frequency is the same as the code chipping rate, and they are synchronized such that the square wave "chops up" each chip equally with $+1$ and -1. Thus, the polarity of the chip makes no difference, and the zero frequency component in the basic code is eliminated entirely by the modulation.

The autocorrelation function for the BOC(1,1) code is shown in Fig. 2.23. Note that the slope of the BOC(1,1) autocorrelation is steeper than that for the reference 31-chip code. This helps with the receiver's timing accuracy.

The spectral functions for both the BOC(1,1) code and the basic reference 31-chip code are shown together in Fig. 2.24. These are shown at baseband and the plots are shown with discrete symbols to emphasize their discrete spectra. Note that when the codes modulate the true high frequency carrier, the baseband spectra are translated symmetrically about the carrier frequency. Clearly, the effect of the BOC(1,1) modulation is to push the signal power out away from the carrier frequency; thus, the term binary offset carrier. This makes for more efficient use of the available bandwidth.

The purpose of this BOC example has been to simply demonstrate one of the many proposed improved codes for use in code-division multiplexed applications.

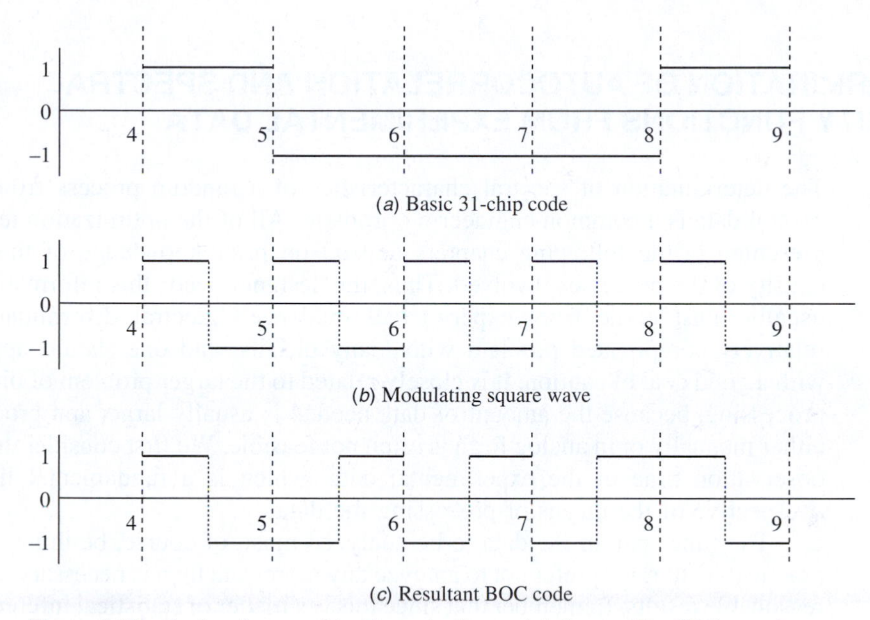

Figure 2.22 Basic reference 31-chip code, the modulating and the BOC waveforms.

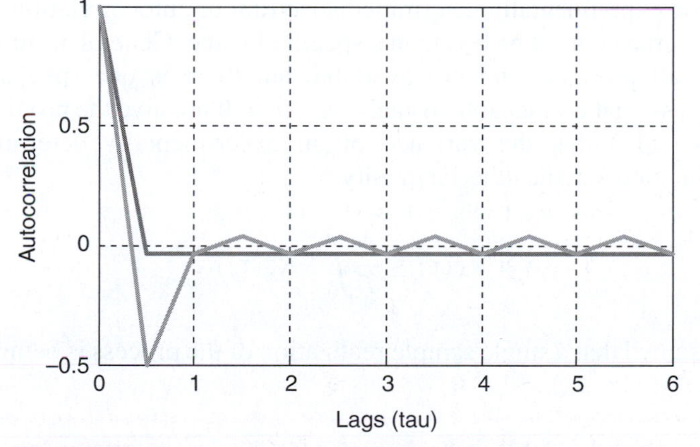

Figure 2.23 Autocorrelation functions for the reference and BOC (1,1) pseudorandom waveform (shown without the periodic extensions that begin at $\tau = 31$).

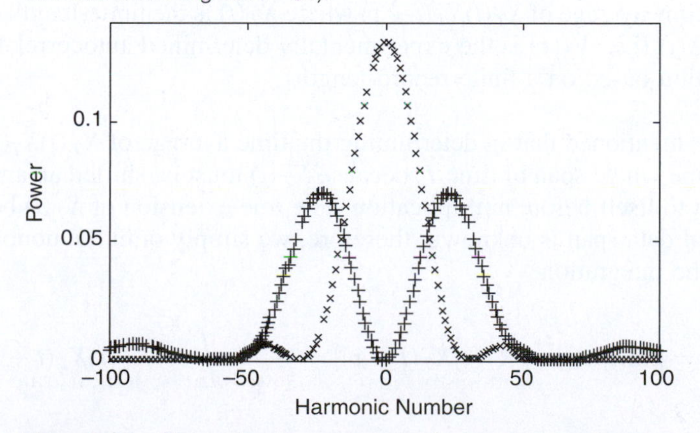

Figure 2.24 Plots of the discrete spectra for the 31-chip reference and BOC(1,1) codes.

2.13 DETERMINATION OF AUTOCORRELATION AND SPECTRAL DENSITY FUNCTIONS FROM EXPERIMENTAL DATA

The determination of spectral characteristics of a random process from experimental data is a common engineering problem. All of the optimization techniques presented in the following chapters depend on prior knowledge of the spectral density of the processes involved. Thus, the designer needs this information and it usually must come from experimental evidence. Spectral determination is a relatively complicated problem with many pitfalls, and one should approach it with a good deal of caution. It is closely related to the larger problem of digital data processing, because the amount of data needed is usually large, and processing it either manually or in analog form is often not feasible. We first consider the span of observation time of the experimental data, which is a fundamental limitation, irrespective of the means of processing the data.

The time span of the data to be analyzed must, of course, be finite; and, as a practical matter, we prefer not to analyze any more data than is necessary to achieve reasonable results. Remember that since this is a matter of statistical inference, there will always remain some statistical uncertainty in the result. One way to specify the accuracy of the experimentally determined spectrum or autocorrelation function is to say that its variance must be less than a specified value. General accuracy bounds applicable to all processes are not available but there is one special case, the Gaussian process, that is amenable to analysis. We will not give the proof here, but it can be shown (11) that the variance of an experimentally determined autocorrelation function satisfies the inequality

$$\text{Var}\, V_X(\tau) \leq \frac{4}{T} \int_0^{\infty} R_X^2(\tau)\, d\tau \tag{2.13.1}$$

where it is assumed that a single sample realization of the process is being analyzed, and

$T =$ time length of the experimental record

$R_X(\tau) =$ autocorrelation function of the Gaussian process under consideration

$V_X(\tau) =$ time average of $X_T(t)X_T(t+\tau)$ where $X_T(t)$ is the finite-length sample of $X(t)$ [i.e., $V_X(\tau)$ is the experimentally determined autocorrelation function based on a finite record length]

It should be mentioned that in determining the time average of $X_T(t)X_T(t+\tau)$, we cannot use the whole span of time T, because $X_T(t)$ must be shifted an amount of τ with respect to itself before multiplication. The true extension of $X_T(t)$ beyond the experimental data span is unknown; therefore, we simply omit the nonoverlapped portion in the integration:

$$V_X(\tau) = [\text{time avg. of } X_T(t)X_T(t+\tau)] = \frac{1}{T-\tau} \int_0^{T-\tau} X_T(t)X_T(t+\tau)\, dt \tag{2.13.2}$$

It will be assumed from this point on that the range of τ being considered is much less than the total data span T, that is, $\tau \ll T$.

We first note that $V_X(\tau)$ is the result of analyzing a single time signal; therefore, $V_X(\tau)$ is itself just a sample function from an ensemble of functions. It is hoped that $V_X(\tau)$ as determined by Eq. (2.13.2) will yield a good estimate of $R_X(\tau)$ and, in order to do so, it should be an unbiased estimator. This can be verified by computing its expectation:

$$\begin{aligned} E[V_X(\tau)] &= E\left[\frac{1}{T-\tau}\int_0^{T-\tau} X_T(t)X_T(t+\tau)\,dt\right] \\ &= \frac{1}{T-\tau}\int_0^{T-\tau} E[X_T(t)X_T(t+\tau)]\,dt \\ &= \frac{1}{T-\tau}\int_0^{T-\tau} R_X(\tau)\,dt = R_X(\tau) \end{aligned} \tag{2.13.3}$$

Thus, $V_X(\tau)$ is an unbiased estimator of $R_X(\tau)$. Also, it can be seen from the equation for Var $V_X(\tau)$, Eq. (2.13.1), that if the integral of R_X^2 converges (e.g., R_X decreases exponentially with τ), then the variance of $V_X(\tau)$ approaches zero as T becomes large. Thus, $V_X(\tau)$ would appear to be a well-behaved estimator of $R_X(\tau)$, that is, $V_X(\tau)$ converges in the mean to $R_X(\tau)$. We will now pursue the estimation accuracy problem further.

Equation (2.13.1) is of little value if the process autocorrelation function is not known. So, at this point, we assume that $X(t)$ is a Gauss–Markov process with an autocorrelation function

$$R_X(\tau) = \sigma^2 e^{-\beta|\tau|} \tag{2.13.4}$$

The σ^2 and β parameters may be difficult to determine in a real-life problem, but we can get at least a rough estimate of the amount of experimental data needed for a given required accuracy. Substituting the assumed Markov autocorrelation function into Eq. (2.13.1) then yields

$$\text{Var}[V_X(\tau)] \leq \frac{2\sigma^4}{\beta T} \tag{2.13.5}$$

We now look at an example illustrating the use of Eq. (2.13.5).

EXAMPLE 2.15

Let us say that the process being investigated is thought to be a Gauss–Markov process with an estimated time constant $(1/\beta)$ of 1 sec. Let us also say that we wish to determine its autocorrelation function within an accuracy of 10 percent, and we want to know the length of experimental data needed. By "accuracy" we mean that the experimentally determined $V(\tau)$ should have a standard deviation less than .1 of the σ^2 of the process, at least for a reasonably small range of τ near zero. Therefore, the ratio of Var$[V(\tau)]$ to $(\sigma^2)^2$ must be less than .01. Using Eq. (2.13.5), we can write

$$\frac{\text{Var}[V(\tau)]}{\sigma^4} \leq \frac{2}{\beta T}$$

Setting the quantity on the left side equal to .01 and using the equality condition yield

$$T = \frac{1}{(.1)^2} \cdot \frac{2}{\beta} = 200 \text{ sec} \tag{2.13.6}$$

A sketch indicating a typical sample experimental autocorrelation function is shown in Fig. 2.25. Note that 10 percent accuracy is really not an especially demanding requirement, but yet the data required is 200 times the time constant of the process. To put this in more graphic terms, if the process under investigation were random gyro drift with an estimated time constant of 10 hours, 2,000 hours of continuous data would be needed to achieve 10 percent accuracy. This could very well be in the same range as the mean time to failure for the gyro. Were we to be more demanding and ask for 1 percent accuracy, about 23 years of data would be required! It can be seen that accurate determination of the autocorrelation function is not a trivial problem in some applications. (This example is pursued further in Problem 2.25.)

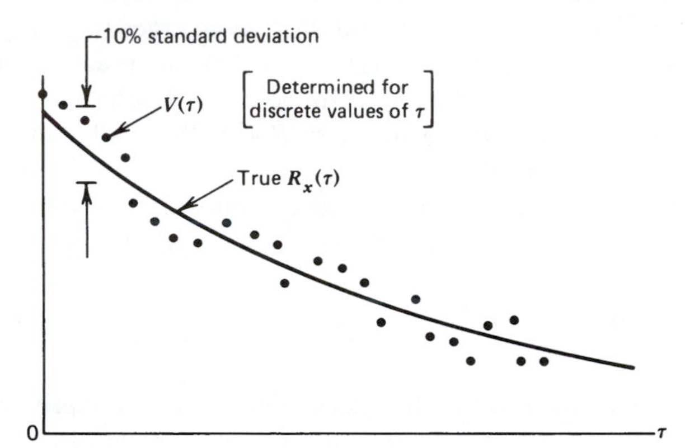

Figure 2.25 Experimental and true autocorrelation functions for Example 2.15.

The main point to be learned from this example is that reliable determination of the autocorrelation function takes considerably more experimental data than one might expect intuitively. The spectral density function is just the Fourier transform of the autocorrelation function, so we might expect a similar accuracy problem in its experimental determination.

As just mentioned, the spectral density function for a given sample signal may be estimated by taking the Fourier transform of the experimentally determined autocorrelation function. This, of course, involves a numerical procedure of some sort because the data describing $V_X(\tau)$ will be in numerical form. The spectral function may also be estimated directly from the periodogram of the sample signal. Recall from Section 2.7 that the average periodogram (the square of the magnitude of the Fourier transform of X_T) is proportional to the spectral density function (for large T). Unfortunately, since we do not usually have the luxury of having a large ensemble of periodograms to average, there are pitfalls in this approach, just as there are in going the autocorrelation route. Nevertheless, modern digital processing methods using fast Fourier transform (FFT) techniques have popularized the periodogram approach. Thus, it is important to understand its limitations (6).

First, there is the truncation problem. When the time record being analyzed is finite in length, we usually assume that the signal will "jump" abruptly to zero outside the valid data interval. This causes frequency spreading and gives rise to high-frequency components that are not truly representative of the process under consideration, which is assumed to ramble on indefinitely in a continuous manner. An extreme case of this would occur if we were to chop up one long record into many very short records and then average the periodograms of the short records. The individual periodograms, with their predominance of high-frequency components due to the truncation, would not be at all representative of the spectral content of the original signal; nor would their average! Thus, the first rule is that we must have a long time record relative to the typical time variations in the signal. This is truc rcgardlcss of the method used in analyzing the data. There is, however, a statistical convergence problem that arises as the record length becomes large, and this will now be examined.

In Section 2.7 it was shown that the expectation of the periodogram approaches the spectral density of the process for large T. This is certainly desirable, because we want the periodogram to be an unbiased estimate of the spectral density. It is also of interest to look at the behavior of the variance of the periodogram as T becomes large. Let us denote the periodogram of $X_T(\tau)$ as $M(\omega, T)$, that is,

$$M(\omega, T) = \frac{1}{T}|\mathfrak{F}\{X_T(t)\}|^2 \tag{2.13.7}$$

Note that the periodogram is a function of the record length T as well as ω. The variance of $M(\omega, T)$ is

$$\operatorname{Var} M = E(M^2) - [E(M)]^2 \tag{2.13.8}$$

Since we have already found $E(M)$ as given by Eqs. (2.7.8) and (2.7.9), we now need to find $E(M)^2$. Squaring Eq. (2.13.7) leads to

$$E(M^2) = \frac{1}{T^2}E\left[\int_0^T\int_0^T\int_0^T\int_0^T X(t)X(s)X(u)X(\upsilon)e^{-j\omega(t-s+u-\upsilon)}\,dt\,ds\,du\,d\upsilon\right] \tag{2.13.9}$$

It can be shown that if $X(t)$ is a Gaussian process,*

$$\begin{aligned} E[X(t)X(s)X(u)X(\upsilon)] = R_X(t-s)R_X(u-\upsilon) \\ + R_X(t-u)R_X(s-\upsilon) \\ + R_X(t-\upsilon)R_X(s-u) \end{aligned} \tag{2.13.10}$$

* See Problem 2.23.

Thus, moving the expectation operator inside the integration in Eq. (2.13.9) and using Eq. (2.13.10) lead to

$$
\begin{aligned}
E(M^2) &= \frac{1}{T^2}\int_0^T\int_0^T\int_0^T\int_0^T [R_X(t-s)R_X(u-v) + R_X(t-u)R_X(s-v) \\
&\quad + R_X(t-v)R_X(s-u)]e^{-j\omega(t-s+u-v)}\,dt\,ds\,du\,dv \\
&= \frac{1}{T^2}\int_0^T\int_0^T R_X(t-s)e^{-j\omega(t-s)}\,dt\,ds \int_0^T\int_0^T R_X(u-v)e^{-j\omega(u-v)}\,du\,dv \\
&\quad + \frac{1}{T^2}\int_0^T\int_0^T R_X(t-v)e^{-j\omega(t-v)}\,dt\,dv \int_0^T\int_0^T R_X(u-u)e^{-j\omega(u-u)}\,ds\,du \\
&\quad + \frac{1}{T^2}\left|\int_0^T\int_0^T R_X(t-u)e^{-j\omega(t+u)}\,dt\,du\right|^2
\end{aligned}
\tag{2.13.11}
$$

Next, substituting Eq. (2.7.3) into (2.13.11) leads to

$$
E(M^2) = 2[E(M)]^2 + \frac{1}{T^2}\left|\int_0^T\int_0^T R_X(t-u)e^{-j\omega(t+u)}\,dt\,du\right|^2 \tag{2.13.12}
$$

Therefore,

$$
\begin{aligned}
\operatorname{Var} M &= E(M^2) - [E(M)]^2 \\
&= [E(M)]^2 + \frac{1}{T^2}\left|\int_0^T\int_0^T R_X(t-u)e^{-j\omega(t-u)}\,dt\,du\right|^2
\end{aligned}
\tag{2.13.13}
$$

The second term of Eq. (2.13.13) is nonnegative, so it should be clear that

$$
\operatorname{Var} M \geq [E(M)]^2 \tag{2.13.14}
$$

But $E(M)$ approaches the spectral function as $T \to \infty$. Thus, the variance of the periodogram does not go to zero as $T \to \infty$ (except possibly at those exceptional points where the spectral function is zero). In other words, the periodogram does not converge in the mean as $T \to \infty$! This is most disturbing, especially in view of the popularity of the periodogram method of spectral determination. The dilemma is summarized in Fig. 2.26. Increasing T will not help reduce the ripples in the individual periodogram. It simply makes M "jump around" faster with ω. This does help, though, with the subsequent averaging that must accompany the spectral analysis. Recall that it is the *average* periodogram that is the measure of the spectral density function. Averaging may not be essential in the analysis of deterministic signals, but it is for random signals. Averaging in both frequency and time is easily accomplished in analog spectrum analyzers by appropriate adjustment of the width of the scanning window and the sweep speed. In digital analyzers, similar averaging over a band of discrete frequencies can be implemented in software. Also, further

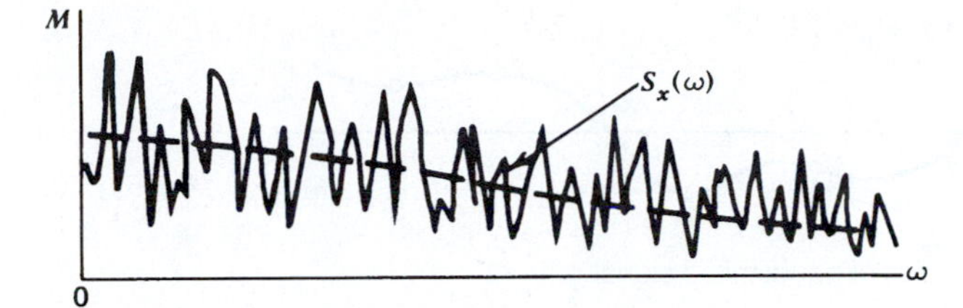

Figure 2.26 Typical periodogram for long record length.

averaging in time may be accomplished by averaging successive periodograms before displaying the spectrum graphically. In either event, analog or digital, some form of averaging is essential when analyzing noise.

Our treatment of the general problem of autocorrelation and spectral determination from experimental data must be brief. However, the message here should be clear. Treat this problem with respect. It is fraught with subtleties and pitfalls. Engineering literature abounds with reports of shoddy spectral analysis methods and the attendant questionable results. Know your digital signal processing methods and recognize the limitations of the results.

We will pursue the subject of digital spectral analysis further in Section 2.15. But first we digress to present Shannon's sampling theorems, which play an important role in digital signal processing.

2.14 SAMPLING THEOREM

Consider a time function $g(t)$ that is bandlimited, that is,

$$\mathfrak{F}[g(t)] = G(\omega) = \begin{cases} \text{Nontrivial}, & |\omega| \le 2\pi W \\ 0, & |\omega| > 2\pi W \end{cases} \tag{2.14.1}$$

Under the conditions of Eq. (2.14.1), the time function can be written in the form

$$g(t) = \sum_{n=-\infty}^{\infty} g\left(\frac{n}{2W}\right) \frac{\sin(2\pi Wt - n\pi)}{2\pi Wt - n\pi} \tag{2.14.2}$$

This remarkable theorem is due to C. E. Shannon (12, 13), and it has special significance when dealing with bandlimited noise.* The theorem says that if one were to specify an infinite sequence of sample values $\ldots, g_1, g_2, g_3, \ldots$, uniformly

*The basic concept of sampling at twice the highest signal frequency is usually attributed to H. Nyquist (14). However, the explicit form of the sampling theorem given by Eq. (2.14.2) and its associated signal bandwidth restriction was first introduced into communication theory by C. E. Shannon. You may wish to refer to Shannon (13) or Black (15) for further comments on the history of sampling theory.

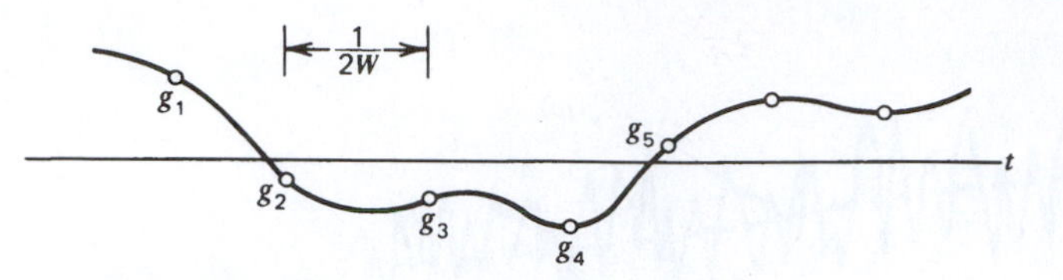

Figure 2.27 Samples of bandlimited signal $g(t)$.

spaced $1/2W$ sec apart as shown in Fig. 2.27, then there would be one and only one bandlimited function that would go through all the sample values. In other words, specifying the signal sample values and requiring $g(t)$ to be bandlimited indirectly specify the signal in between the sample points as well. The sampling rate of $2W$ Hz is known as the *Nyquist rate*. This represents the minimum sampling rate needed to preserve all the information content in the continuous signal. If we sample $g(t)$ at less than the Nyquist rate, some information will be lost, and the original signal cannot be exactly reconstructed on the basis of the sequence of samples. Sampling at a rate higher than the Nyquist rate is not necessary, but it does no harm because this simply extends the allowable range of signal frequencies beyond W Hz. Certainly, a signal lying within the bandwidth W also lies within a bandwidth greater than W.

In describing a stationary random process that is bandlimited, it can be seen that we need to consider only the statistical properties of samples taken at the Nyquist rate of $2W$ Hz. This simplifies the process description considerably. If we add the further requirement that the process is Gaussian and white within the bandwidth W, then the joint probability density for the samples may be written as a simple product of single-variate normal density functions. This simplification is frequently used in noise analysis in order to make the problem mathematically tractable.

Since there is symmetry in the direct and inverse Fourier transforms, we would expect there to be a corresponding sampling theorem in the frequency domain. It may be stated as follows. Consider the time function $g(t)$ to be time limited, that is, nontrivial over a span of time T and zero outside this interval; then its Fourier transform $G(\omega)$ may be written as

$$G(\omega) = \sum_{n=-\infty}^{\infty} G\left(\frac{2\pi n}{T}\right) \frac{\sin\left(\dfrac{\omega T}{2} - n\pi\right)}{\dfrac{\omega T}{2} - n\pi} \tag{2.14.3}$$

All of the previous comments relative to time domain sampling have their corresponding frequency-domain counterparts.

Frequently, it is useful to consider time functions that are limited in both time and frequency. Strictly speaking, this is not possible, but it is a useful approximation. This being the case, the time function can be uniquely represented by $2WT$ samples. These may be specified either in the time or frequency domain.

Sampling theorems have also been worked out for the nonbaseband case (17). These are somewhat more involved than the baseband theorems and will not be given here.

PROBLEMS

2.1 Noise measurements at the output of a certain amplifier (with its input shorted) indicate that the rms output voltage due to internal noise is 100 μV. If we assume that the frequency spectrum of the noise is flat from 0 to 10 MHz and zero above 10 MHz, find:

(a) The spectral density function for the noise.

(b) The autocorrelation function for the noise.

Give proper units for both the spectral density and autocorrelation functions.

2.2 A sketch of a sample realization of a stationary random process $X(t)$ is shown in the figure. The pulse amplitudes a_i are independent samples of a normal random variable with zero mean and variance σ^2. The time origin is completely random. Find the autocorrelation function for the process.

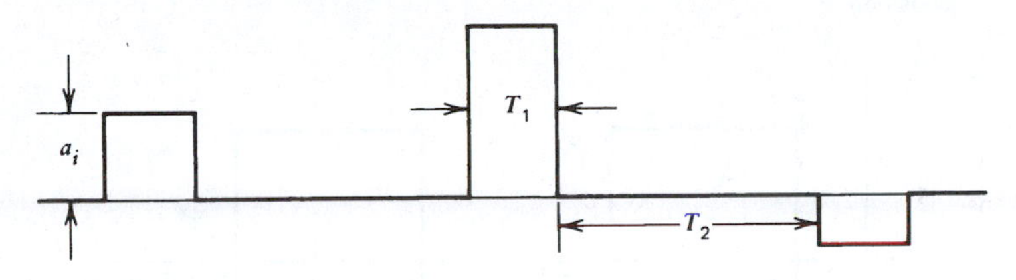

Figure P2.2

2.3 Find the autocorrelation function corresponding to the spectral density function

$$S(j\omega) = \delta(\omega) + \tfrac{1}{2}\delta(\omega - \omega_0) + \tfrac{1}{2}\delta(\omega + \omega_0) + \frac{2}{\omega^2 + 1}$$

2.4 A stationary Gaussian random process $X(t)$ has an autocorrelation function of the form

$$R_X(\tau) = 4e^{-|\tau|}$$

What fraction of the time will $|X(t)|$ exceed four units?

2.5 It is suggested that a certain real process has an autocorrelation function as shown in the figure. Is this possible? Justify your answer.

(*Hint:* Calculate the spectral density function and see if it is plausible.)

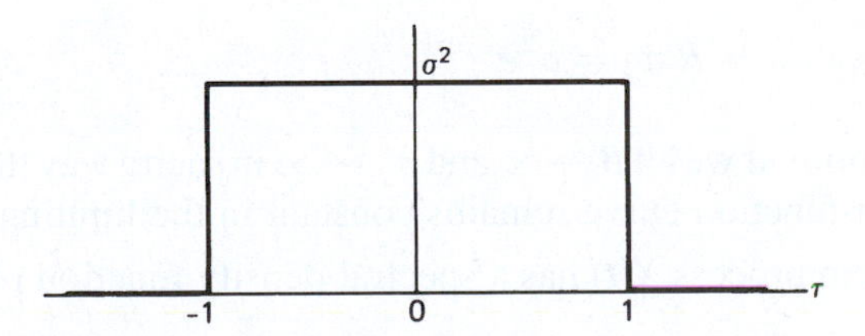

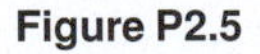

Figure P2.5

2.6 The input to an ideal rectifier (unity forward gain, zero reverse gain) is a stationary Gaussian process.

(a) Is the output stationary?

(b) Is the output a Gaussian process?

Give a brief justification for both answers.

2.7 A random process $X(t)$ has sample realizations of the form

$$X(t) = at + Y$$

where a is a known constant and Y is a random variable whose distribution is $\mathcal{N}(0, \sigma^2)$. Is the process (a) stationary and (b) ergodic? Justify your answers.

2.8 A sample realization of a random process $X(t)$ is shown in the figure. The time t_0 when the transition from the -1 state to the $+1$ state takes place is a random variable that is uniformly distributed between 0 and 2 units.

(a) Is the process stationary?
(b) Is the process deterministic or nondeterministic?
(c) Find the autocorrelation function and spectral density function for the process.

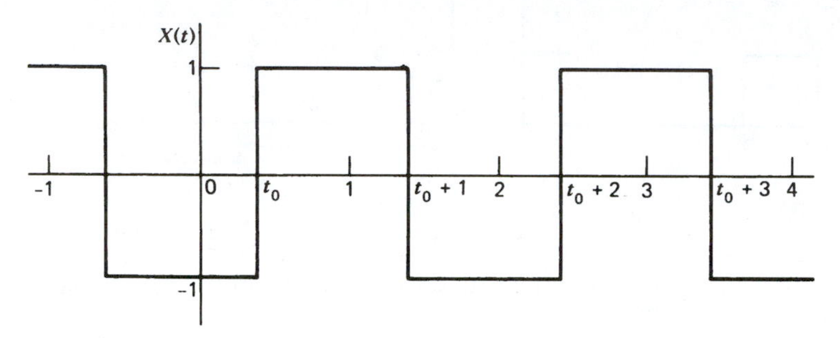

Figure P2.8

2.9 A common autocorrelation function encountered in physical problems is

$$R(\tau) = \sigma^2 e^{-\beta|\tau|} \cos \omega_0 \tau$$

(a) Find the corresponding spectral density function.
(b) $R(\tau)$ will be recognized as a damped cosine function. Sketch both the autocorrelation and spectral density functions for the lightly damped case.

2.10 Show that a Gauss–Markov process described by the autocorrelation function

$$R(\tau) = \sigma^2 e^{-\beta|\tau|}$$

becomes Gaussian white noise if we let $\beta \to \infty$ and $\sigma^2 \to \infty$ in such a way that the area under the autocorrelation-function curve remains constant in the limiting process.

2.11 A stationary random process $X(t)$ has a spectral density function of the form

$$S_X(\omega) = \frac{6\omega^2 + 12}{(\omega^2 + 4)(\omega^2 + 1)}$$

What is the mean-square value of $X(t)$?
(*Hint:* $S_X(\omega)$ may be resolved into a sum of two terms of the form: $[A/(\omega^2 + 4)] + [B/(\omega^2 + 1)]$. Each term may then be integrated using standard integral tables.)

2.12 The stationary process *X*(*t*) has an autocorrelation function of the form

$$R_X(\tau) = \sigma^2 e^{-\beta|\tau|}$$

Another process *Y*(*t*) is related to *X*(*t*) by the deterministic equation

$$Y(t) = aX(t) + b$$

where a and b are known constants.

What is the autocorrelation function for *Y*(*t*)?

2.13 The discrete random walk process is discussed in Section 2.11. Assume each step is of length l and that the steps are independent and equally likely to be positive or negative. Show that the variance of the total distance D traveled in N steps is given by

$$\text{Var}\, D = l^2 N$$

(*Hint:* First write D as the sum $l_1 + l_2 + \dots l_N$ and note that $l_1, l_2, \dots, l_N$ are independent random variables. Then form $E(D)$ and $E(D^2)$ and compute Var D as $E(D^2) - [E(D)]^2$.)

2.14 The Wiener process was discussed in Section 2.11. It is defined to be Gaussian random walk that begins with zero at $t=0$. A more general random walk process can be defined to be similar to the Wiener process except that it starts with a $N(0, \sigma^2)$ random variable where σ is specified.

(a) What is the autocorrelation function for this more general process? Denote the white noise PSD driving the integrator as W.

(b) Write the expression for the mean-square value of the process as a function of W, σ, and the elapsed time from $t=0$.

2.15 Let the process *Z*(*t*) be the product of two independent stationary processes *X*(*t*) and *Y*(*t*). Show that the spectral density function for *Z*(*t*) is given by (in the s domain)

$$S_Z(s) = \frac{1}{2\pi j} \int_{-j\infty}^{j\infty} S_X(w) S_Y(s - w)\, dw$$

[*Hint:* First show that $R_Z(\tau) = R_X(\tau) R_Y(\tau)$.]

2.16 The spectral density function for the stationary process *X*(*t*) is

$$S_X(j\omega) = \frac{1}{(1 + \omega^2)^2}$$

Find the autocorrelation function for *X*(*t*).

2.17 A stationary process *X*(*t*) is Gaussian and has an autocorrelation function of the form

$$R_X(\tau) = 4e^{-|\tau|}$$

Let the random variable X_1 denote $X(t_1)$ and X_2 denote $X(t_1 + 1)$. Write the expression for the joint probability density function $f_{X_1 X_2}(x_1, x_2)$.

2.18 A stationary Gaussian process $X(t)$ has a power spectral density function

$$S_X(j\omega) = \frac{2}{\omega^4 + 1}$$

Find $E(X)$ and $E(X^2)$.

2.19 A typical sample function of a stationary Gauss–Markov process is shown in the sketch. The process has a mean-square value of 9 units, and the random variables X_1 and X_2 indicated on the waveform have a correlation coefficient of 0.5. Write the expression for the autocorrelation function of $X(t)$.

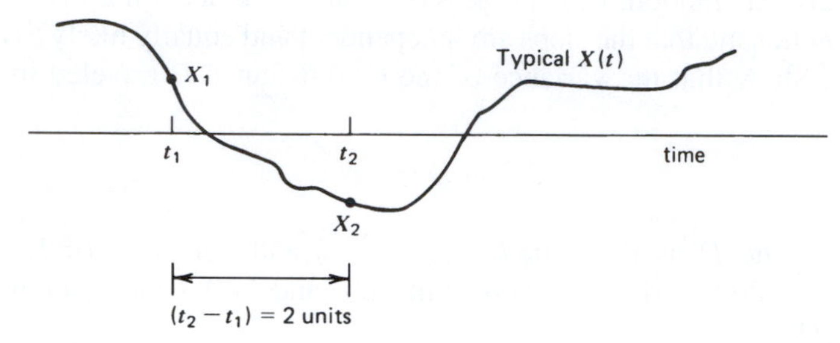

Figure P2.19

2.20 It was stated in Section 2.9 that a first-order Gauss–Markov process has an autocorrelation function of the form

$$R(\tau) = \sigma^2 e^{-\beta|\tau|}$$

It was also stated that a discrete-time version of the process can be generated by the recursive equation

$$X(t_{k+1}) = e^{-\beta\Delta t} X(t_k) + W(t_k)$$

where $W(t_k)$ is a zero-mean white Gaussian sequence that is uncorrelated with $X(t_k)$ and all the preceding X samples. Show that

$$E\left[W^2(t_k)\right] = \sigma^2\left(1 - e^{-2\beta\Delta t}\right)$$

Note that in simulating a first-order Gauss–Markov process, the initial $X(t_0)$ must be a $\mathcal{N}(0, \sigma^2)$ sample in order for the process to be stationary.

2.21 In Example 2.12, it was mentioned that the derivative of a first-order Gauss–Markov process does not exist, and this is certainly true in the continuous-time case. Yet, when we look at adjacent samples of a typical discrete-time first-order process (see Fig. 2.16*a*), it appears that the difference between samples is modest and well-behaved. Furthermore, we know that the $X(t_k)$ samples evolve in accordance with the recursive equation

$$X(t_{k+1}) = e^{-\beta\Delta t} X(t_k) + W(t_k)$$

where $W(t_k)$ is an uncorrelated random sequence with a variance $\sigma^2\left(1 - e^{-2\beta\Delta t}\right)$. (See Problem 2.20.)

Thus, as Δt becomes smaller and smaller, the variance of $W(t_k)$ approaches zero. Thus, it appears that the $X(t_k)$ sequence becomes "smooth" as Δt approaches zero, and one would think intuitively that the derivative (i.e., slope) would exist. This, however, is a mirage, because in forming the slope as the ratio $[X(t_{k+1}) - X(t_k)]/\Delta t$, both numerator and denominator approach zero as $\Delta t \rightarrow 0$.

Show that the denominator in the approximate slope expression approaches zero "faster" than the numerator, with the result that the approximate slope becomes larger as Δt becomes smaller. This confirms that the approximate derivative does not converge as $\Delta t \rightarrow 0$, even in the discrete time case.

2.22 We wish to determine the autocorrelation function a random signal empirically from a single time record. Let us say we have good reason to believe the process is ergodic and at least approximately Gaussian and, furthermore, that the autocorrelation function of the process decays exponentially with a time constant no greater than 10 s. Estimate the record length needed to achieve 5 percent accuracy in the determination of the autocorrelation function. (By 5 percent accuracy, assume we mean that for any τ, the standard deviation of the experimentally determined autocorrelation function will not be more than 5 percent of the maximum value of the true autocorrelation function.)

2.23 Let X_1, X_2, X_3, X_4 be zero-mean Gaussian random variables. Show that

$$E(X_1X_2X_3X_4) = E(X_1X_2)E(X_3X_4) + E(X_1X_3)E(X_2X_4) + E(X_1X_4)E(X_2X_3) \tag{P2.23.1}$$

(*Hint:* The characteristic function was discussed briefly in Section 1.8.)

The multivariate version of the characteristic function is useful here. Let $\psi(\omega_1, \omega_2, \ldots, \omega_n)$ be the multidimensional Fourier transform of $f_{x_1x_2\ldots x_n}(x_1, x_2, \ldots, x_n)$ (but with the signs reversed on $\omega_1, \omega_2, \ldots, \omega_n$). Then it can be readily verified that

$$\begin{aligned}&(-j)^n \frac{\partial^n \psi(\omega_1, \omega_1 \ldots \omega_n)}{\partial\omega_1 \partial\omega_2, \ldots, \partial\omega_n}\Bigg|_{\substack{\omega_1 = 0 \\ \omega_2 = 0 \\ \text{etc.}}} \\ &= \int_{-\infty}^{\infty}\int_{-\infty}^{\infty} \cdots \int_{-\infty}^{\infty} x_1\, x_2, \ldots, \\ &\quad x_n f_{x_1x_2,\ldots,x_n}(x_1, x_2, \ldots, x_n)\, dx_1, dx_2, \ldots, dx_n \\ &= E(X_1, X_2, \ldots, X_n)\end{aligned} \tag{P2.23.2}$$

The characteristic function for a zero-mean, vector Gaussian random variable $\mathbf{X}$ is

$$\psi(\boldsymbol{\omega}) = e^{-\frac{1}{2}\boldsymbol{\omega}^T \mathbf{C}_X \boldsymbol{\omega}} \tag{P2.23.3}$$

where $\mathbf{C}_X$ is the covariance matrix for $\mathbf{X}$. This, along with Eq. (P2.23.2), may now be used to justify the original statement given by Eq. (P2.23.1).

2.24 The accompanying figure shows a means of generating narrowband noise from two independent baseband noise sources. (See Section 2.10.) The bandwidth of the resulting narrowband noise is controlled by the cutoff frequency of the low-pass filters, which are assumed to have identical characteristics. Assume that $F_1(t)$ and $F_2(t)$ are independent white Gaussian noise processes with similar spectral amplitudes. The resulting noise processes after low-pass filtering will then have identical autocorrelation functions that will be denoted $R_X(\tau)$.

(a) Show that the narrowband noise $S(t)$ is a stationary Gaussian random process whose autocorrelation function is

$$R_S(\tau) = R_X(\tau)\cos\omega_c\tau$$

(b) Also show that both the in-phase and quadrature channels are needed to produce stationary narrowband noise. (That is, if either of the $\sin\omega_c t$ or $\cos\omega_c t$ multiplying operations is omitted, the resultant output will not be strictly stationary.)

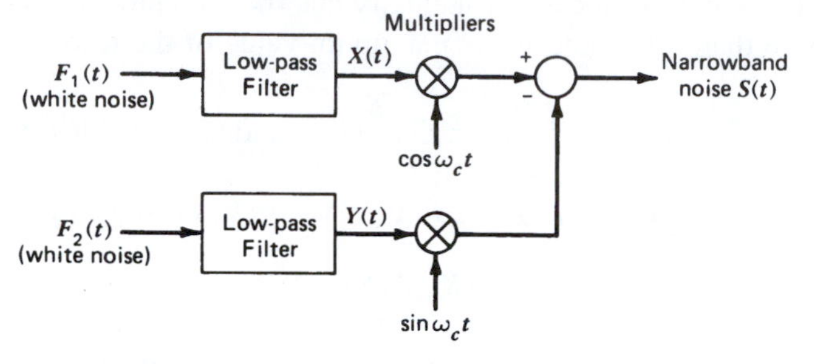

Figure P2.24

2.25 A sequence of discrete samples of a Gauss–Markov process can be generated using the following difference equation (see Section 2.9):

$$X_{k+1} = e^{-\beta\Delta t}X_k + W_k, \quad k = 0, 1, 2, \ldots$$

$W_k =$ white sequence, $\mathcal{N}[0, \sigma^2(1 - e^{-2\beta\Delta t})]$

$\sigma^2 =$ variance of the Markov process

$\beta =$ reciprocal time constant of the process

$\Delta t =$ time interval between samples

If the initial value of the process X_0 is chosen from a population that is $\mathcal{N}(0, \sigma^2)$, then the sequence so generated will be a sample realization of a stationary Gauss–Markov process. Such a sample realization is easily generated with MATLAB's normal random number generator with appropriate scaling of the initial X_0 and the W_k sequence.

(a) Generate 1024 samples of a Gauss–Markov process with $\sigma^2 = 1$, $\beta = 1$, and $\Delta t = .05$ s. As a matter of convenience, let the samples be a 1024-element row vector with a suitable variable name.

(b) Calculate the experimental autocorrelation function for the X_k sequence of part (a). That is, find $V_X(\tau)$ for $\tau = 0, .05, .10, \ldots, 3.0$ (i.e., 60 "lags"). You will find it convenient here to write a general MATLAB program for computing the autocorrelation function for a sequence of length s and for m lags. (This program can then be used in subsequent problems.) Compare your experimentally determined $V_X(\tau)$ with the true autocorrelation function $R_X(\tau)$ by plotting both $V_X(\tau)$ and $R_X(\tau)$ on the same graph. Note that for the relatively short 1024-point time sequence being used here, you should not expect to see a close match between $V_X(\tau)$ and $R_X(\tau)$ (see Example 2.15).

(c) The theory given in Section 2.13 states that the expectation of $V_X(\tau)$ is equal to $R_X(\tau)$ regardless of the length of the sequence. It is also shown that $V_X(\tau)$ converges in the mean for a Gauss–Markov process as T becomes large. One would also expect to see the same sort of convergence when we look at the average of an ensemble of $V_X(\tau)$'s that are generated from "statistically identical," but different, 1024-point sequences. This can be demonstrated (not proved) using a different seed in developing each $V_X(\tau)$. Say we use seeds $1, 2, \ldots, 8$. First plot the $V_X(\tau)$ obtained using seed 1. Next, average the two $V_X(\tau)$'s obtained from seeds 1 and 2, and plot the result. Then average the four $V_X(\tau)$'s for seeds 1, 2, 3, and 4, and plot the result. Finally, average all eight $V_X(\tau)$'s, and plot the result. You should see a general trend toward the true $R_X(\tau)$ as the number of $V_X(\tau)$'s used in the average increases.

2.26 Discrete samples of a Wiener process are easily generated using MATLAB's normal random number generator and implementing the recursion equation:

$$\bar{X}_{k+1} = X_k + W_k, \quad k = 0, 1, 2, \ldots \tag{P2.26}$$

where the subscript k is the time index and the initial condition is set to

$$X_0 = 0$$

Consider a Wiener process where the white noise being integrated has a power spectral density of unity (see Section 2.7), and the sampling interval is 1 s. The increment to be added with each step (i.e., W_k) is a $\mathcal{N}(0, 1)$ random variable, and all the W_k's are independent. [That this will generate samples of a process whose variance is t (in s) is easily verified by working out the variance of X_k for a few steps beginning at $k = 0$.]

(a) Using Eq. (P2.26), generate an ensemble of 50 sample realizations of the Wiener process described above for $k = 0, 1, 2, \ldots, 10$. For convenience, arrange the resulting realizations into a 50×11 matrix, where each row represents a sample realization beginning at $k = 0$.

(b) Plot any 8 sample realizations (i.e., rows) from part (a), and note the obvious nonstationary character of the process.

(c) Form the average squares of the 50 process realizations from part (a), and plot the result vs. time (i.e., k). (The resulting plot should be approximately linear with a slope of unity.)

REFERENCES CITED IN CHAPTER 2

1. E. Wong, *Stochastic Processes in Information and Dynamical Systems*, New York: McGraw-Hill, 1971.
2. A. H. Jazwinski, *Stochastic Processes and Filtering Theory*, New York: Academic Press, 1970.
3. N. Wiener, *Extrapolation, Interpolation, and Smoothing of Stationary Time Series*, Cambridge, MA: MIT Press and New York: Wiley, 1949.
4. S. O. Rice, "Mathematical Analysis of Noise," *Bell System Tech. J.*, 23, 282–332 (1944); 24, 46–256 (1945).
5. W. B. Davenport, Jr.and W. L. Root, *An Introduction to the Theory of Random Signals and Noise*, New York: McGraw-Hill, 1958.
6. R. C. Dixon, *Spread Spectrum Systems*, New York: Wiley, 1976.
7. R. P. Denaro, "Navstar: The All-Purpose Satellite," *IEEE Spectrum*, 18(5), 35–40 (May 1981).
8. B. W. Parkinson and S. W. Gilbert, "NAVSTAR: Global Positioning System—Ten Years Later," *Proc. IEEE, 71*: 10, 1177–1186 (Oct. 1983).
9. R. E. Ziemer and R. L. Peterson, *Digital Communications and Spread Spectrum Systems*, New York: Macmillan, 1985.
10. J. W. Betz, "Binary offset Carrier Modulation for Ratio Navigation," = NAVIGATION, Jour. of the Inst. of Navigation, v. 48, No. 4, Winter 2001–2002, pp. 227–246.
11. J. H. Laning, Jr.and R. H. Battin, *Random Processes in Automatic Control*, New York: McGraw-Hill, 1956.
12. C. E. Shannon, "The Mathematical Theory of Communication," *Bell System Tech. J.* (July and Oct. 1948). (Later reprinted in book form by the University of Illinois Press, 1949.)
13. C. E. Shannon, "Communication in the Presence of Noise," *Proc. Inst. Radio Engr.*, 37: 1, 10–21 (Jan. 1949).
14. H. Nyquist, "Certain Topics in Telegraph Transmission Theory," *Trans. Am. Inst. Elect. Engr.*, 47, 617–644 (April 1928).
15. H. S. Black, *Modulation Theory*, New York: Van Nostrand Co., 1950.
16. S. Goldman, *Information Theory*, Englewood Cliffs, NJ: Prentice-Hall, 1953
17. K. S. Shanmugam, *Digital and Analog Communication Systems*, New York: Wiley, 1979.

Additional References on Probability and Random Signals

18. John A. Gubner, *Probability and Random Processes for Electrical and Computer Engineers*, Combidge University Press, 2006.
19. Scott Miller, Donal Childers, *Probability and Random Processes: With Applications to Signal Processing and Communications*, 2nd Edition, Academic Press, 2004.
20. Geoffrey R. Grimmet, David R. Stirzaker, *Probability and Random Processes*, 3rd Edition, Oxford University Press, 2001.
21. Steven Kay, *Intuitive Probability and Random Processes Using MATLAB*, Springer, 2005.

3

Linear Systems Response, State-Space Modeling, and Monte Carlo Simulation

The central problem of linear systems analysis is: Given the input, what is the output? In the deterministic case, we usually seek an explicit expression for the response or output. In the random-input problem no such explicit expression is possible, except for the special case where the input is a so-called deterministic random process (and not always in this case). Usually, in random-input problems, we must settle for a considerably less complete description of the output than we get for corresponding deterministic problems. In the case of random processes the most convenient descriptors to work with are autocorrelation function, power spectral density function, and mean-square value. We will now examine the input–output relationships of linear systems in these terms.

3.1 INTRODUCTION: THE ANALYSIS PROBLEM

In any system satisfying a set of linear differential equations, the solution may be written as a superposition of an initial-condition part and another part due to the driving or forcing functions. Both the initial conditions and forcing functions may be random; and, if so, the resultant response is a random process. We direct our attention here to such situations, and it will be assumed that the reader has at least an elementary acquaintance with deterministic methods of linear system analysis (1, 2).

With reference to Fig. 3.1, the analysis problem may be simply stated: Given the initial conditions and the input and the system's dynamical characteristics [i.e., $G(s)$ in Fig. 3.1], what is the output? Of course, in the stochastic problem, the input and output will have to be described in probabilistic terms.

We need to digress here for a moment and discuss a notational matter. In Chapters 1 and 2 we were careful to use uppercase symbols to denote random variables and lowercase symbols for the corresponding arguments of their probability density functions. This is the custom in most current books on probability. There is, however, a long tradition in engineering books on automatic control and

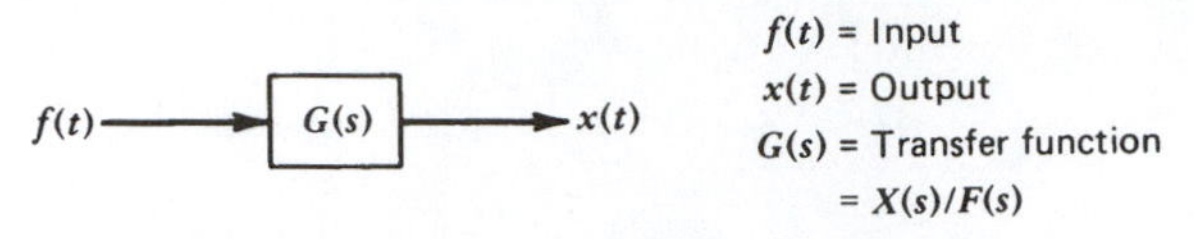

Figure 3.1 Block diagram for elementary analysis problem.

linear systems analysis of using lowercase for time functions and uppercase for the corresponding Laplace or Fourier transforms. Hence, we are confronted with notational conflict. We will resolve this in favor of the traditional linear analysis notation, and from this point on we will use lowercase symbols for time signals—either random or deterministic—and uppercase for their transforms. This seems to be the lesser of the two evils. The reader will simply have to interpret the meaning of symbols such as $x(t)$, $f(t)$, and the like, within the context of the subject matter under discussion. This usually presents no problem. For example, with reference to Fig. 3.1, $g(t)$ would mean inverse transform of $G(s)$, and it clearly is a deterministic time function. On the other hand, the input and output, $f(t)$ and $x(t)$, will usually be random processes in the subsequent material.

Generally, analysis problems can be divided into two categories:

1. Stationary (steady-state) analysis. Here the input is assumed to be time stationary, and the system is assumed to have fixed parameters with a stable transfer function. This leads to a stationary output, provided the input has been present for a long period of time relative to the system time constants.
2. Nonstationary (transient) analysis. Here we usually consider the driving function as being applied at $t = 0$, and the system may be initially at rest or have nontrivial initial conditions. The response in this case is usually nonstationary. We note that analysis of unstable systems falls into this category, because no steady-state (stationary) condition will exist.

The similarity between these two categories and the corresponding ones in deterministic analysis should be apparent. Just as in circuit analysis, we would expect the transient solution to lead to the steady-state response as $t \to \infty$. However, if we are only interested in the stationary result, this is getting at the solution the "hard way." Much simpler methods are available for the stationary solution, and these will now be considered.

3.2 STATIONARY (STEADY-STATE) ANALYSIS

We assume in Fig. 3.1 that $G(s)$ represents a stable, fixed-parameter system and that the input is covariance (wide-sense) stationary with a known power spectral density function (PSD). In deterministic analysis, we know that if the input is Fourier transformable, the input spectrum is simply modified by $G(j\omega)$ in going through the filter. In the random process case, one interpretation of the spectral function is that it is proportional to the magnitude of the *square* of the Fourier transform. Thus, the equation relating the input and output spectral functions is

$$S_x(s) = G(s)G(-s)S_f(s) \tag{3.2.1}$$

Note that Eq. (3.2.1) is written in the s domain where the imaginary axis has the meaning of real angular frequency ω. If you prefer to write Eq. (3.2.1) in terms of ω, just replace s with $j\omega$. Equation (3.2.1) then becomes

$$\begin{aligned} S_x(j\omega) &= G(j\omega)G(-j\omega)S_f(j\omega) \\ &= |G(j\omega)|^2 S_f(j\omega) \end{aligned} \tag{3.2.2}$$

Because of the special properties of spectral functions, both sides of Eq. (3.2.2) work out to be real functions of ω. Also note that the autocorrelation function of the output can be obtained as the inverse Fourier transform of $S_x(j\omega)$. Two examples will now illustrate the use of Eq. (3.2.1).

EXAMPLE 3.1

Consider a first-order low-pass filter with unity white noise as the input. With reference to Fig. 3.1, then

$$\begin{aligned} S_f(s) &= 1 \\ G(s) &= \frac{1}{1+Ts} \end{aligned}$$

where T is the time constant of the filter. The output spectral function is then

$$\begin{aligned} S_x(s) &= \frac{1}{1+Ts} \cdot \frac{1}{1+T(-s)} \cdot 1 \\ &= \frac{(1/T)^2}{-s^2 + (1/T)^2} \end{aligned}$$

Or, in terms of real frequency ω,

$$S_x(j\omega) = \frac{(1/T)^2}{\omega^2 + (1/T)^2}$$

This is sketched as a function of ω in Fig. 3.2. As would be expected, most of the spectral content is concentrated at low frequencies and then it gradually diminishes as $\omega \to \infty$.

It is also of interest to compute the mean-square value of the output. It is given by Eq. (2.7.17).

$$E(x^2) = \frac{1}{2\pi j} \int_{-j\infty}^{j\infty} \frac{1}{1+Ts} \cdot \frac{1}{1+T(-s)} ds \tag{3.2.3}$$

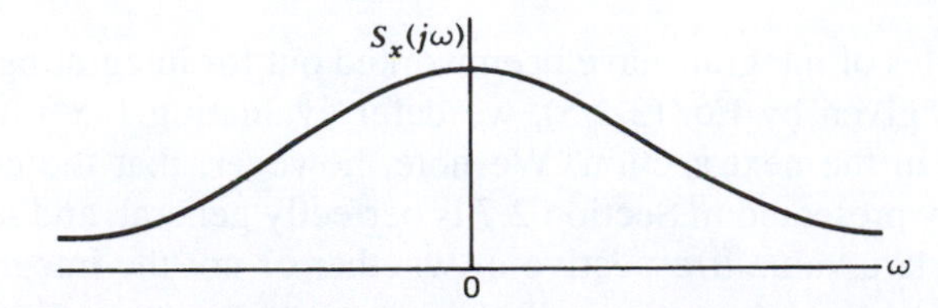

Figure 3.2 Spectral function for low-pass filter output with white noise input.

The integral of Eq. (3.2.3) is easily evaluated in this case by substituting $j\omega$ for s and using a standard table of integrals. This leads to

$$E(x^2) = \frac{1}{2T}$$

The "standard" table-of-integrals approach is of limited value, though, as will be seen in the next example.

EXAMPLE 3.2

Consider the input process to have an exponential autocorrelation function and the filter to be the same as in the previous example. Then

$$R_f(\tau) = \sigma^2 e^{-\beta|\tau|}$$

$$G(s) = \frac{1}{1 + Ts}$$

First, we transform R_f to obtain the input spectral function.

$$\mathfrak{F}\left[R_f(\tau)\right] = \frac{2\sigma^2\beta}{-s^2 + \beta^2}$$

The output spectral function is then

$$S_x(s) = \frac{2\sigma^2\beta}{-s^2 + \beta^2} \cdot \frac{(1/T)}{[s + (1/T)]} \cdot \frac{(1/T)}{[-s + (1/T)]} \tag{3.2.4}$$

Now, if we wish to find $E(x^2)$ in this case, it will involve integrating a function that is fourth-order in the denominator, and most tables of integrals will be of no help. We note, though, that the input spectral function can be factored and the terms of Eq. (3.2.4) can be rearranged as follows:

$$S_x(s) = \left[\frac{\sqrt{2\sigma^2\beta}}{(s + \beta)} \cdot \frac{(1/T)}{[s + (1/T)]}\right]\left[\frac{\sqrt{2\sigma^2\beta}}{(-s + \beta)} \cdot \frac{(1/T)}{[-s + (1/T)]}\right] \tag{3.2.5}$$

The first term has all its poles and zeros in the left half-plane, and the second term has mirror-image poles and zeros in the right half-plane. This regrouping of terms is known as *spectral factorization* and can always be done if the spectral function is rational in form (i.e., if it can be written as a ratio of polynomials in even powers of s).

Since special tables of integrals have been worked out for integrating complex functions of the type given by Eq. (3.2.5), we defer evaluating $E(x^2)$ until these have been presented in the next section. We note, however, that the concept of power spectral density presented in Section 2.7 is perfectly general, and its integral represents a mean-square value irrespective of whether or not the integral can be evaluated in closed form.

3.3 INTEGRAL TABLES FOR COMPUTING MEAN-SQUARE VALUE

In linear analysis problems, the spectral function can often be written as a ratio of polynomials in s^2. If this is the case, spectral factorization can be used to write the function in the form

$$S_x(s) = \frac{c(s)}{d(s)} \cdot \frac{c(-s)}{d(-s)} \tag{3.3.1}$$

where $c(s)/d(s)$ has all its poles and zeros in the left half-plane and $c(-s)/d(-s)$ has mirror-image poles and zeros in the right half-plane. No roots of $d(s)$ are permitted on the imaginary axis. The mean-square value of x can now be written as

$$E(x^2) = \frac{1}{2\pi j} \int_{-j\infty}^{j\infty} \frac{c(s)c(-s)}{d(s)d(-s)} ds \tag{3.3.2}$$

R. S. Phillips (3) was the first to prepare a table of integrals for definite integrals of the type given by Eq. (3.3.2). His table has since been repeated in many texts with a variety of minor modifications (4,5,6). An abbreviated table in terms of the complex s domain follows. An example will now illustrate the use of Table 3.1.

Table 3.1 Table of Integrals

$$I_n = \frac{1}{2\pi j} \int_{-j\infty}^{j\infty} \frac{c(s)c(-s)}{d(s)d(-s)} ds \tag{3.3.3}$$

$$c(s) = c_{n-1}s^{n-1} + c_{n-2}s^{n-2} + \cdots + c_0$$

$$d(s) = d_n s^n + d_{n-1}s^{n-1} + \cdots + d_0$$

$$I_1 = \frac{c_0^2}{2d_0 d_1}$$

$$I_2 = \frac{c_1^2 d_0 + c_0^2 d_2}{2d_0 d_1 d_2}$$

$$I_3 = \frac{c_2^2 d_0 d_1 + (c_1^2 - 2c_0 c_2) d_0 d_3 + c_0^2 d_2 d_3}{2d_0 d_3 (d_1 d_2 - d_0 d_3)}$$

$$I_4 = \frac{c_3^2(-d_0^2 d_3 + d_0 d_1 d_2) + (c_2^2 - 2c_1 c_3) d_0 d_1 d_4 + (c_1^2 - 2c_0 c_2) d_0 d_3 d_4 + c_0^2(-d_1 d_4^2 + d_2 d_3 d_4)}{2d_0 d_4 (-d_0 d_3^2 - d_1^2 d_4 + d_1 d_2 d_3)}$$

EXAMPLE 3.3

The solution in Example 3.2 was brought to the point where the spectral function had been written in the form

$$S_x(s) = \left[\frac{\sqrt{2\sigma^2\beta}\cdot 1/T}{(s+\beta)(s+1/T)}\right]\left[\frac{\sqrt{2\sigma^2\beta}\cdot 1/T}{(-s+\beta)(-s+1/T)}\right] \tag{3.3.4}$$

Clearly, S_x has been factored properly with its poles separated into left and right half-plane parts. The mean-square value of x is given by

$$E(x^2) = \frac{1}{2\pi j}\int_{-j\infty}^{j\infty} S_x(s)ds \tag{3.3.5}$$

Comparing the form of $S_x(s)$ in Eq. (3.3.4) with the standard form given in Eq. (3.3.3), we see that

$$\begin{aligned} c(s) &= \frac{\sqrt{2\sigma^2\beta}}{T} \\ d(s) &= s^2 + (\beta + 1/T)s + \beta/T \end{aligned}$$

Thus, we can use the I_2 integral of Table 3.1. The coefficients for this case are

$$\begin{aligned} c_1 &= 0 & d_2 &= 1 \\ c_0 &= \frac{\sqrt{2\sigma^2\beta}}{T} & d_1 &= (\beta + 1/T) \\ & & d_0 &= \beta/T \end{aligned}$$

and $E(x^2)$ is then

$$E(x^2) = \frac{c_0^2}{2d_0d_1} = \frac{2\sigma^2\beta/T^2}{2(\beta/T)(\beta+1/T)} = \frac{\sigma^2}{1+\beta T}$$

■

3.4 PURE WHITE NOISE AND BANDLIMITED SYSTEMS

We are now in a position to demonstrate the validity of using the pure white noise model in certain problems, even though white noise has infinite variance. This will be done by posing two hypothetical mean-square analysis problems:

1. Consider a simple first-order low-pass filter with bandlimited white noise as the input. Specifically, with reference to Fig. 3.1, let

$$S_f(j\omega) = \begin{cases} A, & |\omega| \le \omega_c \\ 0, & |\omega| > \omega_c \end{cases} \tag{3.4.1}$$

$$G(s) = \frac{1}{1+Ts} \tag{3.4.2}$$

2. Consider the same low-pass filter as in problem 1, but with pure white noise as the input:

$$S_f(j\omega) = A, \quad \text{for all } \omega \tag{3.4.3}$$

$$G(s) = \frac{1}{1 + Ts} \tag{3.4.4}$$

Certainly, problem 1 is physically plausible because bandlimited white noise has finite variance. Conversely, problem 2 is not because the input has infinite variance. The preceding theory enables us to evaluate the mean-square value of the output for both problems. As a matter of convenience, we do this in the real frequency domain rather than the complex s domain.

Problem 1:

$$S_x(j\omega) = \begin{cases} \dfrac{1}{1 + (T\omega)^2}, & |\omega| \le \omega_c \\ 0, & |\omega| > \omega_c \end{cases} \tag{3.4.5}$$

$$E(x^2) = \frac{1}{2\pi} \int_{-\omega_c}^{\omega_c} \frac{A}{1 + (T\omega)^2} d\omega = \frac{A}{\pi T} \tan^{-1}(\omega_c T) \tag{3.4.6}$$

Problem 2:

$$S_x(j\omega) = \frac{A}{1 + (T\omega)^2}, \quad \text{for all } \omega \tag{3.4.7}$$

$$\begin{aligned} E(x^2) &= \frac{1}{2\pi} \int_{-\infty}^{\infty} \frac{A}{1 + (T\omega)^2} d\omega \\ &= \frac{A}{\pi T} \tan^{-1}(\infty) = \frac{A}{2T} \end{aligned} \tag{3.4.8}$$

Now, we see by comparing the results given by Eqs. (3.4.6) and (3.4.8) that the difference is just that between $\tan^{-1}(\omega_c T)$ and $\tan^{-1}(\infty)$. The bandwidth of the input is ω_c and the filter bandwidth is $1/T$. Thus, if their ratio is large, $\tan^{-1}(\omega_c T) \approx \tan^{-1}(\infty)$. For a ratio of 100:1, the error is less than 1 percent. Thus, if the input spectrum is flat considerably out beyond the point where the system response is decreasing at 20 db/decade (or faster), there is relatively little error introduced by assuming that the input is flat out to infinity. The resulting simplification in the analysis is significant.

3.5 NOISE EQUIVALENT BANDWIDTH

In filter theory, it is sometimes convenient to think of an idealized filter whose frequency response is unity over a prescribed bandwidth B (in hertz) and zero

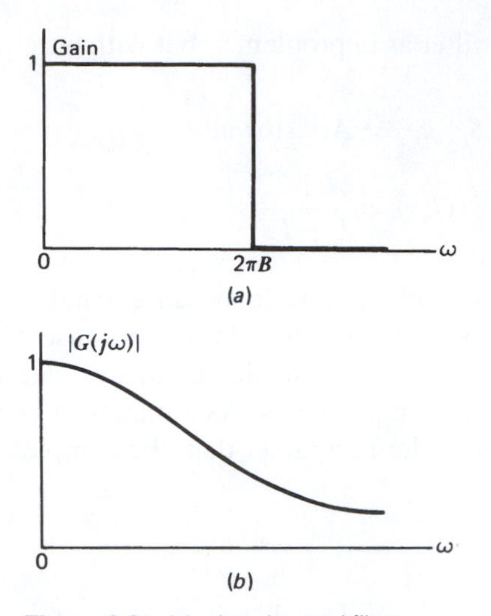

Figure 3.3 Ideal and actual filter responses. (*a*) Ideal. (*b*) Actual.

outside this band. This response is depicted in Fig. 3.3*a*. If this ideal filter is driven by white noise with amplitude A, its mean-square response is

$$E(x^2) \quad \text{(ideal)} = \frac{1}{2\pi}\int_{-2\pi B}^{2\pi B} A d\omega = 2AB \tag{3.5.1}$$

Next, consider an actual filter $G(s)$ whose gain has been normalized to yield a peak response of unity. An example is shown in Fig. 3.3*b*. The mean-square response of the actual filter to white noise of amplitude A is given by

$$E(x^2) \quad \text{(actual)} = \frac{1}{2\pi j}\int_{-j\infty}^{j\infty} AG(s)G(-s)ds \tag{3.5.2}$$

Now, if we wish to find the idealized filter that will yield this same response, we simply equate $E(x^2)$ (ideal) and $E(x^2)$ (actual) and solve for the bandwidth that gives equality. The resultant bandwidth B is known as the *noise equivalent bandwidth.* It may, of course, be written explicitly as

$$B \quad \text{(in hertz)} = \frac{1}{2}\left[\frac{1}{2\pi j}\int_{-j\infty}^{j\infty} G(s)G(-s)ds\right] \tag{3.5.3}$$

EXAMPLE 3.4

Suppose we wish to find the noise equivalent bandwidth of the second-order low-pass filter

$$G(s) = \frac{1}{(1+Ts)^2}$$

Since the peak response of $G(s)$ occurs at zero frequency and is unity, the gain scale factor is set properly. We must next evaluate the integral in brackets in Eq. (3.5.3). Clearly, $G(s)$ is second-order in the denominator, and therefore we use I_2 of the integral tables given in Section 3.3. The coefficients in this case are

$$\begin{aligned} c_1 &= 0 \quad & d_2 &= T^2 \\ c_0 &= 1 \quad & d_1 &= 2T \\ & & d_0 &= 1 \end{aligned}$$

and thus I_2 is

$$I_2 = \frac{c_0^2}{2d_0 d_1} = \frac{1}{4T}$$

The filter's noise equivalent noise bandwidth is then

$$B = \frac{1}{8T} \text{ Hz}$$

This says, in effect, that an idealized filter with a bandwidth of $1/8T$ Hz would pass the same amount of noise as the actual second-order filter.

■

3.6 SHAPING FILTER

With reference to Fig. 3.4, we have seen that the output spectral function can be written as

$$S_x(s) = 1 \cdot G(s)G(-s) \tag{3.6.1}$$

If $G(s)$ is minimum phase and rational in form,* Eq. (3.6.1) immediately provides a factored form for $S_x(s)$ with poles and zeros automatically separated into left and right half-plane parts.

Clearly, we can reverse the analysis problem and pose the question: What minimum-phase transfer function will shape unity white noise into a given spectral function $S_x(s)$? The answer should be apparent. If we can use spectral factorization on $S_x(s)$, the part with poles and zeros in the left half-plane provides the appropriate shaping filter. This is a useful concept, both as a mathematical artifice and also as a physical means of obtaining a noise source with desired spectral characteristics from a wideband source.

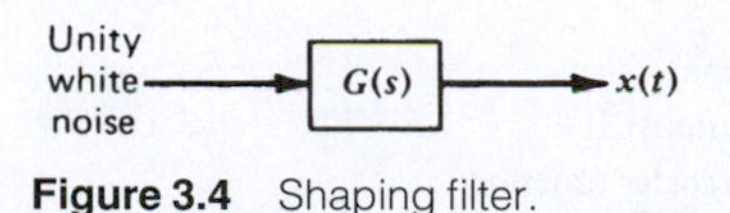

Figure 3.4 Shaping filter.

*This condition requires $G(s)$ to have a finite number of poles and zeros, all of which must be in the left half-plane.

EXAMPLE 3.5

Suppose we wish to find the shaping filter that will shape unity white noise into noise with a spectral function

$$S_x(j\omega) = \frac{16}{\omega^4 + 64} \tag{3.6.2}$$

First, we write S_x in the s domain as

$$S_x(s) = \frac{16}{s^4 + 64} \tag{3.6.3}$$

Next, we find the poles of S_x. (There are no zeros.)

$$\text{Poles} = -2 \pm j2,\ 2 \pm j2$$

Finally, we group together left and right half-plane parts. $S_x(s)$ can then be written as

$$S_x(s) = \frac{4}{s^2 + 4s + 8} \cdot \frac{4}{s^2 - 4s + 8} \tag{3.6.4}$$

The desired shaping filter is then

$$G(s) = \frac{4}{s^2 + 4s + 8} \tag{3.6.5}$$

■

3.7 NONSTATIONARY (TRANSIENT) ANALYSIS

Thus far we have only considered the stationary (steady-state) analysis problem. We will now look at the nonstationary (transient) problem.

The block diagram of Fig. 3.1 is repeated as Fig. 3.5 with the addition of a switch in the input. Imagine the system to be initially at rest, and then close the switch at $t = 0$. A transient response takes place in the stochastic problem just as in the corresponding deterministic problem. If the input $f(t)$ is a nondeterministic random process, we would expect the response also to be nondeterministic, and its autocorrelation function may be computed in terms of the input autocorrelation function. This is done as follows.

The system response can be written as a convolution integral

$$x(t) = \int_0^t g(u) f(t-u)\, du \tag{3.7.1}$$

where $g(u)$ is the inverse Laplace transform of $G(s)$ and is usually referred to as the system weighting function or impulsive response. To find the autocorrelation

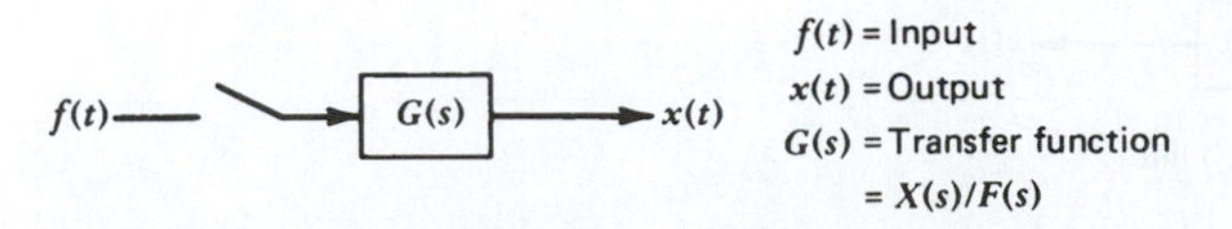

Figure 3.5 Block diagram for nonstationary analysis problem.

function, we simply evaluate $E[x(t_1)x(t_2)]$.

$$
\begin{aligned}
R_x(t_1, t_2) &= E[x(t_1)x(t_2)] \\
&= E\left[\int_0^{t_1} g(u)f(t_1 - u)\,du \int_0^{t_2} g(v)f(t_2 - v)\,dv\right] \\
&= \int_0^{t_2}\int_0^{t_1} g(u)g(v)E[f(t_1 - u)f(t_2 - v)]\,du\,dv
\end{aligned}
\tag{3.7.2}
$$

Now, if $f(t)$ is stationary, Eq. (3.7.2) can be written as

$$
R_x(t_1, t_2) = \int_0^{t_2}\int_0^{t_1} g(u)g(v)R_f(u - v + t_2 - t_1)\,du\,dv \tag{3.7.3}
$$

and we now have an expression for the output autocorrelation function in terms of the input autocorrelation function and system weighting function.

Equation (3.7.3) is difficult to evaluate except for relatively simple systems. Thus, we are often willing to settle for less information about the response and just compute its mean-square value. This is done by letting $t_2 = t_1 = t$ in Eq. (3.7.3) with the result

$$
E[x^2(t)] = \int_0^t\int_0^t g(u)g(v)R_f(u - v)\,du\,dv \tag{3.7.4}
$$

Three examples will now illustrate the use of Eqs. (3.7.3) and (3.7.4).

EXAMPLE 3.6

Let $G(s)$ be a first-order low-pass filter, and let $f(t)$ be white noise with amplitude A. Then

$$
\begin{aligned}
G(s) &= \frac{1}{1 + Ts} \\
S_f(\omega) &= A
\end{aligned}
$$

Taking inverse transforms gives

$$
\begin{aligned}
g(u) &= \frac{1}{T}e^{-u/T} \\
R_f(\tau) &= A\delta(\tau)
\end{aligned}
$$

Next, substituting in Eq. (3.7.4) yields

$$
\begin{aligned}
E[x^2(t)] &= \int_0^t\int_0^t \frac{A}{T^2}e^{-u/T}e^{-v/T}\delta(u - v)\,du\,dv \\
&= \frac{A}{T^2}\int_0^t e^{-2v/T}\,dv \\
&= \frac{A}{2T}\left[1 - e^{-2t/T}\right]
\end{aligned}
\tag{3.7.5}
$$

Note that as $t \to \infty$, the mean-square value approaches $A/2T$, which is the same result obtained in Section 3.2 using spectral analysis methods.

■

EXAMPLE 3.7

Let $G(s)$ be an integrator with zero initial conditions, and let $f(t)$ be a Gauss–Markov process with variance σ^2 and time constant $1/\beta$. We desire the mean-square value of the output x. The transfer function and input autocorrelation function are

$$G(s) = \frac{1}{s} \quad \text{or} \quad g(u) = 1$$

and

$$R_f(\tau) = \sigma^2 e^{-\beta|\tau|}$$

Next, we use Eq. (3.7.4) to obtain $E[x^2(t)]$.

$$E\left[x^2(t)\right] = \int_0^t \int_0^t 1 \cdot 1 \cdot \sigma^2 e^{-\beta|u-v|} \, du \, dv \tag{3.7.6}$$

Some care is required in evaluating Eq. (3.7.6) because one functional expression for $e^{-\beta|u-v|}$ applies for $u > v$, and a different one applies for $u < v$. This is shown in Fig. 3.6. Recognizing that the region of integration must be split into two parts, we have

$$E\left[x^2(t)\right] = \int_0^t \int_0^v \sigma^2 e^{-\beta(u-v)} \, du \, dv + \int_0^t \int_v^t \sigma^2 e^{-\beta(u-v)} \, du \, dv \tag{3.7.7}$$

Since there is symmetry in the two integrals of Eq. (3.7.7), we can simply evaluate the first one and multiply by 2. The mean-square value of x is then

$$\begin{aligned} E[x^2(t)] &= 2 \int_0^t \int_0^v \sigma^2 e^{-\beta(u-v)} \, du \, dv = 2 \int_0^t \sigma^2 e^{-\beta v} \int_0^v e^{\beta u} \, du \, dv \\ &= \frac{2\sigma^2}{\beta} \int_0^t e^{-\beta v}\left(e^{\beta v} - 1\right) dv \\ &= \frac{2\sigma^2}{\beta^2}\left[\beta t - \left(1 - e^{-\beta t}\right)\right] \end{aligned} \tag{3.7.8}$$

Note that $E[x^2(t)]$ increases without bound as $t \to \infty$. This might be expected because an integrator is an unstable system.

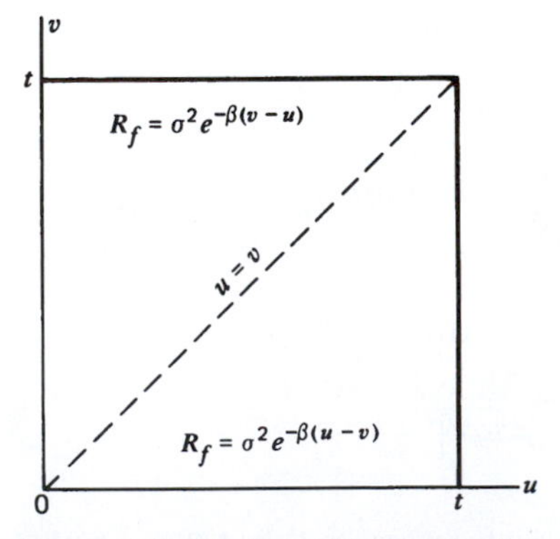

Figure 3.6 Regions of integration for Example 3.7.

■

EXAMPLE 3.8

As our final example, we find the autocorrelation function of the output of a simple integrator driven by unity-amplitude Gaussian white noise. The transfer function and input autocorrelation function are

$$G(s) = \frac{1}{s} \quad \text{or} \quad g(u) = 1$$

$$R_f(\tau) = \delta(\tau)$$

We obtain $R_x(t_1, t_2)$ from Eq. (3.7.3):

$$R_x(t_1, t_2) = \int_0^{t_2} \int_0^{t_1} 1 \cdot 1 \cdot \delta(u - v + t_2 - t_1)\, du\, dv \tag{3.7.9}$$

The region of integration for this double integral is shown in Fig. 3.7 for $t_2 > t_1$. The argument of the Dirac delta function in Eq. (3.7.9) is zero along the dashed line in the figure. It can be seen that it is convenient to integrate first with respect to v if $t_2 > t_1$ as shown in the figure. Considering this case first (i.e., $t_2 > t_1$), we have

$$R_x(t_1, t_2) = \int_0^{t_1} \int_0^{t_2} \delta(u - v + t_2 - t_1)\, dv\, du = \int_0^{t_1} 1 \cdot du = t_1$$

Similarly, when $t_2 < t_1$,

$$R_x(t_1, t_2) = t_2$$

The final result is then

$$R_x(t_1, t_2) = \begin{cases} t_1, & t_2 \geq t_1 \\ t_2, & t_2 < t_1 \end{cases} \tag{3.7.10}$$

Note that this is the same result obtained for the Wiener process in Chapter 2.

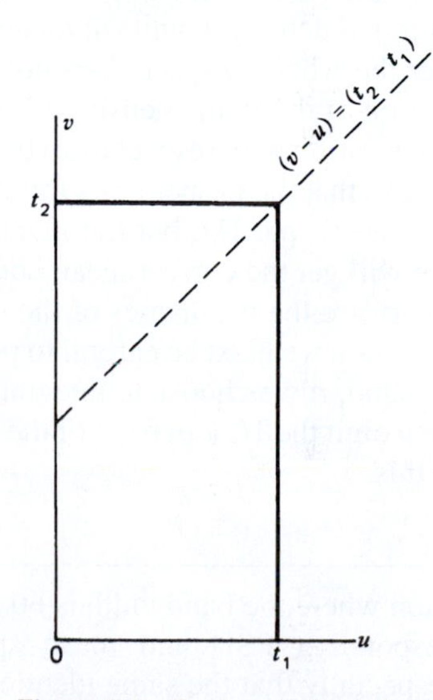

Figure 3.7 Region of integration for Example 3.8.

In concluding this section, we might comment that if the transient response includes both forced and initial-condition components, the total response is just the superposition of the two. The mean-square value must be evaluated with care, though, because the total mean-square value is the sum of the two only when the crosscorrelation is zero. If the crosscorrelation between the two responses is not zero, it must be properly accounted for in computing the mean-square value.

3.8 NOTE ON UNITS AND UNITY WHITE NOISE

The Pesky 2π Problem

We have been a bit cavalier about units in our examples thus far, but this is a good place to pause and consider the matter of units more carefully. First of all, in the term *power spectral density* (PSD) we will always interpret "power" to be mean-square-value in squared units of the random variable under consideration. Also, in our analysis thus far we have dealt with frequency in terms of ω rather than f, so there has been a tacit assumption that the units of PSD in the denominator are radians per second (rad/s). This is where the "units" problem begins.

Say, for example, the random variable x being considered is distance in meters. Our formula for computing mean-square-value is given by Eq. (2.7.16), which is repeated here for convenience:

$$E(x^2) = \frac{1}{2\pi} \int_{-\infty}^{\infty} S_x(j\omega) d\omega \tag{2.7.16}$$

The $S_x(j\omega)$ was rigorously derived as the Fourier transform of the autocorrelation function, and we are tempted to refer to $S_x(j\omega)$ as density in units of meters^2 per rad/s. But yet the direct integral of $S_x(j\omega)$ over the whole ω space does not yield total power. This is in conflict with the basic notion of the term "density"! It is not until we modify the summation of $S_x(j\omega)$ (i.e., integration with respect to ω) by a factor of $1/2\pi$ that we get the true total power. This says that the proper units for $S_x(j\omega)$ must be meters^2 per (2π rad)/s, or equivalently meters^2 per Hz, but not meters^2/(rad/s). Then, when we sum with respect to Hz, we will get the correct mean-square-value. Now, all this being said, we can actually perform the mechanics of the summation with respect to ω if we so choose; but, if we do so, we must be careful to put the $1/2\pi$ factor in front of the integral. On the other hand, if we choose to integrate the *same* $S_x(j\omega)$ as before but with respect to f, we then omit the $1/2\pi$ in front of the integral. A simple numerical example will illustrate this.

EXAMPLE 3.9

Consider a bandlimited white noise situation where the bandwidth is 60 Hz and the mean-square-value is $36\,\text{m}^2$. The corresponding PSD and mean-square-value calculations are shown in Fig. 3.8. Note especially that the same identical value of the PSD magnitude is used in both the (*a*) and (*b*) parts of the figure.

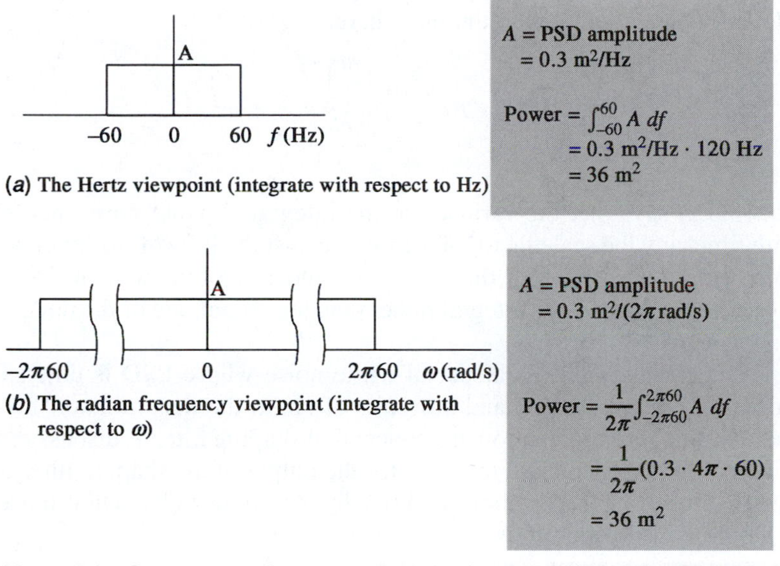

Figure 3.8 Computation of mean-square-value from two different viewpoints.

White Noise And Unity White Noise

Pure white noise is a conceptual idea. It has a spectral density that is flat out to infinity and its power is infinite (whatever that means). Yet the PSD of white noise is finite and has meaning, so we need to consider its units. For example, let y be a velocity random variable that is assumed to be white. From the discussion in the previous paragraphs, the units would then be $(\text{m/s})^2/\text{Hz}$. Then the integral with respect to Hz over any finite bandwidth would give the power in the specified bandwidth. Let A be the amplitude of $S_y(j\omega)$. Then the corresponding autocorrelation function would be

$$R_y(\tau) = A\delta(\tau) \tag{3.8.1}$$

where

$$\delta(\tau) = \text{Dirac delta function (dimensionless)}$$

Now continue this scenario. Let us say we integrate the velocity y to get position. We can now use the methods of Sec. 3.7 to compute the variance of the output of the integrator that we will call x. If the integrator has zero initial conditions:

$$\begin{aligned} E(x^2) &= \int_0^t \int_0^t 1 \cdot 1 \cdot A \cdot \delta(u - v)\,du\,dv \\ &= \int_0^t A\,dv \\ &= At \end{aligned} \tag{3.8.2}$$

We can pause to check on the units here:

$$At \overset{units}{\Rightarrow} \frac{\left(\frac{m}{s}\right)^2}{\frac{cycles}{s}} \cdot s = m^2$$

Eq. (3.8.2) says that the variance of the integrated white noise increases linearly with time, and the scale factor of the ramp is just the PSD of the input white noise in units $(m/s)^2/Hz$. Note that this gives us a more rigorous way of describing white noise, i.e., in terms of its integral rather than the magnitude of the noise itself, which does not exist.

Unity white noise is a special white noise whose PSD is unity. It is dimensionless in the numerator and has units of Hz in the denominator. It is especially useful as the forcing function in a conceptual shaping filter as discuss in Section 3.6. When used in this context, the units for the output of the shaping filter are provided by the filter transfer function, and not by the input unity white noise. A simple example will illustrate this.

EXAMPLE 3.10

Consider the shaping filter shown in Fig. 3.9.

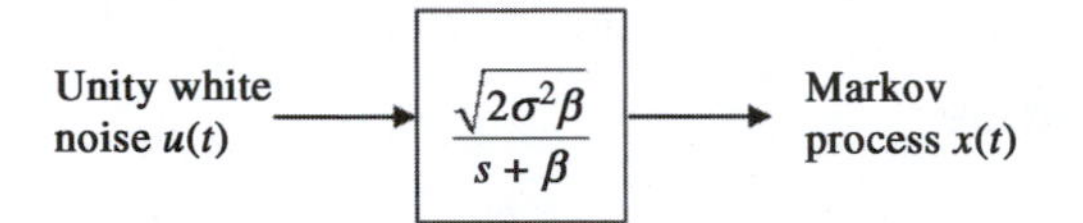

Figure 3.9 Shaping filter for a Markov process.

We wish to shape $u(t)$ into a Markov process whose variance is to be σ^2 in m^2, and β is in the usual rad/s units. The units for the PSD of $u(t)$ are $(Hz)^{-1}$. Using Eq. (3.2.2) we get the relationship between the input and output PSDs.

$$\begin{aligned} S_x(j\omega) &= G(j\omega)G(-j\omega)S_u(j\omega) \\ &= \frac{\sqrt{2\sigma^2\beta}}{j\omega+\beta} \cdot \frac{\sqrt{2\sigma^2\beta}}{-j\omega+\beta} \cdot 1 \\ &= \frac{2\sigma^2\beta}{\omega^2+\beta^2} m^2/Hz \end{aligned}$$

Note that the input PSD is "per Hz," so the output should also be "per Hz." As a check on this we can compute the mean-square-value of x as discussed previously. $S_x(j\omega)$ is in m^2/Hz, but we will be integrating with respect to ω (not f). Therefore, we need $(1/2\pi)$ in front of the integral:

$$\begin{aligned} E(x^2) &= \frac{1}{2\pi} \int_{-\infty}^{\infty} \frac{2\sigma^2\beta}{\omega^2+\beta^2} d\omega \\ &= \frac{2\sigma^2}{2\pi} \int_{-\infty}^{\infty} \frac{\beta}{\omega^2+\beta^2} d\omega \\ &= \sigma^2\, m^2 \text{ (from standard integral tables)} \end{aligned}$$

■

3.9 VECTOR DESCRIPTION OF RANDOM PROCESSES

In the subsequent chapters on Kalman filtering we will have need to put the various random processes being considered into vector state-space format. We will consider the continuous-time model first and then follow this with the discrete-time model.

Continuous-Time Model

Even though the filter measurement stream may consist of discrete-time samples, some of the underlying random processes may be time-continuous (e.g., physical dynamics, Newton's law, etc.). Thus, the continuous-time model is often just as important as the corresponding discrete-time model. So, consider a continuous-time linear state-space model of the form

$$\dot{\mathbf{x}} = \mathbf{F}\mathbf{x} + \mathbf{G}\mathbf{u} \quad \text{(random process differential equation)} \tag{3.9.1}$$

$$\mathbf{z} = \mathbf{H}\mathbf{x} + \mathbf{v} \quad \text{(linear measurement relationship)} \tag{3.9.2}$$

where

$\mathbf{x}$ is the n x 1 state vector.
$\mathbf{u}$ is the vector forcing function whose components are white noise.
$\mathbf{z}$ is the m x 1 noisy measurement vector.
$\mathbf{v}$ is the measurement noise vector.
$\mathbf{F}$, $\mathbf{G}$, $\mathbf{H}$ are matrices that give the linear connections appropriate for the problem at hand.

Eq. (3.9.1) is a linear differential equation, and the components of $\mathbf{x}$ describe the dynamics of the various processes under consideration. Some of these process descriptions may initially be in the form of power spectral densities, so we need to be able to translate these PSD descriptions into differential equation form. So, we will now look at this problem, and we will take advantage of the shaping-filter methods of Section 3.6 in doing so.

Suppose we start with a PSD in rational form where both the numerator and denominator are polynomials in ω^2 (or s^2). There are some restrictions on the orders of the polynomials. These are imposed to assure that the variances of the phase variables of the state vector will be finite. They are fairly obvious, so we will not pursue this further. A simple numerical example will illustrate the procedure for getting a state model from the spectral description of the process.

EXAMPLE 3.11

Suppose that we want to get a state-space model for the second-order Markov process that was discussed in Section 2.9. We will let $\omega_0 = 2$ rad/s and $\sigma^2 = 1\ \text{m}^2$ for this numerical example. Then the PSD is

$$S_x(s) = \frac{16\sqrt{2}}{s^4 + 16} \quad \text{(in terms of } s \text{ rather than } j\omega\text{)}$$

First, we do spectral factorization of $S_x(s)$.

$$S_x(s) = \frac{4\sqrt[4]{2}}{s^2 + 2\sqrt{2}s + 4} \cdot \frac{4\sqrt[4]{2}}{(-s)^2 + 2\sqrt{2}(-s) + 4} \tag{3.9.3}$$

The units of $S_x(s)$ are m^2/Hz. The shaping filter and pole locations for this exercise are shown in Fig. 3.10.

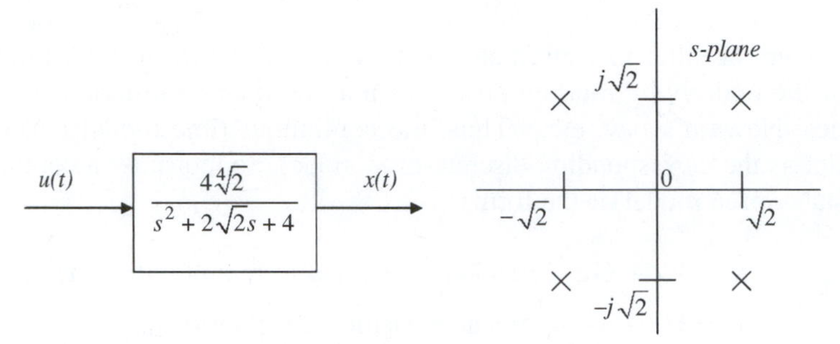

Figure 3.10 Shaping filter and pole locations for $S_x(s)$.

The scalar differential equation relating $x(t)$ to $u(t)$ is now obtained directly from the transfer function of the shaping filter. It is

$$\ddot{x} + 2\sqrt{2}\dot{x} + 4x = 4\sqrt[4]{2}u \tag{3.9.4}$$

We now choose phase variables x and $\dot{x}$ as our two state variables, and the resulting vector differential equation becomes:*

$$\underbrace{\begin{bmatrix} \dot{x}_1 \\ \dot{x}_2 \end{bmatrix}}_{\dot{\mathbf{x}}} = \underbrace{\begin{bmatrix} 0 & 1 \\ -4 & -2\sqrt{2} \end{bmatrix}}_{\mathbf{F}} \underbrace{\begin{bmatrix} x_1 \\ x_2 \end{bmatrix}}_{\mathbf{x}} + \underbrace{\begin{bmatrix} 0 \\ 4\sqrt[4]{2} \end{bmatrix}}_{\mathbf{G}} u \tag{3.9.5}$$

Let us assume that in this example we only have a measurement of position, and not velocity. The measurement equation would then be

$$z = \underbrace{[1 \quad 0]}_{\mathbf{H}} \underbrace{\begin{bmatrix} x_1 \\ x_2 \end{bmatrix}}_{\mathbf{x}} + v \tag{3.9.6}$$

We have now defined the **F**, **G** and **H** parameters for the state model for our second-order Markov process.

■

*In an nth-order differential equation in x, the scalar x and its (n–1) derivatives are often referred to as the phase variables. In state-space notation these then become the n elements of the nxl state vector.

Sampled Continuous-Time Systems

Discrete-time processes may arise in either of two ways. First, there may be applications where a sequence of random events take place naturally in discrete steps, and there is no underlying continuous dynamics to consider. Or, the discrete process may be the result of sampling a continuous-time process such has physical motion governed by Newton's laws. It is such sampled processes that need further discussion here. So, let us now consider a continuous-time random process of the form given by Eq. (3.9.1), but the accompanying measurements come to us sampled in the form

$$\mathbf{z}(t_k) = \mathbf{H}(t_k)\mathbf{x}(t_k) + \mathbf{v}(t_k) \tag{3.9.7}$$

where $\mathbf{H}(t_k)$ is known, and the mean and covariance of $\mathbf{v}(t_k)$ are also known.

We only have measurements at discrete times t_k, t_{k+1}, t_{k+2}, . . . , so we will be primarily interested in the solution of the differential equation, Eq. (3.9.1), at the corresponding times. In linear systems, this solution can always be written as the sum of an initial condition part and a driven-response part. Thus, from basic differential equation theory we have

$$\mathbf{x}(t_{k+1}) = \underbrace{\boldsymbol{\phi}(t_{k+1}, t_k)\mathbf{x}(t_k)}_{\text{Initial condition part}} + \underbrace{\int_{t_k}^{t_{k+1}} \boldsymbol{\phi}(t_{k+1}, \tau)\mathbf{G}(\tau)\mathbf{u}(\tau)d\tau}_{\text{Driven part } \mathbf{w}(t_k)} \tag{3.9.8}$$

Clearly, $\boldsymbol{\phi}(t_{k+1}, t_k)$ is the state transition matrix, and $\mathbf{w}(t_k)$ is the driven response for the (t_{k+1}, t_k) interval. Note that the scale factor for the white noise driving function is accounted for in the $\mathbf{G}$ matrix. Also, we are assured that the $\mathbf{w}(t_k)$, $\mathbf{w}(t_{k+1})$, . . . sequence is a zero-mean white sequence, because we have assumed that the driving function is vector unity white noise. The solution as given by Eq. (3.9.8) is quite formidable if the system is time varying in the intervals between samples. So, we will assume here that the Δt intervals are small, and that the system parameters are constant within the Δt intervals. (More will be said of this assumption in Chapter 4.)

With the constant parameter assumption in place, the state transition matrix can be written out explicitly in terms of Laplace transforms:

$$\boldsymbol{\phi}(\Delta t) = \mathfrak{L}^{-1}\left[(s\mathbf{I} - \mathbf{F})^{-1}\right]_{t=\Delta t} \tag{3.9.9}$$

where t is the dummy inverse Laplace transform variable, and Δt is the (t_{k+1}, t_k) interval. This calculation is quite manageable for lower-order systems; but, if the system order is large, it is next to impossible to do this with "paper-and-pencil" methods. (More will be said about numerical methods presently.)

It will be shown in Chapter 4 that the covariance of $\mathbf{w}(k)$ is one of the key parameters of the Kalman filter, so we need to be able to compute it as well as the state transition matrix. We will call the covariance of $\mathbf{w}(k)$, $\mathbf{Q}(t_k)$ (or sometimes, just

$\mathbf{Q}_k$ for short), and it can be written out explicitly as

$$\begin{aligned}\mathbf{Q}(t_k) &= E[\mathbf{w}(t_k)\mathbf{w}^T(t_k)] \\ &= E\left\{\left[\int_{t_k}^{t_{k+1}} \boldsymbol{\phi}(t_{k+1},\xi)\mathbf{G}(\xi)\mathbf{u}(\xi)d\xi\right]\left[\int_{t_k}^{t_{k+1}} \boldsymbol{\phi}(t_{k+1},\eta)\mathbf{G}(\eta)\mathbf{u}(\eta)d\eta\right]^T\right\} \\ &= \int_{t_k}^{t_{k+1}}\int_{t_k}^{t_{k+1}} \boldsymbol{\phi}(t_{k+1},\xi)\mathbf{G}(\xi)E\left[\mathbf{u}(\xi)\mathbf{u}^T(\eta)\right]\mathbf{G}^T(\eta)\boldsymbol{\phi}^T(t_{k+1},\eta)d\xi d\eta\end{aligned} \tag{3.9.10}$$

The matrix $E[\mathbf{u}(\xi)\mathbf{u}^T(\eta)]$ is a matrix of Dirac delta functions that, presumably, is known from the continuous model. Thus, in principle, $\mathbf{Q}_k$ may be evaluated from Eq. (3.9.10). This is not a trivial task, though, even for low-order systems. If the continuous system giving rise to the discrete situation has constant parameters and if the various white noise inputs have zero crosscorrelation, some simplification is possible and the weighting function methods of Section 3.7 may be applied. This is best illustrated with an example rather than in general terms.

EXAMPLE 3.12

The integrated Gauss–Markov process shown in Fig. 3.11 is frequently encountered in engineering applications. The continuous model in this case is

$$\begin{bmatrix}\dot{x}_1\\ \dot{x}_2\end{bmatrix} = \begin{bmatrix}0 & 1\\ 0 & -\beta\end{bmatrix}\begin{bmatrix}x_1\\ x_2\end{bmatrix} + \begin{bmatrix}0\\ \sqrt{2\sigma^2\beta}\end{bmatrix}u(t) \tag{3.9.11}$$

$$y = [1 \quad 0]\begin{bmatrix}x_1\\ x_2\end{bmatrix} \tag{3.9.12}$$

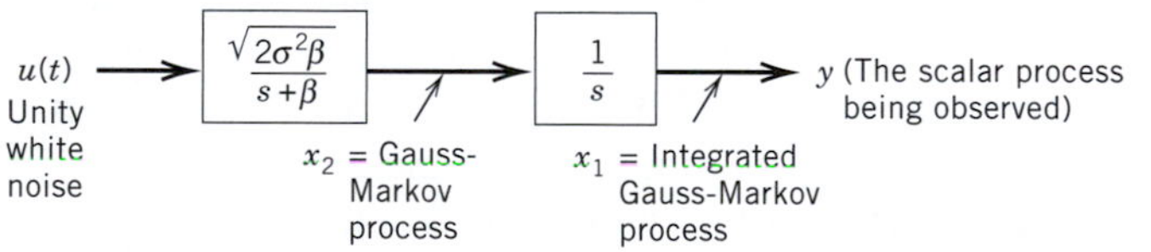

Figure 3.11 Integrated Gauss–Markov process.

Let us say the sampling interval is Δt and we wish to find the corresponding discrete model. The key parameters to be determined are $\boldsymbol{\phi}_k$, and $\mathbf{Q}_k$. The transition matrix is easily determined as

$$\begin{aligned}\boldsymbol{\phi}_k &= \left[\mathfrak{L}^{-1}[(s\mathbf{I}-\mathbf{F})^{-1}]\right]_{t=\Delta t} \\ &= \mathfrak{L}^{-1}\begin{bmatrix}s & -1\\ 0 & s+\beta\end{bmatrix}^{-1} = \mathfrak{L}^{-1}\begin{bmatrix}\dfrac{1}{s} & \dfrac{1}{s(s+\beta)}\\ 0 & \dfrac{1}{s+\beta}\end{bmatrix} \\ &= \begin{bmatrix}1 & \dfrac{1}{\beta}(1-e^{-\beta\Delta t})\\ 0 & e^{-\beta\Delta t}\end{bmatrix}\end{aligned} \tag{3.9.13}$$

Next, rather than using Eq. (3.9.10) directly to determine $\mathbf{Q}_k$, we will use the transfer function approach. From the block diagram of Fig. 3.11, we observe the following transfer functions:

$$G(u \text{ to } x_1) = G_1 = \frac{\sqrt{2\sigma^2\beta}}{s(s+\beta)} \tag{3.9.14}$$

$$G(u \text{ to } x_2) = G_2 = \frac{\sqrt{2\sigma^2\beta}}{s+\beta} \tag{3.9.15}$$

The corresponding weighting functions are

$$g_1(t) = \sqrt{\frac{2\sigma^2}{\beta}}(1 - e^{-\beta t}) \tag{3.9.16}$$

$$g_2(t) = \sqrt{2\sigma^2\beta}\, e^{-\beta t} \tag{3.9.17}$$

We can now use the methods of Section 3.7 to find the needed mean-square responses:

$$\begin{aligned} E[x_1x_1] &= \int_0^{\Delta t}\int_0^{\Delta t} g_1(\xi)g_1(\eta)E[u(\xi)u(\eta)]d\xi\, d\eta \\ &= \int_0^{\Delta t}\int_0^{\Delta t} \frac{2\sigma^2}{\beta}(1-e^{-\beta\xi})(1-e^{-\beta\eta})\delta(\xi-\eta)\, d\xi\, d\eta \\ &= \frac{2\sigma^2}{\beta}\left[\Delta t - \frac{2}{\beta}(1-e^{-\beta\Delta t}) + \frac{1}{2\beta}(1-e^{-2\beta\Delta t})\right] \end{aligned} \tag{3.9.18}$$

$$\begin{aligned} E[x_1x_2] &= \int_0^{\Delta t}\int_0^{\Delta t} g_1(\xi)g_2(\eta)E[u(\xi)u(\eta)]d\xi\, d\eta \\ &= \int_0^{\Delta t}\int_0^{\Delta t} 2\sigma^2 e^{-\beta\xi}(1-e^{-\beta\eta})\delta(\xi-\eta)\, d\xi\, d\eta \\ &= 2\sigma^2\left[\frac{1}{\beta}(1-e^{-\beta\Delta t}) - \frac{1}{2\beta}(1-e^{-2\beta\Delta t})\right] \end{aligned} \tag{3.9.19}$$

$$\begin{aligned} E[x_2x_2] &= \int_0^{\Delta t}\int_0^{\Delta t} g_2(\xi)g_2(\eta)E[u(\xi)u(\eta)]d\xi\, d\eta \\ &= \int_0^{\Delta t}\int_0^{\Delta t} 2\sigma^2\beta e^{-\beta\xi}e^{-\beta\eta}\delta(\xi-\eta)\, d\xi\, d\eta \\ &= \sigma^2(1-e^{-2\beta\Delta t}) \end{aligned} \tag{3.9.20}$$

Thus, the $\mathbf{Q}_k$ matrix is

$$\mathbf{Q}_k = \begin{bmatrix} E[x_1x_1] & E[x_1x_2] \\ E[x_1x_2] & E[x_2x_2] \end{bmatrix} = \begin{bmatrix} \text{Eq. (3.9.18)} & \text{Eq. (3.9.19)} \\ \text{Eq. (3.9.19)} & \text{Eq. (3.9.20)} \end{bmatrix} \tag{3.9.21}$$

The discrete model is now complete with the specification of $\boldsymbol{\phi}_k$, and $\mathbf{Q}_k$, as given by Eqs. (3.9.13) and (3.9.21). Note that the k subscript could have been dropped in this example because the sampling interval is constant.

Numerical Evaluation of ϕ_k and $\mathbf{Q}_k$

Analytical methods for finding $\boldsymbol{\phi}_k$ and $\mathbf{Q}_k$ work quite well for constant parameter systems with just a few elements in the state vector. However, the dimensionality does not have to be very large before it becomes virtually impossible to work out explicit expressions for $\boldsymbol{\phi}_k$ and $\mathbf{Q}_k$. A numerical method for determining $\boldsymbol{\phi}_k$ and $\mathbf{Q}_k$ for large scale systems has been worked out by C. F. van Loan [7], and it is especially convenient when using MATLAB. With reference to the continuous-time model given by Eq. (3.9.1), the van Loan method proceeds as follows:

1. First, form a $2n \times 2n$ matrix that we will call $\mathbf{A}$ (n is the dimension of $\mathbf{x}$ and $\mathbf{W}$ is the power spectral density of u).*

$$\mathbf{A} = \left[\begin{array}{c|c} -\mathbf{F} & \mathbf{GWG}^T \\ \hline \mathbf{0} & \mathbf{F}^T \end{array}\right] \Delta t \tag{3.9.22}$$

2. Using MATLAB (or other software), form $e^{\mathbf{A}}$ and call it $\mathbf{B}$.

$$\mathbf{B} = \text{expm}(\mathbf{A}) = \left[\begin{array}{c|c} \cdots & \boldsymbol{\phi}_k^{-1}\mathbf{Q}_k \\ \hline 0 & \boldsymbol{\phi}_k^T \end{array}\right] \tag{3.9.23}$$

(The upper-left partition of $\mathbf{B}$ is of no concern here.)

3. Transpose the lower-right partition of $\mathbf{B}$ to get $\boldsymbol{\phi}_k$.

$$\boldsymbol{\phi}_k = \text{transpose of lower-right partition of } \mathbf{B} \tag{3.9.24}$$

4. Finally, $\mathbf{Q}_k$ is obtained from a matrix product as follows:

$$\mathbf{Q}_k = \boldsymbol{\phi}_k[\text{upper-right partition of } \mathbf{B}] \tag{3.9.25}$$

The method will now be illustrated with an example.

*In the shaping filter notation used here (see Section 3.6), we usually account for the white noise scale factor within the $\mathbf{G}$ matrix. Thus, $\mathbf{W}$ is an identity matrix with dimensions that are compatible with $\mathbf{G}$. Other authors may prefer to include the noise scaling directly in u, and $\mathbf{W}$ is not trivial in that case. The "bottom line" is simply that the triple product $\mathbf{GWG}^\mathrm{T}$ must properly account for the white noise forcing function, whatever it may be.

EXAMPLE 3.13

Consider the nonstationary harmonic-motion process described by the differential equation

$$\ddot{y} + y = 2u(t) \tag{3.9.26}$$

where $u(t)$ is unity white noise and let $\Delta t = .1$ sec. The continuous state model for this process is then

$$\begin{bmatrix} \dot{x}_1 \\ \dot{x}_2 \end{bmatrix} = \underbrace{\begin{bmatrix} 0 & 1 \\ -1 & 0 \end{bmatrix}}_{\mathbf{F}} \begin{bmatrix} x_1 \\ x_2 \end{bmatrix} + \underbrace{\begin{bmatrix} 0 \\ 2 \end{bmatrix}}_{\mathbf{G}} u(t) \tag{3.9.27}$$

where x_1 and x_2 are the usual phase variables. In this case, $\mathbf{W}$ is

$$\mathbf{W} = 1 \tag{3.9.28}$$

(The scale factor is accounted for in $\mathbf{G}$.) $\mathbf{GWG}^T$ is then

$$\mathbf{GWG}^T = \begin{bmatrix} 0 & 0 \\ 0 & 4 \end{bmatrix} \tag{3.9.29}$$

Now form the partitioned $\mathbf{A}$ matrix. Let $\Delta t = .1$

$$\mathbf{A} = \begin{bmatrix} -\mathbf{F}\Delta t & \mathbf{GWG}^T\Delta t \\ \mathbf{0} & \mathbf{F}^T\Delta t \end{bmatrix} = \left[\begin{array}{cc|cc} 0 & -.1 & 0 & 0 \\ .1 & 0 & 0 & .4 \\ \hline 0 & 0 & 0 & -.1 \\ 0 & 0 & .1 & 0 \end{array}\right] \tag{3.9.30}$$

The next step is to compute $\mathbf{B} = e^{\mathbf{A}}$. The result is (with numerical rounding)

$$\mathbf{B} = \text{expm}(\mathbf{A}) = \left[\begin{array}{cc|cc} .9950 & -.0998 & -.0007 & -.0200 \\ .0998 & .9950 & .0200 & .3987 \\ \hline 0 & 0 & .9950 & -.0998 \\ 0 & 0 & .0998 & .9950 \end{array}\right] \tag{3.9.31}$$

Finally, we get both $\boldsymbol{\phi}_k$ and $\mathbf{Q}_k$ from

$$\begin{aligned} \boldsymbol{\phi}_k &= \text{transpose of lower-right partition of } \mathbf{B} \\ &= \begin{bmatrix} .9950 & .0998 \\ .0998 & .9950 \end{bmatrix} \end{aligned} \tag{3.9.32}$$

$$\begin{aligned} \mathbf{Q}_k &= \boldsymbol{\phi}_k[\text{upper-right partition of } \mathbf{B}] \\ &= \begin{bmatrix} .0013 & .0199 \\ .0199 & .3987 \end{bmatrix} \end{aligned} \tag{3.9.33}$$

■

3.10 MONTE CARLO SIMULATION OF DISCRETE-TIME PROCESSES

Monte Carlo simulation refers to system simulation using random sequences as inputs. Such methods are often helpful in understanding the behavior of stochastic systems that are not amenable to analysis by usual direct mathematical methods. This is especially true of nonlinear filtering problems (considered later in Chapter 7), but there are also many other applications where Monte Carlo methods are useful. Briefly, these methods involve setting up a statistical experiment that matches the physical problem of interest, then repeating the experiment over and over with typical sequences of random numbers, and finally, analyzing the results of the experiment statistically. We are concerned here primarily with experiments where the random processes are Gaussian and sampled in time.

Simulation of Sampled Continuous-Time Random Processes

The usual description of a stationary continuous-time random process is its power spectral density (PSD) function or the corresponding autocorrelation function. It was mentioned in Chapter 2 that the autocorrelation function provides a complete statistical description of the process when it is Gaussian. This is important in Monte Carlo simulation (even though somewhat restrictive), because Gaussian processes have a firm theoretical foundation and this adds credibility to the resulting analysis. In Section 3.9 a general method was given for obtaining a discrete state-space model for a random process, given its power spectral density. Thus, we will begin with a state model of the form given by Eq. (3.9.8) (Here we will use the more compact notation where the subscript k refers to the time t_k.)

$$\mathbf{x}_{k+1} = \boldsymbol{\phi}_k \mathbf{x}_k + \mathbf{w}_k \tag{3.10.1}$$

Presumably, $\boldsymbol{\phi}_k$ is known, and $\mathbf{w}_k$ is a Gaussian white sequence with known covariance $\mathbf{Q}_k$. The problem is to generate an ensemble of random trials of $\mathbf{x}_k$ (i.e., sample realizations of the process) for $k = 0, 1, 2, \ldots, m$.

Equation (3.10.1) is explicit. Thus, once methods are established for generating $\mathbf{w}_k$ for $k = 0, 1, 2, \ldots, (m-1)$ and setting the initial condition for $\mathbf{x}$ at $k = 0$, then programming the few lines of code needed to implement Eq. (3.10.1) is routine. MATLAB is especially useful here because of its "user friendliness" in performing matrix calculations. If $\boldsymbol{\phi}_k$ is a constant, it is simply assigned a variable name and given a numerical value in the MATLAB workspace. If $\boldsymbol{\phi}_k$ is time-variable, it is relatively easy to reevaluate the parameters with each step as the simulation proceeds in time. Generating the $\mathbf{w}_k$ sequence is a bit more difficult, though, because $\mathbf{Q}_k$ is usually not diagonal. Proceeding on this basis, we begin with a vector $\mathbf{u}_k$ whose components are independent samples from an $\mathcal{N}(0, 1)$ population (which is readily obtained in MATLAB), and then operate on this vector with a linear transformation $\mathbf{C}_k$ that is chosen so as to yield a $\mathbf{w}_k$ vector with the desired covariance structure. The desired $\mathbf{C}_k$ is not unique, but a simple way of forming a suitable $\mathbf{C}_k$ is to let it be lower triangular and then solve for the unknown elements. Stated mathematically, we have (temporarily omitting the k subscripts)

$$\mathbf{w} = \mathbf{C}\mathbf{u} \tag{3.10.2}$$

and we demand that

$$E\left[(\mathbf{Cu})(\mathbf{Cu})^T\right] = E\left[\mathbf{ww}^T\right] = \mathbf{Q} \tag{3.10.3}$$

Now, $E[\mathbf{uu}^T]$ is the unitary matrix because of the way we obtain the elements of $\mathbf{u}$ as independent $\mathcal{N}(0, 1)$ samples. Therefore,

$$\mathbf{CC}^T = \mathbf{Q} \tag{3.10.4}$$

We will now proceed to show that the algebra for solving for the elements of $\mathbf{C}$ is simple, provided that the steps are done in the proper order. This will be demonstrated for a 2×2 $\mathbf{Q}$ matrix. Recall that $\mathbf{Q}$ is symmetric and positive definite. For the 2×2 case, we have (with the usual matrix subscript notation)

$$\begin{bmatrix} c_{11} & 0 \\ c_{21} & c_{22} \end{bmatrix} \begin{bmatrix} c_{11} & c_{21} \\ 0 & c_{22} \end{bmatrix} = \begin{bmatrix} q_{11} & q_{12} \\ q_{21} & q_{22} \end{bmatrix}$$

or

$$\begin{bmatrix} c_{11}^2 & c_{11}c_{21} \\ c_{11}c_{21} & c_{21}^2 + c_{22}^2 \end{bmatrix} = \begin{bmatrix} q_{11} & q_{12} \\ q_{21} & q_{22} \end{bmatrix} \tag{3.10.5}$$

We start first with the 11 term.

$$c_{11} = \sqrt{q_{11}} \tag{3.10.6}$$

Next, we solve for the 21 term.

$$c_{21} = \frac{q_{12}}{c_{11}} \tag{3.10.7}$$

Finally, c_{22} is obtained as

$$c_{22} = \sqrt{q_{22} - c_{21}^2} \tag{3.10.8}$$

The preceding 2×2 example is a special case of what is known as Cholesky factorization, and it is easily generalized to higher-order cases. This procedure factors a symmetric, positive definite matrix into upper- and lower-triangular parts, and MATLAB has a built-in function chol to perform this operation. The user defines a matrix variable, say, QUE, with the numerical values of $\mathbf{Q}$, and then chol (QUE) returns the transpose of the desired $\mathbf{C}$ in the notation used here. This is a very nice feature of MATLAB, and it is a valuable timesaver when dealing with higher-order systems.

It should also be clear that if the transformation $\mathbf{C}$ takes a vector of uncorrelated, unit-variance random variables into a corresponding set of correlated random variables, then $\mathbf{C}^{-1}$ will do just the opposite. If we start with a set of random variables with covariance $\mathbf{Q}$, then $\mathbf{C}^{-1}\mathbf{Q}\mathbf{C}^{-1T}$ will be the covariance of the transformed set. This covariance is, of course, just the identity matrix.

One limitation of the Cholesky factorization is that it cannot be used if the covariance matrix is singular. We can overcome this limitation by the use of another factorization method called the *singular value decomposition*, which is given as "svd" in MATLAB. If a variable Q is singular, then a MATLAB command line

$$[U, T, V] = \text{svd}\,(Q)$$

will return three matrices U, T, and V such that $UTV = Q$.

If Q is symmetric, as would be the case if it is a covariance matrix, then we also get $U = V^{\mathrm{T}}$. The T matrix is diagonal with non-negative elements so we can easily compute its square root in MATLAB with $S = \text{sqrt}\ (T)$.

$$T = \begin{bmatrix} t_{11} & 0 & 0 \\ 0 & t_{22} & 0 \\ 0 & 0 & t_{33} \end{bmatrix} \quad \Rightarrow \quad S = \begin{bmatrix} \sqrt{t_{11}} & 0 & 0 \\ 0 & \sqrt{t_{22}} & 0 \\ 0 & 0 & \sqrt{t_{33}} \end{bmatrix}$$

Note that $T = SS^T$. So Q can be factored into $Q = USS^{\mathrm{T}}U^{\mathrm{T}} = (US)(US)^{\mathrm{T}}$ and the desired C factor in Eq. 3.10.4 is simply formed by $C = US$.

Specifying an appropriate initial condition on $\mathbf{x}$ in the simulation can also be troublesome, and each case has to be considered on its own merits. If the process being simulated is nonstationary, there is no "typical" starting point. This will depend on the definition of the process. For example, a Wiener process is defined to have a zero initial condition. All sample realizations must be initialized at zero in this case. On the other hand, a simple one-state random-walk process can be defined to start with any specified $\mathbf{x}_0$, be it deterministic or random.

If the process being considered is stationary, one usually wants to generate an ensemble of realizations that are stationary throughout the time span of the runs. The initial condition on $\mathbf{x}$ must be chosen carefully to assure this. There is one special case where specification of the initial components of $\mathbf{x}$ is relatively easy. If the process is stationary and the state variables are chosen to be phase variables, it works out that the covariance matrix of the state variables is diagonal in the steady-state condition (see Problem 3.18). Thus, one simply chooses the components of $\mathbf{x}$ as independent samples from an $\mathcal{N}(0, 1)$ population appropriately scaled in accordance with the rms values of the process "position," "velocity," "acceleration," and so forth. If the state variables are not phase variables, however, then they will be correlated (in general), and this complicates matters considerably. Sometimes, the most expeditious way of circumventing the problem is to start the simulation with zero initial conditions, then let the process run until the steady-state condition is reached (or nearly so), and finally use just the latter portion of the realization for "serious" analysis. This may not be an elegant solution to the initial-condition problem, but it is effective.

3.11 SUMMARY

In this chapter we have developed various stochastic input/output relationships for linear systems. In steady-state analysis the primary descriptors for the random processes being considered are autocorrelation function and power spectral density

(PSD). They are Fourier transform pairs, so both contain the same information about the process at hand. Thus, whichever description we choose to work with is a matter of mathematical convenience.

The process magnitude description of primary interest in our analysis is mean-square-value, also called power. This is partly a matter of convenience, because it is relatively easy to calculate mean-square-value and not so easy with other measures of amplitude. In Section 3.8 it was mentioned that we must be careful in our interpretation of $S(j\omega)$ as being power density. When we formulate our equations in terms of ω (in contrast to frequency in Hz), and when we integrate $S(j\omega)$ with respect to ω to get power, we must remember to multiply the integral by $1/2\pi$ in order to get the correct power. This little bit of confusion came about because we defined PSD as the Fourier transform of the autocorrelation function, and thus the inversion integral had to contain a $1/2\pi$ factor. There is no confusion, though, when using the integral tables given in Section 3.3, because the $1/2\pi$ factor is already accounted for in the tables.

Input/output calculations for transient problems are considerably more complicated than for the steady-state case. Explicit expressions for mean-square-value can usually be calculated for first- and second-order systems. However, beyond that it becomes quite difficult using "paper-and-pencil" methods. Thus we usually have to resort to numerical methods for higher-order cases. In subsequent chapters we will see that there are two key parameters in the discrete Kalman filter that depend on the step size between measurements. These are the state transition matrix and the process noise covariance matrix. Section 3.9 gives a convenient algorithm for computing these parameters. This is an important algorithm and is now used almost universally, in offline analysis at least.

We have now presented in Chapters 1, 2, and 3 all of the basic background material needed for the study of Kalman filtering. So, the remaining chapters will concentrate on Kalman filtering, as such, and its many variations and applications.

PROBLEMS

3.1 Find the steady-state mean-square value of the output for the following filters. The input is white noise with a spectral density amplitude A.

(a) $G(s) = \dfrac{Ts}{(1+Ts)^2}$

(b) $G(s) = \dfrac{\omega_0^2}{s^2 + 2\zeta\omega_0 s + \omega_0^2}$

(c) $G(s) = \dfrac{s+1}{(s+2)^2}$

3.2 A white noise process having a spectral density amplitude of A is applied to the circuit shown. The circuit has been in operation for a long time. Find the steady-state mean-square value of the output voltage.

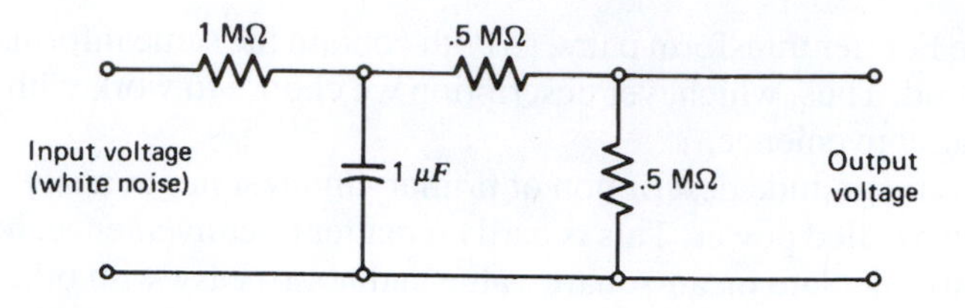

Figure P3.2

3.3 The input to the feedback system shown is a stationary Markov process with an autocorrelation function

$$R_f(\tau) = \sigma^2 e^{-\beta|\tau|}$$

The system is in a stationary condition.

(a) What is the spectral density function of the output?

(b) What is the mean-square value of the output?

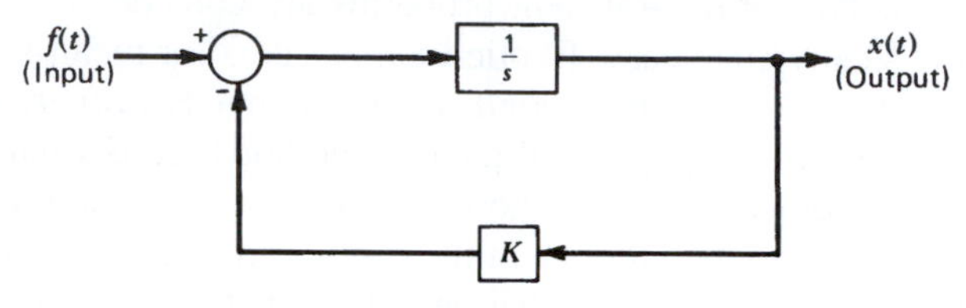

Figure P3.3

3.4 Find the steady-state mean-square value of the output for a first-order low-pass filter, i.e., $G(s) = 1/(1 + Ts)$ if the input has an autocorrelation function of the form

$$R(\tau) = \begin{cases} \sigma^2(1 - \beta|\tau|), & -\dfrac{1}{\beta} \le \tau \le \dfrac{1}{\beta} \\ 0, & |\tau| > \dfrac{1}{\beta} \end{cases}$$

[*Hint:* The input spectral function is irrational so the integrals given in Table 3.1 are of no help here. One approach is to write the integral expression for $E(X^2)$ in terms of real ω rather than s and then use conventional integral tables. Also, those familiar with residue theory will find that the integral can be evaluated by the method of residues.]

3.5 Consider a linear filter whose weighting function is shown in the figure. (This filter is sometimes referred to as a finite-time integrator.) The input to the filter is white noise with a spectral density amplitude A, and the filter has been in operation a long time. What is the mean-square value of the output?

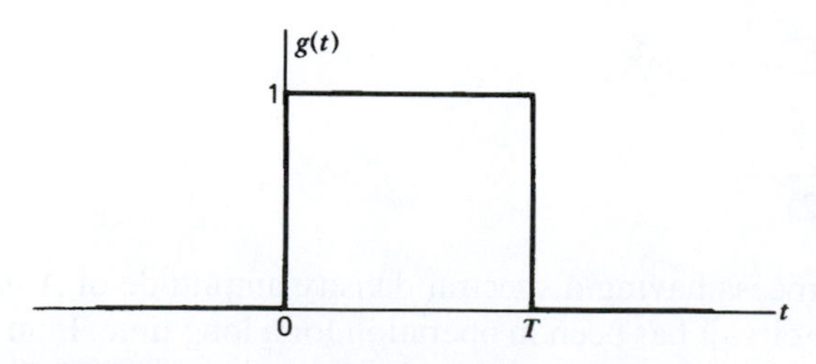

Figure P3.5 Filter weighting function.

3.6 Find the shaping filter that will shape unity white noise into noise with a spectral function

$$S(j\omega) = \frac{\omega^2 + 1}{\omega^4 + 8\omega^2 + 16}$$

3.7 A series resonant circuit is shown in the figure. Let the resistance R be small such that the circuit is sharply tuned (i.e., high Q or very low damping ratio). Find the noise equivalent bandwidth for this circuit and express it in terms of the damping ratio ζ and the natural undamped resonant frequency ω_r (i.e., $\omega_r = 1/\sqrt{LC}$). Note that the "ideal" response in this case is a unity-gain rectangular pass band centered about ω_r. Also find the usual half-power bandwidth and compare this with the noise equivalent bandwidth. (Half-power bandwidth is defined to be the frequency difference between the two points on the response curve that are "down" by a factor of $1/\sqrt{2}$ from the peak value. It is useful to approximate the resonance curve as being symmetric about the peak for this part of the problem.)

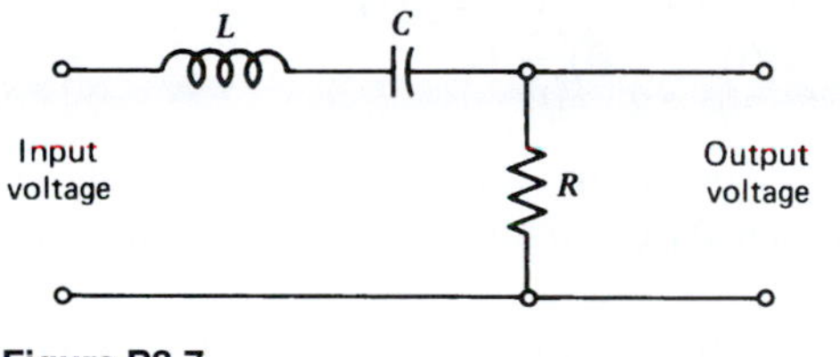

Figure P3.7

3.8 The transfer functions and corresponding bandpass characteristics for first-, second-, and third-order Butterworth filters are shown in the figure below. These filters are said to be "maximally flat" at zero frequency with each successive higher-order filter more nearly approaching the ideal curve than the previous one. All three

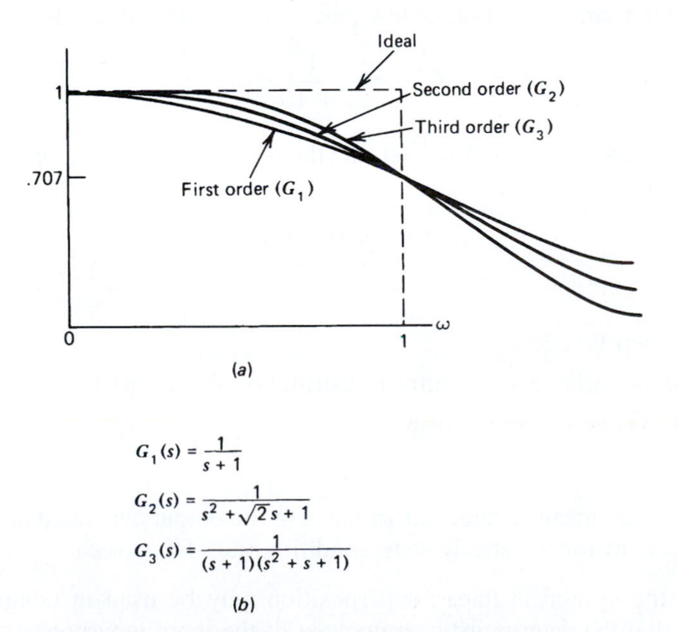

Figure P3.8 (*a*) Responses of three Butterworth filters. (*b*) Transfer functions of Butterworth filters.

filters have been normalized such that all responses intersect the −3-db point at 1 rad/s (or $1/2\pi$ Hz).

(a) Find the noise equivalent bandwidth for each of the filters.

(b) Insofar as noise suppression is concerned, is there much to be gained by using anything higher-order than a third-order Butterworth filter?

3.9 Find the mean-square value of the output (averaged in an ensemble sense) for the following transfer functions. In both cases, the initial conditions are zero and the input $f(t)$ is applied at $t=0$.

$$\text{(a)}\ G(s) = \frac{1}{s^2}, \quad R_f(\tau) = A\delta(\tau)$$

$$\text{(b)}\ G(s) = \frac{1}{s^2 + \omega_0^2}, \quad R_f(\tau) = A\delta(\tau)$$

3.10 A certain linear system is known to satisfy the following differential equation:

$$\ddot{x} + 10\dot{x} + 100x = f(t)$$
$$x(0) = \dot{x}(0) = 0$$

where $x(t)$ is the response (say, position) and the $f(t)$ is a white noise forcing function with a power spectral density of 10 units. (Assume the units are consistent throughout.)

(a) Let x and $\dot{x}$ be state variables x_1 and x_2. Then write out the vector state space differential equations for this random process.

(b) Suppose this process is sampled at a uniform rate beginning at $t=0$. The sampling interval is 0.2 s. Find the $\boldsymbol{\phi}$ and $\mathbf{Q}$ parameters for this situation. (The Van Loan method described in Section 3.9 is recommended.)

(c) What are the mean-square values of $x(t)$ and $\dot{x}(t)$ at $t=0.2$ s?

3.11 Consider a simple first-order low-pass filter whose transfer function is

$$G(s) = \frac{1}{1 + 10s}$$

The input to the filter is initiated at $t=0$, and the filter's initial condition is zero. The input is given by

$$f(t) = Au(t) + n(t)$$

where

$u(t) =$ unit step function

$A =$ random variable with uniform distribution from 0 to 1

$n(t) =$ unity Gaussian white noise

Find:

(a) The mean, mean square, and variance of the output evaluated at $t = .1$ sec.

(b) Repeat (a) for the steady-state condition (i.e., for $t=\infty$).

(*Hint:* Since the system is linear, superposition may be used in computing the output. Note that the deterministic component of the input is written explicitly in functional form. Therefore, deterministic methods may be used to compute the portion of the output due to this component. Also, remember that the mean-square value and the variance are not the same if the mean is nonzero.)

3.12 A signal is known to have the following form:

$$s(t) = a_0 + n(t)$$

where a_0 is an unknown constant and $n(t)$ is a stationary noise process with a known autocorrelation function

$$R_n(\tau) = \sigma^2 e^{-\beta|\tau|}$$

It is suggested that a_0 can be estimated by simply averaging $s(t)$ over a finite interval of time T. What would be the rms error in the determination of a_0 by this method?

(*Note:* The *root* mean square rather than mean square value is requested in this problem.)

3.13 Unity Gaussian white noise $f(t)$ is applied to the cascaded combination of integrators shown in the figure. The switch is closed at $t=0$. The initial condition for the first integrator is zero, and the second integrator has two units as its initial value.

(a) What is the mean-square value of the output at $t=2$ s?

(b) Sketch the probability density function for the output evaluated at $t=2$ s.

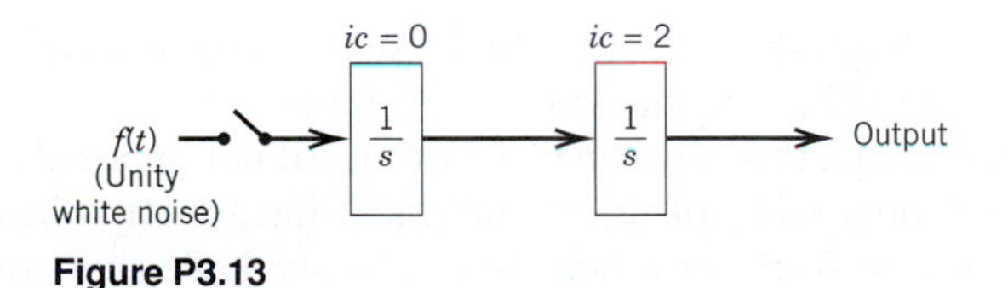

Figure P3.13

3.14 Consider again the filter with a rectangular weighting function discussed in Problem 3.5. Consider the filter to be driven with unity Gaussian white noise, which is initiated at $t=0$ with zero initial conditions.

(a) Find the mean-square response in the interval from 0 to T.

(b) Find the mean-square response for $t \geq T$ and compare the result with that obtained in Problem 3.5.

(c) From the result of (b), would you say the filter's "memory" is finite or infinite?

3.15 The block diagram on the next page describes the error propagation in one channel of an inertial navigation system with external-velocity-reference damping (8). The inputs shown as $f_1(t)$ and $f_2(t)$ are random driving functions due to the accelerometer and external-velocity-reference instrument errors. These will be assumed to be independent white noise processes with spectral amplitudes A_1 and A_2, respectively. The outputs are labeled x_1, x_2, and x_3, and these physically represent the inertial system's platform tilt, velocity error, and position error.

(a) Write out the vector state-space model for this system.

(b) Find the steady-state mean-square values for the three state variables. Express them in terms of A_1, A_2, R, g, and K parameters.

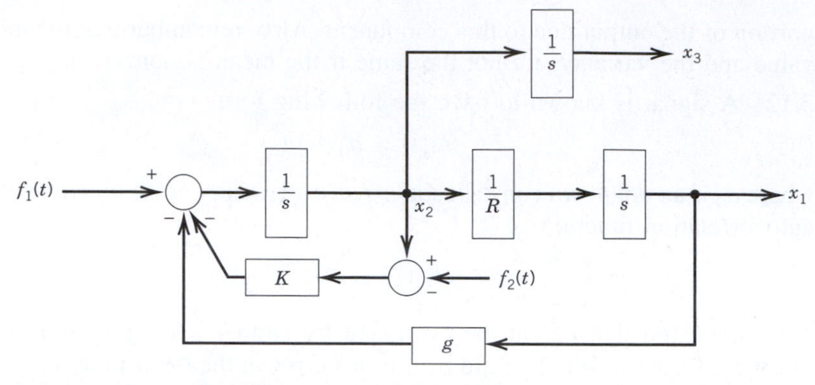

R = earth radius ≈ 2.09×10^7 ft
g = gravitational constant ≈ 32.2 ft/sec^2
K = feedback constant (adjustable to yield the desired damping ratio)

Figure P3.15

3.16 In the nonstationary mean-square-response problem, it is worth noting that the double integral in Eq. (3.7.4) reduces to a single integral when the input is white noise. That is, if $R_f(\tau) = A\delta(\tau)$, then

$$E\left[x^2(t)\right] = A \int_0^t g^2(\upsilon) d\upsilon \tag{P3.16}$$

where $g(\upsilon)$ is the inverse Laplace transform of the transfer function $G(s)$, and A is the power spectral density (PSD) of the white noise input.

Evaluation of integrals analytically can be laborious (if not impossible), so one should not overlook the possibility of using numerical integration when $g^2(\upsilon)$ is either rather complicated in form, or when it is only available numerically. To demonstrate the effectiveness of the numerical approach, consider the following system driven by white noise whose PSD = 10 units:

$$G(s) = \frac{1}{s^2 + \omega_0^2}; \quad \omega_0 = 20\pi$$

Let us say that we want to find $E[x^2(t)]$ for $0 \le t \le .1$ with a sample spacing of .001 sec (i.e., 101 samples including end points). Using MATLAB's numerical integration function *quad* (or other suitable software), find the desired mean-square response numerically. Then plot the result along with the exact theoretical response for comparison [see part (b) of Problem 3.9].

3.17 In the discussion of Markov Processes in Chapter 2 it was mentioned that the derivative of a first-order Markov process does not exist, but the derivative of a second-order Markov does exist. This can be readily demonstrated with a Monte Carlo simulation. For example, consider the same second-order Markov process that was used in Example 2.12. Its power spectral density (PSD) is

$$S_x(j\omega) = \frac{2\sqrt{2}\omega_0^3\sigma^2}{\omega^4 + \omega_0^4} \tag{P3.17.1}$$

where

$$\sigma^2 = 1 \text{ m}^2$$
$$\omega_0 = 0.1 \text{ rad/s}$$

Do a Monte Carlo simulation of this process for 100 steps where the Δt step size is 1 s. To do this you will need to develop a vector state-space model for the process. This is discussed in Section 3.9, and the Van Loan method of computing $\boldsymbol{\phi}_k$ and $\mathbf{Q}_k$ is recommended. Once the key $\boldsymbol{\phi}_k$ and $\mathbf{Q}_k$ parameters are determined, it is relatively easy to implement samples of the process (both position and velocity) using the vector equation

$$\mathbf{x}(t_{k+1}) = \boldsymbol{\phi}_k \mathbf{x}(t_k) + \mathbf{w}(t_k) \tag{P3.17.2}$$

In this demonstration we only want to show that the velocity sequence exists and is well-behaved. So, for simplicity, you may initialize the simulated process with $x_1(0)$ and $x_2(0)$ set at zero. This is somewhat artificial, but legitimate. Plot both position and velocity. We are especially interested in velocity. Note it appears to be well-behaved, and it is approaching a steady-state condition near the end of the run.

3.18 In the preceding Problem 3.17 the process was initialized with both x_1 and x_2 set at zero at $t = 0$. This is artificial and it leads to a transient period during the run. A more natural way to do the simulation would be to begin the run with $x_1(0)$ and $x_2(0)$ chosen at random for stationary conditions as dictated by the PSD of the process.

(a) First show that in steady-state conditions $x_1(t)$ and $x_2(t)$ are uncorrelated when the states are chosen to be position and velocity. The initial 2×2 process covariance matrix will then be diagonal. The two nontrivial elements of the covariance are easily computed using the integral table given in Section 3.3.
Hint: The indefinite integral $\int u \frac{du}{dt} dt = u^2/2$ will be helpful here.

(b) Now re-run the simulation described in Problem 3.17, except start the simulation with typical random initial conditions in accordance with the covariance computed in Part (a).

REFERENCES CITED IN CHAPTER 3

1. J.J. D'Azzo, C.H. Houpis, and S.N. Sheldon, *Linear Control System Analysis and Design*, 5th edition, CRC Press, 2003.
2. B.P. Lathi, *Linear Systems and Signals*, 2nd ed., Oxford University Press, 2004.
3. H.M. James, N.B. Nichols, and R.S. Phillips, *Theory of Servomechanisms*, Radiation Laboratory Series (Vol. 25) New York: McGraw-Hill, 1947.
4. G.C. Newton, L.A. Gould,and J.F. Kaiser, *Analytical Design of Linear Feedback Controls*, New York: Wiley, 1957.
5. G.R. Cooperand C.D. McGillem, *Probabilistic Methods of Signal and System Analysis*, 2nd edition, New York: Holt, Rinehart, and Winston, 1986.
6. K.S. Shanmugan and A.M. Breipohl, *Random Signals: Detection, Estimation, and Data Analysis*, New York: Wiley, 1988.
7. C.F. van Loan, "Computing Integrals Involving the Matrix Exponential," *IEEE Trans. Automatic Control*, AC-23 (3): 395–404 (June 1978).
8. G.R. Pitman, (ed.), *Inertial Guidance*, New York: Wiley, 1962.

PART 2

Kalman Filtering And Applications

4

Discrete Kalman Filter Basics

Modern filter theory began with N. Wiener's work in the 1940s (1). His work was based on minimizing the mean-square error, so this branch of filter theory is sometimes referred to as least-squares filtering. This is an oversimplification though, because a more exact description would be "linear time-domain minimum mean-square error filtering." This is a bit wordy though, so the shortened version will suffice for now. Regardless of what it is called, the central problem is simply a matter of separation of the signal from an additive combination of signal and noise. In hindsight, the Wiener solution turned out to be one of those subjects that was much discussed in textbooks, but little used in practice. Perhaps Wiener's main contribution was the way in which he posed the the problem in terms of minimizing the mean-square error in the time domain. This is in contrast to the frequency-separation methods that were in use at the time. However, in fairness to Wiener's work, the weighting function approach (which is central in the Wiener theory) still has some merit. More is said of this in Section 6.8.

In 1960 R.E. Kalman considered the same problem that Wiener had dealt with earlier, but in his 1960 paper he considered the noisy measurement to be a discrete sequence in time in contrast to a continuous-time signal (2). He also posed the problem in a state-space setting that accommodated the time-variable multiple-input/multiple-output scenario nicely. Engineers, especially in the field of navigation, were quick to see the Kalman technique as a practical solution to some applied filtering problems that were intractable using Wiener methods. Also, the rapid advances in computer technology that occurred in the 1960s certainly contributed to popularizing Kalman filtering as a practical means of separating signal from noise. After some 50 years now, Kalman filtering is still alive and well, and new applications keep appearing on the scene regularly.

4.1 A SIMPLE RECURSIVE EXAMPLE

When working with practical problems involving discrete data, it is important that our methods be computationally feasible as well as mathematically correct. A simple example will illustrate this. Consider the problem of estimating the mean of some random constant based on a sequence of noisy measurements. That is, in filtering terms the true mean is the "signal," and the measurement error is the "noise."

The resulting estimate is the filter output. Now think of processing the data on-line. Let the measurement sequence be denoted as $z_1, z_2, \ldots, z_n$, where the subscript denotes the time at which the measurement is taken. One method of processing the data would be to store each measurement as it becomes available and then compute the sample mean in accordance with the following algorithm (in words):

1. ***First measurement z_1:*** Store z_1 and estimate the mean as

$$\hat{m}_1 = z_1$$

2. ***Second measurement z_2:*** Store z_2 along with z_1 and estimate the mean as

$$\hat{m}_2 = \frac{z_1 + z_2}{2}$$

3. ***Third measurement z_3:*** Store z_3 along with z_1 and z_2 and estimate the mean as

$$\hat{m}_3 = \frac{z_1 + z_2 + z_3}{3}$$

4. And so forth.

Clearly, this would yield the correct sequence of sample means as the experiment progresses. It should also be clear that the amount of memory needed to store the measurements keeps increasing with time, and also the number of arithmetic operations needed to form the estimate increases correspondingly. This would lead to obvious problems when the total amount of data is large. Thus, consider a simple variation in the computational procedure in which each new estimate is formed as a blend of the old estimate and the current measurement. To be specific, consider the following algorithm:

1. ***First measurement z_1:*** Compute the estimate as

$$\hat{m}_1 = z_1$$

   Store $\hat{m}_1$ and discard z_1.

2. ***Second measurement z_2:*** Compute the estimate as a weighted sum of the previous estimate $\hat{m}_1$ and the current measurement z_2:

$$\hat{m}_2 = \tfrac{1}{2}\hat{m}_1 + \tfrac{1}{2}z_2$$

   Store $\hat{m}_2$ and discard z_2 and $\hat{m}_1$.

3. ***Third measurement z_3:*** Compute the estimate as a weighted sum of $\hat{m}_2$ and z_3:

$$\hat{m}_3 = \tfrac{2}{3}\hat{m}_2 + \tfrac{1}{3}z_3$$

   Store $\hat{m}_3$ and discard z_3 and $\hat{m}_2$.

4. And so forth. It should be obvious that at the nth stage the weighted sum is

$$\hat{m}_n = \left(\frac{n-1}{n}\right)\hat{m}_{n-1} + \left(\frac{1}{n}\right)z_n$$

Clearly, the above procedure yields the same identical sequence of estimates as before, but without the need to store all the previous measurements. We simply use the result of the previous step to help obtain the estimate at the current step of the process. In this way, the previous computational effort is used to good advantage and not wasted. The second algorithm can proceed on *ad infinitum* without a growing memory problem. Eventually, of course, as n becomes extremely large, a round-off problem might be encountered. However, this is to be expected with either of the two algorithms.

The second algorithm is a simple example of a *recursive* mode of operation. The key element in any recursive procedure is the use of the results of the previous step to aid in obtaining the desired result for the current step. This is one of the main features of Kalman filtering, and one that clearly distinguishes it from the weight-factor (Wiener) approach.

In order to apply the recursive philosophy to estimation of a random process, it is first necessary that both the process and the measurement noise be modeled in vector form. This was discussed in Section 3.9, and we will proceed on that basis.

4.2 THE DISCRETE KALMAN FILTER

We will now proceed to develop the Kalman filter recursive equations. The optimization criterion used here is minimization of the mean-square estimation error of the random variable x. Then later in Section 4.7 we will show that this same linear estimate also corresponds to the mean of x conditioned on the entire past measurement stream.

We begin by assuming the random process to be estimated can be modeled in the form

$$\mathbf{x}_{k+1} = \boldsymbol{\phi}_k \mathbf{x}_k + \mathbf{w}_k \tag{4.2.1}$$

The observation (measurement) of the process is assumed to occur at discrete points in time in accordance with the linear relationship

$$\mathbf{z}_k = \mathbf{H}_k \mathbf{x}_k + \mathbf{v}_k \tag{4.2.2}$$

and we assume that we know $\boldsymbol{\phi}_k$, $\mathbf{H}_k$, and the covariances describing $\mathbf{w}_k$ and $\mathbf{v}_k$. Also, we will be using the same shortened notation here that was introduced earlier in Section 3.9. It will be repeated here for easy reference:

$\mathbf{x}_k = (n \times 1)$ process state vector at time t_k

$\boldsymbol{\phi}_k = (n \times n)$ matrix relating $\mathbf{x}_k$ to $\mathbf{x}_{k+1}$ in the absence of a forcing function—if $\mathbf{x}_k$ is a sample of continuous process, $\boldsymbol{\phi}_k$ is the usual state transition matrix $\boldsymbol{\phi}(t_{k+1}, t_k)$

$\mathbf{w}_k = (n \times 1)$ vector—assumed to be a white sequence with known covariance structure. It is the input white noise contribution to the state vector for the time interval (t_{k+1}, t_k)

$\mathbf{z}_k = (m \times 1)$ vector measurement at time t_k

$\mathbf{H}_k = (m \times n)$ matrix giving the ideal (noiseless) connection between the measurement and the state vector at time t_k

$\mathbf{v}_k = (m \times 1)$ measurement error—assumed to be a white sequence with known covariance structure and having zero crosscorrelation with the $\mathbf{w}_k$ sequence

The covariance matrices for the $\mathbf{w}_k$ and $\mathbf{v}_k$ vectors are given by

$$E\left[\mathbf{w}_k \mathbf{w}_i^T\right] = \begin{cases} \mathbf{Q}_k, & i = k \\ \mathbf{0}, & i \neq k \end{cases} \tag{4.2.3}$$

$$E\left[\mathbf{v}_k \mathbf{v}_i^T\right] = \begin{cases} \mathbf{R}_k, & i = k \\ \mathbf{0}, & i \neq k \end{cases} \tag{4.2.4}$$

$$E\left[\mathbf{w}_k \mathbf{v}_i^T\right] = \mathbf{0}, \quad \text{for all } k \text{ and } i \tag{4.2.5}$$

We assume at this point that we have an initial estimate of the process at some point in time t_k, and that this estimate is based on all our knowledge about the process prior to t_k. This prior (or *a priori*) estimate will be denoted as $\hat{\mathbf{x}}_k^-$ where the "hat" denotes estimate, and the "super minus" is a reminder that this is our best estimate prior to assimilating the measurement at t_k. (Note that super minus as used here is not related in any way to the super minus notation used in spectral factorization.) We also assume that we know the error covariance matrix associated with $\hat{\mathbf{x}}_k^-$. That is, we define the estimation error to be

$$\mathbf{e}_k^- = \mathbf{x}_k - \hat{\mathbf{x}}_k^- \tag{4.2.6}$$

and the associated error covariance matrix is*

$$\mathbf{P}_k^- = E\left[\mathbf{e}_k^- \mathbf{e}_k^{-T}\right] = E\left[\left(\mathbf{x}_k - \hat{\mathbf{x}}_k^-\right)\left(\mathbf{x}_k - \hat{\mathbf{x}}_k^-\right)^T\right] \tag{4.2.7}$$

In many cases, we begin the estimation problem with no prior measurements. Thus, in this case, if the process mean is zero, the initial estimate is zero, and the associated error covariance matrix is just the covariance matrix of $\mathbf{x}$ itself.

With the assumption of a prior estimate $\hat{\mathbf{x}}_k^-$, we now seek to use the measurement $\mathbf{z}_k$ to improve the prior estimate. We choose a linear blending of the noisy measurement and the prior estimate in accordance with the equation

$$\hat{\mathbf{x}}_k = \hat{\mathbf{x}}_k^- + \mathbf{K}_k\left(\mathbf{z}_k - \mathbf{H}_k \hat{\mathbf{x}}_k^-\right) \tag{4.2.8}$$

where

$\hat{\mathbf{x}}_k =$ updated estimate

$\mathbf{K}_k =$ blending factor (yet to be determined)

The justification of the special form of Eq. (4.2.8) will be deferred until Section 4.7. The problem now is to find the particular blending factor $\mathbf{K}_k$ that yields an updated estimate that is optimal in some sense. Just as in the Wiener solution, we use minimum mean-square error as the performance criterion. Toward this end, we first form the expression for the error covariance matrix associated with the updated (*a posteriori*) estimate.

$$\mathbf{P}_k = E\left[\mathbf{e}_k \mathbf{e}_k^T\right] = E\left[(\mathbf{x}_k - \hat{\mathbf{x}}_k)(\mathbf{x}_k - \hat{\mathbf{x}}_k)^T\right] \tag{4.2.9}$$

* We tacitly assume here that the estimation error has zero mean, and thus, it is proper to refer to $E\left[\mathbf{e}_k^- \mathbf{e}_k^{-T}\right]$ as a covariance matrix. It is also, of course, a moment matrix, but it is usually not referred to as such.

Next, we substitute Eq. (4.2.2) into Eq. (4.2.8) and then substitute the resulting expression for $\hat{\mathbf{x}}_k$ into Eq. (4.2.9). The result is

$$\begin{aligned}\mathbf{P}_k = E\{&\left[\left(\mathbf{x}_k - \hat{\mathbf{x}}_k^-\right) - \mathbf{K}_k\left(\mathbf{H}_k\mathbf{x}_k + \mathbf{v}_k - \mathbf{H}_k\hat{\mathbf{x}}_k^-\right)\right] \\ &\left[\left(\mathbf{x}_k - \hat{\mathbf{x}}_k^-\right) - \mathbf{K}_k\left(\mathbf{H}_k\mathbf{x}_k + \mathbf{v}_k - \mathbf{H}_k\hat{\mathbf{x}}_k^-\right)\right]^T\}\end{aligned} \tag{4.2.10}$$

Now, performing the indicated expectation and noting the $\left(\mathbf{x}_k - \hat{\mathbf{x}}_k^-\right)$ is the *a priori* estimation error that is uncorrelated with the current measurement error $\mathbf{v}_k$, we have

$$\mathbf{P}_k = (\mathbf{I} - \mathbf{K}_k\mathbf{H}_k)\mathbf{P}_k^-(\mathbf{I} - \mathbf{K}_k\mathbf{H}_k)^T + \mathbf{K}_k\mathbf{R}_k\mathbf{K}_k^T \tag{4.2.11}$$

Notice here that Eq. (4.2.11) is a perfectly general expression for the updated error covariance matrix, and it applies for any gain $\mathbf{K}_k$, suboptimal or otherwise.

Returning to the optimization problem, we wish to find the particular $\mathbf{K}_k$ that minimizes the individual terms along the major diagonal of $\mathbf{P}_k$, because these terms represent the estimation error variances for the elements of the state vector being estimated. The optimization can be done in a number of ways. We will do this using a straightforward differential calculus approach, and to do so we need two matrix differentiation formulas. They are

$$\frac{d[\text{trace}(\mathbf{AB})]}{d\mathbf{A}} = \mathbf{B}^T \quad (\mathbf{AB}\text{ must be square}) \tag{4.2.12}$$

$$\frac{d\left[\text{trace}\left(\mathbf{ACA}^T\right)\right]}{d\mathbf{A}} = 2\mathbf{AC} \quad (\mathbf{C}\text{ must be symmetric}) \tag{4.2.13}$$

where the derivative of a scalar with respect to a matrix is defined as

$$\frac{ds}{dA} = \begin{bmatrix} \frac{ds}{da_{11}} & \frac{ds}{da_{12}} & \cdots \\ \frac{ds}{da_{21}} & \frac{ds}{da_{22}} & \cdots \\ & \vdots & \end{bmatrix} \tag{4.2.14}$$

The two matrix differentiation formulas can be easily verified by writing out the indicated traces explicitly and then differentiating the results term by term. (This will be left as an exercise.)

We will now expand the general form for $\mathbf{P}_k$, Eq. (4.2.11), and rewrite it in the form:

$$\mathbf{P}_k = \mathbf{P}_k^- - \mathbf{K}_k\mathbf{H}_k\mathbf{P}_k^- - \mathbf{P}_k^-\mathbf{H}_k^T\mathbf{K}_k^T + \mathbf{K}_k\left(\mathbf{H}_k\mathbf{P}_k^-\mathbf{H}_k^T + \mathbf{R}_k\right)\mathbf{K}_k^T \tag{4.2.15}$$

Notice that the second and third terms are linear in $\mathbf{K}_k$ and that the fourth term is quadratic in $\mathbf{K}$. The two matrix differentiation formulas may now be applied to Eq. (4.2.15). We wish to minimize the trace of $\mathbf{P}$ because it is the sum of the mean-square errors in the estimates of all the elements of the state vector. We can use the argument here that the individual mean-square errors are also minimized when the total is minimized, provided that we have enough degrees of freedom in the variation of $\mathbf{K}_k$, which we do in this case. We proceed now to differentiate the trace of $\mathbf{P}_k$ with respect to $\mathbf{K}_k$, and we note that the trace of $\mathbf{P}_k^- \mathbf{H}_k^T \mathbf{K}_k^T$ is equal to the trace of its transpose $\mathbf{K}_k \mathbf{H}_k \mathbf{P}_k^-$. The result is

$$\frac{d(\text{trace}\,\mathbf{P}_k)}{d\mathbf{K}_k} = -2\left(\mathbf{H}_k \mathbf{P}_k^-\right)^T + 2\mathbf{K}_k\left(\mathbf{H}_k \mathbf{P}_k^- \mathbf{H}_k^T + \mathbf{R}_k\right) \tag{4.2.16}$$

We now set the derivative equal to zero and solve for the optimal gain. The result is

$$\mathbf{K}_k = \mathbf{P}_k^- \mathbf{H}_k^T \left(\mathbf{H}_k \mathbf{P}_k^- \mathbf{H}_k^T + \mathbf{R}_k\right)^{-1} \tag{4.2.17}$$

This particular $\mathbf{K}_k$, namely, the one that minimizes the mean-square estimation error, is called the *Kalman gain*.

The covariance matrix associated with the optimal estimate may now be computed. Referring to Eq. (4.2.11), we have

$$\mathbf{P}_k = (\mathbf{I} - \mathbf{K}_k \mathbf{H}_k)\mathbf{P}_k^- (\mathbf{I} - \mathbf{K}_k \mathbf{H}_k)^T + \mathbf{K}_k \mathbf{R}_k \mathbf{K}_k^T \tag{4.2.18}$$

$$= \mathbf{P}_k^- - \mathbf{K}_k \mathbf{H}_k \mathbf{P}_k^- - \mathbf{P}_k^- \mathbf{H}_k^T \mathbf{K}_k^T + \mathbf{K}_k\left(\mathbf{H}_k \mathbf{P}_k^- \mathbf{H}_k^T + \mathbf{R}_k\right)\mathbf{K}_k^T \tag{4.2.19}$$

Routine substitution of the optimal gain expression, Eq. (4.2.17), into Eq. (4.2.19) leads to

$$\mathbf{P}_k = \mathbf{P}_k^- - \mathbf{P}_k^- \mathbf{H}_k^T \left(\mathbf{H}_k \mathbf{P}_k^- \mathbf{H}_k^T + \mathbf{R}_k\right)^{-1} \mathbf{H}_k \mathbf{P}_k^- \tag{4.2.20}$$

or

$$\mathbf{P}_k = \mathbf{P}_k^- - \mathbf{K}_k\left(\mathbf{H}_k \mathbf{P}_k^- \mathbf{H}_k^T + \mathbf{R}_k\right)\mathbf{K}_k^T \tag{4.2.21}$$

or

$$\mathbf{P}_k = (\mathbf{I} - \mathbf{K}_k \mathbf{H}_k)\mathbf{P}_k^- \tag{4.2.22}$$

Note that we have four expressions for computing the updated $\mathbf{P}_k$ from the prior $\mathbf{P}_k^-$. Three of these, Eqs. (4.2.20), (4.2.21), and (4.2.22), are only valid for the optimal gain condition. However, Eq. (4.2.18) is valid for any gain, optimal or suboptimal. All four equations yield identical results for optimal gain with perfect arithmetic. We note, though, that in the real engineering world Kalman filtering is a numerical procedure, and some of the **P**-update equations may perform better numerically than others under unusual conditions. More will be said of this later in Section 4.9. For now, we will list the simplest update equation, that is, Eq. (4.2.22), as the usual way to update the error covariance.

We now have a means of assimilating the measurement at t_k by the use of Eq. (4.2.8) with $\mathbf{K}_k$ set equal to the Kalman gain as given by Eq. (4.2.17). Note that we need $\hat{\mathbf{x}}_k^-$ and $\mathbf{P}_k^-$ to accomplish this, and we can anticipate a similar need at the next step in order to make optimal use of the measurement $\mathbf{z}_{k+1}$. The updated estimated $\hat{\mathbf{x}}_k$ is easily projected ahead via the transition matrix. We are justified in ignoring the contribution of $\mathbf{w}_k$ in Eq. (4.2.1) because it has zero mean and is not correlated with any of the previous $\mathbf{w}$'s.* Thus, we have

$$\hat{\mathbf{x}}_{k+1}^- = \boldsymbol{\phi}_k \hat{\mathbf{x}}_k \tag{4.2.23}$$

The error covariance matrix associated with $\hat{\mathbf{x}}_{k+1}^-$ is obtained by first forming the expression for the *a priori* error

$$\begin{aligned} \mathbf{e}_{k+1}^- &= \mathbf{x}_{k+1} - \hat{\mathbf{x}}_{k+1}^- \\ &= (\boldsymbol{\phi}_k \mathbf{x}_k + \mathbf{w}_k) - \boldsymbol{\phi}_k \hat{\mathbf{x}}_k \\ &= \boldsymbol{\phi}_k \mathbf{e}_k + \mathbf{w}_k \end{aligned} \tag{4.2.24}$$

We now note that $\mathbf{w}_k$ and $\mathbf{e}_k$ have zero crosscorrelation, because $\mathbf{w}_k$ is the process noise for the step ahead of t_k. Thus, we can write the expression for $\mathbf{P}_{k+1}^-$ as

$$\begin{aligned} \mathbf{P}_{k+1}^- &= E\left[\mathbf{e}_{k+1}^- \mathbf{e}_{k+1}^{-T}\right] = E\left[(\boldsymbol{\phi}_k \mathbf{e}_k + \mathbf{w}_k)(\boldsymbol{\phi}_k \mathbf{e}_k + \mathbf{w}_k)^T\right] \\ &= \boldsymbol{\phi}_k \mathbf{P}_k \boldsymbol{\phi}_k^T + \mathbf{Q}_k \end{aligned} \tag{4.2.25}$$

We now have the needed quantities at time t_{k+1}, and the measurement $\mathbf{z}_{k+1}$ can be assimilated just as in the previous step.

Equations (4.2.8), (4.2.17), (4.2.22), (4.2.23), and (4.2.25) comprise the Kalman filter recursive equations. It should be clear that once the loop is entered, it can be continued *ad infinitum*. The pertinent equations and the sequence of computational steps are shown pictorially in Fig. 4.1. This summarizes what is now known as the *Kalman filter*.

Before we proceed to some examples, it is interesting to reflect on the Kalman filter in perspective. If you were to stumble onto the recursive process of Fig. 4.1

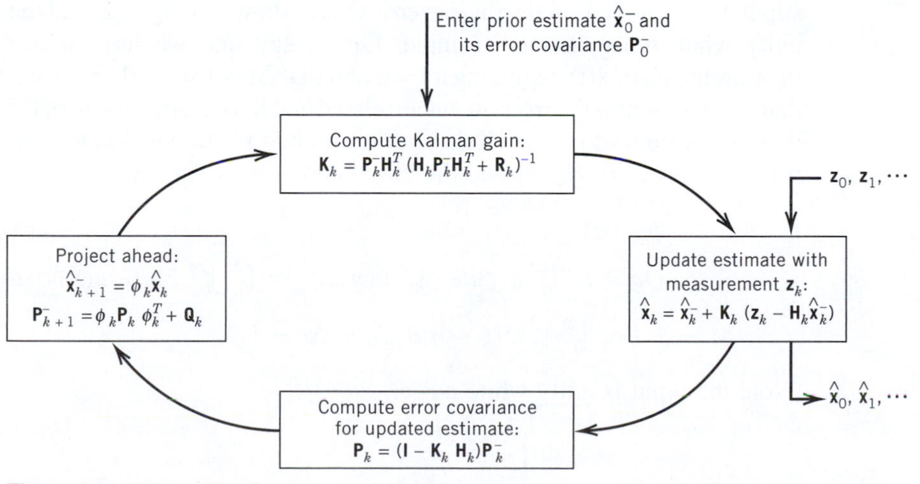

Figure 4.1 Kalman filter loop.

* Recall that in our notation $\mathbf{w}_k$ is the process noise that accumulates during the step ahead from t_k to t_{k+1}. This is purely a matter of notation (but an important one), and in some books it is denoted as $\mathbf{w}_{k+1}$ rather than $\mathbf{w}_k$. Consistency in notation is the important thing here. Conceptually, we are thinking of doing real-time filtering in contrast to smoothing, which we usually think of doing off-line (see Chapter 6).

without benefit of previous history, you might logically ask, "Why in the world did somebody call that a filter? It looks more like a computer algorithm." You would, of course, be quite right in your observation. The Kalman filter is just a computer algorithm for processing discrete measurements (the input) into optimal estimates (the output). Its roots, though, go back to the days when filters were made of electrical elements wired together in such a way as to yield the desired frequency response. The design was often heuristic. Wiener then came on the scene in the 1940s and added a more sophisticated type of filter problem. The end result of his solution was a filter weighting function or a corresponding transfer function in the complex domain. Implementation in terms of electrical elements was left as a further exercise for the designer. The discrete-time version of the Wiener problem remained unsolved (in a practical sense, at least) until Kalman's paper of 1960 (2). Even though his presentation appeared to be quite abstract at first glance, engineers soon realized that this work provided a practical solution to a number of unsolved filtering problems, especially in the field of navigation.

4.3 SIMPLE KALMAN FILTER EXAMPLES AND AUGMENTING THE STATE VECTOR

The basic recursive equations for the Kalman filter were presented in Section 4.2. We will now illustrate the use of these equations with two simple examples. The emphasis here is on modeling. Monte Carlo examples with simulated measurements will be considered later.

EXAMPLE 4.1 WIENER (BROWNIAN MOTION) PROCESS

The Wiener process is defined as integrated Gaussian white noise with the additional stipulation that the initial value is zero. This is shown in Fig. 4.2, and the input u(t) is unity white noise in this example. Let us say that we have uniformly spaced measurements of x(t) beginning a t = 0 and that Δt is 1 second. Also, we will assume that the measurement errors are uncorrelated and have an rms value of 0.5 m. The first thing we need to do in any Kalman filter problem is to develop the four key model parameters ϕ_k, Q_k, H_k, R_k and the initial conditions. They are:

$$\begin{aligned}
\phi_k &= 1 \\
Q_k &= E\left[\int_0^{\Delta t} u(\xi)d\xi \cdot \int_0^{\Delta t} u(\eta)d\eta\right] = \int_0^{\Delta t}\int_0^{\Delta t} E[u(\xi)u(\eta)]d\xi d\eta \\
&= \int_0^{\Delta t}\int_0^{\Delta t} \delta(\xi - \eta)d\xi d\eta = \Delta t = 1
\end{aligned}$$

(Note the input is unity white noise.)

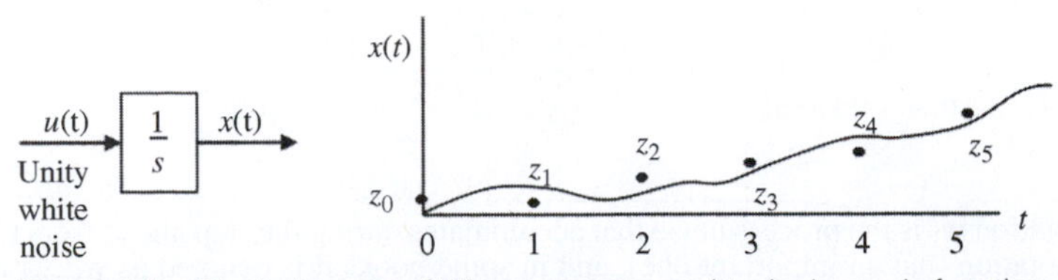

Figure 4.2 Block diagram of Wiener process and typical sample function of the process.

$$\begin{aligned} H_k &= 1 \\ R_k &= (0.5)^2 = 0.25 \end{aligned}$$

(All of these parameters are constant in this example so we will omit their subscripts from here on.)

Next, we need the initial conditions. From the problem statement we can say:

$$\begin{aligned} \hat{x}_0^- &= 0 \\ P_0^- &= 0 \end{aligned}$$

Now we can do the Kalman filter steps as indicated in Fig. 4.1. Begin with the first measurement z_0 at t = 0.

Step 1 Compute gain

$$K_0 = P_0^- H \left(H P_0^- H + R \right)^{-1} = 0 \cdot (R)^{-1} = 0$$

Step 2 Update estimate

$$\hat{x}_0 = \hat{x}_0^- + K_0 \left(z_0 - H \hat{x}_0^- \right) = 0 + 0 \cdot (z_0) = 0$$

Step 3 Update error covariance

$$P_0 = (I - K_0 H) P_0^- = (1 - 0) \cdot 0 = 0$$

(Of course, in this simple example we could have guessed the results of Steps 2 and 3 from the trivial initial conditions.)

Step 4 Project ahead to t = 1

$$\begin{aligned} \hat{x}_1^- &= \phi \hat{x}_0 = \phi \cdot 0 = 0 \\ P_1^- &= \phi P_0 \phi + Q = 1 \cdot 0 \cdot 1 + 1 = 1 \end{aligned}$$

The above steps can now be repeated at t = 1, and the z_1 measurement will be assimilated accordingly. This is left as an exercise. (Answer: $\hat{x}_1 = \frac{4}{5} z_1, P_1 = \frac{1}{5}$).

The above recursive process can now be repeated indefinitely as long as the measurement sequence continues. Note especially that the identities of the $z_1, z_2, \ldots$ measurements are not kept as we proceed in the recursive process. They are "thrown away," so to speak, after the information so contained is assimilated into the estimate.

■

EXAMPLE 4.2 AUGMENTING THE STATE VECTOR

As a variation on the scenario of Example 4.1 consider a situation where the integrator is driven by colored noise in contrast to white noise. To keep it simple let the input to the integrator be a Gauss-Markov process as shown in Fig. 4.3. This

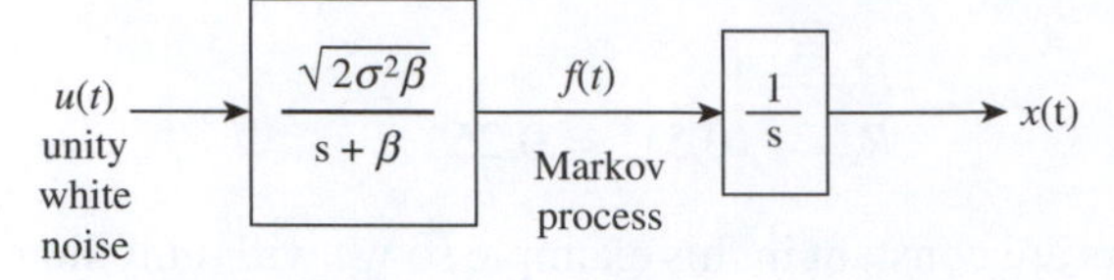

Figure 4.3 Integrator driven by Markov noise.

presents a problem, because if we try to model the Kalman filter with just one state variable, say $x(t)$, we find that the corresponding discrete-model w_k sequence will have nontrivial correlation in time. This violates one of the assumptions in the derivation of the Kalman filter recursive equations. Remember, there are very strict constraints in the Kalman filter model (Eqs. 4.2.3 through 4.2.5), and these must be followed carefully.

To obviate the difficulty just cited we can augment the original one-state model with an additional state variable, namely the Markov state. The resulting two-state model will then meet the model requirements. A proper model can then be developed as follows. Define the state variables (Fig. 4.3):

$$\begin{aligned} x_1 &= x(t) \\ x_2 &= f(t) \end{aligned}$$

Then the vector differential equation for the process is

$$\underbrace{\begin{bmatrix} \dot{x}_1 \\ \dot{x}_2 \end{bmatrix}}_{\dot{\mathbf{x}}} = \underbrace{\begin{bmatrix} 0 & 1 \\ 0 & -\beta \end{bmatrix}}_{\mathbf{F}} \underbrace{\begin{bmatrix} x_1 \\ x_2 \end{bmatrix}}_{\mathbf{x}} + \underbrace{\begin{bmatrix} 0 \\ \sqrt{2\sigma^2\beta} \end{bmatrix}}_{\mathbf{G}} u(t), \quad u(t) = \text{unity white noise}$$

and the discrete measurement equation is:

$$z_k = \underbrace{\begin{bmatrix} 1 & 0 \end{bmatrix}}_{\mathbf{H}} \begin{bmatrix} x_1 \\ x_2 \end{bmatrix}_k + v_k, \quad \mathrm{E}\left[v_k^2\right] = R_k$$

To complete the discrete process model we need to compute $\boldsymbol{\phi}$ and $\mathbf{Q}$. These are easily computed using the Van Loan method (see Section 3.9, Chapter 3). Using the notation in the referenced Section 3.9, the $\mathbf{A}$ matrix is:

$$\mathbf{A} = \Delta t \left[\begin{array}{c|c} -\mathbf{F} & \mathbf{GWG}^{\mathrm{T}} \\ \hline \mathbf{0} & \mathbf{F}^{\mathrm{T}} \end{array} \right]$$

where $\mathbf{W}$ is unity. To get a random walk process that is similar to the Wiener process of Example 4.1 (but not exactly the same) we could try letting;

$$\begin{aligned} \sigma^2 &= 1\ \mathrm{m}^2 \\ \beta &= 1\ \mathrm{rad/s} \\ \Delta t &= 1\ \mathrm{s} \end{aligned}$$

The Van Loan method then yields:

$$\boldsymbol{\phi} = \begin{bmatrix} 1.0000 & 0.6321 \\ 0 & 0.3679 \end{bmatrix}, \quad \mathbf{Q} = \begin{bmatrix} 0.3362 & 0.3996 \\ 0.3996 & 0.8647 \end{bmatrix}$$

The final step in model development is to specify filter initial conditions. Let us say that the integrator is zeroed initially, but the Markov process is in steady-state condition at $t = 0$. In that case:

$$\hat{\mathbf{x}}_0^- = \begin{bmatrix} 0 \\ 0 \end{bmatrix}, \quad \mathbf{P}_0^- = \begin{bmatrix} 0 & 0 \\ 0 & \sigma^2 \end{bmatrix}$$

If we were to now generate a sequence of measurements in accordance with the stated parameters using Monte Carlo methods, we could process the measurements recursively and obtain a corresponding sequence of estimates of both states x_1 and x_2.

4.4 MARINE NAVIGATION APPLICATION WITH MULTIPLE-INPUTS/ MULTIPLE-OUTPUTS

It was mentioned earlier that one of the advantages of the Kalman filter over Wiener methods lies in the convenience of handling multiple-input/multiple-output applications. We will now look at an example of this. Just as in Example 4.2, it will be seen that it is sometimes necessary to expand the original state model in order to achieve a suitable model that will satisfy the strict requirements for a Kalman filter. The example that we will consider is based on a paper by B. E. Bona and R. J. Smay that was published in 1966 (3). This paper is of some historical importance, because it was one of the very early applications of real-time Kalman filtering in a terrestial navigation setting. For tutorial purposes, we will consider a simplified version of the state model used in the Bona–Smay paper.

In marine applications, and especially in the case of submarines, the mission time is usually long, and the ship's inertial navigation system (INS) must operate for long periods without the benefit of position fixes. The major source of position error during such periods is gyro drift. This, in turn, is due to unwanted biases on the axes that control the orientation of the platform (i.e., the inertial instrument cluster). These "biases" may change slowly over long time periods, so they need to be recalibrated occasionally. This is difficult to do at sea, because the biases are hidden from direct one-to-one measurement. One must be content to observe them indirectly through their effect on the inertial system's outputs. Thus, the main function of the Kalman filter in this application is to estimate the three gyro biases and platform azimuth error, so they can be reset to zero. (In this application, the platform tilts are kept close to zero by the gravity vector and by damping the Schuler oscillation with external velocity information from the ship's log.)

In our simplified model, the measurements will be the inertial system's two horizontal position errors, that is, latitude error (N-S direction) and longitude error (E-W direction). These are to be obtained by comparing the INS output with position as determined independently from other sources such as a satellite navigation system, or perhaps from a known (approximately) position at dockside. The mean-square errors associated with external reference are assumed to be known, and they determine the numerical values assigned to the $\mathbf{R}_k$ matrix of the Kalman filter.

The applicable error propagation equations for a damped inertial navigation system in a slow-moving vehicle are*

$$\dot{\psi}_x - \Omega_z \psi_y = \varepsilon_x \tag{4.4.1}$$

$$\dot{\psi}_y + \Omega_z \psi_x - \Omega_x \psi_z = \varepsilon_y \tag{4.4.2}$$

$$\dot{\psi}_z + \Omega_x \psi_y = \varepsilon_z \tag{4.4.3}$$

where x, y, and z denote the platform coordinate axes in the north, west, and up directions, and

$\psi_x =$ inertial system's west position error (in terms of great circle arc distance in radians)

$\psi_y =$ inertial system's south position error (in terms of great circle arc distance in radians)

$\psi_z =$ [platform azimuth error] – [west position error] · [tan (latitude)]

Also,

$$\Omega_x = x \text{ component of earth rate } \Omega \quad [\text{i.e., } \Omega_x = \Omega \cos(\text{lat.})]$$

$$\Omega_z = z \text{ component of earth rate } \Omega \quad [\text{i.e., } \Omega_z = \Omega \sin(\text{lat.})]$$

and

$$\varepsilon_x, \varepsilon_y, \varepsilon_z = \text{gyro drift rates for the } x, y, \text{ and } z \text{ axis gyros}$$

We assume that the ship's latitude is known approximately; therefore, Ω_x and Ω_z are known and may be assumed to be constant over the observation interval.

Nonwhite Forcing Functions

The three differential equations, Eqs. (4.4.1) through (4.4), represent a third-order linear system with the gyro drift rates, ε_x, ε_y, ε_z, as the forcing functions. These will be assumed to be random processes. However, they certainly are not white noises in this application. Quite to the contrary, they are processes that vary very slowly with

* Equations (4.4.1) to (4.4.3) are certainly not obvious, and a considerable amount of background in inertial navigation theory is needed to understand the assumptions and approximations leading to this simple set of equations (4,5). We do not attempt to derive the equations here. For purposes of understanding the Kalman filter, simply assume that these equations do, in fact, accurately describe the error propagation in this application and proceed on to the details of the Kalman filter.

time—just the opposite of white noise. Thus, if we were to discretize this third-order system of equations using relatively small sampling intervals, we would find that the resulting sequence of $\mathbf{w}_k$'s would be highly correlated. This would violate the white-sequence assumption that was used in deriving the filter recursive equations (see Eq. 4.2.3). The solution for this is to expand the size of the model and include the forcing functions as part of the state vector. They can be thought of as the result of passing fictitious white noises through a system with linear dynamics. This yields an additional system of equations that can be appended to the original set; in the expanded or augmented set of equations, the new forcing functions will be white noises. We are assured then that when we discretize the augmented system of equations, the resulting $\mathbf{w}_k$ sequence will be white.

In the interest of simplicity, we will model; ε_x, ε_y, and ε_z as Gaussian random-walk processes. This allows the "biases" to change slowly with time. Each of the gyro biases can then be thought of as the output of an integrator as shown in Fig. 4.4. The three differential equations to be added to the original set are then

$$\dot{\varepsilon}_x = f_x \tag{4.4.4}$$

$$\dot{\varepsilon}_y = f_y \tag{4.4.5}$$

$$\dot{\varepsilon}_z = f_z \tag{4.4.6}$$

where f_x, f_y, and f_z are independent white noise processes with power spectral densities equal to W.

We now have a six-state system of linear equations that can be put into the usual state-space form.

$$\begin{bmatrix} \dot{x}_1 \\ \dot{x}_2 \\ \dot{x}_3 \\ \hline \dot{x}_4 \\ \dot{x}_5 \\ \dot{x}_6 \end{bmatrix} = \left[\begin{array}{ccc|c} 0 & \Omega_z & 0 & \\ -\Omega_z & 0 & \Omega_x & \mathbf{I} \\ 0 & -\Omega_x & 0 & \\ \hline & \mathbf{0} & & \mathbf{0} \end{array}\right] \begin{bmatrix} x_1 \\ x_2 \\ x_3 \\ \hline x_4 \\ x_5 \\ x_6 \end{bmatrix} + \left[\begin{array}{c|c} \mathbf{0} & \mathbf{0} \\ \hline \mathbf{0} & \mathbf{I} \end{array}\right] \begin{bmatrix} 0 \\ 0 \\ 0 \\ \hline f_x \\ f_y \\ f_z \end{bmatrix} \tag{4.4.7}$$

The process dynamics model is now in the proper form for a Kalman filter. It is routine to convert the continuous model to discrete form for a given Δt step size. The key parameters in the discrete model are $\boldsymbol{\phi}_k$ and $\mathbf{Q}_k$, and methods for calculating these are given in Section 3.9.

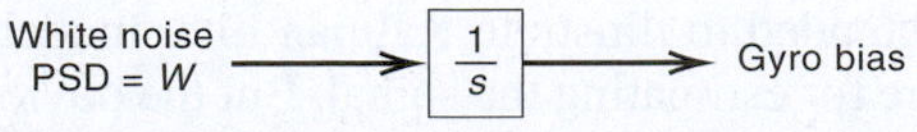

White noise PSD = W → $\boxed{\frac{1}{s}}$ → Gyro bias

Figure 4.4 Random walk model for gyro bias.

As mentioned previously, we will assume that there are only two measurements available to the Kalman filter at time t_k. They are west position error ψ_x and south position error ψ_y. The matrix measurement equation is then

$$\begin{bmatrix} z_1 \\ z_2 \end{bmatrix}_k = \underbrace{\begin{bmatrix} 1 & 0 & 0 & 0 & 0 & 0 \\ 0 & 1 & 0 & 0 & 0 & 0 \end{bmatrix}}_{\mathbf{H}_k} \begin{bmatrix} x_1 \\ x_2 \\ x_3 \\ x_4 \\ x_5 \\ x_6 \end{bmatrix} + \begin{bmatrix} v_1 \\ v_2 \end{bmatrix}_k \qquad (4.4.8)$$

The measurement model is now complete except for specifying $\mathbf{R}_k$ that describes the mean-square errors associated with the external position fixes. The numerical values will, of course, depend on the particular reference system being used for the fixes.

Nonwhite Measurement Noise

We have just seen an example where it was necessary to expand the state model because the random forcing functions were not white. A similar situation can also occur when the measurement noise is not white. This would also violate one of the assumptions used in the derivation of the Kalman filter equations (see Eq. 4.2.4). The correlated measurement-error problem can also be remedied by augmenting the state vector, just as was done in the preceding gyro-calibration application. The correlated part of the measurement noise is simply moved from $\mathbf{v}_k$ into the state vector, and $\mathbf{H}_k$ is changed accordingly. It should be noted, though, that if the white noise part of $\mathbf{v}_k$ was zero in the original model, then in the new model, after augmentation, the measurement noise will be zero. In effect, the model is saying that there exists a perfect measurement of certain linear combinations of state variables. The $\mathbf{R}_k$ matrix will then be singular. Technically, this is permissible in the discrete Kalman filter, provided that the $\mathbf{P}_k^-$ matrix that has been projected ahead from the previous step is positive definite, and the measurement situation is not trivial. The key requirement for permitting a zero $\mathbf{R}_k$ is that $\left(\mathbf{H}_k\mathbf{P}_k^-\mathbf{H}_k^T + \mathbf{R}_k\right)$ be invertible in the gain-computation step. Even so, there is some risk of numerical problems when working with "perfect" measurements. In off-line analysis this problem can be easily avoided by simply letting $\mathbf{R}_k$ be small relative to the $\mathbf{H}_k\mathbf{P}_k^-\mathbf{H}_k^T$ term.

4.5 GAUSSIAN MONTE CARLO EXAMPLES

The previous examples were intended to illustrate Kalman filter modeling and the step-by-step recursive procedure for estimating the signal. But the obvious question still remains: "Does the filter really do a good job of separating the signal from the noise?" The answer to this question is a resounding yes, and we will now present two examples to demonstrate this. In these examples the filter operates on simulated measurement data that is generated with Monte Carlo methods. This is relatively easy to do with MATLAB, which has an excellent normal random number generator. Also, once the MATLAB code is written for a particular scenario, it is then easy to make various runs with different numerical values for selected parameters such as R_k. This kind of simulation can be most helpful in gaining some insight into the filter's effectiveness under various conditions. Our first example will be for a signal that is first-order Gauss-Markov.

EXAMPLE 4.3 FIRST-ORDER GAUSS–MARKOV PROCESS

Consider a stationary first-order Gauss–Markov process whose autocorrelation function is

$$R_x(\tau) = \sigma^2 e^{-\beta|\tau|}$$

where

$$\sigma^2 = 1\,\text{m}^2$$
$$\beta = 0.1\,\text{rad/s}$$

This is the same process that is illustrated in Chapter 2, Fig. 2.16. Using the methods discussed in Section 3.9, we see that the continuous state equation is:

$$\dot{x} = -0.1x + \sqrt{0.2}u(t) \tag{4.5.2}$$

where $u(t)$ is unity white noise.

Let us say that we have a sequence of 51 discrete noisy measurements of the process that begin at $t = 0$ and end at $t = 50$ s, and they are all equally spaced. Therefore,

$$\Delta t = 1.0\,\text{s} \tag{4.5.3}$$

We will further assume that the measurement errors are uncorrelated and their standard deviation is 0.5 m (a relatively coarse measurement for this example). The key process parameters can then be written out explicitly for this scalar example, and they are (with reference to Eq. (3.9.20), Section 3.9):

$$\phi_k = e^{-\beta\Delta t} = e^{-0.1} \approx 0.9048 \tag{4.5.4}$$

$$Q_k = E\left(w_k^2\right) = \sigma^2\left(1 - e^{-2\beta\Delta t}\right) = \left(1 - e^{-0.2}\right) \approx 0.1813 \tag{4.5.5}$$

Clearly, the measurement parameters are obvious from the problem statement:

$$H_k = 1 \tag{4.5.6}$$

$$R_k = (0.5)^2 = 0.25\,\text{m}^2 \tag{4.5.7}$$

We now have the key filter parameters and are ready to begin the recursive process at $t = 0$. The *a priori* estimate of x is just its mean, because we are assuming that the only prior knowledge about x is that it is Gauss–Markov with known σ and β parameters. Therefore,

$$\hat{x}_0^- = 0 \tag{4.5.8}$$

The error covariance associated with this initial estimate is just the variance of x, which in this case is

$$P_0^- = \sigma^2 = 1\ \text{m}^2 \tag{4.5.9}$$

Presumably, the Monte Carlo samples of the true x and the corresponding z_k samples were generated separately and stored before running the filter. So, we now have everything needed to run the filter through 51 steps begin at $t = 0$ and ending at $t = 50$. The results are shown in Fig. 4.5. Statistically, about 95% of the estimates should fall within the $\pm 2\sigma$ bounds, and that is consistent with what we see in the upper plot in Fig. 4.5. The lower plot in the figure shows a relatively rapid convergence of the rms estimation error to its steady-state value of about 0.37 m. This is an improvement over the raw rms measurement error of 0.5 m, but the improvement is not dramatic. This is due to the relatively large Q_k, which represents the variance of new uncertainty that is introduced into the x process with each 1-s step. The only solution to this (all other parameters being held constant) is to increase the measurement sampling rate.

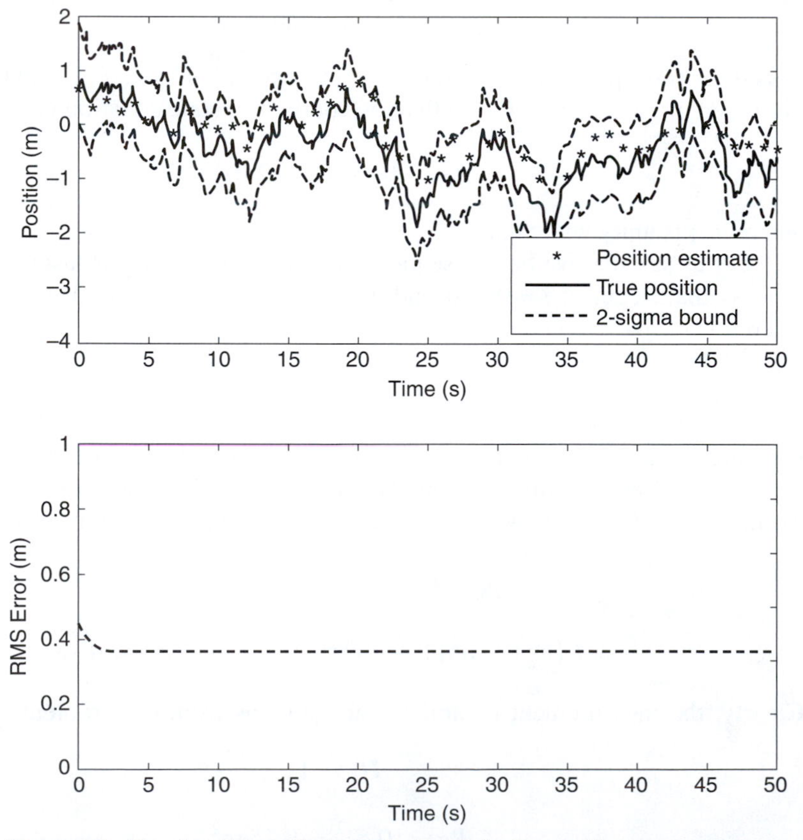

Figure 4.5 Results of Monte Carlo run for first-order Gauss–Markov process.

■

EXAMPLE 4.4 SECOND-ORDER GAUSS–MARKOV PROCESS.
The second-order Gauss–Markov process described in Example 2.12, Chapter 2, will be used for our next Kalman filter example. The parameters for the process have been chosen to make the process similar to the first-order process used for Example 4.3. This was done for the purpose of comparison. It should be remembered, though, that there is no way to make the processes identical in all respects. Their sigmas were chosen to be the same, and their low-frequency spectral characteristics are similar. Also, the measurement situation is the same for both examples.

Recall from Example 2.12 that the second-order process had a PSD of the form:

$$S_x(j\omega) = \frac{b^2\sigma^2}{\omega^4 + \omega_0^4} \tag{4.5.10}$$

where

$$\begin{aligned}
\sigma^2 &= 1\ \text{m}^2 (\text{mean-square-value of the } x \text{ process}) \\
\omega_0 &= 0.1\ \text{rad/s} \\
b^2 &= 2\sqrt{2}\omega_0^3
\end{aligned}$$

We can do spectral factorization and rewrite $S_x(j\omega)$ in product form (after replacing $j\omega$ with s).

$$S_x(s) = \frac{b\sigma}{s^2 + \sqrt{2}\omega_0 s + \omega_0^2} \cdot \frac{b\sigma}{(-s)^2 + \sqrt{2}\omega_0(-s) + \omega_0^2} \tag{4.5.11}$$

The shaping filter that shapes unity white noise into x is obtained by inspection of the positive-time part of $S_x(s)$ and this is shown in Fig. 4.6 (see Section 3.6).

From the transfer function shown in the figure, we get the second-order differential equation:

$$\ddot{x} + \sqrt{2}\omega_0\dot{x} + \omega_0^2 x = b\sigma u \tag{4.5.12}$$

We now choose phase variables for our state variables (i.e., position and velocity). This leads to the continuous-time state model:

$$\begin{bmatrix} \dot{x}_1 \\ \dot{x}_2 \end{bmatrix} = \underbrace{\begin{bmatrix} 0 & 1 \\ -\omega_0^2 & -\sqrt{2}\omega_0 \end{bmatrix}}_{\mathbf{F}} \begin{bmatrix} x_1 \\ x_2 \end{bmatrix} + \underbrace{\begin{bmatrix} 0 \\ b\sigma \end{bmatrix}}_{\mathbf{G}} u \tag{4.5.13}$$

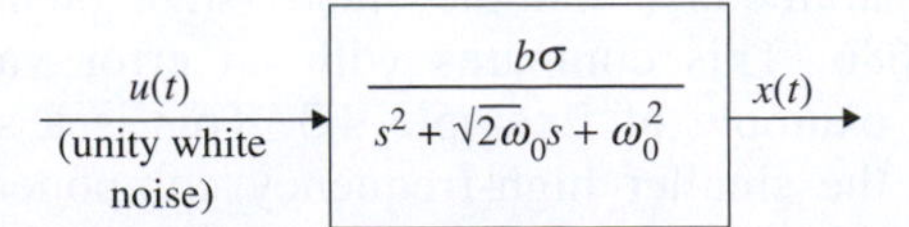

Figure 4.6 Shaping filter for Example 4.4.

The differential equation in this example is second-order, so writing out explicit equations for $\boldsymbol{\phi}_k$ and $\mathbf{Q}_k$ in terms of Δt is a bit messy. So, instead we will use the Van Loan method of evaluating $\boldsymbol{\phi}_k$ and $\mathbf{Q}_k$ (see Section 3.9). The required "inputs" for the Van Loan method are $\mathbf{F}$, $\mathbf{G}$, $\mathbf{W}$, and Δt. $\mathbf{F}$ and $\mathbf{G}$ are shown in Eq. (4.3.18), and W is just the PSD amplitude of the white noise forcing function. This is unity. Also, we will let $\Delta t = 1$ s just as in Example 4.3. So we now use these input values in the Van Loan algorithm, and $\boldsymbol{\phi}_k$ and $\mathbf{Q}_k$ work out to be (with rounding):

$$\boldsymbol{\phi}_k = \begin{bmatrix} 0.9952 & 0.9310 \\ -0.0093 & 0.8638 \end{bmatrix} \tag{4.5.14}$$

$$\mathbf{Q}_k = \begin{bmatrix} 0.000847 & 0.001226 \\ 0.001226 & 0.002456 \end{bmatrix} \tag{4.5.15}$$

In our Kalman filter we will assume that the noisy measurement is of $\mathbf{x}_1$ (i.e., position), and the measurement error sigma is 0.5 m just as in Example 4.3 Therefore, the $\mathbf{H}$ and $\mathbf{R}$ parameters are:

$$\mathbf{H}_k = [1 \quad 0] \tag{4.5.16}$$

$$\mathbf{R}_k = (0.5)^2 = 0.25 \tag{4.5.17}$$

We have now determined the four key parameters for our 2-state Kalman filter. The only remaining items to be specified before running a Monte Carlo simulation are the initial conditions. Presumably, we know nothing about the x process initially other than its spectral function and that it is Gauss–Markov. Therefore, the initial estimate vector is just zero, i.e.,

$$\hat{\mathbf{x}}_0^- = \begin{bmatrix} 0 \\ 0 \end{bmatrix} \tag{4.5.18}$$

Also, because of our choice of phase variables as the state variables, we can say with confidence that x and $\dot{x}$ are uncorrelated (see Problem 3.18). Therefore,

$$\mathbf{P}_0^- = \begin{bmatrix} \sigma^2 & 0 \\ 0 & (\omega_0 \sigma)^2 \end{bmatrix} \tag{4.5.19}$$

It is now a routine matter to run a Monte Carlo simulation for 50 steps. The results for a typical run are shown in Fig. 4.7. The filter reaches steady-state fairly rapidly for the chosen parameters, and the steady-state position error variance is approximately 0.0686. This compares with an error variance of 0.1346 for the first-order filter example of Example 4.3. This is a significant improvement, and it is due to the smaller high-frequency components in the second-order process as compared with those in the first-order Gauss–Markov process.

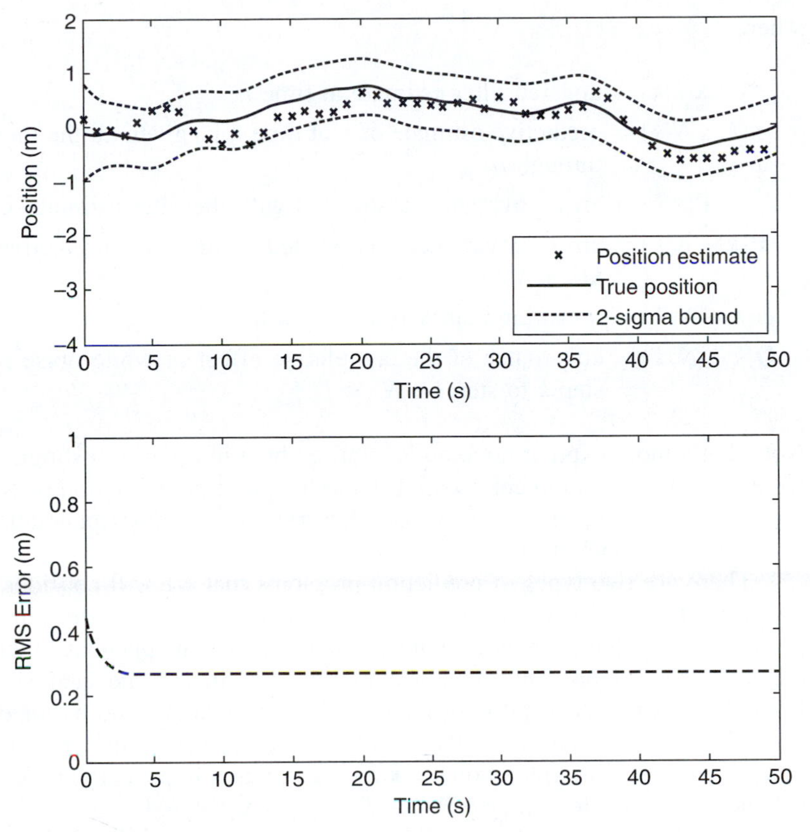

Figure 4.7 Sample Monte Carlo run of the simulated Kalman filter for the second-order Gauss–Markov process.

4.6 PREDICTION

In the older Wiener theory the basic theoretical problem was posed in rather general terms. Say $x(t)$ was the random process to be estimated. Then the problem to be solved was to find the least-squares estimate of $x(t+\alpha)$ in the presence of noise. The α parameter could be either positive (prediction), zero (filtering), or negative (smoothing). The parenthetical words here are the descriptive terms that were used for the three respective cases of the α parameter, and the same terminology continues today. So far, in this chapter we have only considered filtering (i.e., $\alpha=0$). We will now look at prediction as viewed in the context of Kalman filtering. (We will consider smoothing later in Chapter 6).

Extension of Kalman filtering to prediction is straightforward. We first note that the projection step in the filter loop shown in Fig. 4.1 is, in fact, one-step prediction. This was justified on the basis of the white noise assumption for the w_k sequence in the process model (Eq. 4.2.1). We can use the same identical argument for projecting (i.e., predicting) N steps ahead of the current measurement. The obvious equations for N-step prediction are then:

$$\hat{\mathbf{x}}(k+N|k) = \boldsymbol{\phi}(k+N,k)\hat{\mathbf{x}}(k|k) \tag{4.6.1}$$

$$\mathbf{P}(k+N|k) = \boldsymbol{\phi}(k+N,k)\mathbf{P}(k|k)\boldsymbol{\phi}^T(k+N,k) + \mathbf{Q}(k+N,k) \tag{4.6.2}$$

where

$$
\begin{aligned}
\hat{\mathbf{x}}(k|k) &= \text{updated filter estimate at time } t_k \\
\hat{\mathbf{x}}(k+N|k) &= \text{predictive estimate of } \mathbf{x} \text{ at time } t_{k+N} \text{ given all the measurements through } t_k \\
\mathbf{P}(k|k) &= \text{error covariance associated with the filter estimate } \hat{\mathbf{x}}(k|k) \\
\mathbf{P}(k+N|k) &= \text{error covariance associated with the predictive estimate } \hat{\mathbf{x}}(k+N|k) \\
\boldsymbol{\phi}(k+N,k) &= \text{transition matrix from step } k \text{ to } k+N \\
\mathbf{Q}(k+N,k) &= \text{covariance of the cumulative effect of white noise inputs from step } k \text{ to step } k+N
\end{aligned}
$$

Note that a more explicit notation is required here in order to distinguish between the end of the measurement stream (k) and the point of estimation ($k+N$). (These were the same in the filter problem, and thus a shortened subscript notation could be used without ambiguity.)

There are two types of prediction problems that we will consider:

Case 1: Case 1 is where N is fixed and k evolves in integer steps in time just as in the filter problem. In this case, the predictor is just an appendage that we add to the usual filter loop. This is shown in Fig. 4.8. In off-line analysis work, the $\mathbf{P}(k+N|k)$ matrix is of primary interest. The terms along the major diagonal of $\mathbf{P}(k+N|k)$ give a measure of the quality of the predictive state estimate. On the other hand, in on-line prediction it is $\hat{\mathbf{x}}(k+N|k)$ that is of primary interest. Note that it is not necessary to compute $\mathbf{P}(k+N|k)$ to get $\hat{\mathbf{x}}(k+N|k)$.

Case 2: Case 2 is where we fix k and then compute $\hat{\mathbf{x}}(k+N|k)$ and its error covariance for ever-increasing prediction times, that is, $N=1, 2, 3, \ldots$, etc. The error covariance is of special interest here, because it tells us how the predictive estimate degrades as we reach out further and further into the future. We will now consider an example that illustrates this kind of prediction problem.

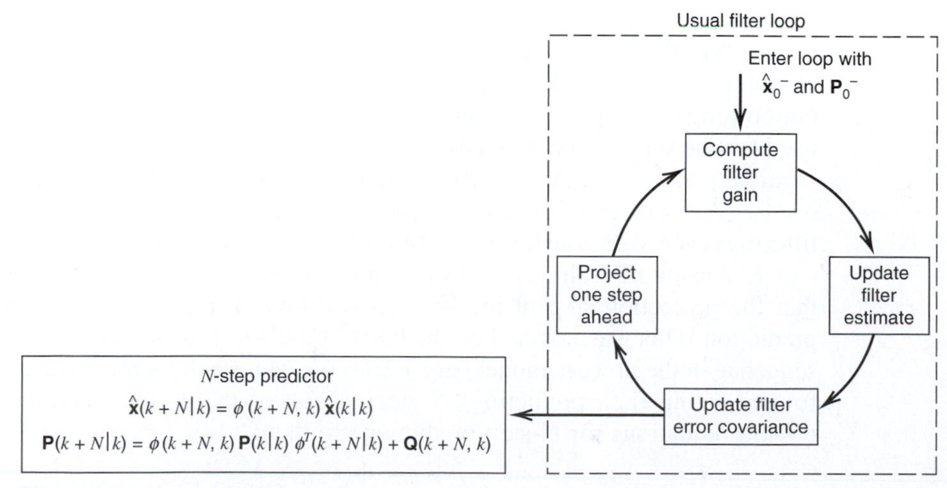

Figure 4.8 *N*-step prediction.

EXAMPLE 4.5 A PREDICTION SIMULATION

Let us return to the second-order Markov process that was considered in Example 4.4. In the Monte Carlo simulation there, the filter had clearly reached a steady-state condition at the end of the run (i.e., 50 sec). Suppose at that point in time we wish to predict ahead and get a position estimate and its error covariance for another 50 steps. This would be a Case 2 situation. However, rather than use the method shown in Fig. 4.8 (where we would recompute $\hat{\mathbf{x}}_k$ and $\mathbf{P}_k$ for ever-increasing time spans), we can use an alternative approach for off-line simulation. We can accomplish the desired result using the usual filter equations for the steps after $t = 50$, but with $\mathbf{R}_k$ set at an extremely large value, rather than the 0.25 value used earlier in the filter part of the run where there were legitimate measurements available for processing. This is equivalent to saying that the measurements after $t = 50$ are worthless, and the filter will automatically give them zero weight during the prediction period. This is what prediction is all about—making do with past, but not current, information.

The results of such a simulation are shown in Fig. 4.9. There the same Monte Carlo run that was used in Example 4.4 was simply extended 50 more steps with $\mathbf{R}_k$ switched to a very large number at t=51 and beyond. As one would expect, the rms estimation error increases during the prediction period, and the simulated estimate relaxes exponentially to zero (the unconditional mean of $\mathbf{x}$) during the same period.

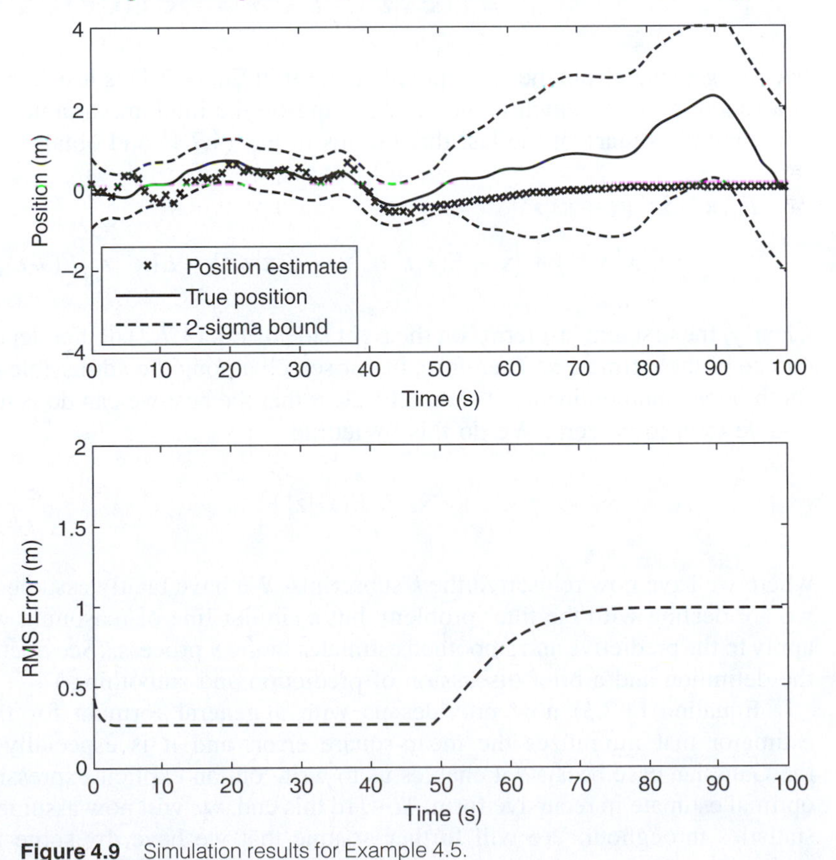

Figure 4.9 Simulation results for Example 4.5.

4.7 THE CONDITIONAL DENSITY VIEWPOINT

In our discussion thus far, we have used minimum mean-square error as the performance criterion, and we have assumed a linear form for the filter. This was partly a matter of convenience, but not entirely so, because we will see presently that the resulting linear filter has more far-reaching consequences than are apparent at first glance. This is especially so in the Gaussian case. We now elaborate on this.

We first show that if we choose as our estimate the mean of $\mathbf{x}_k$ conditioned on the available measurement stream, then this estimate will minimize the mean-square error. This is a somewhat restrictive form of what is sometimes called the fundamental theorem of estimation theory (6,7). The same notation and model assumptions that were used in Section 4.2 will be used here, and our derivation follows closely that given by Mendel (6). Also, to save writing, we will temporarily drop the k subscripts, and we will denote the complete measurement stream $\mathbf{z}_0$, $\mathbf{z}_1, \ldots, \mathbf{z}_k$ simply as $\mathbf{z}^*$. We first write the mean-square estimation error of $\mathbf{x}$, conditioned on $\mathbf{z}^*$, as

$$\begin{aligned} E\left[(\mathbf{x}-\hat{\mathbf{x}})^T(\mathbf{x}-\hat{\mathbf{x}})|\mathbf{z}^*\right] &= E\left[\left(\mathbf{x}^T\mathbf{x}-\mathbf{x}^T\hat{\mathbf{x}}-\hat{\mathbf{x}}^T\mathbf{x}+\hat{\mathbf{x}}^T\hat{\mathbf{x}}\right)|\mathbf{z}^*\right] \\ &= E(\mathbf{x}^T\mathbf{x}|\mathbf{z}^*)-E(\mathbf{x}^T|\mathbf{z}^*)\hat{\mathbf{x}}-\hat{\mathbf{x}}^T E(\mathbf{x}|\mathbf{z}^*)+\hat{\mathbf{x}}^T\hat{\mathbf{x}} \end{aligned} \tag{4.7.1}$$

Factoring $\hat{\mathbf{x}}$ away from the expectation operator in Eq. (4.7.1) is justified, because $\hat{\mathbf{x}}$ is a function of $\mathbf{z}^*$, which is the conditioning on the random variable $\mathbf{x}$. We now complete the square of the last three terms in Eq. (4.7.1) and obtain

$$\begin{aligned} &E\left[(\mathbf{x}-\hat{\mathbf{x}})^T(\mathbf{x}-\hat{\mathbf{x}})|\mathbf{z}^*\right] \\ &\quad = E(\mathbf{x}^T\mathbf{x}|\mathbf{z}^*)+[\hat{\mathbf{x}}-E(\mathbf{x}|\mathbf{z}^*)]^T[\hat{\mathbf{x}}-E(\mathbf{x}|\mathbf{z}^*)]-E(\mathbf{x}^T|\mathbf{z}^*)E(\mathbf{x}|\mathbf{z}^*) \end{aligned} \tag{4.7.2}$$

Clearly, the first and last terms on the right side of Eq. (4.7.2) do not depend on our choice of the estimate $\hat{\mathbf{x}}$. Therefore, in our search among the admissible estimators (both linear and nonlinear), it should be clear that the best we can do is to force the middle term to be zero. We do this by letting

$$\hat{\mathbf{x}}_k = E\left(\mathbf{x}_k|\mathbf{z}_k^*\right) \tag{4.7.3}$$

where we have now reinserted the k subscripts. We have tacitly assumed here that we are dealing with the filter problem, but a similar line of reasoning would also apply to the predictive and smoothed estimates of the $\mathbf{x}$ process. (See Section 4.6 for the definition and a brief discussion of prediction and smoothing.)

Equation (4.7.3) now provides us with a general formula for finding the estimator that minimizes the mean-square error, and it is especially useful in the Gaussian case because it enables us to write out an explicit expression for the optimal estimate in recursive form. Toward this end, we will now assume Gaussian statistics throughout. We will further assume that we have, by some means, an optimal prior estimate $\hat{\mathbf{x}}_k^-$ and its associated error covariance $\mathbf{P}_k^-$. Now, at this point we will stretch our notation somewhat and let $\mathbf{x}_k$ denote the $\mathbf{x}$ random variable at t_k

conditioned on the measurement stream $\mathbf{z}_{k-1}^*$. We know that the form of the probability density of $\mathbf{x}_k$ is then

$$f_{x_k} \sim \mathcal{N}\left(\hat{\mathbf{x}}_k^-, \mathbf{P}_k^-\right) \tag{4.7.4}$$

Now, from our measurement model we know that $\mathbf{x}_k$ is related to $\mathbf{z}_k$ by

$$\mathbf{z}_k = \mathbf{H}_k\mathbf{x}_k + \mathbf{v}_k \tag{4.7.5}$$

Therefore, we can immediately write the density function for $\mathbf{z}_k$ as

$$f_{\mathbf{z}_k} \sim \mathcal{N}\left(\mathbf{H}_k\hat{\mathbf{x}}_k^-, \mathbf{H}_k\mathbf{P}_k^-\mathbf{H}_k^T + \mathbf{R}_k\right) \tag{4.7.6}$$

(Again, remember that conditioning on $\mathbf{z}_{k-1}^*$ is implied.) Also, from Eq. (4.7.5) we can write out the form for the conditional density of $\mathbf{z}_k$, given $\mathbf{x}_k$. It is

$$f_{\mathbf{z}_k|\mathbf{x}_k} \sim \mathcal{N}(\mathbf{H}_k\mathbf{x}_k, \mathbf{R}_k) \tag{4.7.7}$$

Finally, we can now use Bayes formula and write

$$f_{\mathbf{x}_k|\mathbf{z}_k} = \frac{f_{\mathbf{z}_k|\mathbf{x}_k} f_{\mathbf{x}_k}}{f_{\mathbf{z}_k}} \tag{4.7.8}$$

where the terms on the right side of the equation are given by Eqs. (4.7.4), (4.7.6), and (4.7.7). But recall that $\mathbf{x}_k$ itself was conditioned on $\mathbf{z}_0, \mathbf{z}_1, \ldots, \mathbf{z}_{k-1}$. Thus, the density function on the left side of Eq. (4.7.8) is actually the density of the usual random variable $\mathbf{x}_k$, conditioned on the whole measurement stream up through $\mathbf{z}_k$. So, we will change the notation slightly and rewrite Eq. (4.7.8) as

$$f_{\mathbf{x}_k|\mathbf{z}_k} = \frac{[\mathcal{N}(\mathbf{H}_k\mathbf{x}_k, \mathbf{R}_k)]\left[\mathcal{N}\left(\hat{\mathbf{x}}_k^-, \mathbf{P}_k^-\right)\right]}{\left[\mathcal{N}\left(\mathbf{H}_k\mathbf{x}_k^-, \mathbf{H}_k\mathbf{P}_k^-\mathbf{H}_k^T + \mathbf{R}_k\right)\right]} \tag{4.7.9}$$

where it is implied that we substitute the indicated normal functional expressions into the right side of the equation (see Section 1.1.4 for the vector normal form). It is a routine matter now to make the appropriate substitutions in Eq. (4.7.9) and determine the mean and covariance by inspection of the exponential term. The algebra is routine, but a bit laborious, so we will not pursue it further here. The resulting mean and covariance for $\mathbf{x}_k|\mathbf{z}_k^*$ are

$$\text{Mean} = \hat{\mathbf{x}}_k^- + \mathbf{P}_k^-\mathbf{H}_k^T\left(\mathbf{H}_k\mathbf{P}_k^-\mathbf{H}_k^T + \mathbf{R}_k\right)^{-1}\left(\mathbf{z}_k - \mathbf{H}_k\hat{\mathbf{x}}_k^-\right) \tag{4.7.10}$$

$$\text{Covariance} = \left[\left(\mathbf{P}_k^-\right)^{-1} + \mathbf{H}_k^T\mathbf{R}_k^{-1}\mathbf{H}_k\right]^{-1} \tag{4.7.11}$$

Note that the expression for the mean is identical to the optimal estimate previously derived by other methods. The expression for the covariance given by Eq. (4.7.11) may not look familiar, but in Chapter 5 it will be shown to be identically equal to the usual $\mathbf{P}_k = (\mathbf{I} - \mathbf{K}_k\mathbf{H}_k)\mathbf{P}_k^-$ expression, provided that $\mathbf{K}_k$ is the Kalman gain.

We also note by comparing Eq. (4.7.10) with Eq. (4.2.8), which was used in the minimum-mean-square-error approach, that the form chosen for the update in Eq. (4.2.8) was correct (for the Gaussian case, at least). Note also that Eq. (4.2.8) can be written in the form

$$\hat{\mathbf{x}}_k = (\mathbf{I} - \mathbf{K}_k\mathbf{H}_k)\hat{\mathbf{x}}_k^- + \mathbf{K}_k\mathbf{z}_k \tag{4.7.12}$$

When the equation is written this way, we see that the updated estimate is formed as a weighted linear combination of two independent measures of $\mathbf{x}_k$; the first is the prior estimate that is the cumulative result of all the past measurements and the prior knowledge of the process statistics, and the second is the new information about $\mathbf{x}_k$ as viewed in the measurement space. Thus, the effective weight factor placed on the new information is $\mathbf{K}_k\mathbf{H}_k$. From Eq. (4.7.12) we see that the weight factor placed on the old information about $\mathbf{x}_k$ is $(\mathbf{I} - \mathbf{K}_k\mathbf{H}_k)$, and thus the sum of the weight factors is $\mathbf{I}$ (or just unity in the scalar case). This implies that $\hat{\mathbf{x}}_k$ will be an unbiased estimator, provided, of course, that the two estimates being combined are themselves unbiased estimates. [An estimate is said to be unbiased if $E(\hat{\mathbf{x}}) = \mathbf{x}$.] Note, though, in the Gaussian case we did not start out demanding that the estimator be unbiased. This fell out naturally by simply requiring the estimate to be the mean of the probability density function of $\mathbf{x}_k$, given the measurement stream and the statistics of the process.

In summary, we see that in the Gaussian case the conditional density viewpoint leads to the same identical result that was obtained in Section 4.2, where we assumed a special linear form for our estimator. There are some far-reaching conclusions that can be drawn from the conditional density viewpoint:

1. Note that in the conditional density function approach, we did not need to *assume* a linear relationship between the estimate and the measurements. Instead, this came out naturally as a consequence of the Gaussian assumption and our choice of the conditional mean as our estimate. Thus, in the Gaussian case, we know that we need not search among nonlinear filters for a better one; it cannot exist. Thus, our earlier linear assumption in the derivation of both the Wiener and Kalman filters turns out to be a fortuitous one. That is, in the Gaussian case, the Wiener–Kalman filter is not just best within a class of linear filters; it is best within a class of all filters, linear or nonlinear.
2. For the Gaussian case, the conditional mean is also the "most likely" value in that the maximum of the density function occurs at the mean. Also, it can be shown that the conditional mean minimizes the expectation of almost any reasonable nondecreasing function of the magnitude of the error (as well as the squared error). [See Meditch (7) for a more complete discussion of this.] Thus, *in the Gaussian case, the Kalman filter is best by almost any reasonable criterion.*
3. In physical problems, we often begin with incomplete knowledge of the process under consideration. Perhaps only the covariance structure of the process is known. In this case, we can always imagine a corresponding Gaussian process with the same covariance structure. This process is then completely defined and conclusions for the equivalent Gaussian process can be drawn. It is, of course, a bit risky to extend these conclusions to the original process, not knowing it to be Gaussian. However, even risky conclusions are better than none if viewed with proper caution.

4.8 RE-CAP AND SPECIAL NOTE ON UPDATING THE ERROR COVARIANCE MATRIX

The Kalman filter basics were presented in this chapter in the simplest terms possible. The recursive equations for the filter are especially simple when viewed in state-space form. So one might logically ask, "How can whole books be written about such a simple subject?" The answer lies in the mathematical modeling and the frailties of the model in fitting the physical situation at hand. Remember, the physical problem must be made to fit a very special stochastic format, namely the process equation

$$\mathbf{x}_{k+1} = \boldsymbol{\phi}_k \mathbf{x}_k + \mathbf{w}_k \tag{4.8.1}$$

and the measurement equation

$$\mathbf{z}_k = \mathbf{H}_k \mathbf{x}_k + \mathbf{v}_k \tag{4.8.2}$$

This is usually the most difficult part of any applied Kalman filter problem. "Turning the crank," so to speak, is easy once a good model is developed. We never expect the mathematical model to fit the physical situation perfectly, though, so much of the associated analysis has to do with the effect of the misfit.

The filter equations were first derived using minimum-mean-square-error as the performance criterion. Then in a later section it was shown that the same equations can be deduced from a conditional density viewpoint. Perhaps the most important conclusion from that section is that for Gaussian statistics, the Kalman filter is best by almost any reasonable performance criterion. This is a remarkable and far-reaching conclusion.

A special comment is in order about updating the filter error covariance matrix. Four different equations (Eqs. 4.2.18, 4.2.20, 4.2.21, 4.2.22) are provided for this in the filter derivation. All give the same result for optimal gain and with perfect arithmetic. So, should we just use the simplest form, Eq. (4.2.22) (i.e., $\mathbf{P} = (\mathbf{I} - \mathbf{KH})\mathbf{P}^-$), and let it go at that? The answer to this is no. We need to be a little more discriminating about this, depending on the situation at hand. For example, in a typical textbook example that only runs for a few hundred steps (or less), we would not expect to see any appreciable difference in the results in using any of the four mentioned update formulas. On the other hand, suppose we had a surveying problem where we are processing millions of measurements taken over a long time span, and we are trying to squeeze the last little bit of accuracy out of the data. In this case the filter gain will be approaching zero near the end of the run, and we might see a small error in the gain computation. If this gain is used in the short-form update (i.e., $\mathbf{P} = (\mathbf{I} - \mathbf{KH})\mathbf{P}^-$), the computed $\mathbf{P}$ will be slightly in error and not representative of the true error covariance for the suboptimal gain. The effect of this can be cumulative step-after-step. Now consider using the "longer" update formula (i.e., $\mathbf{P} = (\mathbf{I} - \mathbf{KH})\mathbf{P}^-(\mathbf{I} - \mathbf{KH})^T + \mathbf{KRK}^T$), which is valid for any gain, optimal or suboptimal. The updated $\mathbf{P}$ so computed will at least be consistent with the suboptimal gain being used in the update. Thus the "crime" of using slightly suboptimal gain is mitigated by properly accounting for the suboptimality. Thus, the longer-form update equation is a much safer update equation when large amounts of data are being processed.

There is another fringe benefit in using the longer **P** update formula.* Each of the two additive terms in the update have natural symmetry, which is not true with the short-form update formula. Thus, if the $\mathbf{P}^-$ that is projected from the previous step is symmetric, then we are assured that the updated **P** will also be symmetric. There are some researchers in the field who say that the longer form should always be used, especially in off-line analysis where computational efficiency is usually not important. After all, the longer form is not that difficult to program.

PROBLEMS

4.1 A Wiener (or Brownian motion) process is a random walk process, which, by definition, is exactly zero at $t=0$. (This process is discussed in detail in Section 2.11.) Now consider a variation on this process where the initial value is a zero-mean Gaussian random variable, rather than the deterministic value of zero. We will say that the variance of the initial value is known. Let us now imagine that two systems analysts are having an argument over lunch about how to model this process for a Kalman filter application.

Analyst A says that the best way to model this situation is to let the initial value be a random bias state (i.e., constant with time), and then add to this a second state variable that is a Wiener process. Thus the state vector in this model would be a 2-tuple, and the H matrix would be a 1×2 row vector $[1 \quad 1]$.

But Analyst B says that two states are unnecessary; one will do just fine! All we need to do is model the Wiener process with a one-state model as usual (see Section 2.11), and then simply start the Kalman filter with a nontrivial initial P_0^- that properly accounts for the initial uncertainty in the conceptual integrator that forms the Wiener process. Who is right, Analyst A or Analyst B?

(a) Student exercise: Implement both the one-state and two-state models and compare the results. Covariance analysis only will suffice (i.e., no Monte Carlo measurement sequences need to be generated for the comparison.) The suggested numerical values for this study are:

Run length: 51 steps, $k=0,1,2 \ldots, 50$
Step size: $\Delta t = 1$ s
Random walk process: $Q = 1\,m^2$
Measurement noise: $R = 4\,m^2$
Initial random bias: Variance $= 100\,m^2$ (zero-mean Gaussian)

(b) Discuss briefly the relative merits of each implementation.

4.2 The accompanying figure shows a generic model for random motion (in one dimension). This model is one of a family of PVA models where PVA stands for position, velocity, and acceleration, the three state variables in the model. Note especially that the acceleration variable is modeled as random walk, i.e., the integral of white noise (see Section 2.11). It should be apparent that all three states in this model are nonstationary, and the model is only useful where there are measurements to bound the states, and that is where the Kalman filter comes into play. But, to

* The longer form for the **P** update equation is sometimes referred to as the Joseph equation. (See Reference (8), Section 4.2, p. 136.)

design the Kalman filter for this model, we first need to know the $\boldsymbol{\phi}_k$ and the $\mathbf{Q}_k$ parameters that describe the statistics of the process in the time interval between the discrete measurements. This problem is directed just toward the $\boldsymbol{\phi}_k$ and the $\mathbf{Q}_k$ parameter determination, and not the filter in its totality.

(a) For a step size of Δt (in general), first show that $\boldsymbol{\phi}_k$ and the $\mathbf{Q}_k$ are given by

$$\boldsymbol{\phi}_k = \begin{bmatrix} 1 & \Delta t & \Delta t^2/2 \\ 0 & 1 & \Delta t \\ 0 & 0 & 1 \end{bmatrix}, \quad \mathbf{Q}_k = \begin{bmatrix} \frac{W}{20}\Delta t^5 & \frac{W}{8}\Delta t^4 & \frac{W}{6}\Delta t^3 \\ \frac{W}{8}\Delta t^4 & \frac{W}{3}\Delta t^3 & \frac{W}{2}\Delta t^2 \\ \frac{W}{6}\Delta t^3 & \frac{W}{2}\Delta t^2 & W\,\Delta t \end{bmatrix}$$

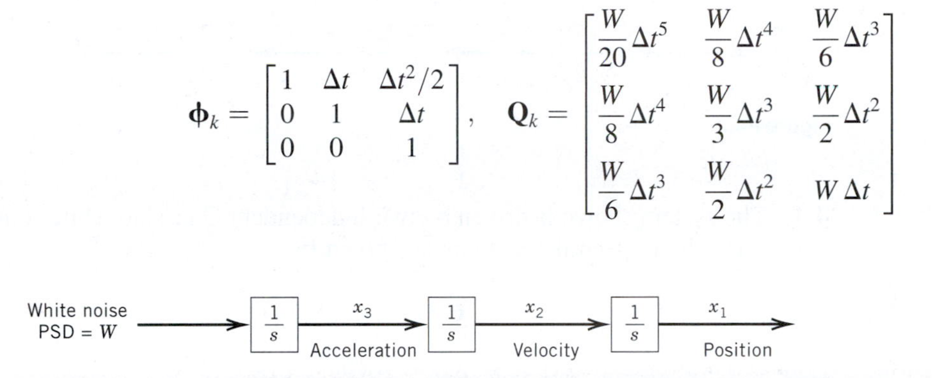

Figure P4.2

(b) Note that there are extreme variations in the powers of Δt in the terms of the $\mathbf{Q}_k$ matrix. This suggests the possibility of making some approximations when Δt is small. One suggestion (rather gross) is to ignore all of the terms except for the (3,3) term, which is first order in Δt. Using MATLAB (or other suitable software) run the following numerical experiment:

Numerical Experiment

Let W = 1 and $\Delta t = 0.1$. Then consider a three-state trivial no-measurement Kalman filter which is initialized with P^- as a 3×3 zero matrix at $t = 0$, and then run the filter for 200 steps (i.e., from $t = 0.1$ to and include $t = 20$). Do this first with the $\mathbf{Q}_k$ terms set at the exact values as given in Part (a). Then repeat the experiment with all the terms of $\mathbf{Q}_k$ set to zero except the (3,3) term, which is set to $\mathrm{W}\Delta t$. Now compare the two runs and make a statement about the validity of the approximation.

(Note that we cannot draw any firm conclusions from just this one numerical experiment. It does, however, provide some insight as to whether this crude approximation is worthy of more careful evaluation.)

4.3 A variation on the dynamic position-velocity-acceleration (PVA) model given in Problem 4.2 is obtained by modeling acceleration as a Markov process rather than random walk. The model is shown in block-diagram form in the accompanying figure. The linear differential equation for this model is of the form

$$\dot{\mathbf{x}} = \mathbf{F}\mathbf{x} + \mathbf{G}\mathbf{f}$$

(a) Write out the **F**, **G**, and $\mathbf{GWG}^T$ matrices for this model, showing each term in the respective matrices explicitly.

(b) The exact expressions for the terms of $\boldsymbol{\phi}_k$ and $\mathbf{Q}_k$ are considerably more difficult to work out in this model than they were in Problem 4.2. However, their numerical values for any reasonable values of W and β can be found readily using the method referred to as the van Loan method discussed in

Section 3.9. Using MATLAB (or other suitable software), find $\boldsymbol{\phi}_k$ and $\mathbf{Q}_k$ for $\beta = 0.2\,\text{s}^{-1}$, $W = 10\ (\text{m/s}^3)^2/\text{Hz}$, and $\Delta t = .1$ s.

(*Note:* These numerical values yield a relatively high-dynamics model where the sigma of the Markov process is about .5 g with a time constant of 5 s.

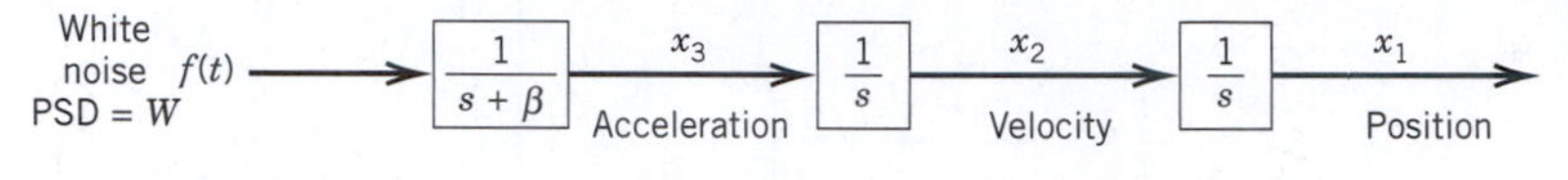

Figure P4.3

4.4 The system shown is driven by two independent Gaussian white sources $u_1(t)$ and $u_2(t)$. Their spectral functions are given by

$$S_{a1} = 4(\text{ft/s}^2)^2/\text{Hz}$$
$$S_{a2} = 16(\text{ft/s})^2/\text{Hz}$$

Let the state variables be chosen as shown on (he diagram, and assume that noisy measurements of x_1, are obtained at unit intervals of time. A discrete Kalman filter model is desired. Find $\boldsymbol{\phi}_k$ and $\mathbf{Q}_k$ for this model.

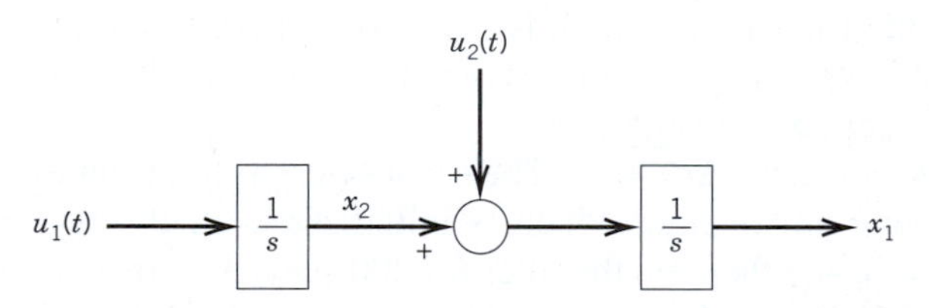

Figure P4.4

4.5 A stationary Gaussian random process is known to have a power spectral density function of the form

$$S_y(j\omega) = \frac{\omega^2 + 1}{\omega^4 + 20\omega^2 + 64}$$

Assume that discrete noisy measurements of y are available at $t = 0, 1, 2, 3, \ldots$ and that the measurement errors are uncorrelated and have a variance of two units. Develop a Kalman filter model for this process. That is, find $\boldsymbol{\phi}_k$, $\mathbf{Q}_k$, $\mathbf{H}_k$, $\mathbf{R}_k$, and the initial conditions $\hat{\mathbf{x}}_0^-$ and $\mathbf{P}_0^-$. Note that numerical answers are requested, so MATLAB and the algorithms for determining $\boldsymbol{\phi}_k$ and $\mathbf{Q}_k$ given in Section 3.9 will be helpful. You may assume that in (the stationary condition, the x_1 and x_2 state variables are uncorrelated. Thus, $\mathbf{P}_0^-$ will be diagonal. (See Problem 3.18 for more on this.)

Hint: It should be apparent that $\dot{y}$ does not exist in this example because of the ω^2 term in the numerator of $S_y(j\omega)$. Thus, we cannot choose y and $\dot{y}$ as state variables in our Kalman filter model. However, a suitable model can be developed as follows:

First, do spectral factorization of $S_y(j\omega)$ as discussed in Section 3.9. This leads to a shaping filter of the form:

$$\text{Shaping filter} \Rightarrow \frac{as+b}{s^2+cs+d}$$

A block diagram for this shaping filter is then as shown in the accompanying figure. Now, let the Kalman filter states be r and $\dot{r}$. The Kalman filter should be fairly obvious now by inspection of the block diagram.

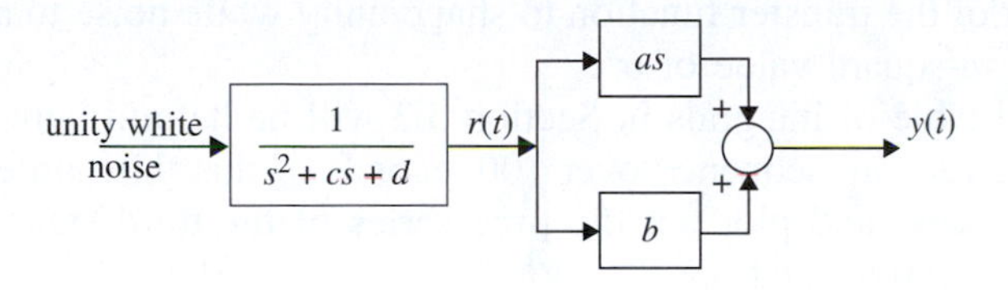

Figure P4.5

4.6 A stationary Gaussian random process is known to have a power spectral density function of the form (same as in Problem 4.5):

$$S_y(j\omega) = \frac{\omega^2+1}{\omega^4+20\omega^2+64}$$

As a variation on the Kalman filter development used in Problem 4.5, do the following:

(a) First, do the spectral factorization of $S_y(j\omega)$ just as in Problem 4.5. This will produce a shaping filter that will convert unity white noise into the $y(t)$ process.

(b) Instead of following the hint in Problem 4.5, do a partial fraction expansion of the shaping filter (just as taught in elementary linear systems analysis). This leads to a block diagram as shown in the figure accompanying this problem.

(c) Now develop a Kalman filter model based on this block diagram. The state variables are shown as x_1 and x_2 in the figure. The same numerical values given in Problem 4.5 apply here also.

(d) Are the Kalman filters of Problem 4.5 and 4.6 equivalent in the steady-state condition? (To be equivalent in every sense of the word, the estimates of the two filters must be identical.)

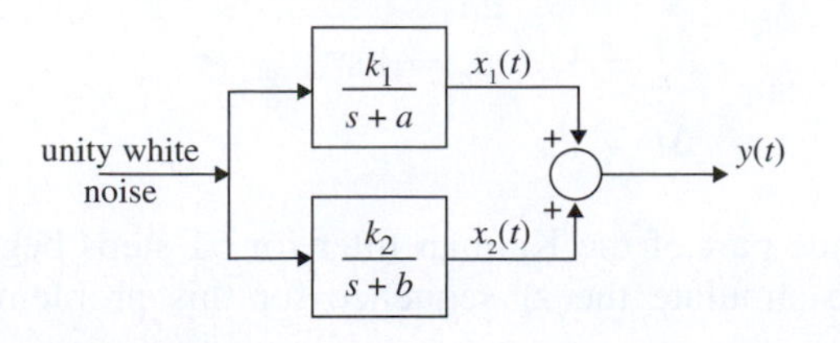

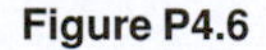

Figure P4.6

4.7 It has been pointed out in this chapter that one of the main features of a Gauss-Markov process is its *boundedness*, unlike a random walk process that will wander off *ad infinitum*. In Example 4.4, we introduced the second-order Gauss-Markov model for processes where the derivative state (e.g., the rate part of a phase/rate pair of states) plays a prominent role in the problem such as if it is strongly connected to a measurement. In some cases, if the second derivative state has similar prominence, we may need to resort to a third-order Gauss-Markov model. Derive such a model as an extension of the second-order Gauss-Markov model given in Example 4.4, by multiplying another factor $(s + \omega_0)$ to the denominator of the "shaping filter" transfer function.

(a) Using methods of Chapter 3, determine what needs to be modified in the numerator of the transfer function to shape unity white noise to an output with a mean-square value of σ^2.
(Hint: The table of integrals in Section 3.3 will be helpful here.)

(b) Generate a random sequence over 100 seconds of this third-order Gauss-Markov process and plot out the time series of the third state (i.e., the "second-derivative" state).

4.8 A classical problem in Wiener-filter theory is one of separating signal from noise when both the signal and noise have exponential autocorrelation functions. Let the noisy measurement be

$$z(t) = s(t) + n(t)$$

and let signal and noise be stationary independent processes with autocorrelation functions

$$R_s(\tau) = \sigma_s^2 e^{-\beta_s|\tau|}$$
$$R_n(\tau) = \sigma_n^2 e^{-\beta_n|\tau|}$$

(a) Assume that we have discrete samples of $z(t)$ spaced Δt apart and wish to form the optimal estimate of $s(t)$ at the sample points. Let $s(t)$ and $n(t)$ be the first and second elements of the process state vector, and then find the parameters of the Kalman filter for this situation. That is, find $\boldsymbol{\phi}_k$, $\mathbf{H}_k$, $\mathbf{Q}_k$, $\mathbf{R}_k$, and the initial conditions $\hat{\mathbf{x}}_0^-$ and $\mathbf{P}_0^-$. Assume that the measurement sequence begins at $t = 0$, and write your results in general terms of σ_s, σ_n, β_s, β_n, and Δt.

(b) To demonstrate that the discrete Kalman filter can be run with $\mathbf{R}_k = \mathbf{0}$ (for a limited number of steps, at least), use the following numerical values in the model developed in part (a):

$$\sigma_s^2 = 9, \quad \beta_s = .1\ \text{s}^{-1}$$
$$\sigma_n^2 = 1, \quad \beta_n = 1\ \text{s}^{-1}$$
$$\Delta t = 1\ \text{s}$$

Then run the error covariance part of the Kalman filter for 51 steps beginning at $t = 0$. (You do not need to simulate the $\mathbf{z}_k$ sequence for this problem. Simple covariance analysis will do.)

4.9 Suppose that we make the following linear transformation on the process state vector of Problem 4.8:

$$\begin{bmatrix} x_1' \\ x_2' \end{bmatrix} = \begin{bmatrix} 1 & 1 \\ 0 & 1 \end{bmatrix} \begin{bmatrix} x_1 \\ x_2 \end{bmatrix}$$

This transformation is nonsingular, and hence we should be able to consider $\mathbf{x}'$ as the state vector to be estimated and write out the Kalman filter equations accordingly. Specify the Kalman filter parameters for the transformed problem. (*Note:* The specified transformation yields a simplification of the measurement matrix, but this is at the expense of complicating the model elsewhere.)

4.10 It is almost self-evident that if the estimation errors are minimized in one set of state variables, this also will minimize the error in any linear combination of those state variables. This can be shown formally by considering a new slate vector $\mathbf{x}'$ to be related to (the original state vector $\mathbf{x}$ via a general nonsingular transformation $\mathbf{x}' = \mathbf{A}\mathbf{x}$. Proceeding in this manner, show that the Kalman estimate obtained in the transformed domain is the same as would be obtained by performing the update (i.e., Eq. 4.2.8) in the original $\mathbf{x}$ domain and then transforming this estimate via the $\mathbf{A}$ matrix.

4.11 Consider two different measurement situations for the same random-walk dynamical process:

Process model:

$$x_{k+1} = x_k + w_k$$

Measurement model 1:

$$z_k = .5x_k + v_k$$

Measurement model 2:

$$z_k = (\cos\theta_k)x_k + v_k, \quad \theta_k = 1 + \frac{k}{120} \text{ rad}$$

Using $Q = 4$, $R = 1$, and $P_0^- = 100$, run error covariance analyses for each measurement model for $k = 0, 1, 2, \ldots, 200$. Plot the estimation error variance for the scalar state x against the time index k for each case. Explain the difference seen between the two plots, particularly as the recursive process approaches and passes $k \approx 70$.

REFERENCES CITED IN CHAPTER 4

1. N. Wiener, *Extrapolation, Interpolation, and Smoothing of Stationary Time Series*, New York: Wiley, 1949.
2. R.E. Kalman, "A New Approach to Linear Filtering and Prediction Problems," *Trans. ASME – J. Basic Eng.*, 35–45 (March 1960).
3. B.E. Bona and R.J. Smay, "Optimum Reset of Ship's Inertial Navigation System," *IEEE Trans. Aerospace Electr. Syst.*, AES-2 (4): 409–414 (July 1966).

4. G.R. Pitman (ed.), *Inertial Guidance*, New York: Wiley, 1962.
5. J.C. Pinson,"Inertial Guidance for Cruise Vehicles," in C.T. Leondes (ed.), *Guidance and Control of Aerospace Vehicles*, New York: McGraw-Hill, 1963.
6. J.M. Mendel, *Optimal Seismic Deconvolution*, New York: Academic Press, 1983.
7. J.S. Meditch, *Stochastic Optimal Linear Estimation and Control*, New York: McGraw-Hill, 1969.
8. M.S. Grewal and A.P. Andrews, *Kalman Filtering Theory and Practice*, Englewood Cliffs, NJ: Prentice-Hall, 1993.

Additional References

9. M.S. Grewal and A.P. Andrews, *Kalman Filtering: Theory and Practice Using MATLAB*, 3rd ed., New York: Wiley, 2008.
10. A. Gelb (ed.), *Applied Optimal Estimation*, Cambridge, MA: MIT Press, 1974.
11. P.S. Maybeck, *Stochastic Models, Estimation and Control* (Vol. 1), New York: Academic Press, 1979.

5

Intermediate Topics on Kalman Filtering

5.1 ALTERNATIVE FORM OF THE DISCRETE KALMAN FILTER–THE INFORMATION FILTER

The Kalman filter equations given in Chapter 4 can be algebraically manipulated into a variety of forms. An alternative form that is especially useful will now be presented (1). We begin with the expression for updating the error covariance, Eq. (4.2.22), and we temporarily omit the subscripts to save writing:

$$\begin{aligned} \mathbf{P} &= (\mathbf{I} - \mathbf{KH})\mathbf{P}^- \\ \mathbf{KHP}^- &= \mathbf{P}^- - \mathbf{P} \end{aligned} \tag{5.1.1}$$

Recall that the Kalman gain is given by Eq. (4.2.17):

$$\begin{aligned} \mathbf{K} &= \mathbf{P}^-\mathbf{H}^T\left(\mathbf{HP}^-\mathbf{H}^T + \mathbf{R}\right)^{-1} \\ \mathbf{K}\left(\mathbf{HP}^-\mathbf{H}^T + \mathbf{R}\right) &= \mathbf{P}^-\mathbf{H}^T \\ \mathbf{KHP}^-\mathbf{H}^T + \mathbf{KR} &= \mathbf{P}^-\mathbf{H}^T \end{aligned} \tag{5.1.2}$$

Substituting Eq. (5.1.1) into Eq. (5.1.2),

$$\begin{aligned} (\mathbf{P}^- - \mathbf{P})\mathbf{H}^T + \mathbf{KR} &= \mathbf{P}^-\mathbf{H}^T \\ \mathbf{KR} &= \mathbf{PH}^T \\ \mathbf{K} &= \mathbf{PH}^T\mathbf{R}^{-1} \end{aligned} \tag{5.1.3}$$

Going back to the error covariance update:

$$\begin{aligned} \mathbf{P} &= (\mathbf{I} - \mathbf{KH})\mathbf{P}^- \\ \mathbf{I} - \mathbf{KH} &= \mathbf{P}(\mathbf{P}^-)^{-1} \end{aligned} \tag{5.1.4}$$

Substituting Eq. (5.1.3) for the gain $\mathbf{K}$ in Eq. (5.1.4), we get

$$\mathbf{I} - \mathbf{P}\mathbf{H}^T\mathbf{R}^{-1}\mathbf{H} = \mathbf{P}(\mathbf{P}^-)^{-1}$$
$$\mathbf{P}(\mathbf{P}^-)^{-1} + \mathbf{P}\mathbf{H}^T\mathbf{R}^{-1}\mathbf{H} = \mathbf{I}$$

Factoring out $\mathbf{P}$ and rearranging the equation in terms of $(\mathbf{P})^{-1}$, we get

$$(\mathbf{P})^{-1} = (\mathbf{P}^-)^{-1} + \mathbf{H}^T\mathbf{R}^{-1}\mathbf{H} \tag{5.1.5}$$

For the error covariance projection equation, we start with the usual prediction stage:

$$\mathbf{P}^- = \boldsymbol{\phi}\mathbf{P}\boldsymbol{\phi}^T + \mathbf{Q}$$
$$(\mathbf{P}^-)^{-1} = (\boldsymbol{\phi}\mathbf{P}\boldsymbol{\phi}^T + \mathbf{Q})^{-1} \tag{5.1.6}$$

Using the following version of the matrix inversion lemma,

$$(\mathbf{A} + \mathbf{B})^{-1} = \mathbf{A}^{-1} - \mathbf{A}^{-1}\mathbf{B}(\mathbf{I} + \mathbf{A}^{-1}\mathbf{B})^{-1}\mathbf{A}^{-1}$$

where

$$\mathbf{A} = \boldsymbol{\phi}\mathbf{P}\boldsymbol{\phi}^T \text{ and } \mathbf{B} = \mathbf{Q},$$

and, defining $\mathbf{M} = \left(\boldsymbol{\phi}\mathbf{P}\boldsymbol{\phi}^T\right)^{-1} = \left(\boldsymbol{\phi}^T\right)^{-1}(\mathbf{P})^{-1}\boldsymbol{\phi}^{-1}$, Eq. (5.1.6) becomes

$$\begin{aligned}(\mathbf{P}^-)^{-1} &= \left(\boldsymbol{\phi}\mathbf{P}\boldsymbol{\phi}^T + \mathbf{Q}\right)^{-1} \\ &= \mathbf{M} - \mathbf{M}\mathbf{Q}(\mathbf{I} + \mathbf{M}\mathbf{Q})^{-1}\mathbf{M}\end{aligned} \tag{5.1.7}$$

If $(\mathbf{Q})^{-1}$ exists, then Eq. (5.1.7) becomes

$$(\mathbf{P}^-)^{-1} = \mathbf{M} - \mathbf{M}\left(\mathbf{Q}^{-1} + \mathbf{M}\right)^{-1}\mathbf{M} \tag{5.1.8a}$$

In the special case where $\mathbf{Q} = \mathbf{0}$, then Eq. (5.1.7) degenerates into

$$(\mathbf{P}^-)^{-1} = \mathbf{M} \tag{5.1.8b}$$

Eq. (5.1.5), Eqs. (5.1.8a) and (5.1.8b) provide the necessary update and projection mechanisms for the inverse of the error covariance matrix $\mathbf{P}$, the so-called *information matrix*. Note that in these equations, we are no longer dealing with the *a priori* error covariance nor the *a posteriori* error covariance, but only with their inverses, $(\mathbf{P}^-)^{-1}$ and $(\mathbf{P})^{-1}$.

Now let us look at the state update and projection equations that go along with this alternative form of the Kalman filter. Let us revisit the original state estimate update equation from Eq. (4.2.8):

$$\hat{\mathbf{x}} = \hat{\mathbf{x}}^- + \mathbf{K}(\mathbf{z} - \mathbf{H}\hat{\mathbf{x}}^-)$$

If we premultiply the inverse of $\mathbf{P}$ on both sides of this equation, we get

$$(\mathbf{P})^{-1}\hat{\mathbf{x}} = (\mathbf{P})^{-1}\hat{\mathbf{x}}^{-} + (\mathbf{P})^{-1}\mathbf{K}(\mathbf{z} - \mathbf{H}\hat{\mathbf{x}}^{-})$$

Simplifying further, and redefining a new variable, $\mathbf{y}$ (to be called the information state vector), related to the state estimate via the inverse of the error covariance, we get

$$\begin{aligned}
\underbrace{(\mathbf{P})^{-1}\hat{\mathbf{x}}}_{\hat{\mathbf{y}}} &= (\mathbf{P})^{-1}\hat{\mathbf{x}}^{-} + \mathbf{H}^T\mathbf{R}^{-1}(\mathbf{z} - \mathbf{H}\hat{\mathbf{x}}^{-}) \\
&= (\mathbf{P})^{-1}\hat{\mathbf{x}}^{-} - \mathbf{H}^T\mathbf{R}^{-1}\mathbf{H}\hat{\mathbf{x}}^{-} + \mathbf{H}^T\mathbf{R}^{-1}\mathbf{z} \\
&= \left[(\mathbf{P})^{-1} - \mathbf{H}^T\mathbf{R}^{-1}\mathbf{H}\right]\hat{\mathbf{x}}^{-} + \mathbf{H}^T\mathbf{R}^{-1}\mathbf{z} \\
&= \underbrace{(\mathbf{P}^{-})^{-1}\hat{\mathbf{x}}^{-}}_{\hat{\mathbf{y}}^{-}} + \mathbf{H}^T\mathbf{R}^{-1}\mathbf{z}
\end{aligned} \tag{5.1.9}$$

This gives us a relationship between two modified quantities related to the original *a priori* and *a posteriori* state estimates through the inverses of the respective error covariances.

If we carry out the same kind of transformation on the state projection equation, by premultiplying with the inverse of $\mathbf{P}^{-}$, we get

$$(\mathbf{P}^{-})^{-1}\hat{\mathbf{x}}^{-} = (\mathbf{P}^{-})^{-1}\boldsymbol{\phi}\hat{\mathbf{x}}$$

If $(\mathbf{Q})^{-1}$ exists, we can substitute the $(\mathbf{P}^{-})^{-1}$ term on the right hand side with Eq. (5.1.8a):

$$\begin{aligned}
\underbrace{(\mathbf{P}^{-})^{-1}\hat{\mathbf{x}}^{-}}_{\hat{\mathbf{y}}^{-}} &= \left[\mathbf{M} - \mathbf{M}(\mathbf{Q}^{-1} + \mathbf{M})^{-1}\mathbf{M}\right]\boldsymbol{\phi}\hat{\mathbf{x}} \\
&= \left[\mathbf{I} - \mathbf{M}(\mathbf{Q}^{-1} + \mathbf{M})^{-1}\right](\boldsymbol{\phi}^T)^{-1}(\mathbf{P})^{-1}\boldsymbol{\phi}^{-1}\boldsymbol{\phi}\hat{\mathbf{x}} \\
&= \left[\mathbf{I} - \mathbf{M}(\mathbf{Q}^{-1} + \mathbf{M})^{-1}\right](\boldsymbol{\phi}^T)^{-1}\underbrace{(\mathbf{P})^{-1}\hat{\mathbf{x}}}_{\hat{\mathbf{y}}}
\end{aligned} \tag{5.1.10a}$$

If $(\mathbf{Q})^{-1}$ does not exist, we substitute for that same term with Eq. (5.1.8b) instead:

$$\underbrace{(\mathbf{P}^{-})^{-1}\hat{\mathbf{x}}^{-}}_{\hat{\mathbf{y}}^{-}} = \mathbf{M}\boldsymbol{\phi}\hat{\mathbf{x}} = (\boldsymbol{\phi}^T)^{-1}(\mathbf{P})^{-1}\boldsymbol{\phi}^{-1}\boldsymbol{\phi}\hat{\mathbf{x}} = (\boldsymbol{\phi}^T)^{-1}\underbrace{(\mathbf{P})^{-1}\hat{\mathbf{x}}}_{\hat{\mathbf{y}}} \tag{5.1.10b}$$

Finally, we bring together the relevant equations, restore the time step indices, and summarize the new algorithm for the Information Filter in the flow diagram of Fig. 5.1.

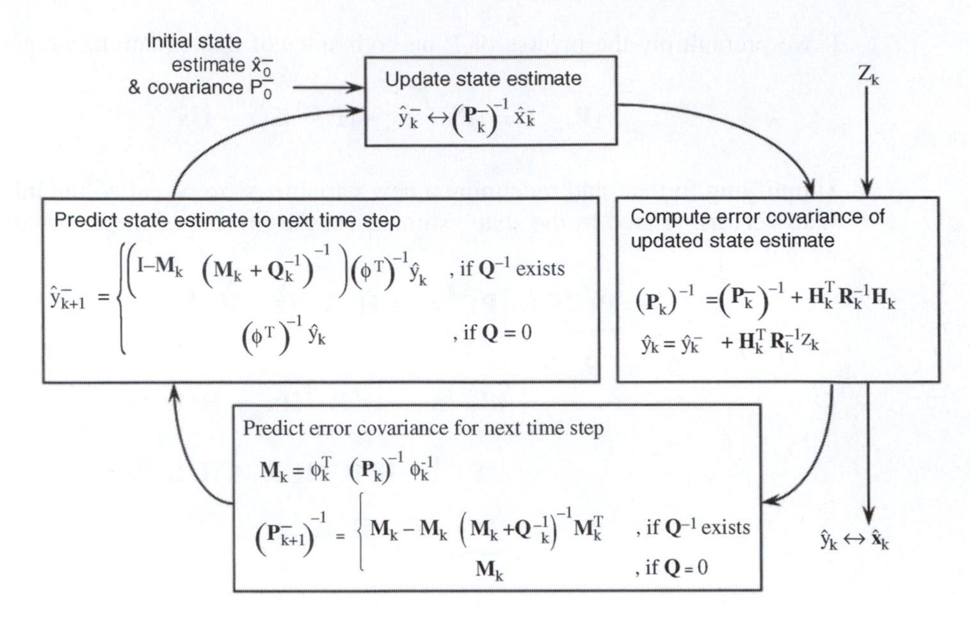

Figure 5.1 The Information filter.

5.2
PROCESSING THE MEASUREMENTS ONE AT A TIME

We now have two different Kalman filter algorithms for processing the measurement information. The first is the usual error-covariance algorithm that is shown pictorially in Fig. 4.1, Chapter 4. Then in the preceding Section 5.1 we presented an alternative algorithm that was centered around the inverse error covariance that has a physical interpretation as the information matrix. Both algorithms yield identical results with perfect arithmetic and proper interpretation. Which one should be used is a matter of convenience in programming and online implementation. We will now show that, under certain circumstances, the components of the measurement vector can be processed one at a time.

We begin with the information matrix update formula that was derived as Eq. 5.1.5 in Section 5.1 (repeated here for convenience).

$$(\mathbf{P})^{-1} = (\mathbf{P}^-)^{-1} + \mathbf{H}^T \mathbf{R}^{-1} \mathbf{H} \tag{5.2.1}$$

We note first that this update equation was derived with the assumption of optimal gain. Let us assume that the **R** matrix is block diagonal. Then Eq. (5.2.1) can be rewritten in partitioned form as (with the k subscripts reinserted):

$$(\mathbf{P}_k)^{-1} = (\mathbf{P}_k^-)^{-1} + \left[\mathbf{H}_k^{aT} \mid \mathbf{H}_k^{bT} \mid \cdots \right] \begin{bmatrix} (\mathbf{R}_k^a)^{-1} & \mathbf{0} & \mathbf{0} \\ \mathbf{0} & (\mathbf{R}_k^b)^{-1} & \mathbf{0} \\ \mathbf{0} & \mathbf{0} & \ddots \end{bmatrix} \begin{bmatrix} \mathbf{H}_k^a \\ \mathbf{H}_k^b \\ \vdots \end{bmatrix} \tag{5.2.2}$$

Physically, the partitioned **R** matrix means that the measurement suite at time t_k can be arranged in blocks such that the measurement errors among the a, b, . . . blocks are uncorrelated. This is often the case when the measurements come from different sensors or instrumentation sources. We next expand Eq. (5.2.2) to get:

$$\left(\mathbf{P}_k\right)^{-1} = \underbrace{\underbrace{\left(\mathbf{P}_k^-\right)^{-1} + \mathbf{H}_k^{aT}\left(\mathbf{R}_k^a\right)^{-1}\mathbf{H}_k^a}_{\substack{(\mathbf{P}_k)^{-1} \text{ after assimilating} \\ \text{block } a \text{ measurements}}} + \mathbf{H}_k^{bT}\left(\mathbf{R}_k^b\right)^{-1}\mathbf{H}_k^b}_{(\mathbf{P}_k)^{-1} \text{ after assimilating block } b \text{ measurements}} + \cdots \tag{5.2.3}$$

Note that the sum of the first two terms in Eq. (5.2.3) is just the $(\mathbf{P}_k)^{-1}$ one would obtain after assimilating the "block *a*" measurement, just as if no further measurements were available. The Kalman gain associated with this block (Eq. 5.1.3, Section 5.1) may now be used to update the state estimate accordingly. Now think of making a trivial projection through zero time, and then starting through the update loop again with the "block *b*" measurements. This is shown in Fig. 5.2 with the projection step shown bypassed. In effect, we can use the usual error covariance update loop over and over again until all of the measurement blocks at time t_k have been assimilated. The end result is then the same as if we had processed the measurements all at once as a vector quantity. We then step ahead with the usual nontrivial projection to t_{k+1}, and there we start all over again with the one-at-a-time measurement assimilation. It is as simple as that.

From a programming viewpoint one-at-a-time measurement processing may, or may not, be advantageous depending on the application at hand. It is presented here without specific recommendations, simply as a useful option in system programming. It was probably noticed that we justified the one-at-a-time idea with information filter concepts, and then we proceeded to illustrate the mechanics of carrying out the procedure with the usual error-covariance filter. This is perfectly legitimate, because the only thing that really matters is the end result after all the measurements at t_k have been assimilated. And the equivalence of the information

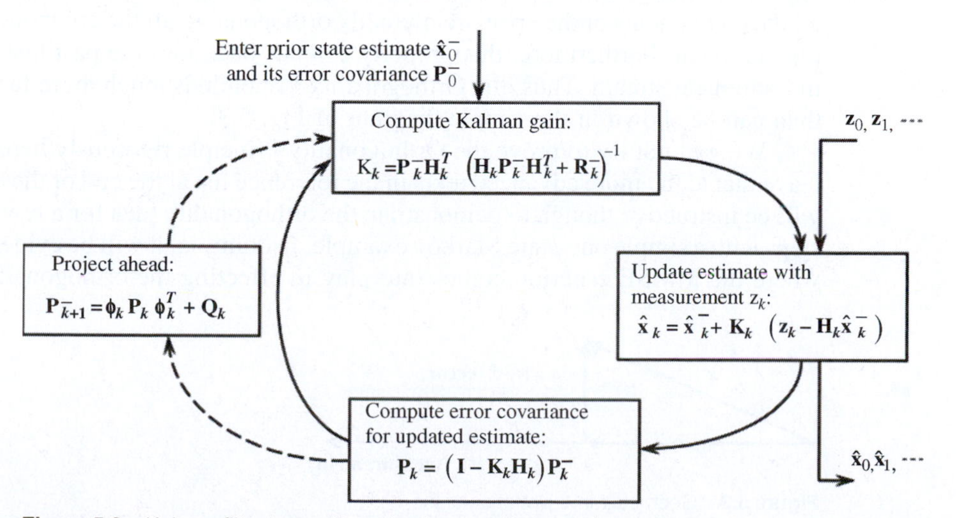

Figure 5.2 Kalman filter loop with projection step bypassed.

and error-covariance filters was shown in Section 5.1, and this is enough justification for the discussion here.

5.3 ORTHOGONALITY PRINCIPLE

In linear minimum-mean-square-error (MMSE) estimation theory there is a remarkable connection among the estimation error, the measurement sequence and the random process being estimated. This relationship is known as the Orthogonality Principle, and it dates back to the early days of Wiener filtering (16). As applied to discrete Kalman filtering, this principle simply says that the estimation error is orthogonal to the measurement sequence, provided the filter gain is set to yield the minimum mean-square error. This is not at all obvious at the outset, but it is a direct consequence of the MMSE criterion. It is also easy to show that the filter estimation error is also orthogonal to the estimate, because it is a linear function of the measurement stream.

The Orthogonality Principle is often explained geometrically with a triangular diagram as shown in Fig. 5.3. The estimate $\hat{x}$ lies in the measurement space z, and it is adjustable, depending on how the measurement is to be weighted to yield the desired estimate.

Clearly, the minimum error e occurs when $\hat{x}$ is adjusted such that e is orthogonal to z. This geometric picture is helpful in understanding the Orthogonality Principle and its connection to minimizing the error (in some sense), but one should not take the picture too literally. First of all, the principle is a vector concept, and it is not easily described in a simple two-dimensional figure. One should also remember that it is the expectation of the squared error that is minimized, not the error itself. Also, it works out that the principle is much more general than the figure would indicate. When the principle is written out mathematically, i.e.,

$$E\left[(\hat{x} - x)z^T\right] = 0 \tag{5.3.1}$$

we note that it is the outer product that is zero (not an inner product). This means that all the components of the error are mutually orthogonal to all the components of the measurement. Furthermore, this property extends back into the past history of the measurement stream. Thus, the Orthogonality Principle is much more far-reaching than can be shown in the simple diagram of Fig. 5.3.

We will not try to prove the Orthogonality Principle rigorously here. We will leave that to the more advanced texts in the reference list at the end of the chapter. It will be instructive though to demonstrate the orthogonality idea for a few recursive steps with a simple one-state Markov example. In doing so, it will be evident exactly where the MMSE criterion comes into play in effecting the orthogonality.

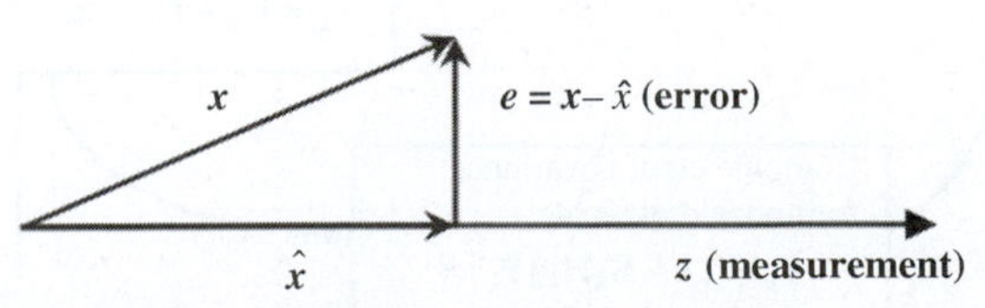

Figure 5.3 Geometric interpretation of the Orthogonality Principle.

EXAMPLE 5.1

Consider a first-order Gauss Markov process x with known σ, and β, and say we have a direct one-to-one measurement of x with an rms error of σ_v. We wish to investigate the orthogonal relationships that exist right from the very start, beginning at $k=0$. The filter demands that we specify an initial $\hat{x}_0^-$ and P_0^-. These must be consistent with the prior assumption about the x process, which in this case is the known autocorrelation function. Therefore,

$$\hat{x}_0^- = 0,\ P_0^- = \sigma_x^2 \tag{5.3.2}$$

Note also that the initial estimation error is

$$e_0^- = \hat{x}_0^- - x_0 = -x_0 \tag{5.3.3}$$

The first measurement z_0 is to occur at $k=0$, and the filter will process this measurement and update the initial estimate in accordance with the usual Kalman filter equations as described in Chapter 4. This leads to:

$$\text{Gain } K_0 = \frac{P_0^-}{P_0^- + \sigma_v^2} = \frac{\sigma_x^2}{\sigma_x^2 + \sigma_v^2} \tag{5.3.4}$$

Also,

$$(1 - K_0) = \frac{\sigma_v^2}{\sigma_x^2 + \sigma_v^2} \tag{5.3.5}$$

The updated estimate is given by

$$\hat{x}_0 = \hat{x}_0^- + K_0\left(z_0 - \hat{x}_0^-\right) = K_0 z_0 \tag{5.3.6}$$

and the estimation error and its variance are

$$e_0 = \hat{x}_0 - x_0 = K_0 z_0 - x_0 = K_0(x_0 + v_0) - x_0 = -(1 - K_0)x_0 + K_0 v_0 \tag{5.3.7}$$

Now, according to the Orthogonality Principle the estimation error at $k=0$ should be orthogonal to z_0, so we need to verify this:

$$E[e_0 z_0] = E[\{-(1 - K_0)x_0 + K_0 v_0\}\{x_0 + v_0\}] \tag{5.3.8}$$

We note here that x_0 and v_0 are uncorrelated. Therefore,

$$E[e_0 z_0] = -(1 - K_0)E\left[x_0^2\right] + K_0 E\left[v_0^2\right] \tag{5.3.9}$$

If we now substitute the optimal values for K_0 and $(1 - K_0)$ we get

$$E[e_0 z_0] = -\frac{\sigma_v^2 \sigma_x^2}{\sigma_x^2 + \sigma_v^2} + \frac{\sigma_x^2 \sigma_v^2}{\sigma_x^2 + \sigma_v^2} = 0 \tag{5.3.10}$$

It is important to note here that the gain K_0 has to be the optimal gain that minimizes the mean-square error in order to get cancellation of the two terms in Eq. (5.3.9). Otherwise, e_0 will not be orthogonal to z_0!

Let us now go one step further and look at the *a priori* estimate at the next recursive step (i.e., at $k=1$).

$$\hat{x}_1^- = \phi \hat{x}_0 \tag{5.3.11}$$

and, similarly, for the process model:

$$x_1 = \phi x_0 + w_0 \tag{5.3.12}$$

Therefore, the *a priori* error is

$$e_1^- = \hat{x}_1^- - x_1 = \phi e_0 - w_0 \tag{5.3.13}$$

Now, the e_1^- is also a minimum mean-square error, so we might expect $E\left[e_1^- z_0\right]$ to be zero, just as we found that $E[e_0 z_0]$ was zero.

To verify this we can write out $E\left[e_1^- z_0\right]$ explicitly as

$$E\left[e_1^- z_0\right] = E[(\phi e_0 - w_0)z_0] = E[\phi e_0 z_0] - E[w_0 z_0] \tag{5.3.14}$$

We have just shown that $E[e_0 z_0]$ is zero. Also $E[w_0 z_0]$ must be zero, because w_0 is the additive white noise that comes into the x process after $k=0$. Therefore, e_1^- is orthogonal to the previous measurement z_0.

We could now go further and show that the updated estimation error e_1 is orthogonal to both z_1 and z_0. The algebra is a bit messy, though, so we will leave this as an exercise (see Problem 5.3).

■

Application of the Orthogonality Principle to the Measurement Residuals

The term $\left(\mathbf{z}_k - \mathbf{H}_k \hat{\mathbf{x}}_k^-\right)$ that appears in the measurement update equation is usually referred to as the measurement residual. To be more precise, perhaps it should be called the *a priori* measurement residual, because it is formed by differencing the current measurement $\mathbf{z}_k$ with the *a priori* estimate of $\hat{\mathbf{x}}_k^-$ (not $\hat{\mathbf{x}}_k$) as it reflects into the measurement space. However, it is customary to omit the "a priori," so we will do likewise here. Note that the $\mathbf{H}_k \hat{\mathbf{x}}_k^-$ quantity in the residual is the filter's best estimate of $\mathbf{z}_k$ before it gets to "see" the actual $\mathbf{z}_k$. It is conditioned on the measurement stream up through $\mathbf{z}_{k-1}$, just like $\hat{\mathbf{x}}_k^-$, so it is also minimum-mean-square-error (in the Gaussian case, at least). Thus, the measurement residual is an estimation error of sorts. So, from the Orthogonality Principle, it should be orthogonal to the measurement stream from $\mathbf{z}_{k-1}$ clear back to $\mathbf{z}_0$. In mathematical terms we then have:

$$\left(\mathbf{z}_k - \mathbf{H}_k \hat{\mathbf{x}}_k^-\right)\mathbf{z}_{k-1}^T = 0 \tag{5.3.15}$$

$$\begin{gathered} \left(\mathbf{z}_k - \mathbf{H}_k \hat{\mathbf{x}}_k^-\right)\mathbf{z}_{k-2}^T = 0 \\ \vdots \\ \text{etc.} \end{gathered} \tag{5.3.16}$$

Next, consider the expectation of the outer product of any adjacent pair of measurement residuals, say, for example, the most "recent" pair:

$$E[\text{"}adjacent\ pair\text{"}] = E\left[\left(\mathbf{z}_k - \mathbf{H}_k \hat{\mathbf{x}}_k^-\right)\left(\mathbf{z}_{k-1} - \mathbf{H}_{k-1} \hat{\mathbf{x}}_{k-1}^-\right)^T\right] \tag{5.3.17}$$

The first parenthetical quantity in Eq. (5.3.17) will be recognized as the estimation error at t_k, and the $\mathbf{H}_{k-1}\hat{\mathbf{x}}_{k-1}^-$ in the second parenthetical term can be written as a linear function of $\mathbf{z}_{k-2}, \mathbf{z}_{k-3}, \ldots, \mathbf{z}_0$. Therefore, the Orthogonality Principle says that the expectation of the sum of the terms within the brackets of Eq. (5.3.17) must be zero. A similar argument can be applied to any pair of residuals, so we can say that the sequence of measurement residuals is a white sequence. This is also known as an innovations sequence. It is also worth noting that the residuals are uncorrelated (i.e., "white") in an outer-product sense, not just an inner-product sense. This is to say that any of the elements of a residual sampled at a particular time are orthogonal to any of the elements of another residual sampled at a different time. This is a very far-reaching form of orthogonality.

5.4 DIVERGENCE PROBLEMS

Since the discrete Kalman filter is recursive, the looping can be continued indefinitely, in principle, at least. There are practical limits, though, and under certain conditions divergence problems can arise. We elaborate briefly on three common sources of difficulty.

Roundoff Errors

As with any numerical procedure, roundoff error can lead to problems as the number of steps becomes large. There is no one simple solution to this, and each case has to be examined on its own merits. Fortunately, if the system is observable and process noise drives each of the state variables, the Kalman filter has a degree of natural stability. In this case a stable, steady-state solution for the **P** matrix will normally exist, even if the process is nonstationary. If the **P** matrix is perturbed from its steady-state solution in such a way as not to lose positive definiteness, then it tends to return to the same steady-state solution. This is obviously helpful, provided **P** does not lose its symmetry and positive definiteness. (See Section 5.8 for more on filter stability.)

Some techniques that have been found useful in preventing, or at least forestalling, roundoff error problems are:

1. Use high-precision arithmetic, especially in off-line analysis work.
2. If possible, avoid deterministic (undriven) processes in the filter modeling. (*Example*: a random constant.) These usually lead to a situation where the **P** matrix approaches a semidefinite condition as the number of steps becomes large. A small error may then trip the **P** matrix into a non-positive definite condition, and this can then lead to divergence. A good solution is to add (if necessary) small positive quantities to the major diagonal terms of the **Q** matrix. This amounts to inserting a small amount of process noise to each of the states. This leads to a degree of suboptimality, but that is better than having the filter diverge!
3. Symmetrize the error covariance matrix with each step. This is easily done by forming the average of the matrix and its transpose. This is probably the

most important way of avoiding divergence due to computational frailties. It might appear at first glance that triple product operations such as $\boldsymbol{\phi}\mathbf{P}\boldsymbol{\phi}^T$ or $(\mathbf{I} - \mathbf{KH})\mathbf{P}^-(\mathbf{I} - \mathbf{KH})^T$ will always produce symmetric results. Be wary of this. Symmetry is only assured if the inner matrice in the product is, in its self, symmetric. Thus, it is important to keep both $\mathbf{P}$ and $\mathbf{P}^-$ symmetric on long runs.

4. Large uncertainty in the initial estimate can sometimes lead to numerical problems. For example, in a navigation situation, if we start the filter's $\mathbf{P}_0^-$ matrix with very large values along the major diagonal, and if we then follow this with a very precise measurement at $t = 0$, the $\mathbf{P}$ matrix must transition from a very large value to a value close to zero in one step. It can be seen from the $\mathbf{P}$-update equation

$$\mathbf{P}_k = (\mathbf{I} - \mathbf{K}_k\mathbf{H}_k)\mathbf{P}_k^- \tag{5.4.1}$$

that this situation approximates the indeterminate form $0 \times \infty$. One should always be cautious in this kind of numerical situation. One possible solution is to make the elements of $\mathbf{P}_0^-$ artificially smaller and simply recognize that the filter will be suboptimal for the first few steps. Another possibility is to use one of the forms of square-root filtering. This mitigates the problem of extremely large swings in the error covariance from step to step. (See Section 5.7 for more on square root filtering.)

Gross Modeling Errors

Another type of divergence may arise because of inaccurate modeling of the process being estimated. This has nothing to do with numerical roundoff; it occurs simply because the designer (engineer) "told" the Kalman filter that the process behaved one way, whereas, in fact, it behaves another way. As a simple example, if you tell the filter that the process is a random constant (i.e., zero slope), and the actual process is a random ramp (nonzero slope), the filter will be continually trying to fit the wrong curve to the measurement data! This can also occur with nondeterministic as well as deterministic processes, as will now be demonstrated.

EXAMPLE 5.2

Consider a process that is actually random walk but is incorrectly modeled as a random constant. We have then (with numerical values inserted to correspond to a subsequent simulation):

(a) The "truth model":

$$\dot{x} = u(t),\ u(t) = \text{unity Gaussian white noise, and } \mathrm{Var}[x(0)] = 1$$

$$z_k = x_k + v_k, \quad \text{measurement samples at } t = 0,\ 1,\ 2, \ldots$$

$$\text{and } \mathrm{Var}(v_k) = .1$$

(b) Incorrect Kalman filter model:

$$x = \text{constant}, \quad \text{where } x \sim N(0,\ 1)$$

$$z_k = x_k + v_k \quad \text{(same as for truth model)}$$

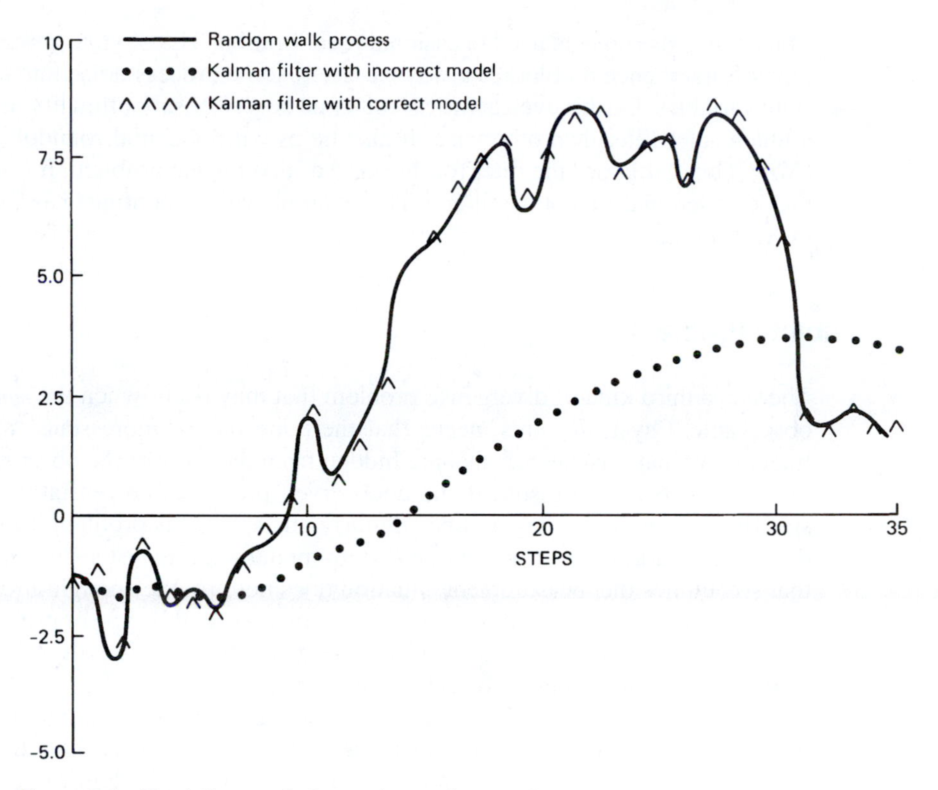

Figure 5.4 Simulation results for random walk example.

The Kalman filter parameters for the incorrect model (b) are: $\phi_k = 1$, $Q_k = 0$, $H_k = 1$, $R_k = .1$, $\hat{x}_0 = 0$, and $P_0^- = 1$. For the truth model the parameters are the same except that $Q_k = 1$, rather than zero.

The random walk process (a) was simulated using Gaussian random numbers with zero mean and unity variance. The resulting sample process for 35 sec is shown in Fig. 5.4. A measurement sequence z_k of this sample process was also generated using another set of $\mathcal{N}(0, 1)$ random numbers for u_k. This measurement sequence was first processed using the incorrect model (i.e., $Q_k = 0$), and again with the correct model (i.e., $Q_k = 1$). The results are shown along with the sample process in Fig. 5.4. In this case, the measurement noise is relatively small ($\sigma \approx .3$), and we note that the estimates of the correctly modeled filter follow the random walk quite well. On the other hand, the incorrectly modeled filter does very poorly after the first few steps. This is due to the filter's gain becoming less and less with each succeeding step. At the 35th step the gain is almost two orders of magnitude less than at the beginning. Thus, the filter becomes very sluggish and will not follow the random walk. Had the simulation been allowed to go on further, it would have become even more sluggish.

The moral to Example 5.2 is simply this. Any model that assumes the process, or any facet of the process, to be absolutely constant forever and ever is a risky model. In the physical world, very few things remain absolutely constant. Instrument biases, even though called "biases," have a way of slowly changing with time.

Thus, most instruments need occasional recalibration. The obvious remedy for this type of divergence problem is always to insert some process noise into each of the state variables. Do this even at the risk of some degree of suboptimality; it makes for a much safer filter than otherwise. It also helps with potential roundoff problems. (*Note*: Don't "blame" the filter for this kind of divergence problem. It is the fault of the designer/analyst, not the filter!). This example will be continued as Example 5.5 in Section 5.5.

Observability Problem

There is a third kind of divergence problem that may occur when the system is not observable. Physically, this means that there are one or more state variables (or linear combinations thereof) that are hidden from the view of the observer (i.e., the measurements). As a result, if the unobserved processes are unstable, the corresponding estimation errors will be similarly unstable. This problem has nothing to do with roundoff error or inaccurate system modeling. It is just a simple fact of life that sometimes the measurement situation does not provide enough information to estimate all the state variables of the system. In a sense, this type of problem should not even be referred to as divergence, because the filter is still doing the best estimation possible under adverse circumstances.

There are formal tests of observability that may be applied to systems of low dimensionality, These tests are not always practical to apply, though, in higher-order systems. Sometimes one is not even aware that a problem exists until after extensive error analysis of the system. If unstable estimation errors exist, this will be evidenced by one or more terms along the major diagonal of $\mathbf{P}$ tending to increase without bound. If this is observed, and proper precautions against roundoff error have been taken, the analyst knows an observability problem exists. The only really good solution to this kind of divergence is to improve the observability situation by adding appropriate measurements to make the system completely observable.

5.5 SUBOPTIMAL ERROR ANALYSIS

A Kalman filter is optimal in a minimum mean square error sense if the filter model matches the physical situation perfectly. However, in practical real-life applications the filter actually being implemented will always be suboptimal to some degree or other, so it is important to be able to assess the degree of suboptimality. In this section we will assume that the structure of the filter is correct (i.e., the order of the state vector and measurement relationship are correct), but some of the filter parameters may not match the physical situation perfectly. To some extent, the effect of such mismodeling errors is amenable to analysis, and this will be discussed in this section. Then the analysis of reduced-order (i.e., state vector) systems will be considered in Section 5.6.

Fig. 5.5 shows two conceptual suboptimal filters that will be helpful in our analysis. The top filter, No. 1, represents the actual filter being implemented with the wrong parameters (which can be $\mathbf{R}$, $\mathbf{Q}$ or $\mathbf{P}_0^-$, or any combination of these). We first note that the usual online nonadaptive Kalman filter is really not very smart. The

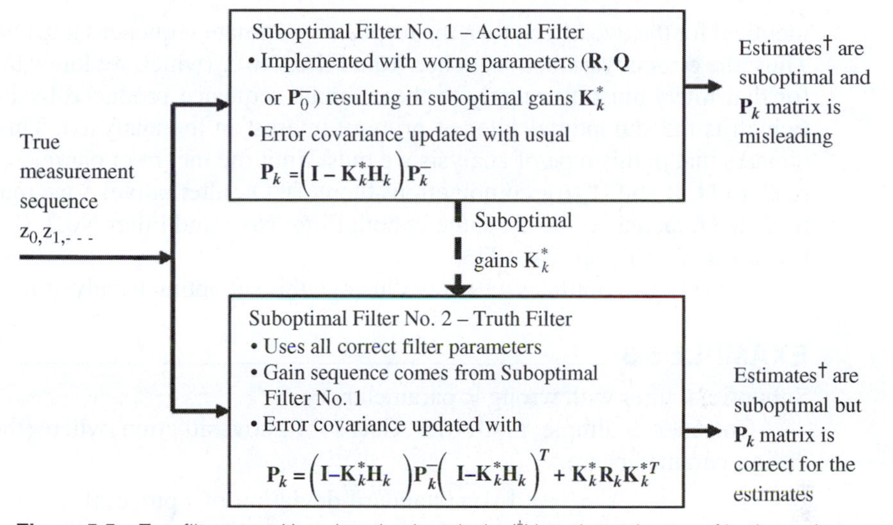

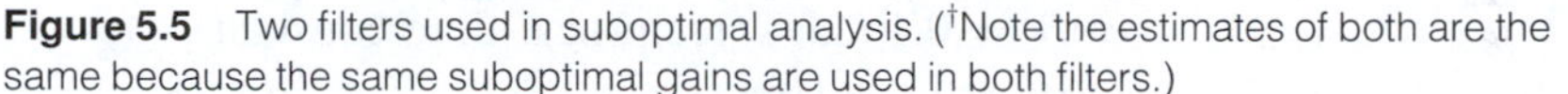

Figure 5.5 Two filters used in suboptimal analysis. (†Note the estimates of both are the same because the same suboptimal gains are used in both filters.)

only thing it knows is what the designer tells it in terms of the model parameters. It does not know truth from fiction or optimality from suboptimality. It simply takes the given information and follows the rules as faithfully as possible numerically. It was pointed out in the filter derivation in Chapter 4 that there are a number of equations for updating the **P** matrix. The two most frequently used equations are a "short form" Eq. (4.2.22) and a "long form" Eq. (4.2.18). These are repeated in Fig. 5.5. One is for optimal gain only and the other applies to any gain. We also note that the "long form" is often referred to as the Joseph update equation.

Now the online Filter No. 1 wants to be optimal (and it does not know that it is using the wrong **R**, for example); so it can update its computed **P** with the short-form equation. But the resulting **P** will not be truly representative of the actual estimation error covariance, because the wrong **R** affects the gain calculation; which, in turn, affects the **P** calculation. Thus, the **P** so computed by Filter No. 1 is misleading. Remember, though, the online filter has no way of knowing this, because it thinks the given parameters are true and accurate. Note, though, that Filter No. 1 is suboptimal because it is using the wrong parameter, **R** for example, not because it is using the short-form update equation.

Consider Filter No. 2 next. It is entirely conceptual, and it is "implemented" offline for analysis purposes only. Its parameters represent truth, including the correct **R** parameter. However, Filter No. 2 does not calculate its gain using the usual formula, i.e., $\mathbf{K}_k = \mathbf{P}_k^- \mathbf{H}_k^T \left(\mathbf{H}_k \mathbf{P}_k^- \mathbf{H}_k^T + \mathbf{R}_k \right)^{-1}$. Rather, the filter is programmed to use a numerical gain sequence from another source—namely, Filter No. 1. Thus, Filter No. 2 is also suboptimal because it is using a gain sequence that is inconsistent with the truth model. Now the theory says that the **P** produced by the Joseph update formula will be the true error covariance for Filter No. 1.

To justify the foregoing statement we need to look carefully at the estimates as well as the error covariances for the two filters of Fig. 5.5. In general, the Kalman filter estimate sequence is only affected directly by the **K**, **H**, $\hat{\mathbf{x}}_0^-$, and $\boldsymbol{\phi}$ combination of parameters. Therefore, if the gains **K** are the same for both Filter No. 1 and Filter No. 2 (which they are according to Fig. 5.5), and if the **H**, $\boldsymbol{\phi}$, and $\hat{\mathbf{x}}_0^-$ parameters are identical for the two filters, then the respective estimate sequences must be identical. Thus, the error covariance sequence from Filter No. 2 (which we know to be correct for that filter) must also apply to the estimate sequence produced by Filter No. 1 (which is the suboptimal filter of primary interest in the analysis). This then also dictates that in this type of analysis we must limit the incorrect parameters in Filter No.1 to $\mathbf{Q}$, $\mathbf{R}$ and $\mathbf{P}_0^-$ (or combinations thereof). Or, alternatively, we must demand that the $\mathbf{H}$, $\boldsymbol{\phi}$, and $\hat{\mathbf{x}}_0^-$ be the same in both Filter No.1 and Filter No.2. (For more on this see Gelb (3), pp. 246–273).

Two simple examples will now illustrate this suboptimal analysis methodology.

EXAMPLE 5.3

Suboptimal filter with wrong R parameter.

Consider a simple first-order Gauss–Markov situation where the nominal design parameters are:

$$\begin{aligned}
\sigma_x &= 1.0\,\text{m (standard deviation of } x \text{ process)}\\
\beta &= 0.1\,\text{rad/s (reciprocal time constant)}\\
\Delta t &= 1.0\,\text{s (step size)}\\
\phi &= e^{-\beta\Delta t}\\
Q &= \sigma_x^2\left(1 - e^{-2\beta\Delta t}\right)\,\text{m}^2\\
H &= 1.0\\
R &= 1.0\,\text{m}^2\\
P_0^- &= \sigma_x^2 \text{ (Initial error covariance)}
\end{aligned}$$

The online filter tries to be optimal, and it is programmed accordingly using the nominal parameters which it thinks represent truth. For reference purposes, let us first look at the updated values of P produced by this filter at its first step at $t = 0$:

Computed gain at $t = 0$

$$K = \frac{P_0^-}{P_0^- + R} = \frac{1.0}{1.0 + 1.0} = 0.5$$

$$P_0 = (\text{I} - KH)P_0^- = (1 - 0.5) \cdot 1.0 = 0.5\,\text{m}^2$$

Let us now speculate about a suboptimal scenario where we say the true R in this situation is $2.0\,\text{m}^2$ rather than 1.0 as implemented in Filter No. 1. This filter is now suboptimal, and we wish to assess the effect of the mismodeling. We can do this simply by cycling the suboptimal gain from Filter No.1 into the truth model, which is Filter No. 2. We will now look at the result at the very first update at $t = 0$. (Note especially that we use the Joseph update, rather than the short form).

$$\begin{aligned}
\text{P}_0(\textit{realistic}) &= (\text{I} - KH)P_0^-(\text{I} - KH)^T + KRK^T\\
&= (1 - 0.5)(1.0)(1 - 0.5) + (0.5)(2.0)(0.5)\\
&= 0.75\,\text{m}^2
\end{aligned}$$

Note that there is a significant difference between what Filter No.1 thinks is its estimation accuracy ($0.5\,\text{m}^2$) and its actual accuracy ($0.75\,\text{m}^2$). Of course, this comparison is only shown here for the first recursive step. It will change as the recursive process proceeds onwards. We will leave this analysis as an exercise (see Problem 5.5). ■

EXAMPLE 5.4

Suboptimal Filter with Wrong Q Parameter

This example is a continuation of the scenario given in Example 5.2 in Section 5.4. There the emphasis was on the divergence problem, and this was demonstrated with a Monte Carlo simulation. Here we will analyze the mean-square error growth quantitatively using suboptimal analysis methods. With reference to Fig. 5.5, the two filter models are as follows:

Suboptimal Filter No. 1: (Wrong Q model)

$$\begin{aligned}
\Delta t &= 1.0\,\text{s} \\
\phi &= 1.0 \\
Q &= 0 \\
R &= 0.1\,\text{m}^2 \\
H &= 1.0 \\
\text{P}_0^- &= 1.0\,\text{m}^2 \text{ (Initial } x \text{ is a } N(0,1) \text{ random variable)}
\end{aligned}$$

Suboptimal Filter No. 2: (Truth model)

(a) Same parameters as Filter No.1, except $Q = 1.0\,\text{m}^2$;

(b) Gains from Filter No.1 are recycled through Filter No. 2; and

(c) Joseph P update equation is used.

The suboptimal error analysis proceeds as follows. We first need the suboptimal gain sequence that comes from Filter No. 1. These gains are computed using the usual recursive equations shown in Fig. 4.1 of Chapter 4. This is easily done with pencil-and-paper methods in this example. The resulting sequence beginning with $t = 0$ is:

$$\text{Gain } K \text{ sequence for 36 cycles (Filter No. 1)} : \frac{10}{11}, \frac{10}{21}, \frac{10}{31}, \frac{10}{41}, \cdots \frac{10}{361}$$

This gain sequence is then recycled through Filter No. 2, and the error variance is shown in the plot of Fig. 5.6. The error variance growth appears to be linear near the end of the run, which might have been expected intuitively for random walk. However, the suboptimal analysis methodology used here provides a quantitative measure of growth rate that is not so obvious intuitively.

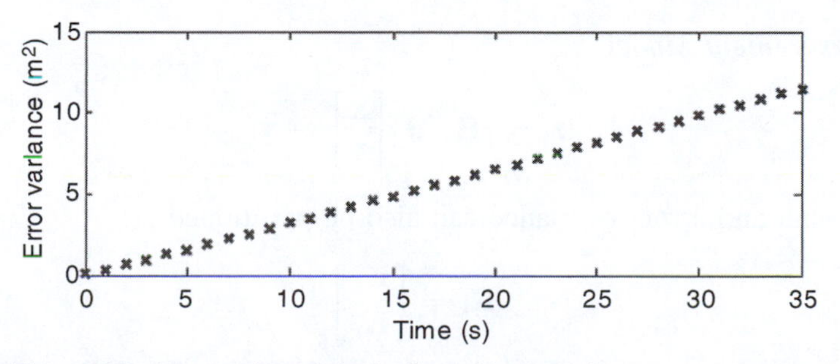

Figure 5.6 Error variance plot for filter with wrong Q.

5.6 REDUCED-ORDER SUBOPTIMALITY

In many real-life applications the size of the state vector can be a problem, and it becomes necessary to eliminate some of the elements to make the filter computationally feasible in real time. This is especially true in integrated navigation systems where there is often a wealth of measurements involved. Recall that when non-white measurement noises are present, they must be absorbed as elements of the state vector to fit the required format for the Kalman filter (see Section 4.2). The estimates of these states are usually not of primary interest, and they are often only weakly observable. Thus, they immediately become candidates for elimination if the need arises. If this is done, the reduced-order system becomes suboptimal, and it is important to be able to assess the degree of suboptimality induced by this choice. The analysis methods discussed in the preceding section were restricted to situations where the suboptimal filter model and the truth model were of the same order, so some special remodeling must be done before applying these methods to the reduced-order filter problem. We will now look at two reduced-order scenarios where exact suboptimal analyses are possible.

The Schmidt Kalman Filter

In the early days of Kalman filtering, Stanley F. Schmidt suggested a method of reducing the order of the state vector, but at the same time still accounting for the deleted states (4). Of course, in general, there is no way to eliminate states and maintain true optimality. But the Schmidt scheme does partially account for the "eliminated" states to some extent at least. Thus, these states in the Schmidt filter are sometimes referred to as consider states, because they are considered but not implemented literally in the filter.

We begin by assuming that we start with a finite-state truth model. Usually the states to be eliminated are in the measurement part of the model, but this is not necessary. The truth model can then be partitioned as follows. (The $\mathbf{y}$ states are the ones to be eliminated.)

Process Model

$$\begin{bmatrix} \mathbf{x} \\ \hline \mathbf{y} \end{bmatrix}_{k+1} = \left[\begin{array}{c|c} \boldsymbol{\phi}_x & \mathbf{0} \\ \hline \mathbf{0} & \boldsymbol{\phi}_y \end{array}\right] \begin{bmatrix} \mathbf{x} \\ \hline \mathbf{y} \end{bmatrix}_k + \begin{bmatrix} \mathbf{w_x} \\ \hline \mathbf{w_y} \end{bmatrix}_k \tag{5.6.1}$$

Measurement Model

$$\mathbf{z}_k = [\mathbf{H} \quad \mathbf{J}] \begin{bmatrix} \mathbf{x} \\ \hline \mathbf{y} \end{bmatrix}_k + \mathbf{v}_k \tag{5.6.2}$$

And the gain and error covariance can also be partitioned as:

$$\mathbf{K}_k = \begin{bmatrix} \mathbf{K}_x \\ \hline \mathbf{K}_y \end{bmatrix}_k \tag{5.6.3}$$

$$\mathbf{P}_k = \left[\begin{array}{c|c} \mathbf{P}_x & \mathbf{P}_{xy} \\ \hline \mathbf{P}_{yx} & \mathbf{P}_y \end{array}\right] \tag{5.6.4}$$

Now, it is the lower partitioned part of the state vector that is to be eliminated and not estimated. So, in the suboptimal model we will arbitrarily set the **y** estimate and its associated gain equal to zero. That is,

$$\hat{\mathbf{y}}_k = \mathbf{0} \tag{5.6.5}$$

$$\mathbf{K}_y = \mathbf{0} \tag{5.6.6}$$

The suboptimal filter is then implemented (conceptually) in partitioned form using the usual matrix equations shown in Fig. 4.1 of Chapter 4, except that the error covariance matrix must be updated using the "Joseph form" of the **P** update equation which also must be expanded out into partitioned form. The partitioning is important if there is to be any computational savings in the Schmidt scheme. (Most of the savings is due to setting $\mathbf{K}_y = \mathbf{0}$.)

In summary, it is the upper-left partitioned part of **P** that gives meaningful error variances (along the major diagonal) for the states of primary interest in the Schmidt-Kalman filter. It is worth noting that the off-diagonal parts of the **P** matrix are not zero. This indicates that the "consider" states have played a role in estimating states of primary interest.

The Pure Reduced-Order Kalman Filter

If the states to be eliminated are only weakly observable, one might logically ask, "Why not just ignore the **y** states completely and shorten the state vector accordingly?" This is certainly a reasonable suggestion, and it would be much simpler than the partitioned Schmidt solution. The difficulty with this suggestion is not, of course, with implementation. (It is easy to "throw away" states.) Rather, the difficulty lies in the analysis of the effect of simply eliminating the unwanted states.

The reduced-order filter, like any suboptimal filter, cannot count on its error covariance matrix **P** to provide the "correct story" about its true error. In the previous section, Section 5.5, we introduced a reasonably doable way of analyzing the true error of a suboptimal filter by way of making two passes, the first time through the suboptimal model, and then sequencing those collected suboptimal gains to run through a second time through the optimal model. One major requirement of this particular method of analysis is that both the "optimal" and "suboptimal" models must have the same dimensionality and the same parameters of $\boldsymbol{\phi}$ and **H**. In order to use this same method of analysis for our reduced-order filter model, we need to satisfy those very requirements.

By definition, a reduced-order suboptimal filter will be estimating a set of states, call it $\hat{\mathbf{x}}_R$, that is a subset of the states of the full-order filter it is being "reduced" from. Therefore, we must first seek out how to represent the reduced-order filter in the same structure of the full-order filter. It turns out that if the set of filter states to be "eliminated," call it $\hat{\mathbf{x}}_E$, is rigged such that the filter "thinks" that it perpetually knows these states perfectly, i.e., the covariance matrix associated with $\hat{\mathbf{x}}_E$ is a zero matrix, then this Kalman filter model behaves exactly like the reduced-order filter with the $\hat{\mathbf{x}}_R$ states only.

We begin with the partitioning the full-order model into the two subspaces associated with $\mathbf{x}_R$ and $\mathbf{x}_E$.

$$\begin{bmatrix} \mathbf{x}_R \\ \mathbf{x}_E \end{bmatrix}_{k+1} = \begin{bmatrix} \boldsymbol{\phi}_{RR} & \boldsymbol{\phi}_{RE} \\ \boldsymbol{\phi}_{ER} & \boldsymbol{\phi}_{EE} \end{bmatrix} \begin{bmatrix} \mathbf{x}_R \\ \mathbf{x}_E \end{bmatrix}_k + \begin{bmatrix} \mathbf{w}_R \\ \mathbf{w}_E \end{bmatrix}_k \tag{5.6.7}$$

$$\mathbf{z}_k = \begin{bmatrix} \mathbf{H}_R & \mathbf{H}_E \end{bmatrix} \begin{bmatrix} \mathbf{x}_R \\ \mathbf{x}_E \end{bmatrix}_k + \mathbf{v}_k \tag{5.6.8}$$

The Kalman filter model for the full-order model would be specified along with the following parameters:

$$\hat{\mathbf{x}}_0^- = \begin{bmatrix} (\hat{\mathbf{x}}_R)_0^- \\ (\hat{\mathbf{x}}_E)_0^- \end{bmatrix} \qquad \mathbf{P}_0^- = \begin{bmatrix} (\mathbf{P}_{RR})_0^- & (\mathbf{P}_{RE})_0^- \\ (\mathbf{P}_{RE}^T)_0^- & (\mathbf{P}_{EE})_0^- \end{bmatrix} \tag{5.6.9}$$

To contrive a Kalman filter such that the error covariance associated with $\hat{\mathbf{x}}_E$ to be a zero matrix, we need to start out with the following representation for an initial error covariance matrix $\mathbf{P}_0^-$:

$$\hat{\mathbf{x}}_0^- = \begin{bmatrix} (\hat{\mathbf{x}}_R)_0^- \\ \mathbf{0} \end{bmatrix} \qquad \mathbf{P}_0^- = \begin{bmatrix} (\mathbf{P}_R)_0^- & \mathbf{0} \\ \mathbf{0} & \mathbf{0} \end{bmatrix} \tag{5.6.10}$$

Let us next examine how the state vector and error covariance matrix progresses through the update processes of the Kalman filter.

$$\begin{aligned} \mathbf{K}_0 &= \mathbf{P}_0^- \mathbf{H}_0^T \left(\mathbf{H}_0 \mathbf{P}_0^- \mathbf{H}_0^T + \mathbf{R}_0 \right)^{-1} \\ &= \begin{bmatrix} (\mathbf{P}_R)_0^- & \mathbf{0} \\ \mathbf{0} & \mathbf{0} \end{bmatrix} \begin{bmatrix} \mathbf{H}_R \\ \mathbf{H}_E \end{bmatrix} \left(\begin{bmatrix} \mathbf{H}_R & \mathbf{H}_E \end{bmatrix} \begin{bmatrix} (\mathbf{P}_R)_0^- & \mathbf{0} \\ \mathbf{0} & \mathbf{0} \end{bmatrix} \begin{bmatrix} \mathbf{H}_R \\ \mathbf{H}_E \end{bmatrix} + \mathbf{R}_0 \right)^{-1} \\ &= \begin{bmatrix} (\mathbf{P}_R)_0^- \mathbf{H}_R \left(\mathbf{H}_R (\mathbf{P}_R)_0^- \mathbf{H}_R + \mathbf{R}_0 \right)^{-1} \\ \mathbf{0} \end{bmatrix} \triangleq \begin{bmatrix} \mathbf{K}_R \\ \mathbf{0} \end{bmatrix} \end{aligned} \tag{5.6.11}$$

$$\begin{aligned} \hat{\mathbf{x}}_0 &= \hat{\mathbf{x}}_0^- + \mathbf{K} \left(\mathbf{z}_0 - \mathbf{H}_0 \hat{\mathbf{x}}_0^- \right) \\ &= \begin{bmatrix} (\hat{\mathbf{x}}_R)_0^- \\ \mathbf{0} \end{bmatrix} + \begin{bmatrix} \mathbf{K}_R \\ \mathbf{0} \end{bmatrix} \left(\mathbf{z}_0 - \begin{bmatrix} \mathbf{H}_R & \mathbf{H}_E \end{bmatrix} \begin{bmatrix} (\hat{\mathbf{x}}_R)_0^- \\ \mathbf{0} \end{bmatrix} \right) \\ &= \begin{bmatrix} (\hat{\mathbf{x}}_R)_0^- + \mathbf{K}_R \left(\mathbf{z}_0 - \mathbf{H}_R (\hat{\mathbf{x}}_R)_0^- \right) \\ \mathbf{0} \end{bmatrix} \triangleq \begin{bmatrix} (\hat{\mathbf{x}}_R)_0 \\ \mathbf{0} \end{bmatrix} \end{aligned} \tag{5.6.12}$$

$$
\begin{aligned}
\mathbf{P}_0 &= (\mathbf{I} - \mathbf{K}_0\mathbf{H}_0)\mathbf{P}_0^- \\
&= \left(\begin{bmatrix} \mathbf{I}_R & \mathbf{0} \\ \mathbf{0} & \mathbf{I}_E \end{bmatrix} - \begin{bmatrix} \mathbf{K}_R \\ \mathbf{0} \end{bmatrix} [\mathbf{H}_R \mid \mathbf{H}_E] \right) \begin{bmatrix} (\mathbf{P}_R)_0^- & \mathbf{0} \\ \mathbf{0} & \mathbf{0} \end{bmatrix} \\
&= \begin{bmatrix} (\mathbf{I}_R - \mathbf{K}_R\mathbf{H}_R)(\mathbf{P}_R)_0^- & \mathbf{0} \\ \mathbf{0} & \mathbf{0} \end{bmatrix} \triangleq \begin{bmatrix} (\mathbf{P}_R)_0 & \mathbf{0} \\ \mathbf{0} & \mathbf{0} \end{bmatrix}
\end{aligned} \tag{5.6.13}
$$

From the update equations of Eqs. (5.6.11)-(5.6.13), we see that the upper partition essentially maintains what the reduced-order state estimates would undergo while the lower partition associated with the eliminated states remains zeroed. Now, to complete the cycle, we next examine the projection step, where we assert that the partition of the **Q** matrix associated with the eliminated states must also be zeroed to preserve the required condition we are seeking:

$$
\begin{aligned}
\mathbf{P}_1^- &= \boldsymbol{\phi}_0\mathbf{P}_0\boldsymbol{\phi}_0^T + \mathbf{Q}_0 \\
&= \begin{bmatrix} \boldsymbol{\phi}_{RR} & \boldsymbol{\phi}_{RE} \\ \boldsymbol{\phi}_{ER} & \boldsymbol{\phi}_{EE} \end{bmatrix} \begin{bmatrix} \mathbf{P}_R & \mathbf{0} \\ \mathbf{0} & \mathbf{0} \end{bmatrix} \begin{bmatrix} \boldsymbol{\phi}_{RR} & \boldsymbol{\phi}_{RE} \\ \boldsymbol{\phi}_{ER} & \boldsymbol{\phi}_{EE} \end{bmatrix}^T + \begin{bmatrix} \mathbf{Q}_R & \mathbf{0} \\ \mathbf{0} & \mathbf{0} \end{bmatrix} \\
&= \begin{bmatrix} \boldsymbol{\phi}_{RR}\mathbf{P}_R\boldsymbol{\phi}_{RR}^T + \mathbf{Q}_R & \mathbf{0} \\ \mathbf{0} & \mathbf{0} \end{bmatrix} \triangleq \begin{bmatrix} (\mathbf{P}_R)_1 & \mathbf{0} \\ \mathbf{0} & \mathbf{0} \end{bmatrix}
\end{aligned} \tag{5.6.14}
$$

Thus, we have seen that the initialization of Eq. (5.6.10) and the modification of **Q** in Eq. (5.6.14) are able to ensure the retention of a zeroing of the "eliminated" states and its associated error covariance, the sole purpose of which is to allow the filter model with the full-order structure of Eqs. (5.6.7)–(5.6.8) to mimic the reduced-order filter containing the subset states of $\hat{\mathbf{x}}_R$.

By doing so, we are now able to exploit the method outline in Section 5.5 where we make two passes, the first time through the suboptimal model, and then sequencing those collected suboptimal gains to run through a second time through the optimal model, in order to analyze the true error of reduced-order filter. Such an exercise is often made mostly in tradeoff studies to determine, when one starts out with a rather high-dimensionality filter state vector, how much compromise in the true error covariance performance would be made by eliminating certain states in the final reduced-order filter implementation.

The contrived full-order filter may seem counterintuitive at first glance. But look at it this way. Suppose you had a measurement of some unknown quantity, voltage for example. And also say you know that the voltmeter has a known bias. Certainly, a prudent person (without the benefit of Kalman filter background) would simply subtract the bias from the voltmeter reading, and then have faith that the compensated reading is in fact the correct voltage at hand. This is exactly what the Kalman filter does when it is told that one of the state variables is known perfectly. It estimates the other states just as if the one that is known perfectly did not even exist. Of course, this is all fiction in the problem at hand, because it does not match truth. Thus, the contrived filter is, in fact, suboptimal.

EXAMPLE 5.5

We will now look at a numerical example where we compare the estimation error variances for three different filters: (1) optimal; (2) Schmidt-Kalman; and (3) pure reduced-order. The scenario is briefly this: The x process is Gaussian random walk where the initial condition is, itself, a Gaussian random variable. The measurement is one-to-one of the x process with additive Markov and white noises.

The numerical values for the various models are:

Truth Model (also the Optimal Filter)

$$\sigma_x^2 = 1.0\,\text{m}^2/\text{s} \quad \text{(Rate of growth of the } x \text{ process)}$$
$$\sigma_0^2 = 1.0\,\text{m}^2 \quad \text{(Initial variance of the } x \text{ process)}$$
$$\sigma_y^2 = 1.0\,\text{m}^2 \quad \text{(Variance of the Markov measurement noise component)}$$
$$\beta = 0.5\,\text{s}^{-1} \quad \text{(Reciprocal time constant of Markov noise process)}$$
$$\Delta t = 1.0\,\text{s} \quad \text{(Sampling interval)}$$
$$\boldsymbol{\phi} = \begin{bmatrix} 1 & 0 \\ 0 & e^{-\beta\Delta t} \end{bmatrix} \quad \text{(State transition matrix)}$$
$$\mathbf{Q} = \begin{bmatrix} \sigma_x^2 & 0 \\ 0 & \sigma_y^2\left(1 - e^{-2\beta\Delta t}\right) \end{bmatrix} \quad \text{(Process noise covariance matrix)}$$
$$\mathbf{H} = [1 \quad 1] \quad \text{(Measurement connection matrix)}$$
$$\mathbf{R} = 1.0\,\text{m}^2 \quad \text{(Measurement noise covariance matrix)}$$
$$\mathbf{P}_0^- = \begin{bmatrix} \sigma_0^2 & 0 \\ 0 & \sigma_y^2 \end{bmatrix} \quad \text{(Process noise covariance matrix)}$$

Schmidt-Kalman Model

Same model as the optimal filter except $\mathbf{K_y}$ is set to zero for each step, and all the add and multiply operations are done in partitioned form. The Joseph form **P** update formula is used, and this is also expanded out in partitioned form (with $\mathbf{K_y} = \mathbf{0}$, of course).

Pure Reduced-Order Kalman Filter

$$\boldsymbol{\phi} = [1]$$
$$\mathbf{Q} = [1]$$
$$\mathbf{H} = [1]$$
$$\mathbf{R} = 1.0\,\text{m}^2$$
$$\mathbf{P}_0^- = [\sigma_0^2]$$

(Full-Order suboptimal model that mimics the Reduced-Order model):

$$\boldsymbol{\phi}_{sub} = \begin{bmatrix} 1 & 0 \\ 0 & e^{-\beta\Delta t} \end{bmatrix}$$
$$\mathbf{Q}_{sub} = \begin{bmatrix} 1 & 0 \\ 0 & 0 \end{bmatrix}$$
$$\mathbf{H}_{sub} = [1 \quad 1]$$
$$\mathbf{R} = 1.0\,\text{m}^2$$
$$\mathbf{P}_0^- = \begin{bmatrix} \sigma_0^2 & 0 \\ 0 & 0 \end{bmatrix}$$

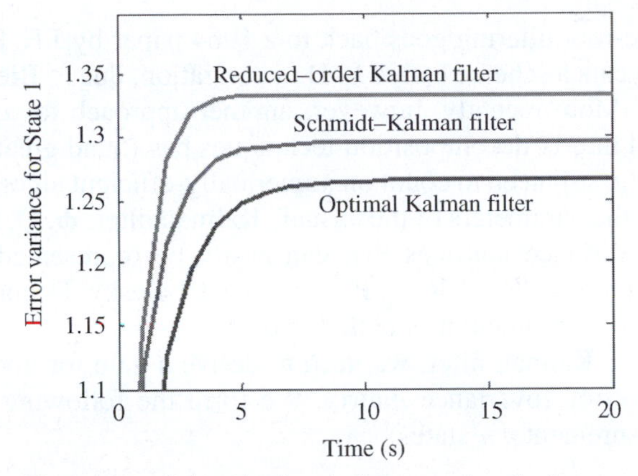

Figure 5.7 Suboptimal analysis comparison for three different filters.

Note:
The contrived $\mathbf{x}_E$ state is assumed to be the known constant zero. This is assured by letting $\mathbf{Q}_{sub}(2, 2) = 0$ and $\mathbf{P}_0^-(2,2) = 0$, even though we have left $\boldsymbol{\phi}_{sub}(2,2) = e^{-\beta\Delta t}$ in order to match the $\boldsymbol{\phi}$ in the truth model.

The three filters just described were programmed for 21 steps doing covariance analysis only. The results are shown in Fig. 5.7. The only surprise in the results is in the comparison of the Schmidt-Kalman and the reduced-order filters. When put in rms terms, the improvement of the Schmidt-Kalman over the reduced-order filter is only about 2%. This is not very much when one considers the extra computational effort involved in implementing the Schmidt-Kalman filter. Of course, this comparison is just for one numerical example. The important message here is the methodology of analysis, not the numerical results.

■

5.7 SQUARE-ROOT FILTERING AND U-D FACTORIZATION

Due to its numerical nature, the Kalman filter may be saddled with computational problems under certain circumstances. But in truth, such problems are far less worrisome today as they were in the early 1960s because of the spectacular progress in computer technology over the decades past. Also, problems of divergence are better understood now than they were in the early days of Kalman filtering. Even so, there are occasional applications where roundoff errors can be a problem, and one must take all possible precautions against divergence. A class of Kalman filter algorithms known as square-root filtering was developed out of necessity in the early days of computer infancy, and they have somewhat better numerical behavior than the "usual" algorithm given in Chapter 4. The basic idea is to propagate something analogous to $\sqrt{\mathbf{P}}$ (standard deviation) rather than $\mathbf{P}$ (variance). One of the critical situations is where the elements of $\mathbf{P}$ go through an extremely wide dynamical range in the course of the filter operation. For example, if the dynamical range of $\mathbf{P}$ is 20 orders of magnitude, the corresponding range of $\sqrt{\mathbf{P}}$ will be 10 orders of magnitude; and it does not take much imagination to see that we would be better off, numerically, manipulating $\sqrt{\mathbf{P}}$ rather than $\mathbf{P}$.

The idea of square-root filtering goes back to a 1964 paper by J.E. Potter (5), but in the intervening years a scheme called U-D factorization, due to Bierman (6), became quite popular. More recently, however, another approach to square-root filtering using standard matrix decomposition techniques has found greater acceptance among the few who still need to count on numerically-efficient algorithms (7). We begin by revisiting the parameters of the "usual" Kalman filter: $\boldsymbol{\phi}, \mathbf{Q}, \mathbf{H}, \mathbf{R}, \mathbf{P}_0^-$. Of these, three are covariance matrices that can easily be represented by their "square root" counterparts, $\sqrt{\mathbf{Q}}$, $\sqrt{\mathbf{R}}$, $\sqrt{\mathbf{P}_0^-}$ via the Cholesky Decomposition technique (In MATLAB, the function is called *chol*).

As a first step in the Kalman filter, we need to derive a gain for updating the state estimate and the error covariance matrix. We form the following $(m+n)$ x $(m+n)$ matrix (m measurements, n states):

$$\mathbf{A} = \begin{bmatrix} \left(\sqrt{\mathbf{R}_k}\right)^T & \mathbf{0} \\ \left(\sqrt{\mathbf{P}_k^-}\right)^T \mathbf{H}_k^T & \left(\sqrt{\mathbf{P}_k^-}\right)^T \end{bmatrix} \tag{5.7.1}$$

If we perform a QR decomposition of $\mathbf{A}$ (MATLAB function is called *qr*), the resultant factors $\mathbf{T}_\mathrm{A}$ and $\mathbf{U}_\mathrm{A}$ are such that $\mathbf{T}_\mathrm{A}\mathbf{U}_\mathrm{A} = \mathbf{A}$. Without proof, we state that

$$\mathbf{U}_\mathrm{A} = \begin{bmatrix} \left(\sqrt{\mathbf{B}_k}\right)^T & \mathbf{W}_k^T \\ \mathbf{0} & \left(\sqrt{\mathbf{P}_k}\right)^T \end{bmatrix} \tag{5.7.2}$$

But we then verify next this by forming $\mathbf{A}^T\mathbf{A} = (\mathbf{T}_\mathrm{A}\mathbf{U}_\mathrm{A})^T\mathbf{T}_\mathrm{A}\mathbf{U}_\mathrm{A} = \mathbf{U}_\mathrm{A}^T\mathbf{T}_\mathrm{A}^T\mathbf{T}_\mathrm{A}\mathbf{U}_\mathrm{A}$:

$$\begin{aligned} &\begin{bmatrix} \sqrt{\mathbf{R}_k} & \mathbf{H}_k\sqrt{\mathbf{P}_k^-} \\ \mathbf{0} & \sqrt{\mathbf{P}_k^-} \end{bmatrix} \begin{bmatrix} \left(\sqrt{\mathbf{R}_k}\right)^T & \mathbf{0} \\ \left(\sqrt{\mathbf{P}_k^-}\right)^T \mathbf{H}_k^T & \left(\sqrt{\mathbf{P}_k^-}\right)^T \end{bmatrix} \\ &= \begin{bmatrix} \sqrt{\mathbf{B}_k} & \mathbf{0} \\ \mathbf{W}_k & \sqrt{\mathbf{P}_k} \end{bmatrix} \mathbf{T}_\mathrm{A}^T\mathbf{T}_\mathrm{A} \begin{bmatrix} \left(\sqrt{\mathbf{B}_k}\right)^T & \mathbf{W}_k^T \\ \mathbf{0} & \left(\sqrt{\mathbf{P}_k}\right)^T \end{bmatrix} \end{aligned} \tag{5.7.3}$$

Since $\mathbf{T}_\mathrm{A}^T\mathbf{T}_\mathrm{A} = \mathbf{I}$ (the *idempotent* property of this factor of the QR decomposition), we can rewrite Eq. (5.7.3) as:

$$\begin{bmatrix} \mathbf{R}_k + \mathbf{H}_k\mathbf{P}_k^-\mathbf{H}_k^T & \mathbf{H}_k\mathbf{P}_k^- \\ \mathbf{P}_k^-\mathbf{H}_k^T & \mathbf{P}_k^- \end{bmatrix} = \begin{bmatrix} \mathbf{B}_k & \sqrt{\mathbf{B}_k}\mathbf{W}_k^T \\ \mathbf{W}_k\left(\sqrt{\mathbf{B}_k}\right)^T & \mathbf{W}_k\mathbf{W}_k^T + \mathbf{P}_k \end{bmatrix} \tag{5.7.4}$$

The (2, 1) and (1, 1) submatrix elements of Eq. (5.7.4) can therefore be combined to form the Kalman gain:

$$\mathbf{K}_k = \mathbf{W}_k\left(\sqrt{\mathbf{B}_k}\right)^T\mathbf{B}_k^{-1} = \mathbf{W}_k\left(\sqrt{\mathbf{B}_k}\right)^T\left[\sqrt{\mathbf{B}_k}\left(\sqrt{\mathbf{B}_k}\right)^T\right]^{-1} = \mathbf{W}_k\left(\sqrt{\mathbf{B}_k}\right)^{-1} \tag{5.7.5}$$

Finally, the updated square-root of the error covariance matrix is simply obtained from the transpose of the (2, 2) submatrix element of $\mathbf{U}_A$. This completes the gain computation and the error covariance update portion of the Kalman filter algorithm.

Equating the (2, 2) terms in Eq. (5.7.4), we get

$$\begin{aligned}\mathbf{P}_k &= \mathbf{P}_k^- - \mathbf{W}_k\mathbf{W}_k^T \\ &= \mathbf{P}_k^- - \mathbf{K}_k\sqrt{\mathbf{B}_k}\left(\sqrt{\mathbf{B}_k}\right)^T\mathbf{K}_k^T \\ &= \mathbf{P}_k^- - \mathbf{K}_k\mathbf{B}_k\mathbf{K}_k^T \\ &= \mathbf{P}_k^- - \mathbf{K}_k\left(\mathbf{R}_k + \mathbf{H}_k\mathbf{P}_k^-\mathbf{H}_k^T\right)\mathbf{K}_k^T \\ &= \mathbf{P}_k^- - \mathbf{P}_k^-\mathbf{H}_k^T\mathbf{K}_k^T \\ &= (\mathbf{I} - \mathbf{K}_k\mathbf{H}_k)\mathbf{P}_k^-\end{aligned} \tag{5.7.6}$$

This relationship of Eq. (5.7.6) is, in fact, the covariance update equation of the Kalman filter, which verifies the conjecture of Eq. (5.7.2).

Now, for the projection step, we form a $2n \times n$ matrix

$$\mathbf{C} = \begin{bmatrix} \left(\sqrt{\mathbf{P}_k}\right)^T \boldsymbol{\phi}_k^T \\ \left(\sqrt{\mathbf{Q}_k}\right)^T \end{bmatrix} \tag{5.7.7}$$

Here again, we perform a QR decomposition of $\mathbf{C}$ to obtain the resultant factors $\mathbf{T}_C$ and $\mathbf{U}_C$ are such that $\mathbf{T}_C\mathbf{U}_C = \mathbf{C}$. And, without proof, we state that

$$\mathbf{U}_C = \begin{bmatrix} \left(\sqrt{\mathbf{P}_{k+1}^-}\right)^T \\ \mathbf{0} \end{bmatrix} \tag{5.7.8}$$

We verify this by forming $\mathbf{C}^T\mathbf{C} = (\mathbf{T}_C\mathbf{U}_C)^T\mathbf{T}_C\mathbf{U}_C = \mathbf{U}_C^T\mathbf{T}_C^T\mathbf{T}_C\mathbf{U}_C$:

$$\begin{bmatrix} \boldsymbol{\phi}_k\sqrt{\mathbf{P}_k} & \sqrt{\mathbf{Q}_k} \end{bmatrix} \begin{bmatrix} \left(\sqrt{\mathbf{P}_k}\right)^T \boldsymbol{\phi}_k^T \\ \left(\sqrt{\mathbf{Q}_k}\right)^T \end{bmatrix} = \begin{bmatrix} \sqrt{\mathbf{P}_{k+1}^-} & \mathbf{0} \end{bmatrix} \mathbf{T}_C^T\mathbf{T}_C \begin{bmatrix} \left(\sqrt{\mathbf{P}_{k+1}^-}\right)^T \\ \mathbf{0} \end{bmatrix} \tag{5.7.9}$$

Again, since $\mathbf{T}_C^T\mathbf{T}_C = \mathbf{I}$, equating the left and right hand sides of Eq. (5.7.9), we get the covariance projection equation of the Kalman filter, which verifies the conjecture of Eq. (5.7.8):

$$\boldsymbol{\phi}_k\mathbf{P}_k\boldsymbol{\phi}_k^T + \mathbf{Q}_k = \mathbf{P}_{k+1}^- \tag{5.7.10}$$

The square-root Kalman filter algorithm is summarized in Fig. 5.8.

Note, in Fig. 5.8, that the entire Kalman filter algorithm can be run recursively by maintaining the state estimate vector $\hat{\mathbf{x}}_k^-$ and $\hat{\mathbf{x}}_k$ and the square roots of the covariance matrices $\sqrt{\mathbf{P}_k^-}$ and $\sqrt{\mathbf{P}_k}$. At any point in time, the full covariance matrices may be reconstituted from their respective square root forms, but they are not needed or used anywhere in the recursive loop.

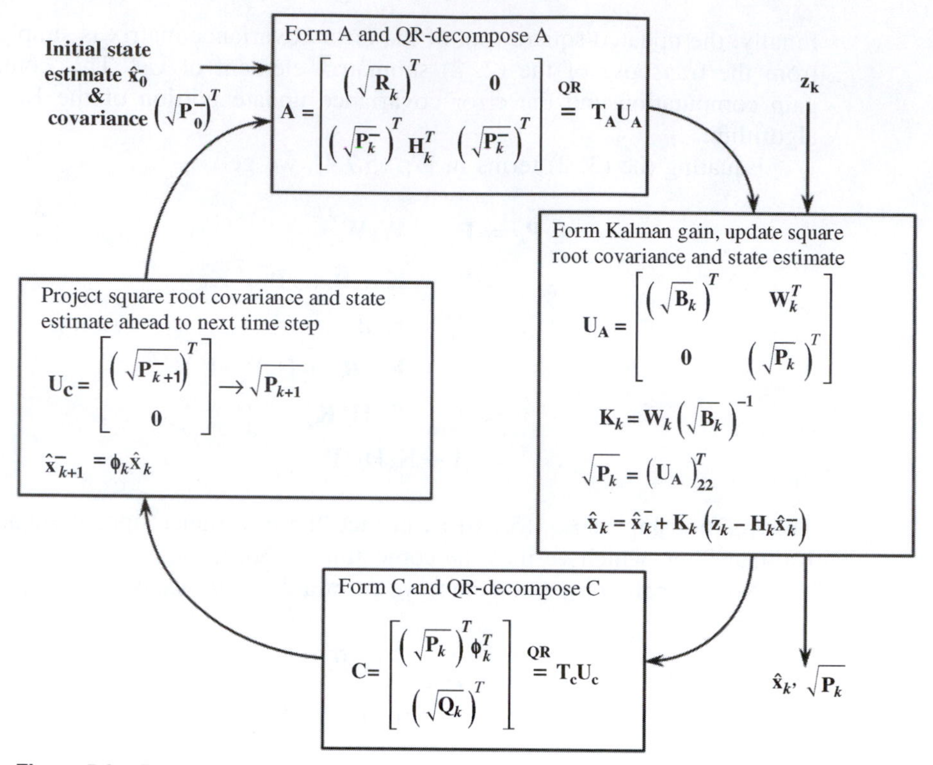

Figure 5.8 Flow chart for a square-root Kalman filter.

U-D Factorization

U-D factorization, due to Bierman, was one of several alternatives available to those in need of computationally efficient and stable algorithms in the early days of Kalman filtering. One of the distinct features of the U-D factorization method was, when decomposing the error covariance matrix into $\mathbf{P} = \mathbf{UDU}^{\mathrm{T}}$, there is no need to solve for simultaneous equations (much like the Cholesky Decomposition method), and also there is no need for any square root numerical operation, which was deemed to be exceedingly burdensome in the early dark ages of computers. Clearly, many such computational constraints have gone by the wayside with the advances of the past couple of decades, and even if there remains any tangible reason to look towards square-root filtering to solve certain limited problems under the appropriate circumstances, the square-root covariance filter method presented earlier in this section now has clear advantages over the Bierman U-D factorization method. For one, the square-root method uses a standard QR factorization operation that is readily available in MATLAB and other mathematical tools, and it can clearly account for the square-root form of the covariance matrix throughout the computational cycle without much fuss. Historically, the U-D factorization method played its part in computational implementations of the Kalman filter and it still lingers on in legacy software. If needed, the reader can consult the third edition of this reference textbook (2), Maybeck (8), or Bierman (6) for details of the many variations of this algorithm, but we will leave this subject without further fanfare.

5.8 KALMAN FILTER STABILITY

A Kalman filter is sometimes referred to as a time-domain filter, because the design is done in the time domain rather than the frequency domain; of course, one of the beauties of the Kalman filter is its ability to accommodate time-variable parameters. However, there are some applications where the filter, after many recursive steps, approaches a steady-state condition. When this happens and the sampling rate is fixed, the Kalman filter behaves much the same as any other digital filter (9), the main difference being the vector input/output property of the Kalman filter. The stability of conventional digital filters is easily analyzed with $\mathscr{z}$-transform methods. We shall proceed to do the same for the Kalman filter.

We begin by assuming that the Kalman filter under consideration has reached a constant-gain condition. The basic estimate update equation is repeated here for convenience:

$$\hat{\mathbf{x}}_k = \hat{\mathbf{x}}_k^- + \mathbf{K}_k\left(\mathbf{z}_k - \mathbf{H}_k\hat{\mathbf{x}}_k^-\right) \tag{5.8.1}$$

We first need to rewrite Eq. (5.8.1) as a first-order vector difference equation. Toward this end, we replace $\hat{\mathbf{x}}_k^-$ with $\boldsymbol{\phi}_{k-1}\hat{\mathbf{x}}_{k-1}$ in Eq. (5.8.2). The result is

$$\hat{\mathbf{x}}_k = (\boldsymbol{\phi}_{k-1} - \mathbf{K}_k\mathbf{H}_k\boldsymbol{\phi}_{k-1})\hat{\mathbf{x}}_{k-1} + \mathbf{K}_k\mathbf{z}_k \tag{5.8.2}$$

We now take the $\mathscr{z}$-transform of both sides of Eq. (5.8.2) and note that retarding $\hat{\mathbf{x}}_k$ by one step in the time domain is the equivalent of multiplying by $\mathscr{z}^{-1}$ in the $\mathscr{z}$-domain. This yields {in the $\mathscr{z}$-domain)

$$\hat{\mathbf{X}}_k(\mathscr{z}) = (\boldsymbol{\phi}_{k-1} - \mathbf{K}_k\mathbf{H}_k\boldsymbol{\phi}_{k-1})\mathscr{z}^{-1}\hat{\mathbf{X}}_k(\mathscr{z}) + \mathbf{K}_k\mathbf{Z}_k(\mathscr{z}) \tag{5.8.3}$$

Or, after rearranging terms, we have

$$\left[\mathscr{z}\mathbf{I} - (\boldsymbol{\phi}_{k-1} - \mathbf{K}_k\mathbf{H}_k\boldsymbol{\phi}_{k-1})\right]\hat{\mathbf{X}}(\mathscr{z}) = \mathscr{z}\mathbf{K}_k\mathbf{Z}_k(\mathscr{z}) \tag{5.8.4}$$

[We note that in Eqs. (5.8.3) and (5.8.4), italic script $\mathscr{z}$ denotes the usual z-transform variable, whereas boldface $\mathbf{Z}_k$ refers to the $\mathscr{z}$-transformed measurement vector.]

We know from linear system theory that the bracketed quantity on the left side of Eq. (5.8.4) describes the natural modes of the system. The determinant of the bracketed $n \times n$ matrix gives us the characteristic polynomial for the system, that is,

$$\text{Characteristic polynomial} = \left|\mathscr{z}\mathbf{I} - (\boldsymbol{\phi}_{k-1} - \mathbf{K}_k\mathbf{H}_k\boldsymbol{\phi}_{k-1})\right| \tag{5.8.5}$$

and the roots of this polynomial provide information about the filter stability. If all the roots lie inside the unit circle in the $\mathscr{z}$-plane, the filter is stable; conversely, if any root lies on or outside the unit circle, the filter is unstable. [As a matter of terminology, the roots of the characteristic polynomial are the same as the eigenvalues of $(\boldsymbol{\phi}_{k-1} - \mathbf{K}_k\mathbf{H}_k\boldsymbol{\phi}_{k-1})$]. A simple example will illustrate the usefulness of the stability concept.

EXAMPLE 5.6

Let us return to the random walk problem of Example 5.2 (Section 5.4) and investigate the stability of the filter in the steady-state condition. The discrete model in this example is

$$x_{k+1} = x_k + w_k \tag{5.8.6}$$

$$z_k = x_k + v_k \tag{5.8.7}$$

and the discrete filter parameters are

$$\phi_k = 1, \quad H_k = 1, \quad Q_k = 1, \quad R_k = .1, \quad P_0^- = 1, \quad \hat{x}_0^- = 0$$

In this example the gain reaches steady state in just a few steps, and it is easily verified that its steady-state value is

$$K_k \approx .916$$

We can now form the characteristic polynomial from Eq. (5.8.5)

$$\begin{aligned} \text{Characteristic polynomial} &= z - [1 - (.916)(1)(1)] \\ &= z - .084 \end{aligned} \tag{5.8.8}$$

The characteristic root is at .084, which is well within the unit circle in the z-plane. Thus we see that the filter is highly stable in this case.

Note that even though the input in this case is nonstationary, the filter itself is intrinsically stable. Furthermore, the filter pole location tells us that any small perturbation from the steady-state condition (e.g., due to roundoff error) will damp out quickly. Any such perturbations will be attenuated by a factor of .084 with each step in this case, so their effect "vaporizes" rapidly. This same kind of reasoning can be extended to the vector case, provided that the **P** matrix is kept symmetric in the recursive process and that it is never allowed to lose its positive definiteness. Thus, we see that we can gain considerable insight into the filter operation just by looking at its characteristic poles in the steady-state condition, provided, of course, that a steady-state condition exists.

5.9 RELATIONSHIP TO DETERMINISTIC LEAST SQUARES ESTIMATION

Kalman filtering is sometimes loosely referred to as least-squares filtering (10, 11, 12). It was mentioned earlier in Chapter 4 that this is an oversimplification, because the criterion for optimization in Kalman filtering is minimum mean-square error and not just the squared error in a deterministic sense. There is, however, a coincidental connection between Kalman filtering and classical deterministic least squares, and this will now be demonstrated (13).

Consider an algebraic problem where we have a set of m linear equations in n unknowns, and the unknowns are overdetermined, i.e., $m > n$. In vector form the equations are

$$\mathbf{Mx} = \mathbf{b} \tag{5.9.1}$$

And we think of $\mathbf{x}$ as the unknown vector, and $\mathbf{M}$ and $\mathbf{b}$ are given. Clearly, $\mathbf{x}$ is $(n \times 1)$, $\mathbf{M}$ is $(m \times n)$, and $\mathbf{b}$ is $(m \times 1)$. This situation arises frequently in physical experiements where we have redundant noisy measurements of linear combinations of certain variables of interest, and there is an abundance of measurement information. In such cases where the system of equations is overdetermined, there is no "solution" for $\mathbf{x}$ that will satisfy all the equations. So, it is reasonable to ask, "What solution will best fit all the equations?" The term best must be defined, of course, and it is frequently defined to be the particular $\mathbf{x}$ that will minimize the sum of the squared residuals. That is, think of $\mathbf{x}$ in Eq. (5.9.1) as a test of $\mathbf{x}$ in our search for the best $\mathbf{x}$. And, in our search, move $\mathbf{b}$ in Eq. (5.9.1) over to the left side of the equation and form the residual expression

$$\mathbf{Mx} - \mathbf{b} = \boldsymbol{\varepsilon} \tag{5.9.2}$$

We now seek the particular $\mathbf{x}$ that will minimize the sum of residuals given by $\boldsymbol{\varepsilon}^{\mathbf{T}}\boldsymbol{\varepsilon}$. Or, we can generalize at this point and say that we wish to minimize the sum of weighted residuals. That is,

$$[\text{Weighted sum of residuals}] = (\mathbf{Mx} - \mathbf{b})^T\mathbf{W}(\mathbf{Mx} - \mathbf{b}) \tag{5.9.3}$$

We assume that the weighting matrix $\mathbf{W}$ is symmetric and positive definite and, hence, so is its inverse. If we wish equal weighting of the residuals, we simply let $\mathbf{W}$ be the identity matrix. The problem now is to find the particular $\mathbf{x}$ (i.e., $\mathbf{x}_{\text{opt}}$) that minimizes the weighted sum of the residuals. Toward this end, the expression given by Eq. (5.9.3) may be expanded and differentiated term by term and then set equal to zero.[†]

[†]The derivative of a scalar s with respect to a vector $\mathbf{x}$ is defined to be

$$\frac{ds}{d\mathbf{x}} = \begin{bmatrix} \dfrac{ds}{dx_1} \\ \dfrac{ds}{dx_2} \\ \vdots \\ \dfrac{ds}{dx_n} \end{bmatrix}$$

The two matrix differentiation formulas used to arrive at Eq. (5.9.4) are

$$\frac{d(\mathbf{x}^T\mathbf{A}\mathbf{x})}{d\mathbf{x}} = 2\mathbf{A}\mathbf{x} \quad (\text{for symmetric } \mathbf{A})$$

and

$$\frac{d(\mathbf{a}^T\mathbf{x})}{d\mathbf{x}} = \frac{d(\mathbf{x}^T\mathbf{a})}{d\mathbf{x}} = \mathbf{a}$$

Both of these formulas can be verified by writing out a few scalar terms of the matrix expressions and using ordinary differentiation methods.

This leads to

$$\begin{aligned}&\frac{d}{d\mathbf{x}_{\text{opt}}}\left[\mathbf{x}_{\text{opt}}^T(\mathbf{M}^T\mathbf{W}\mathbf{M})\mathbf{x}_{\text{opt}} - \mathbf{b}^T\mathbf{W}\mathbf{M}\mathbf{x}_{\text{opt}} - \mathbf{x}_{\text{opt}}^T\mathbf{M}^T\mathbf{W}\mathbf{b} + \mathbf{b}^T\mathbf{b}\right] \\ &= 2(\mathbf{M}^T\mathbf{W}\mathbf{M})\mathbf{x}_{\text{opt}} - (\mathbf{b}^T\mathbf{W}\mathbf{M})^T - \mathbf{M}^T\mathbf{W}\mathbf{b} = \mathbf{0}\end{aligned} \tag{5.9.4}$$

Equation (5.9.4) may now be solved for $\mathbf{x}_{\text{opt}}$. The result is

$$\mathbf{x}_{\text{opt}} = \left[(\mathbf{M}^T\mathbf{W}\mathbf{M})^{-1}\mathbf{M}^T\mathbf{W}\right]\mathbf{b} \tag{5.9.5}$$

and this is the solution of the deterministic least-squares problem.

Next consider the Kalman filter solution for the same measurement situation. The vector $\mathbf{x}$ is assumed to be a random constant, so the differential equation for $\mathbf{x}$ is

$$\dot{\mathbf{x}} = \mathbf{0} \tag{5.9.6}$$

The corresponding discrete model is then

$$\mathbf{x}_{k+1} = \mathbf{x}_k + \mathbf{0} \tag{5.9.7}$$

The measurement equation is

$$\mathbf{z}_k = \mathbf{H}_k\mathbf{x}_k + \mathbf{v}_k \tag{5.9.8}$$

where $\mathbf{z}_k$ and $\mathbf{H}_k$ play the same roles as $\mathbf{b}$ and $\mathbf{M}$ in the deterministic problem. Since time is of no consequence, we assume that all measurements occur simultaneously. Furthermore, we assume that we have no *a priori* knowledge of $\mathbf{x}$, so the initial $\hat{\mathbf{x}}_0^-$ will be zero and its associated error covariance will be infinity. Therefore, using the alternative form of the Kalman filter (Section 5.1), we have

$$\begin{aligned}\mathbf{P}_0^{-1} &= (\infty)^{-1} + \mathbf{H}_0^T\mathbf{R}_0^{-1}\mathbf{H}_0 \\ &= \mathbf{H}_0^T\mathbf{R}_0^{-1}\mathbf{H}_0\end{aligned} \tag{5.9.9}$$

The Kalman gain is then

$$\mathbf{K}_0 = (\mathbf{H}_0^T\mathbf{R}_0^{-1}\mathbf{H}_0)\mathbf{H}_0^T\mathbf{R}_0^{-1}$$

and the Kalman filter estimate of $\mathbf{x}$ at $t = 0$ is

$$\hat{\mathbf{x}}_0 = \left[(\mathbf{H}_0^T\mathbf{R}_0^{-1}\mathbf{H}_0)\mathbf{H}_0^T\mathbf{R}_0^{-1}\right]\mathbf{z}_0 \tag{5.9.10}$$

This is the same identical expression obtained for $\mathbf{x}_{\text{opt}}$ in the deterministic least-squares problem with $\mathbf{R}_0^{-1}$ playing the role of the weighting matrix $\mathbf{W}$.

Let us now recapitulate the conditions under which the Kalman filter estimate coincides with the deterministic least-squares estimate. First, the system state vector was assumed to be a random constant (the dynamics are thus trivial). Second, we assumed the measurement sequence was such as to yield an overdetermined system of linear equations [otherwise $\left(\mathbf{H}_0^T\mathbf{R}_0^{-1}\mathbf{H}_0\right)^{-1}$ will not exist]. And, finally, we assumed that we had no prior knowledge of the process being estimated. This is one of the things that distinguishes the Kalman filter from the least squares estimator and this is an important distinction.

In summary, we should always remember that least squares estimation is basically a deterministic problem. In the beginning we make certain assumptions about the linear measurement structure, and we assume that the system of equations is overdetermined. We then look for a best-fit solution and call this the least-squares estimate. At this point we say nothing whatsoever about the statistics of the best-fit solution. This comes later. Now, if we go further and make some assumptions about the measurement noise, then we can infer something about the statistics of the best-fit solution.

Now, contrast this with the Kalman filter methodology. There, the initial structure of the **x** process and the measurements are put into a probabilistic setting right at the beginning, and the estimation statistics follow automatically with no further assumptions. Furthermore, the Kalman filter allows considerably more flexibility in the probabilistic setting than does the least-squares method. In short, the Kalman filter can do everything that least-squares can do—and much more!

5.10 DETERMINISTIC INPUTS

In many situations the random processes under consideration are driven by deterministic as well as random inputs. That is, me process equation may be of the form

$$\dot{\mathbf{x}} = \mathbf{F}\mathbf{x} + \mathbf{G}\mathbf{u} + \mathbf{B}\mathbf{u}_d \tag{5.10.1}$$

where $\mathbf{B}\mathbf{u}_d$ is the additional deterministic input. Since the system is linear, we can use superposition and consider the random and deterministic responses separately. Thus, the discrete Kalman filter equations are modified only slightly. The only change required is in the estimate projection equation. In this equation the contribution due to $\mathbf{B}\mathbf{u}_d$ must be properly accounted for. Using the same zero-mean argument as before, relative to the random response, we then have

$$\hat{\mathbf{x}}_{k+1}^- = \boldsymbol{\phi}_k\hat{\mathbf{x}}_k + \mathbf{0} + \int_{t_k}^{t_{k+1}} \boldsymbol{\phi}(t_{k+1},\ \tau)\mathbf{B}(\tau)\mathbf{u}_d(\tau)d\tau \tag{5.10.2}$$

where the integral term is the contribution due to $\mathbf{B}\mathbf{u}_d$ in the interval $(t_k,\ t_{k+1})$. The associated equation for $\mathbf{P}_{k+1}^-$ is $\left(\boldsymbol{\phi}_k\mathbf{P}_k\boldsymbol{\phi}_k^T + \mathbf{Q}_k\right)$, as before, because the uncertainty in the deterministic term is zero. Also, the estimate update and associated covariance expressions (see Fig. 4.1) are unchanged, provided the deterministic contribution has been properly accounted for in computing the a priori estimate $\hat{\mathbf{x}}_k^-$.

Another way of accounting for the deterministic input is to treat the problem as a superposition of two entirely separate estimation problems, one deterministic and the other random. The deterministic one is trivial, of course, and the random one is not trivial. This complete separation approach is not necessary, though, provided one properly accounts for the deterministic contribution in the projection step.

PROBLEMS

5.1 In the Information Filter shown in Fig. 5.1, the measurement update of the state and error covariance matrix is quite simple as compared to the time projection of the same quantities. In this problem, we examine the latter with a special case.

Consider the following Kalman filter model that consists of two states, a random walk, and a random constant.

Process Model:

$$\begin{bmatrix} x_1 \\ x_2 \end{bmatrix}_{k+1} = \begin{bmatrix} 1 & 0 \\ 0 & 1 \end{bmatrix} \begin{bmatrix} x_1 \\ x_2 \end{bmatrix}_k + \begin{bmatrix} w_1 \\ w_2 \end{bmatrix}_k$$

$$\mathbf{Q}_k = \begin{bmatrix} 10^2 & 0 \\ 0 & 0 \end{bmatrix}$$

$$\mathbf{P}_0^- = \begin{bmatrix} 10^6 & 0 \\ 0 & 10^6 \end{bmatrix}$$

Measurement Model:

$$z_k = \begin{bmatrix} \cos\left(\dfrac{2\pi k}{600}\right) & 1 \end{bmatrix} \begin{bmatrix} x_1 \\ x_2 \end{bmatrix}_k + v_k$$

$$R_k = 4^2$$

(a) Run a covariance analysis for 600 steps with the usual Kalman filter, save the variance associated with the first state from the updated **P** matrix, and plot the square root of the saved items.

(b) Replicate this analysis with the Information Filter. What makes this a special case is the nature of the **Q** matrix, which is neither a zero matrix nor an invertible one. Find a simple numerical approximation to overcome this problem.

(c) Using this Information Filter with a mechanism to handle the special case of **Q**, starting with the given initial error covariance $\mathbf{P}_0^-$, save the inverted updated **P** matrix, invert the saved matrices, and finally plot the square root of the variance term associated with the first state.

(d) Generate simulated measurements with random numbers for a Monte Carlo simulation that are processed by the Information Filter formulated in (b). Plot the state estimate errors along with the rms error computed in (c).

5.2 Consider the measurement to be a two-tuple $[z_1, z_2]^T$, and assume that the measurement errors are correlated such that the **R** matrix is of the form

$$\mathbf{R} = \begin{bmatrix} r_{11} & r_{12} \\ r_{12} & r_{22} \end{bmatrix}$$

(a) Form a new measurement pair, z_1' and z_2', as a linear combination of the originai pair such that the errors in the new pair are uncorrelated. (*Hint:* First, let $z_1' = z_1$ and then assume $z_2' = c_1 z_1 + c_2 z_2$ and choose the constants c_1 and c_2 such that the new measurement errors are uncorrelated.)

(b) Find the **H** and **R** matrices associated with the new z' measurement vector.

(c) Note that there is no uniqueness in the choice of c_1 and c_2, but rather the specific combination that results in the desired outcome. A good alternative for this choice is the use of the Cholesky Decomposition described in Section 3.10. Show how the Cholesky method can be used to form the new measurement pair that meets the condition desired in (a).

5.3 In Example 5.1, it was shown that the *a priori* estimation error is orthogonal to the previous measurement, i.e., $E\left[e_k^- z_{k-1}\right] = 0$. Proceed further to show that the updated estimation error e_k is orthogonal to both z_k and z_{k-1}.

5.4 The accompanying block diagram shows two cascaded integrators driven by white noise. The two state variables x_1 and x_2 can be thought of as position and velocity for convenience, and the forcing function is acceleration. Let us suppose that we have a noisy measurement of velocity, but there is no direct observation of position. From linear control theory, we know that this system is not observable on the basis of velocity measurements alone. (This is also obvious from the ambiguity in initial position, given only the integral of velocity.) Clearly, there will be no divergence of the estimation error in x_2, because we have a direct measurement of it. However, divergence of the estimation error of x_1 is not so obvious.

The question of divergence of the error in estimating x, is easily answered empirically by cycling through the Kalman filter error covariance equations until either (a) a stationary condition for p_{11} is reached, or (b) divergence becomes obvious by continued growth of p_{11} with each recursive step. Perform the suggested experiment using appropriate covariance analysis software. You will find the following numerical values suitable for this exercise:

$$\text{Power spectral density of } f(t) = .1\ (\text{m/s}^2)^2/(\text{rad/s})$$

$$\text{Step size } \Delta t = 1\ \text{s}$$

$$\text{Measurement error variance} = .01\ (\text{m/s})^2$$

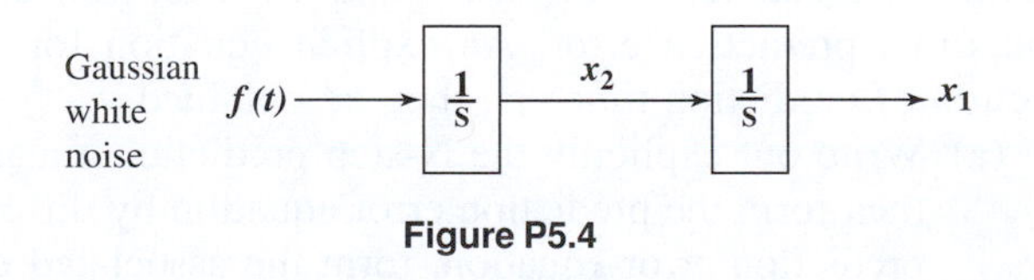

Figure P5.4

Note that if divergence is found, it is not the "fault" of the filter. It simply reflects an inadequate measurement situation. This should not be confused with *computational* divergence.

5.5 In Example 5.3, we considered a suboptimal filter when the R parameter is incorrectly chosen to be 1 instead of its correct value of 2. Extend the analysis begun for the first step in the example to 100 steps. Plot the "realistic" variance P_k over this time duration. Compare this against a plot of the optimal variance of P_k if the parameter $R = 2$ had been correctly chosen.

5.6 In Example 4.5, the concept of prediction was demonstrated for a second-order Gauss-Markov process (derived in Example 4.4). The end result of the example was a plot of the rms prediction error for the range correction for the optimal predictor. It is also of interest to compare the optimal results with corresponding results for a suboptimal predictor that is being considered for this application (14). The suboptimal predictor simply takes the range and range-rate corrections, as provided at the start time, and projects these ahead with a constant rate (much as is done in dead reckoning). This, of course, does not take advantage of any prior knowledge of the spectral characteristics of this process.

The continuous dynamic model for the suboptimal predictor is

$$\ddot{x} = w(t) \tag{P5.6-1}$$

where x is range and $w(t)$ is white noise. [The PSD of $w(t)$ does not affect the projection of x.] If we choose range and range rate as our state variables, the continuous state model becomes

$$\begin{bmatrix} \dot{x}_1 \\ \dot{x}_2 \end{bmatrix} = \underbrace{\begin{bmatrix} 0 & 1 \\ 0 & 0 \end{bmatrix}}_{\mathbf{F}_{\text{subopt}}} \begin{bmatrix} x_1 \\ x_2 \end{bmatrix} + \begin{bmatrix} 0 \\ 1 \end{bmatrix} w(t) \tag{P5.6-2}$$

The **F** matrix for the suboptimal model can now be compared with **F** for the optimal model from Example 4.4. It is

$$\mathbf{F}_{\text{opt}} = \begin{bmatrix} 0 & 1 \\ -\omega_0^2 & -\sqrt{2}\,\omega_0 \end{bmatrix} \quad \text{(optimal model)} \tag{P5.6-3}$$

The optimal system is the truth model in this example, and clearly, $\mathbf{F}_{\text{opt}}$ and $\mathbf{F}_{\text{subopt}}$ are quite different. This means then that the $\boldsymbol{\phi}_k$ matrices for the two models will be different, and this precludes the use of the "recycling suboptimal gains" method of analyzing the suboptimal system performance. All is not lost, though. In this simple situation we can return to basics and write out an explicit expression for the suboptimal prediction error, An explicit equation for the error covariance as a function of prediction time can then be obtained.

(a) Write out explicitly the N-step prediction equations for both models and then form the prediction error equation by differencing the two. From the prediction error equation, form the associated error covariance matrix.

(b) Plot the RMS error of the first state, i.e., the square root of the (1,1) term of the covariance matrix for a prediction time interval from 0 to 200 seconds. From visual inspection, is this error bounded?

5.7 Consider an elementary physics experiment that is intended to measure the gravity constant g. A mass is released at $t = 0$ in a vertical, evacuated column, and multiple-exposure photographs of the failing mass are taken at .05-s intervals beginning at $t = .05$-s. A sequence of N such exposures is taken, and then the position of the mass at each time is read from a scale in the photograph. There will be experimental errors for a number of reasons; let us assume that they are random (i.e., not systematic) and are such that the statistical uncertainties in all position readings are the same and that the standard deviation of these is σ.

Consider g to be an unknown constant, and suppose we say that we have no prior knowledge about its value.

Develop a Kalman filter for processing the noisy position measurements by letting g be the single state variable x. Form the necessary process and measurement equations thereby specifying the relevant Kalman filter parameters.

Writing out the Kalman filter recursive equations for two steps through the second measurement update, obtain an explicit expression for $\hat{x}$ in terms of' z_1 and z_2. Next, compute the ordinary least-squares estimate of g using just two measurements z_1 and z_2. Do this on a batch basis using the equation

$$\hat{g} = \left(\mathbf{H}^T\mathbf{H}\right)^{-1}\mathbf{H}^T\mathbf{z}$$

where

$$\mathbf{H} = \begin{bmatrix} \frac{1}{2}t_1^2 \\ \frac{1}{2}t_2^2 \end{bmatrix} \quad \text{and} \quad \mathbf{z} = \begin{bmatrix} z_1 \\ z_2 \end{bmatrix}$$

Compare the least-squares result with that obtained after carrying the Kalman filter recursive process through two steps.

5.8 In Problem 5.7, the initial position and velocity for the falling mass were assumed to be known, that is, they were exactly zero; the gravity constant was presumed to be unknown and to be estimated. Bozic (15) presents an interesting variation on this problem where the situation is reversed; g is assumed to be known perfectly and the initial position and velocity are random variables with known Gaussian distribution, Fven though the trajectory obeys a known deterministic equation of motion, the random initial conditions add sufficient uncertainty to the motion to make the trajectory a legitimate random process. Assume that the initial position and velocity are normal random variables described by $\mathcal{N}(0, \sigma_p^2)$ and $\mathcal{N}(0, \sigma_v^2)$. Let state variables x_1 and x_2 be position and velocity measured downward, and let the measurements take place at uniform intervals Δt beginning at $t = 0$. The measurement error variance is R. Work out the key parameters for the Kalman filter model for this situation. That is, find $\boldsymbol{\phi}_k$, $\mathbf{Q}_k$, $\mathbf{H}_k$, $\mathbf{R}_k$ and the initial $\hat{\mathbf{x}}_0^-$ and $\mathbf{P}_0^-$. (Note that a deterministic forcing function has to be accounted for in this example. The effect of this forcing function, though, will appear in projecting $\hat{\mathbf{x}}_{k+1}^-$, but not in the model parameters.)

REFERENCES CITED IN CHAPTER 5

1. H.F. Durrant-Whyte, "Equations for the Prediction Stage of the Information Filter," August 2000.
2. R.G. Brown and P.Y.C. Hwang, *Introduction to Random Signals and Applied Kalman Filtering*, 3rd edition, New York: Wiley, 1997.
3. A. Gelb (ed.), *Applied Optimal Estimation*, Cambridge, MA: MIT Press, 1974.
4. S.F. Schmidt, "Application of State-Space Methods to Navigation Problems," C.T. Leondes (ed.), *Advances in Control Systems*, Vol. 3, New York: Academic Press, 1966.

5. R.H. Battin, *Astronautical Guidance*, New York: McGraw-Hill, 1964, pp. 338–340.
6. G.J. Bierman, *Factorization Methods for Discrete Sequential Estimation*, New York: Academic Press, 1977.
7. T. Prvan and M.R. Osborne, "A Square-Root Fixed-Interval Discrete-Time Smoother," *J. Australian Math. Soc. Series B*, 30: 57–68 (1988).
8. P.S. Maybeck, *Stochastic Models, Estimation and Control*, Vol. 1, New York: Academic Press, 1979.
9. A.V. Oppenheim and R.W. Schafer, *Discrete-Time Signal Processing,* 3rd edition, Prentice-Hall, 2009.
10. H.W. Bode and C.E. Shannon, "A Simplified Derivation of Linear Least Squares Smoothing and Prediction Theory," *Proc. I.R.E.*, 38: 417–424 (April 1950).
11. H.W. Sorenson, "Least-Squares Estimation: From Gauss to Kalman," *IEEE Spectrum*, 7: 63–68 (July 1970).
12. T. Kailath, "A View of Three Decades of Linear Filtering Theory," *IEEE Trans. Information Theory*, IT-20(2): 146–181 (March 1974).
13. H.W. Sorenson, "Kalman Filtering Techniques," C.T. Leondes (ed.), *Advances in Control Systems*, Vol. 3, New York: Academic Press, 1966.
14. P.Y. Hwang, "Recommendation for Enhancement of RTCM-104 Differential Standard and Its Derivatives," *Proceedings of ION-GPS-93*, The Institute of Navigation, Sept. 22–24, 1993, pp. 1501–1508.
15. S.M. Bozic, *Digital and Kalman Filtering*, London: E. Arnold, Publisher, 1979.
16. M.S. Grewal and A.P. Andrews, *Kalman Filtering: Theory and Practice Using MATLAB*, 3rd edition, New York: Wiley, 2008.

6

Smoothing and Further Intermediate Topics

Chapters 4 and 5 were devoted entirely to Kalman filtering and prediction. We will now look at the *smoothing* problem. This is just the opposite of prediction, because in smoothing we are concerned with estimating the random process x in the past, rather than out into the future as was the case with prediction. To be more exact, we will be seeking the minimum-mean-square-error (MMSE) estimate of $x\ (t+\alpha)$ where α is negative. (Note that $\alpha = 0$ is filtering and $\alpha > 0$ is prediction.) The three classifications of smoothing will be discussed in detail in Sections 6.1 through 6.4. Then, in Sections 6.5 through 6.9 we will take a brief look at five other intermediate topics which have relevance in applications. There are no special connections among these topics, so they may be studied in any desired order.

6.1 CLASSIFICATION OF SMOOTHING PROBLEMS

A great deal has been written about the smoothing problem, especially in the early years of Kalman filtering. An older text by Meditch (1) is still a fine reference on the subject. A more recent book by Grewal and Andrews (2) is also an excellent reference. The early researchers on the subject were generally searching for efficient algorithms for the various types of smoothing problems. Computational efficiency was especially important with the computer facilities of the 1960s, but perhaps efficiency is not so much of an issue today. Precision and algorithm stability in processing large amounts of data are the more pressing concerns nowadays.

Meditch (1) classified smoothing into three categories:

1. ***Fixed-interval smoothing.*** Here the time interval of measurements (i.e., the data span) is fixed, and we seek optimal estimates at some, or perhaps all, interior points. This is the typical problem encountered when processing noisy measurement data off-line.
2. ***Fixed-point smoothing.*** In this case, we seek an estimate at a single fixed point in time, and we think of the measurements continuing on indefinitely

ahead of the point estimation. An example of this would be the estimation of initial conditions based on noisy trajectory observations after $t=0$. In fixed-point smoothing there is no loss of generality in letting the fixed point be at the beginning of the data stream, i.e., $t=0$.

3. ***Fixed-lag smoothing.*** In this problem, we again envision the measurement information proceeding on indefinitely with the running time variable t, and we seek an optimal estimate of the process at a fixed length of time back in the past. Clearly, the Wiener problem with α negative is fixed-lag smoothing. It is of interest to note that the Wiener formulation will not accommodate either fixed-interval or fixed-point smoothing without using multiple sweeps through the same data with different values of α. This would be a most awkward way to process measurement data.

We will begin our discussion with the fixed-interval problem. We will see presently that the algorithm for it can be used, with some modifications, as a starting point for the solutions for other two categories. Thus, the fixed-interval algorithm as presented here is especially important.

6.2 DISCRETE FIXED-INTERVAL SMOOTHING

The algorithm to be presented here is due to Rauch, Tung, and Striebel (3, 4) and its derivation is given in Meditch (1) as well as the referenced papers. In the interest of brevity, the algorithm will be subsequently referred to as the RTS algorithm. Consider a fixed-length interval containing $N+1$ measurements. These will be indexed in ascending order $\mathbf{z}_0, \mathbf{z}_1, \ldots, \mathbf{z}_N$. The assumptions relative to the process and measurement models are the same as for the filter problem. The computational procedure for the RTS algorithm consists of a forward recursive sweep followed by a backward sweep. This is illustrated in Fig. 6.1. We enter the algorithm as usual at $k=0$ with the initial conditions $\hat{\mathbf{x}}_0^-$ and $\mathbf{P}_0^-$. We then sweep forward using the conventional filter algorithm. With each step of the forward sweep, we must save the computed a priori and a posteriori estimates and their associated $\mathbf{P}$ matrices. These are needed for the backward sweep. After completing the forward sweep, we

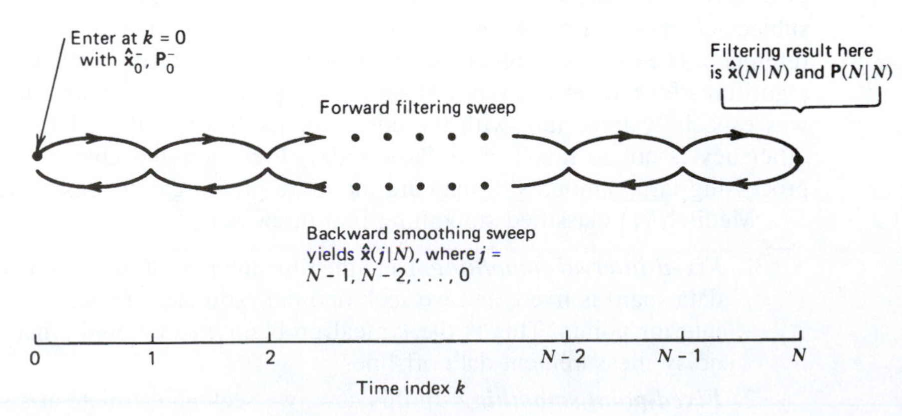

Figure 6.1 Procedure for fixed-interval smoothing.

begin the backward sweep with "initial" conditions $\hat{\mathbf{x}}(N|N)$ and $\mathbf{P}(N|N)$ obtained as the final computation in the forward sweep.* With each step of the backward sweep, the old *filter* estimate is updated to yield an improved smoothed estimate, which is based on all the measurement data. The recursive equations for the backward sweep are

$$\hat{\mathbf{x}}(k|N) = \hat{\mathbf{x}}(k|k) + \mathbf{A}(k)[\hat{\mathbf{x}}(k+1|N) - \hat{\mathbf{x}}(k+1|k)] \tag{6.2.1}$$

where the smoothing gain $\mathbf{A}(k)$ is given by

$$\mathbf{A}(k) = \mathbf{P}(k|k)\boldsymbol{\phi}^T(k+1,k)\mathbf{P}^{-1}(k+1|k) \tag{6.2.2}$$

and

$$k = N-1, N-2, \ldots, 0$$

The error covariance matrix for the smoothed estimates is given by the recursive equation

$$\mathbf{P}(k|N) = \mathbf{P}(k|k) + \mathbf{A}(k)[\mathbf{P}(k+1|N) - \mathbf{P}(k+1|k)]\mathbf{A}^T(k) \tag{6.2.3}$$

It is of interest to note that the *smoothing* error covariance matrix is not needed for the computation of the estimates in the backward sweep. This is in contrast to the situation in the filter (forward) sweep where the **P**-matrix sequence is needed for the gain and associated estimate computations. An example illustrating the use of the RTS algorithm is now in order.

EXAMPLE 6.1

Consider a first-order Gauss–Markov Kalman filter scenario for position x whose autocorrelation function is $R_x(\tau) = \sigma^2 e^{-\beta|\tau|}$ and the filter parameters are:

$$\begin{aligned}\sigma^2 &= 1\,\text{m}^2\\ \beta &= 1\,\text{rad/s}\\ \Delta t &= 0.02\,\text{s}\\ H_k &= 1.0\\ R_k &= 1.0\,\text{m}^2\end{aligned}$$

Now suppose we wish to find the 51-point fixed-interval smoothing solution for all the sample points from $t=0$ to $t=1$ s. Suppose we also wish to look at a Monte Carlo trial solution, so it can be compared with the corresponding filter solution. This is easily programmed with the RTS algorithm, and typical results are shown in Figs. 6.2 and 6.3. Also, for future reference the numerical error covariances are given in Table 6.1. (We will defer the step-by-step details of the RTS recursive procedure until the next example. For now, we will just look at the results of the simulation.)

* The notation used here is the same as that used in the prediction problem. See Chapter 4, Section 4.6.

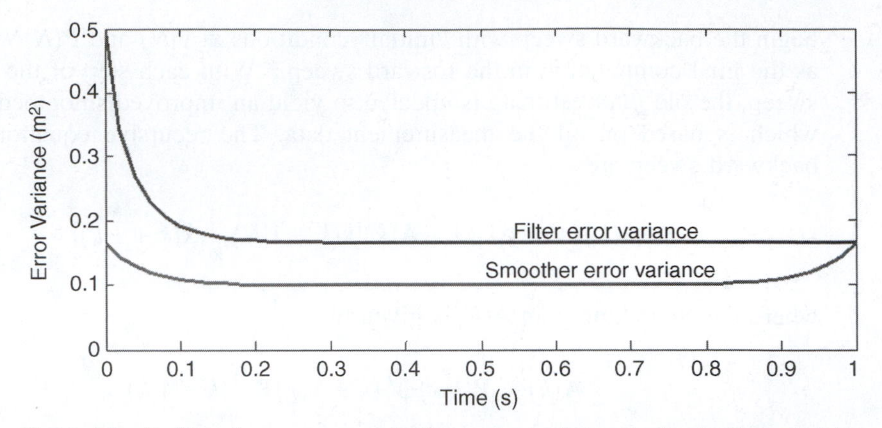

Figure 6.2 Profiles of error variances for filter and smoother estimates for Example 6.1.

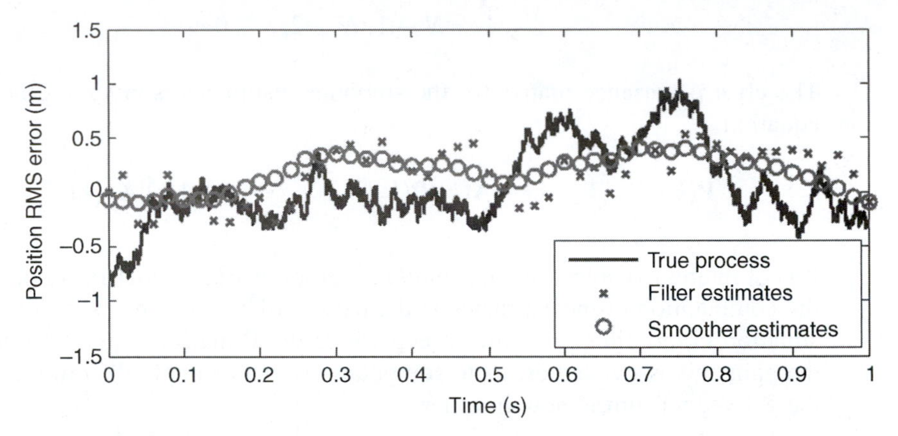

Figure 6.3 Monte Carlo plot of estimates and the true process for Example 6.1.

Note first in Table 6.1, and also from the "Smoother" plot of Fig. 6.2, that the error covariances are symmetric about the midpoint of the fixed interval. This is to be expected because the autocorrelation function is an even function. This is to say that the smoother, in weighting the measurements to obtain the estimate, is indifferent as to the "direction" of time away from the point of estimation. This is as it must be in the fixed-interval solution. Also note that in this example the smoother error variance is considerably less in the middle of the fixed-interval than at the end points. This is also as it should be. In estimating the x process in the middle of the interval, the estimator gets to look at the measurements in "both directions," whereas it only gets to look in one direction at either end point. This improvement in the smoothing estimate versus the filter estimate might lead one to think that smoothing is always better than filtering. However, be wary of this conclusion. It is not always true. For example, when the random variable of interest is a constant, the filter and smoother estimates are one and the same once the filter has the opportunity of observing all the measurements in the fixed interval of the smoother. This is demonstrated in Fig. 6.4. Note that it is only at the end point where we have equivalence.

Before leaving this example, it is worth mentioning that it is obvious from the plots in Fig. 6.3 why this type of estimation is called smoothing. Clearly, the time-domain smoother plot is conspicuously "smoother" than the filter plot!

Table 6.1 Fixed-interval error covariance for Example 6.1

k	P(k|50)
0	0.1653
1	0.1434
2	0.1287
3	0.1189
4	0.1123
5	0.1079
⋮	⋮
25	0.0990
⋮	⋮
45	0.1079
46	0.1123
47	0.1189
48	0.1287
49	0.1434
50	0.1653

Also, it should be remembered that the fixed-interval smoothing problem is an off-line problem, in contrast to "running" real-time problems. Thus, issues other than computational efficiency, such as ease of programming and concern for numerical stability, are usually more pertinent to consider in such cases. The RTS algorithm presented here was developed early in the history of Kalman filtering (circa 1965), and it is still the much-used tried-and-true algorithm. However, other algorithms have also appeared on the scene since 1965, and perhaps the most notable of these are the 1969 solution by Fraser and Potter (19, 20), and the complementary model scheme by Wienert, et al, in 1983 (21). All of the mentioned solutions are equivalent in accuracy performance, so they cannot be compared on that basis. Rather, the

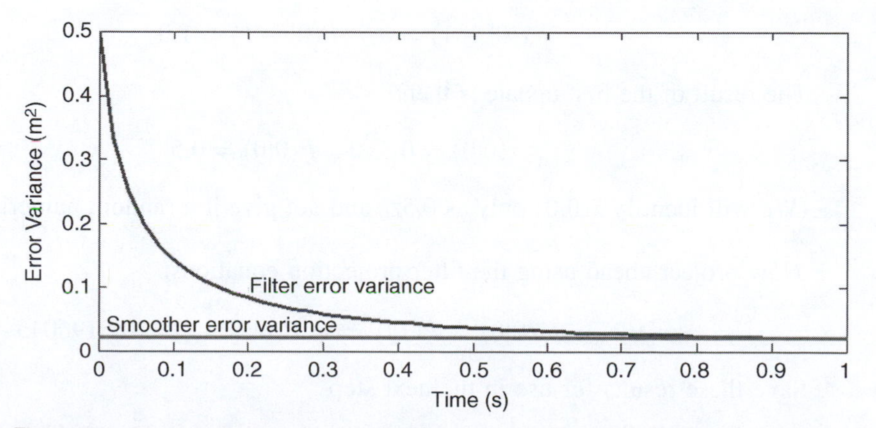

Figure 6.4 Profiles of error variances for filter and smoother estimates for a random bias process (with measurement noise variance of 1 m^2).

comparison has to be made on the basis of ease of conceptual understanding and difficulty in implementation. Arguably, the RTS algorithm is conceptually the simplest of the three, and its inclusion here follows a recommendation that the beginning student should concentrate on it first. After all, what could be simpler than extending the usual Kalman filter (as learned from Chapter 4), save the results, and then do a second (backward) pass using those results in the two simple recursive equations, Eq. (6.2.1) and Eq. (6.2.3).

6.3 DISCRETE FIXED-POINT SMOOTHING

Algorithms for the fixed-point smoothing problem are given in the Meditch (1) and Grewal/Andrews (2) references. They are somewhat more complicated than the straightforward RTS algorithm, so we will be content here to present a simpler (but not quite as general) method than those in the cited references. In our solution the fixed point will always be at $t=0$, and we will let the measurement stream progress positively, with no particular upper bound. With each new measurement we wish to recompute the optimal estimate at $t=0$ based on the whole measurement stream from $t=0$ up to and including the measurement at the end point. Clearly, this problem can be solved treating it as a sequence of RTS probems with ever-increasing upper end points. An example will illustrate the RTS procedure step-by-step; and, as a reminder, in using the RTS algorithm we must be very careful about observing the "parenthesis" notation, because both *a priori* and *a posteriori* variables appear in Eqs. (6.2.1), (6.2.2), and (6.2.3).

EXAMPLE 6.2

In this example, we will demonstrate the step-by-step procedure for solving the fixed-point smoothing problem with the RTS algorithm. The parameters are the same as those in Example 6.1.

Step 1. Assimilate the first measurement at $t=0$ (i.e., z_0). Use the usual Kalman filter algorithm. The initial conditions at $t=0$ (in the absence of prior measurements) are:

$$\hat{x}(0|-1)=0; \quad P(0|-1)=1.0$$

The result of the first update is then:

$$\hat{x}(0|0)=0.5z_0; \quad P(0|0)=0.5$$

(We will identify $\hat{x}(0|0)$ only as $0.5z_0$ and not give it a random numerical value.)

Now project ahead using the filter projection equations:

$$\hat{x}(1|0)=\phi\hat{x}(0|0); \quad P(1|0)=\phi^2P(0|0)+Q=0.5196045$$

Save these results for use in the next step.

Step 2. Assimilate the second measurement at $k=1$ (i.e., z_1). This becomes a two-point fixed-interval smoothing problem. We can now use the RTS directly here.

We must first update the prior state estimate and its error covariance at k = 1 using the usual filter equations. The results are:

$$\hat{x}(1|1) = \hat{x}(1|0) + K_1[z_1 - \hat{x}(1|0)]$$
$$P(1|1) = [1 - K_1]P(1|0) = 0.341934$$

These now become the "initial" conditions for the backward sweep to $k=0$.

To do the backward sweep and get the smoother estimate at $k=0$, we first need to compute the smoother gain. It is, from Eq. (6.2.2):

$$A(0) = P(0|0)\phi/P(1|0) = 0.5 \cdot 0.980199/0.5196045 = 0.9432164$$

The smoother error covariance is given by Eq. (6.2.3) and is:

$$P(0|1) = P(1|1) + A(0)[P(1|1) - P(1|0)]A(0) = 0.341934$$

and the smoother estimate is given by Eq. (6.2.1):

$$\hat{x}(0|1) = \hat{x}(0|0) + A(0)[\hat{x}(1|1) - \hat{x}(1|0)]$$

where $\hat{x}(1|0), \hat{x}(0|0)$, and $\hat{x}(1|1)$ have all been computed previously and saved for the backward sweep.

Step 3. We are now ready to move on to the third measurement at $k=2$. Here we repeat the filter projection and update steps that assimilate the z_2 measurement. We are then ready for the backward sweep to $k=0$. Here, though, we will need to make the sweep in two stages rather than just one. The first stage takes us from $k=2$ to $k=1$; then the second stage takes us from $k=1$ to $k=0$, which is the point of special interest. On the way back to $k=0$ we must compute the smoother estimate at $k=1$, whether we want it or not; it is needed in the final stage of the sweep.

Steps 4, 5, . . . etc. This procedure can then be continued indefinitely to assimilate the measurements z_3, z_4, . . . , etc. Obviously the computational effort keeps increasing without bound.

6.4 DISCRETE FIXED-LAG SMOOTHING

The fixed-lag smoothing problem was originally solved by Wiener in the 1940s. However, he considered only the stationary case; that is, the smoother was assumed to have the entire past history of the input available for weighting in its determination of the estimate. This steady-state solution is certainly interesting in its own right, but it lacks generality. The transient part of the solution is also very much of interest.

Of the three smoothing categories mentioned in Section 6.1, the fixed-lag problem is generally considered to be the most complicated (1, 2). This is mostly due to the start-up problem. If the measurement stream begins at t = 0, we have no nontrivial measurements or corresponding estimates for a few steps back in time where we want to do the estimation. Thus some of the essential variables are missing at the start-up when we try to use the RTS algorithm. An effective (but not very general) solution to the start-up problem is simple avoidance. Say we want to do two-steps-back fixed-lag smoothing, and the measurements begin at $t=0$. Then all we have to do is run the filter for three steps (i.e., $k=0, 1, 2$) without any backwards sweeps. This will "fill up" the missing filter quantities that are needed to compute $\hat{x}(0|2)$ using the RTS algorithm. This can then be repeated when z_3 arrives, and $\hat{x}(1|3)$ can be computed, and so forth. Also, it is obvious that the "older" filtering results that are no longer needed in the two-steps-back sweep can be discarded, so there is no growing memory problem. This process using the RTS algorithm can be continued indefinitely without any special problems. This solution does, of course, beg the issue of the start-up smoother results, but they may not be of great interest in some applications anyway.

There is an alternative to the RTS algorithm that can be applied to the fixed-lag problem when the number of lags is not excessive. In this method we augment the original state vector $x(k)$ with the delayed (i.e., lagged) states $x(k-1)$, $x(k-2)$, . . . $x(k-m)$ where m is the lag of primary interest. The new augmented state vector then becomes

$$\mathbf{X}(k) = \begin{bmatrix} x(k) \\ x(k-1) \\ x(k-2) \\ \vdots \\ x(k-m) \end{bmatrix}$$

We then treat the m-lag smoothing problem as an enlarged filter problem, and no new algorithms are needed. Example 6.3 will illustrate the procedure in detail.

EXAMPLE 6.3 AUGMENTED-STATE METHOD FOR FIXED-LAG SMOOTHING

Let us return to the Kalman filter model used for Example 6.1. The filter parameters will be repeated for easy reference:

$$\begin{aligned}
\sigma &= 1.0\,\text{m } (x \text{ process sigma}) \\
\beta &= 1.0\,\text{rad/s (reciprocal time constant)} \\
\Delta t &= 0.02\text{ s (step size)} \\
h &= 1 \text{ (measurement connection)} \\
r &= 1 \text{ (measurement error variance)} \\
q &= \sigma^2\,(1 - e^{-2\beta\Delta t}) \text{ (variance of process } w_k) \\
\phi &= \mathrm{e}^{-\beta\Delta t}
\end{aligned}$$

Now let us say that we want the solution for the two-steps-back smoothing problem. The measurement stream begins at $k=0$ (i.e., $t=0$). First, we define a new "super" state vector as the three-tuple:

$$\mathbf{X}(k) = \begin{bmatrix} x(k) \\ x(k-1) \\ x(k-2) \end{bmatrix}$$

Then, using the scalar difference equation for x and the definitions of the second and third elements of the super state $\mathbf{X}$, we have the discrete process equation:

$$\underbrace{\begin{bmatrix} x(k+1) \\ x(k) \\ x(k-1) \end{bmatrix}}_{\mathbf{X}(k+1)} = \underbrace{\begin{bmatrix} \phi & 0 & 0 \\ 1 & 0 & 0 \\ 0 & 1 & 0 \end{bmatrix}}_{\boldsymbol{\Phi}(k)} \underbrace{\begin{bmatrix} x(k) \\ x(k-1) \\ x(k-2) \end{bmatrix}}_{\mathbf{X}(k)} + \underbrace{\begin{bmatrix} w(k) \\ 0 \\ 0 \end{bmatrix}}_{\mathbf{W}(k)} \tag{6.4.1}$$

The measurements have a direct connection to only the first element of $\mathbf{X}(k)$, so the measurement equation for the super filter is:

$$z(k) = \underbrace{\begin{bmatrix} 1 & 0 & 0 \end{bmatrix}}_{\mathbf{H}(k)} \begin{bmatrix} x(k) \\ x(k-1) \\ x(k-2) \end{bmatrix} + v(k) \tag{6.4.2}$$

Note that Eqs. (6.4.1) and (6.4.2) are in the exact form required for a Kalman filter. (see Eqs. 4.2.1 and 4.2.2 in Chapter 4.) The $\boldsymbol{\Phi}$ and $\mathbf{H}$ parameters for the super filter are specified in Eqs. (6.4.1) and (6.4.2). It should also be apparent that the $\mathbf{Q}$ and $\mathbf{R}$ parameters for the super filter are:

$$\mathbf{Q}(k) = E\left[\mathbf{W}(k)\mathbf{W}^T(k)\right] = \begin{bmatrix} q & 0 & 0 \\ 0 & 0 & 0 \\ 0 & 0 & 0 \end{bmatrix} \tag{6.4.3}$$

$$\mathbf{R}(k) = E\left[v(k)v^T(k)\right] = r$$

We need one more parameter before running a filter covariance analysis in MATLAB. This is the initial *a priori* error covariance. Here we can use the same arguments for the super filter that we would use for any Kalman filter. Initially, we have no knowledge about x, other than its autocorrelation function and that it is Gaussian. In the example at hand the initial estimate is zero, and the error covariance is just the same as for the process itself. Therefore,

$$\mathbf{P}^-(0) = \sigma^2 \begin{bmatrix} 1 & e^{-\beta\Delta t} & e^{-2\beta\Delta t} \\ e^{-\beta\Delta t} & 1 & e^{-\beta\Delta t} \\ e^{-2\beta\Delta t} & e^{-\beta\Delta t} & 1 \end{bmatrix} \tag{6.4.4}$$

The augmented filter just described was run for 31 steps in MATLAB, and the results are shown in Table 6.2. Perhaps the first thing to notice in the table is that the entries in the first two rows contain nontrivial values for negative time (i.e., when $k-1$ and $k-2$ are negative). This is the start-up region that was alluded to in the discussion of the RTS solution. Here, the augmented filter yields meaningful results in

Table 6.2 Error Covariance Results for Fixed-Lag Smoother Exercise

k	P(k|k)	P(k – 1|k)	P(k – 2|k)
0	0.5000	0.5196	0.5384
1	0.3419	0.3419	0.3677
2	0.2689	0.2589	0.2689
3	0.2293	0.2153	0.2153
4	0.2061	0.1892	0.1839
⋮	⋮	⋮	⋮
15	0.1657	0.1439	0.1293
⋮	⋮	⋮	⋮
26	0.1653	0.1434	0.1287
27	0.1653	0.1434	0.1287
28	0.1653	0.1434	0.1287
29	0.1653	0.1434	0.1287
30	0.1653	0.1434	0.1287

the start-up region automatically without any special effort. For example, it is interesting to note that for $k = 0$ the one-step-back smoother estimate is poorer than the filter estimate. One might then ask, "How can that be? Isn't smoothing supposed to be better than filtering?" Answer: The most relevant measurement occurs at a later time $t = 0$, and its connection to x at an earlier time is only statistical, not direct one-to-one. Thus, the $t = 0$ filter estimate is better than the corresponding one-step-back estimate.

Note also that the augmented filter has converged to steady-state nicely by the end of the 31-step run. It is worthwhile mentioning that the three entries in the last row of Table 6.2 correspond exactly to the last three entries in the P(k|50) column of Table 6.1, which was for the RTS fixed-interval run. This is as it should be, because the parameters are the same for both examples, and both runs have reached steady-state by the ends of their respective runs.

■

6.5 ADAPTIVE KALMAN FILTER (MULTIPLE MODEL ADAPTIVE ESTIMATOR)

In the usual Kalman filter we assume that all the process parameters, that is, $\boldsymbol{\phi}_k$, $\mathbf{H}_k$, $\mathbf{R}_k$, and $\mathbf{Q}_k$, are known. They may vary with time (index k) but, if so, the nature of the variation is assumed to be known. In physical problems this is often a rash assumption. There may be large uncertainty in some parameters because of inadequate prior test data about the process. Or some parameter might be expected to change slowly with time, but the exact nature of the change is not predictable. In such cases, it is highly desirable to design the filter to be self-learning, so that it can adapt itself to the situation at hand, whatever that might be. This problem has received considerable attention since Kalman's original papers of the early 1960s. However, it is not an easy problem with one simple solution. This is evidenced by the fact that 50 years later we still see occasional papers on the subject in current control system journals.

We will concentrate our attention here on an adaptive filter scheme that was first presented by D. T. Magill (5). We will see presently that Magill's adaptive filter is not just one filter but, instead, is a whole bank of Kalman filters running in parallel. At the time that this scheme was first suggested in 1965, it was considered to be impractical for implementation on-line. However, the spectacular advances in computer technology over the past few decades have made Magill's parallel-filter scheme quite feasible in a number of applications (6, 7, 8, 9, 10, 11). Because of the parallel bank of filters, this scheme is usually referred to as the multiple model adaptive estimator (MMAE). In the interest of simplicity, we will confine our attention here to Magill's original MMAE scheme in its primitive form. It is worth mentioning that there have been many extensions and variations on the original scheme since 1965, including recent papers by Caputi (12) and Blair and Bar-Shalom (11). (These are interesting papers, both for their technical content and the references contained therein.) We will now proceed to the derivation that leads to the bank of parallel filters.

We begin with the simple statement that the desired estimator is to be the conditional mean given by

$$\hat{\mathbf{x}}_k = \int_{\mathbf{x}} \mathbf{x}\, p\left(\mathbf{x}|\mathbf{z}_k^*\right) d\mathbf{x} \tag{6.5.1}$$

where $\mathbf{z}_k^*$ denotes all the measurements up to and including time t_k (i.e., $\mathbf{z}_1, \mathbf{z}_2, \ldots, \mathbf{z}_k$), and $p\left(\mathbf{x}|\mathbf{z}_k^*\right)$ is the probability density function of $\mathbf{x}_k$ with the conditioning shown in parentheses.* The indicated integration is over the entire $\mathbf{x}$ space. If the $\mathbf{x}$ and $\mathbf{z}$ processes are Gaussian, we are assured that the estimate given by Eq. (6.5.1) will be optimal by almost any reasonable criterion of optimality, least-mean-square or otherwise (1). We also wish to assume that some parameter of the process, say, α, is unknown to the observer, and that this parameter is a random variable (not necessarily Gaussian). Thus, on any particular sample run it will be an unknown constant, but with a known statistical distribution. Hence, rather than beginning with $p\left(\mathbf{x}|\mathbf{z}_k^*\right)$, we really need to begin with the joint density $p\left(\mathbf{x}, \alpha|\mathbf{z}_k^*\right)$ and sum out on α to get $p\left(\mathbf{x}|\mathbf{z}_k^*\right)$. Thus, we will rewrite Eq. (6.5.1) in the form

$$\hat{\mathbf{x}}_k = \int_{\mathbf{x}} \mathbf{x} \int_{\alpha} p\left((\mathbf{x}, \alpha)|\mathbf{z}_k^*\right) d\alpha\, d\mathbf{x} \tag{6.5.2}$$

But the joint density in Eq. (6.5.2) can be written as

$$p\left(\mathbf{x}, \alpha|\mathbf{z}_k^*\right) = p\left(\mathbf{x}|\alpha, \mathbf{z}_k^*\right) p\left(\alpha|\mathbf{z}_k^*\right) \tag{6.5.3}$$

* Throughout this section we will use a looser notation than that used in Chapter 1 in that p will be used for both probability density and discrete probability. In this way we avoid the multitudinous subscripts that would otherwise be required for conditioned multivariate random variables. However, this means that the student must use a little imagination and interpret the symbol p properly within the context of its use in any particular derivation.

Substituting Eq. (6.5.3) into (6.5.2) and interchanging the order of integration lead to

$$\hat{\mathbf{x}}_k = \int_\alpha p(\alpha|\mathbf{z}_k^*) \int_{\mathbf{x}} \mathbf{x}\, p(\mathbf{x}|\alpha, \mathbf{z}_k^*)\, d\mathbf{x}\, d\alpha \tag{6.5.4}$$

The inner integral will be recognized as just the usual Kalman filter estimate for a given α. This is denoted as $\hat{\mathbf{x}}_k(\alpha)$ where α shown in parentheses is intended as a reminder that there is α dependence. Equation (6.5.4) may now be rewritten as

$$\hat{\mathbf{x}}_k = \int_\alpha \hat{\mathbf{x}}_k(\alpha) p(\alpha|\mathbf{z}_k^*)\, d\alpha \tag{6.5.5}$$

Or the discrete random variable equivalent to Eq. (6.5.5) would be

$$\hat{\mathbf{x}}_k = \sum_{i=1}^{L} \hat{\mathbf{x}}_k(\alpha_i) p(\alpha_i|\mathbf{z}_k^*) \tag{6.5.6}$$

where $p(\alpha_i|\mathbf{z}_k^*)$ is the discrete probability for α_i, conditioned on the measurement sequence $\mathbf{z}_k^*$. We will concentrate on the discrete form from this point on in our discussion.

Equation (6.5.6) simply says that the optimal estimate is a weighted sum of Kalman filter estimates with each Kalman filter operating with a separate assumed value of α. This is shown in Fig. 6.5. The problem now reduces to one of determining the weight factors $p(\alpha_1|\mathbf{z}_k^*)$, $p(\alpha_2|\mathbf{z}_k^*)$, etc. These, of course, change with each recursive step as the measurement process evolves in time. Presumably, as more and more measurements become available, we learn more about the state of the process and the unknown parameter α. (Note that it is constant for any particular sample run of the process.)

We now turn to the matter of finding the weight factors indicated in Fig. 6.5. Toward this end we use Bayes' rule:

$$p(\alpha_i|\mathbf{z}_k^*) = \frac{p(\mathbf{z}_k^*|\alpha_i)p(\alpha_i)}{p(\mathbf{z}_k^*)} \tag{6.5.7}$$

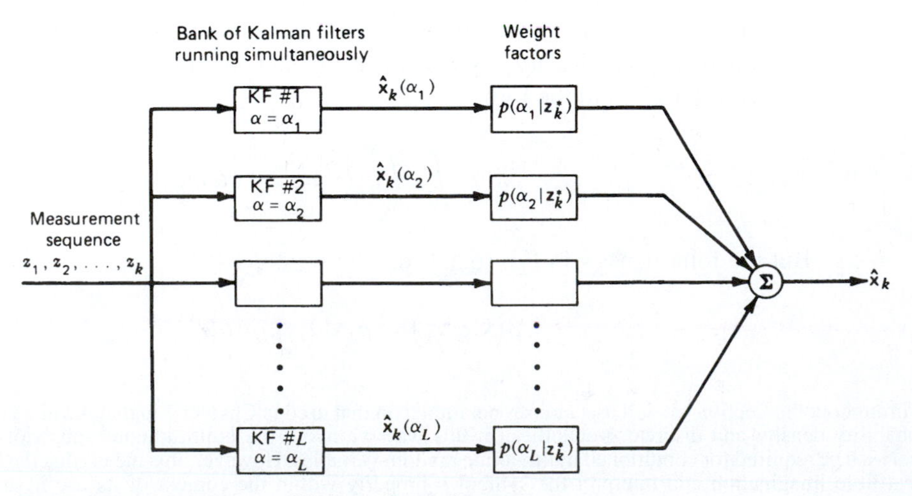

Figure 6.5 Weighted sum of Kalman filter estimates.

But

$$\begin{aligned} p(\mathbf{z}_k^*) &= \sum_{j=1}^{L} p(\mathbf{z}_k^*, \alpha_j) \\ &= \sum_{j=1}^{L} p(\mathbf{z}_k^* | \alpha_j) p(\alpha_j) \end{aligned} \tag{6.5.8}$$

Equation (6.5.8) may now be substituted into Eq. (6.5.7) with the result

$$p(\alpha_1 | \mathbf{z}_k^*) = \left[\frac{p(\mathbf{z}_k^* | \alpha_i) p(\alpha_i)}{\sum_{j=1}^{L} p(\mathbf{z}_k^* | \alpha_j) p(\alpha_j)} \right], \quad i = 1, 2, \ldots, L \tag{6.5.9}$$

The distribution $p(\alpha_i)$ is presumed to be known, so it remains to determine $p(\mathbf{z}_k^* | \alpha_i)$ in Eq. (6.5.9). Toward this end we will write $p(\mathbf{z}_k^* | \alpha_i)$ as a product of conditional density functions. Temporarily omitting the α_i, conditioning (just to save writing), we have

$$\begin{aligned} p(\mathbf{z}_k^*) &= p(\mathbf{z}_k, \mathbf{z}_{k-1}, \ldots \mathbf{z}_0) \\ &= p(\mathbf{z}_k, \mathbf{z}_{k-1}, \ldots \mathbf{z}_1 | \mathbf{z}_0) p(\mathbf{z}_0) \\ &= p(\mathbf{z}_k, \mathbf{z}_{k-1}, \ldots \mathbf{z}_2 | \mathbf{z}_1, \mathbf{z}_0) p(\mathbf{z}_1 | \mathbf{z}_0) p(\mathbf{z}_0) \\ &\;\;\vdots \\ &= p(\mathbf{z}_k | \mathbf{z}_{k-1}, \mathbf{z}_{k-2}, \ldots \mathbf{z}_0) p(\mathbf{z}_{k-1} | \mathbf{z}_{k-2}, \mathbf{z}_{k-3}, \ldots \mathbf{z}_0) \ldots p(\mathbf{z}_1 | \mathbf{z}_0) p(\mathbf{z}_0), k = 1, 2, \ldots \end{aligned} \tag{6.5.10}$$

We now note that the first term in the product string of Eq. (6.5.10) is just $p(\hat{\mathbf{z}}_k^-)$, and that the remaining product is just $p(\mathbf{z}_{k-1}^*)$. Thus, we can rewrite Eq. (6.5.10) in the form

$$p(\mathbf{z}_k^*) = p(\hat{\mathbf{z}}_k^-) p(\mathbf{z}_{k-1}^*) \tag{6.5.11}$$

We now make the Gaussian assumption for the $\mathbf{x}$ and $\mathbf{z}$ processes (but not for α). Also, to simplify matters, we will assume $\mathbf{z}_k^*$ to be a sequence of scalar measurements $z_0, z_1, \ldots z_k$. Equation (6.5.11) then becomes

$$p(\mathbf{z}_k^*) = \frac{1}{(2\pi)^{1/2} \left(\mathbf{H}_k \mathbf{P}_k^- \mathbf{H}_k^T + R_k \right)^{1/2}} \exp \left[-\frac{1}{2} \frac{\left(z_k - \mathbf{H}_k \hat{\mathbf{x}}_k^- \right)^2}{\left(\mathbf{H}_k \mathbf{P}_k^- \mathbf{H}_k^T + R_k \right)} \right] p(\mathbf{z}_{k-1}^*), \quad k = 1, 2, \ldots \tag{6.5.12}$$

Bear in mind that $p(\mathbf{z}_k^*)$ will, in general, be different for each α_i. For example, if the unknown parameter is R_k, each filter in the bank of filters will be modeled around a different value for R_k.

It should be helpful at this point to go through an example step by step (in words, at least) to see how the parallel bank of filters works.

1. We begin with the prior distribution of α and set the filter weight factors accordingly. Frequently, we have very little prior knowledge of the unknown parameter α. In this case we would assume a uniform probability

distribution and set all the weight factors equal initially. This does not have to be the case in general, though.

2. The initial prior estimates $\hat{\mathbf{x}}_0^-$ for each filter are set in accordance with whatever prior information is available. Usually, if $\mathbf{x}$ is a zero-mean process, $\hat{\mathbf{x}}_0^-$ is simply set equal to zero for each filter. We will assume that this is true in this example.
3. Usually, the initial estimate uncertainty does not depend on α, so the initial $\mathbf{P}_0^-$ for each filter will just be the covariance matrix of the $\mathbf{x}$ process.
4. Initially then, before any measurements are received, the prior estimates from each of the filters are weighted by the initial weight factors and summed to give the optimal prior estimate from the bank of filters. In the present example this is trivial, because the initial estimates for each filter were set to zero.
5. At $k=0$ the bank of filters receives the first measurement z_0, and the unconditional $p(z_0)$ must be computed for each permissible α_i. We note that $z = \mathbf{Hx} + v$, so $p(z_0)$ may be written as

$$p(z_0) = \frac{1}{(2\pi)^{1/2}\left(\mathbf{H}_0\mathbf{C}_x\mathbf{H}_0^T + R_0\right)^{1/2}} \exp\left[-\frac{1}{2}\frac{z_0^2}{\left(\mathbf{H}_0\mathbf{C}_x\mathbf{H}_0^T + R_0\right)}\right] \tag{6.5.13}$$

where $\mathbf{C}_x$ is the covariance matrix of $\mathbf{x}$. We note again that one or more of the parameters in Eq. (6.5.13) may have α dependence; thus, in general, $p(z_0)$ will be different for each α_i.

6. Once $p(z_0)$ for each α_i has been determined, Eq. (6.5.9) may be used to find $p(\alpha_i | z_0)$. These are the weight factors to be used in summing the updated estimates $\hat{\mathbf{x}}_0$ that come out of each of the filters in the bank of filters. This then yields the optimal adaptive estimate, given the measurement z_0, and we are ready to project on to the next step.
7. Each of the individual Kalman filter estimates and their error covariances is projected ahead to $k=1$ in the usual manner. The adaptive filter must now compute $p\left(\mathbf{z}_1^*\right)$ for each α_i, and it uses the recursive formula, Eq. (6.5.12), in doing so. Therefore, for $p\left(\mathbf{z}_1^*\right)$ we have

$$p\left(\mathbf{z}_1^*\right) = \frac{1}{(2\pi)^{1/2}\left(\mathbf{H}_1\mathbf{P}_1^-\mathbf{H}_1^T + R_1\right)^{1/2}} \exp\left[-\frac{1}{2}\frac{\left(z_1 - \mathbf{H}_1\hat{\mathbf{x}}_1^-\right)^2}{\left(\mathbf{H}_1\mathbf{P}_1^-\mathbf{H}_1^T + R_1\right)}\right] p\left(\mathbf{z}_0^*\right) \tag{6.5.14}$$

Note that $p\left(\mathbf{z}_0^*\right)$ was computed in the previous step, and the prior $\hat{\mathbf{x}}_1^-$ and $\mathbf{P}_1^-$ for each α_i are obtained from the projection step.

8. Now, the $p\left(\mathbf{z}_1^*\right)$ determined in Step 7 can be used in Bayes' formula, Eq. (6.5.9), and the weight factors $p\left(\alpha_i | \mathbf{z}_1^*\right)$ for $k=1$ are thus determined. It should be clear now that this recursive procedure can be carried on *ad infinitum*.

We are now in a position to reflect on the whole adaptive filter in perspective. At each recursive step the adaptive filter does three things: (1) Each filter in the bank of filters computes its own estimate, which is hypothesized on its own model; (2) the system computes the a posteriori probabilities for each of the hypotheses; and (3) the scheme forms the adaptive optimal estimate of **x** as a weighted sum of the estimates produced by each of the individual Kalman filters. As the measurements evolve with time, the adaptive scheme learns which of the filters is the correct one, and its weight factor approaches unity while the others are going to zero. The bank of filters accomplishes this, in effect, by looking at sums of the weighted squared measurement residuals. The filter with the smallest residuals "wins," so to speak.

The Magill scheme just described is not without some problems and limitations (12). It is still important, though, because it is optimum (within the various assumptions that were made), and it serves as a point of departure for other less rigorous schemes. One of the problems of this technique has to do with numerical behavior as the number of steps becomes large. Clearly, as the sum of the squared measurement residuals becomes large, there is a possibility of computer underflow. Also, note that the unknown parameter was assumed to be constant with time. Thus, there is no way this kind of adaptive filter can readjust if the parameter actually varies slowly with time. This adaptive scheme, in its purest form, never forgets; it tends to give early measurements just as much weight as later ones in its determination of the a posteriori probabilities. Some *ad hoc* procedure, such as periodic reinitialization, has to be used if the scheme is to adapt to a slowly varying parameter situation.

By its very nature, the Magill adaptive filter is a transient scheme, and it converges to the correct α_i in an optimal manner (provided, of course, that the various assumptions are valid). This is one of the scheme's strong points, and it gives rise to a class of applications that are usually not thought of as adaptive filter problems. These are applications where we are more interested in the *a posteriori* probabilities than in the optimal estimate of the vector **x** process. The Magill scheme is an excellent multiple-hypothesis testor in the presence of Gauss–Markov noise. In this setting the main objective of the bank of filters is to determine which hypothesized model in the bank of filters is the correct one, and there have been a number of applications of this type reported in the literature (7, 8, 9). We will discuss one of these briefly at the end of this section. Before doing so, though, we wish to emphasize the assumptions that were used in deriving Magill's parallel filter scheme. These assumptions are most important in understanding the limitations of the scheme. The key assumptions are as follows:

1. We assume Gaussian statistics for the **x** and **z** processes, but not for α.
2. The unknown parameter α is assumed to be a discrete random variable with a known distribution. Thus, for any sample run, α will be constant with time. If, in fact, α is an unknown *deterministic* parameter (in contrast to being random), then we must be cautious about our conclusions relative of optimally.
3. It is tacitly assumed that the process and measurement models are of the proper form for a linear Kalman filter for each allowable α_i. (Note that finite-order state models will not exist for all random processes, band-limited white noise being a good example.)

We will now look at an application of the Magill adaptive filter where both the hypothesis and the optimal estimate are of interest.

EXAMPLE 6.4

The use of carrier phase measurements for accurate relative positioning has brought about one of the more exciting uses of GPS in its three decades of existence thus far. One of the main challenges that come with using the carrier phase measurement is the introduction of an integer ambiguity. Fig. 6.6 shows the notional picture of the measurement geometry that relates the difference between carrier phase measurements to the physical separation between the antennas of the two receivers.

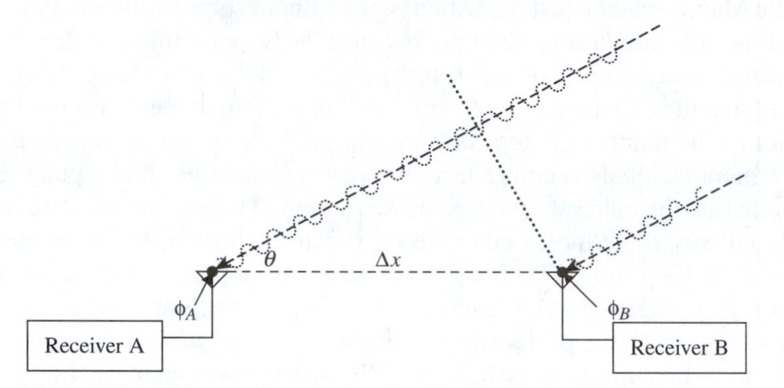

Figure 6.6 Difference between carrier phases measured between receivers, with an additional an integer cycle ambiguity, is an accurate representation of slant range difference that can be related to Δx via the geometry.

If we simplify this illustration to be a one-dimensional problem, the measurement model may be written as follows (measurements are in units of cycles):

$$\phi_A - \phi_B = h \cdot \frac{1}{\lambda} \Delta x + N + v_\phi \tag{6.5.15}$$

where $h = \cos\theta$ and we choose θ, the satellite elevation angle, to be $\frac{\pi}{6}$ radians (30°).

Assume also for this simplified example that the receivers have perfect clocks. The unknown states to solve for in Eq. (6.5.15) are the one-dimensional physical baseline separation Δx and some ambiguity N that is an unknown integer number of cycles. Clearly, this one measurement is not enough to resolve the two associated states. In GPS, carrier phase measurements made at its two carrier frequencies can be combined to form a lower beat frequency with a longer wavelength than either of the two individual ones—this hybrid carrier is often referred to as having a *widelane* wavelength. At the same time, coarser code phase (pseudorange) measurements that are not complicated by any cycle ambiguities are also used to bound the range of ambiguity uncertainties.

Our combined measurement model now becomes:

$$\begin{bmatrix} \rho_A - \rho_B \\ \phi_A^{(L1L2)} - \phi_B^{(L1L2)} \end{bmatrix} = \begin{bmatrix} h \\ \dfrac{h}{\lambda_{L1L2}} \end{bmatrix} \Delta x + \begin{bmatrix} 0 \\ N_{L1L2} \end{bmatrix} + \begin{bmatrix} v_\rho \\ v_\phi \end{bmatrix}$$

$$= \begin{bmatrix} h & 0 \\ \dfrac{h}{\lambda_{L1}} & 1 \end{bmatrix} \begin{bmatrix} \Delta x \\ N_{L1L2} \end{bmatrix} + \begin{bmatrix} v_\rho \\ v_\phi \end{bmatrix} \tag{6.5.16}$$

In the state vector, we have an integer ambiguity state N_{L1L2} that is associated with the widelane carrier phase wavelength. N_{L1L2} is constant and so is modeled as a random bias. For this example, we allow the two receivers to wander so the Δx state is modeled as a random walk with a one-step uncertainty of $(1.0\,\text{m})^2$. Therefore, the process noise covariance matrix is given by:

$$\mathbf{Q} = \begin{bmatrix} (1.0\,m)^2 & 0 \\ 0 & 0 \end{bmatrix}$$

The measurement noise ν vector is associated with the error covariance matrix given by:

$$\mathbf{R} = \begin{bmatrix} (1\,m)^2 & 0 \\ 0 & (0.02\,cycle)^2 \end{bmatrix}$$

The discrete nature of the multiple hypothesis space of the Magill scheme lends itself well to solving problems with integer constraints. The approach we take in this example is to use the Magill adaptive filter to process the available set of noisy measurements with various hypotheses of the correct integer ambiguity. Eventually, one particular hypothesis will win out and its filter estimate will represent the correct answer as well. As we have seen earlier, the α parameter (see Fig. 6.5) of the Magill adaptive filter represents the discrete hypothesis. So here, in this example, α represents an integer ambiguity, N_{L1L2}. Each hypothesis in the parallel bank of Kalman filters will assume a specific integer value. What is interesting is that the error covariance is common to all the hypotheses so the gain and the error covariance computations do not need to be replicated among the various hypotheses, but rather computed just once for each measurement cycle.

The state updates and projection, however, are different for each hypothesis since each ambiguity modifies the measurements to give rise to residuals differently. Fig. 6.7 shows a representation of this particular Magill adaptive filter. The *a priori* and *a posteriori* state estimates must be computed for each hypothesis. Or so it seems!

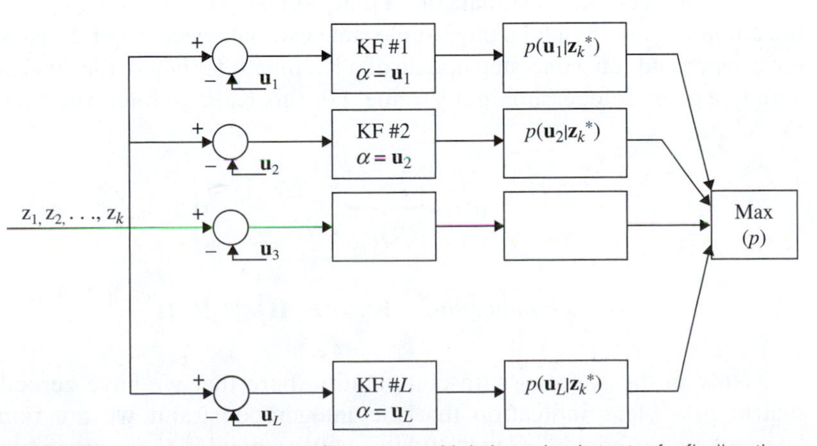

Figure 6.7 The Magill adaptive filter as a multiple hypothesis tester for finding the correct set of integer ambiguities.

It turns out that there is a very clear connectedness that exists between all the hypotheses as a result of the linear connection of the ambiguity within the measurement model. In other words, even the state updates and projections need not be separately accounted for in each hypothesis. A maximum likelihood methodology is outlined in Ref. (13) for a similar problem that involves integer ambiguities in GPS carrier phase relative positioning. In this example, we will consider a slightly different approach that expands the Kalman filter to take the place of the maximum likelihood estimator.

The associated **P** matrix of this Kalman filter is a 2×2 matrix:

$$\mathbf{P} = \left[\begin{array}{c|c} \mathbf{P}_{\Delta x} & \mathbf{P}_{\Delta x,N} \\ \hline \mathbf{P}_{\Delta x,N}^{T} & \mathbf{P}_{N} \end{array}\right] \tag{6.5.17}$$

The lower-right term represents the error variance of the ambiguity state estimate. Together with the relevant state from the updated state estimate vector, we can reconstruct these relationships:

$$\mathbf{W} = \left[\begin{array}{c|c} \mathbf{P}_{N}^{-1} & \mathbf{P}_{N}^{-1}\hat{N}_{L1L2} \\ \hline \hat{N}_{L1L2}\mathbf{P}_{N}^{-1} & w_{nn} \end{array}\right] \tag{6.5.18}$$

where w_{nn} is arbitrary and thus can be set to zero.

The *a priori* conditional probability $p(\mathbf{z}^*|\alpha)$ is computed for each sample in the α-space, which in this case is the ambiguity N_{L1L2}. Let us designate **u** to represent the ambiguity. Then, we have

$$p(\mathbf{z}^*|\mathbf{u}) = \exp\left(-\frac{1}{2}\mathbf{u}_a^{\mathrm{T}}\mathbf{W}\mathbf{u}_a\right) \tag{6.5.19}$$

where $\mathbf{u}_a = \begin{bmatrix} \mathbf{u} \\ 1 \end{bmatrix}$ and $\mathbf{u} = \ldots -1, 0, +1, +2, \ldots$

The *a posteriori* conditional probability $p(\alpha|\mathbf{z}^*)$ still follows Eq. (6.5.9).

For each ambiguity **u**, we can also extract a unique updated state estimate that is constrained by the given **u**. Note that the state estimate vector from the model given in Eq. (6.5.2) is a 2-tuple consisting of $[\Delta\hat{x} \quad \hat{N}_{L1L2}]^T$.

To obtain the state estimate of Δx that has been constrained with fixed values of the ambiguity, we can take the 2-tuple state estimate vector and its associated error covariance and run a one-step update of a Kalman filter where the "measurement" is simply a given integer ambiguity value. For this Kalman filter, the "measurement" model is:

$$\mathbf{u} = \underbrace{[0 \quad 1]}_{\mathbf{H}_c}\begin{bmatrix} \Delta\hat{x} \\ \hat{N}_{L1L2} \end{bmatrix} + 0 \tag{6.5.20}$$

$$\textit{Kalman gain:} \quad \mathbf{K}_c = \mathbf{P}^{-}\mathbf{H}_c^{\mathrm{T}}(\mathbf{H}_c\mathbf{P}^{-}\mathbf{H}_c^{T})^{-1}$$

Note in the Kalman gain computation here that we have zeroed out the **R** matrix in a clear indication that the integer constraint we are imposing is a "perfect measurement." Generally, it is never a good idea to zero out the **R** matrix in a Kalman filter as it tends to lead to numerical instability. Here, however, the problem is diminished as we are only carrying this out for only the state estimate update step.

$$\textit{State update:} \quad \hat{\mathbf{x}}_c = (\mathbf{I}\text{-}\mathbf{K}_c\mathbf{H}_c)\hat{\mathbf{x}}^- + \mathbf{K}_c\mathbf{z}_c$$

The second term of the updated state estimate vector will end up with the exact value of the integer ambiguity input as a "perfect measurement" in $\mathbf{z}_c$. Only the first term of the updated state estimate vector is of interest since this represents the Δx estimate for the given integer ambiguity that represents this hypothesis.

$$\textit{Covariance update:} \quad \mathbf{P}_c = (\mathbf{I}\text{-}\mathbf{K}_c\mathbf{H}_c)\mathbf{P}^-$$

The updated error covariance only needs to be computed if one is interested in the error variance associated with the first term. The variances associated with the second term is nominally zero as might be expected from this process of constraining the solution to the specific integer ambiguity associated with the given hypothesis.

For the two GPS frequencies at 1575.42 MHz and 1227.6 MHz, the widelane wavelength is given by:

$$\lambda_{L1L2} = \frac{c}{f_{L1} - f_{L2}} = 0.861918\,\text{m}$$

Fig. 6.8 shows the weight factor (or probability) associated with the correct ambiguity in this example. After a few processing steps, the probability quickly approaches one indicating overwhelming certainty on the correctness of this hypothesis. During this time, the probabilities associated with all other hypotheses correspondingly diminish to zero.

Fig. 6.9 shows the weighted state estimate of Δx. For this example, the true value of Δx is 100 m. Quite clearly, the solution becomes dominated by the weight of the state estimate from the correct hypothesis as it converges to the true value over time.

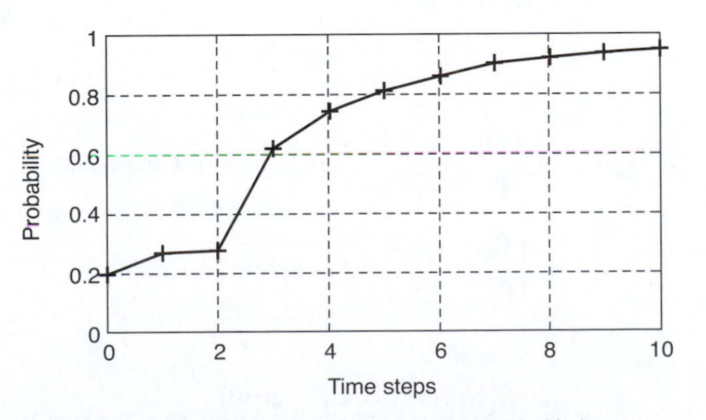

Figure 6.8 Probability or weight factor associated with the correct hypothesis rightly converges to one after processing several measurements.

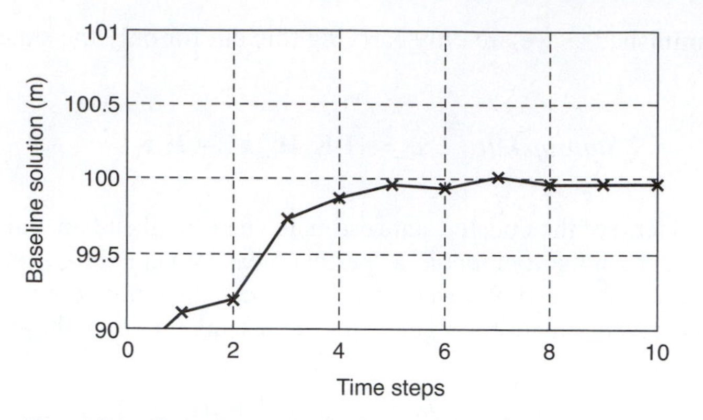

Figure 6.9 State estimate associated with the correct hypothesis converges to the true value of $\Delta x = 100$ m.

6.6 CORRELATED PROCESS AND MEASUREMENT NOISE FOR THE DISCRETE FILTER—DELAYED-STATE FILTER ALGORITHM

Physical situations occasionally arise where the discrete process and measurement noises are correlated. This correlation is not accounted for in the usual Kalman filter equations given in Chapter 4. It can be, though, so this modification and an application are considered further in this section.

Discrete Filter—Correlated Process and Measurement Noise

Just as for the continuous filter, we first define the process and measurement models. They are as follows:

$$\left.\begin{aligned} \mathbf{x}_{k+1} &= \boldsymbol{\phi}_k \mathbf{x}_k + \mathbf{w}_k \\ \mathbf{z}_k &= \mathbf{H}_k \mathbf{x}_k + \mathbf{v}_k \end{aligned}\right\} \text{(as before in Chapter 4)} \qquad \begin{aligned} (6.6.1) \\ (6.6.2) \end{aligned}$$

where

$$E\left[\mathbf{w}_k \mathbf{w}_i^T\right] = \begin{cases} \mathbf{Q}_k, & i = k \\ \mathbf{0}, & i \neq k \end{cases} \quad \text{(as before in Chapter 4)} \tag{6.6.3}$$

$$E\left[\mathbf{v}_k \mathbf{v}_i^T\right] = \begin{cases} \mathbf{R}_k, & i = k \\ \mathbf{0}, & i \neq k \end{cases} \quad \text{(as before in Chapter 4)} \tag{6.6.4}$$

and

$$E\left[\mathbf{w}_{k-1} \mathbf{v}_k^T\right] = \mathbf{C}_k \quad \text{(new)} \tag{6.6.5}$$

Before we proceed, an explanation is in order as to why we are concerned with the crosscorrelation of $\mathbf{v}_k$ with $\mathbf{w}_{k-1}$ rather than $\mathbf{w}_k$, which one might expect from just a casual look at the problem. Rewriting Eq. (6.6.1) with k retarded one step should help in this regard:

$$\mathbf{x}_k = \boldsymbol{\phi}_{k-1}\mathbf{x}_{k-1} + \mathbf{w}_{k-1} \tag{6.6.6}$$

Note that it is $\mathbf{w}_{k-1}$ (and not $\mathbf{w}_k$) that represents the cumulative effect of the white forcing function in the continuous model in the interval (t_{k-1}, t_k). Similarly, $\mathbf{v}_k$ represents the cumulative effect of the white measurement noise in the continuous model when averaged over the same interval (t_{k-1}, t_k) (provided, of course, that we begin with a continuous model). Therefore, if we wish to have a correspondence between the continuous and discrete models for small Δt, it is the crosscorrelation between $\mathbf{v}_k$ and $\mathbf{w}_{k-1}$ that we need to include in the discrete model. This is, of course, largely a matter of notation, but an important one. We will continue with the assumption that we have the discrete model given by Eqs. (6.6.1 through 6.6.5), and we begin with the usual update equation

$$\hat{\mathbf{x}}_k = \hat{\mathbf{x}}_k^- + \mathbf{K}_k\left(\mathbf{z}_k - \mathbf{H}_k\hat{\mathbf{x}}_k^-\right) \tag{6.6.7}$$

Next, we form the expression for the estimation error.

$$\begin{aligned}
\mathbf{e}_k &= \mathbf{x}_k - \hat{\mathbf{x}}_k \\
&= \mathbf{x}_k - \left[\hat{\mathbf{x}}_k^- + \mathbf{K}_k\left(\mathbf{z}_k - \mathbf{H}_k\hat{\mathbf{x}}_k^-\right)\right] \\
&= (\mathbf{I} - \mathbf{K}_k\mathbf{H}_k)\mathbf{e}_k^- - \mathbf{K}_k\mathbf{v}_k
\end{aligned} \tag{6.6.8}$$

We anticipate now that $\mathbf{e}_k^-$ and $\mathbf{v}_k$ will be correlated, so we will work this out as a side problem:

$$\begin{aligned}
E\left[\mathbf{e}_k^-\mathbf{v}_k^T\right] &= E\left[\left(\mathbf{x}_k - \hat{\mathbf{x}}_k^-\right)\mathbf{v}_k^T\right] \\
&= E\left[(\boldsymbol{\phi}_{k-1}\mathbf{x}_{k-1} + \mathbf{w}_{k-1} - \boldsymbol{\phi}_{k-1}\hat{\mathbf{x}}_{k-1})\mathbf{v}_k^T\right]
\end{aligned} \tag{6.6.9}$$

Note that $\mathbf{v}_k$ will not be correlated with either $\mathbf{x}_{k-1}$, or $\hat{\mathbf{x}}_{k-1}$ because of its whiteness. Therefore, Eq. (6.6.9) reduces to

$$E\left[\mathbf{e}_k^-\mathbf{v}_k^T\right] = E\left[\mathbf{w}_{k-1}\mathbf{v}_k^T\right] = \mathbf{C}_k \tag{6.6.10}$$

We now return to the main derivation. By using Eq. (6.6.8), we form the expression for the $\mathbf{P}_k$ matrix:

$$\begin{aligned}
\mathbf{P}_k &= E\left[\mathbf{e}_k\mathbf{e}_k^T\right] \\
&= E\left\{\left[(\mathbf{I} - \mathbf{K}_k\mathbf{H}_k)\mathbf{e}_k^- - \mathbf{K}_k\mathbf{v}_k\right]\left[(\mathbf{I} - \mathbf{K}_k\mathbf{H}_k)\mathbf{e}_k^- - \mathbf{K}_k\mathbf{v}_k\right]^T\right\}
\end{aligned} \tag{6.6.11}$$

Now, expanding Eq. (6.6.11) and taking advantage of Eq. (6.6.10) lead to

$$\begin{aligned}
\mathbf{P}_k &= (\mathbf{I} - \mathbf{K}_k\mathbf{H}_k)\mathbf{P}_k^-(\mathbf{I} - \mathbf{K}_k\mathbf{H}_k)^T + \mathbf{K}_k\mathbf{R}_k\mathbf{K}_k^T \\
&\quad -(\mathbf{I} - \mathbf{K}_k\mathbf{H}_k)\mathbf{C}_k\mathbf{K}_k^T - \mathbf{K}_k\mathbf{C}_k^T(\mathbf{I} - \mathbf{K}_k\mathbf{H}_k)^T
\end{aligned} \tag{6.6.12}$$

This is a perfectly general expression for the error covariance and is valid for any gain $\mathbf{K}_k$. The last two terms in Eq. (6.6.12) are "new" and involve the crosscorrelation parameter $\mathbf{C}_k$.

We now follow the same procedure used in Section 4.2 to find the optimal gain. We differentiate trace $\mathbf{P}_k$ with respect to $\mathbf{K}_k$ and set the result equal to zero. The necessary matrix differentiation formulas are given in Section 4.2, and the resulting optimal gain is

$$\mathbf{K}_k = \left(\mathbf{P}_k^- \mathbf{H}_k^T + \mathbf{C}_k\right)\left[\mathbf{H}_k \mathbf{P}_k^- \mathbf{H}_k^T + \mathbf{R}_k + \mathbf{H}_k \mathbf{C}_k + \mathbf{C}_k^T \mathbf{H}_k^T\right]^{-1} \tag{6.6.13}$$

Note that this expression is similar to the gain formula of Chapter 4 except for the additional terms involving $\mathbf{C}_k$. Let $\mathbf{C}_k$ go to zero, and Eq. (6.6.13) reduces to the same gain as in the zero crosscorrelation model, which is as it should be.

We can now substitute the optimal gain expression, Eq. (6.6.13), into the general $\mathbf{P}_k$ equation, Eq. (6.6.12), to get the a posteriori $\mathbf{P}_k$ equation. After some algebraic manipulation, this leads to either of the two forms:

$$\mathbf{P}_k = \mathbf{P}_k^- - \mathbf{K}_k\left[\mathbf{H}_k \mathbf{P}_k^- \mathbf{H}_k^T + \mathbf{R}_k + \mathbf{H}_k \mathbf{C}_k + \mathbf{C}_k^T \mathbf{H}_k^T\right]\mathbf{K}_k^T \tag{6.6.14}$$

or

$$\mathbf{P}_k = (\mathbf{I} - \mathbf{K}_k \mathbf{H}_k)\mathbf{P}_k^- - \mathbf{K}_k \mathbf{C}_k^T \tag{6.6.15}$$

The projection equations are not affected by the crosscorrelation between $\mathbf{w}_{k-1}$ and $\mathbf{v}_k$ because of the whiteness property of each. Therefore, the projection equations are (repeated here for completeness)

$$\hat{\mathbf{x}}_{k+1}^- = \boldsymbol{\phi}_k \hat{\mathbf{x}}_k \tag{6.6.16}$$

$$\mathbf{P}_{k+1}^- = \boldsymbol{\phi}_k \mathbf{P}_k \boldsymbol{\phi}_k^T + \mathbf{Q}_k \tag{6.6.17}$$

Equations (6.6.7), (6.6.13), (6.6.15), (6.6.16), and (6.6.17) now comprise the complete set of recursive equations for the correlated process and measurement noise case.

Delayed-State Measurement Problem

There are numerous dynamical applications where position and velocity are chosen as state variables. It is also common to have *integrated* velocity over some Δt interval as one of the measurements. In some applications the integration is an intrinsic part of the measurement mechanism, and an associated accumulative "count" is the actual measurement that is available to the Kalman filter (e.g., integrated doppler in a digital GPS receiver. Other times, integration may be performed on the velocity measurement to presmooth the high-frequency noise. In either case, these measurement situations are described by (in words):

$$\begin{aligned}
&(\text{Discrete measurement observed at time } t_k)\\
&= \int_{t_{k-1}}^{t_k} (\text{velocity})dt + (\text{discrete noise})\\
&= (\text{position at } t_k) - (\text{position at } t_{k-1}) + (\text{discrete noise})
\end{aligned} \tag{6.6.18}$$

Or, in general mathematical terms, the measurement equation is of the form

$$\mathbf{z}_k = \mathbf{H}_k\mathbf{x}_k + \mathbf{J}_k\mathbf{x}_{k-1} + \mathbf{v}_k \tag{6.6.19}$$

This, of course, does not fit the required format for the usual Kalman filter because of the $\mathbf{x}_{k-1}$ term. In practice, various approximations have been used to accommodate the delayed-state term: some good, some not so good. (One of the poorest approximations is simply to consider the integral of velocity divided by Δt to be a measure of the instantaneous velocity at the end point of the Δt interval.) The correct way to handle the delayed-state measurement problem, though, is to modify the recursive equations so as to accommodate the $\mathbf{x}_{k-1}$, term exactly (14). This can be done with only a modest increase in complexity, as will be seen presently.

We begin by noting that the recursive equation for $\mathbf{x}_k$ can be shifted back one step, that is,

$$\mathbf{x}_k = \boldsymbol{\phi}_{k-1}\mathbf{x}_{k-1} + \mathbf{w}_{k-1} \tag{6.6.20}$$

Equation (6.6.20) can now be rewritten as

$$\mathbf{x}_{k-1} = \boldsymbol{\phi}_{k-1}^{-1}\mathbf{x}_k - \boldsymbol{\phi}_{k-1}^{-1}\mathbf{w}_{k-1} \tag{6.6.21}$$

and this can be substituted into the measurement equation, Eq. (9.2.19), that yields

$$\mathbf{z}_k = \underbrace{\left(\mathbf{H}_k + \mathbf{J}_k\boldsymbol{\phi}_{k-1}^{-1}\right)}_{\text{New } \mathbf{H}_k}\mathbf{x}_k + \underbrace{\left(-\mathbf{J}_k\boldsymbol{\phi}_{k-1}^{-1}\mathbf{w}_{k-1} + \mathbf{v}_k\right)}_{\text{New } \mathbf{v}_k} \tag{6.6.22}$$

Equation (6.6.22) now has the proper form for a Kalman filter, but the new $\mathbf{v}_k$ term is obviously correlated with the process $\mathbf{w}_{k-1}$, term. We can now take advantage of the correlated measurement-process noise equations that were derived in the first part of this section. Before doing so, though, we need to work out the covariance expression for the new $\mathbf{v}_k$ term and also evaluate $\mathbf{C}_k$ for this application.

We will temporarily let the covariance associated with new $\mathbf{v}_k$ be denoted as "New $\mathbf{R}_k$," and it is

$$\text{New } \mathbf{R}_k = E\left[\left(-\mathbf{J}_k\boldsymbol{\phi}_{k-1}^{-1}\mathbf{w}_{k-1} + \mathbf{v}_k\right)\left(-\mathbf{J}_k\boldsymbol{\phi}_{k-1}^{-1}\mathbf{w}_{k-1} + \mathbf{v}_k\right)^T\right] \tag{6.6.23}$$

We note now that $\mathbf{v}_k$ and $\mathbf{w}_{k-1}$ are uncorrelated. Therefore,

$$\text{New } \mathbf{R}_k = \mathbf{J}_k\boldsymbol{\phi}_{k-1}^{-1}\mathbf{Q}_{k-1}\boldsymbol{\phi}_{k-1}^{-1}\mathbf{J}_k^T + \mathbf{R}_k \tag{6.6.24}$$

Also, with reference to Eq. (6.6.5), we can write $\mathbf{C}_k$ as

$$\mathbf{C}_k = E\left[\mathbf{w}_{k-1}\left(-\mathbf{J}_k\boldsymbol{\phi}_{k-1}^{-1}\mathbf{w}_{k-1} + \mathbf{v}_k\right)^T\right] = -\mathbf{Q}_{k-1}\boldsymbol{\phi}_{k-1}^{-1}\mathbf{J}_k^T \tag{6.6.25}$$

In this application, we can now make the following replacements in Eqs. (6.6.7), (6.6.13), and (6.6.14):

$$\mathbf{H}_k \rightarrow \mathbf{H}_k + \mathbf{J}_k \boldsymbol{\phi}_{k-1}^{-1} \tag{6.6.26}$$

$$\mathbf{R}_k \rightarrow \mathbf{R}_k + \mathbf{J}_k \boldsymbol{\phi}_{k-1}^{-1} \mathbf{Q}_{k-1} \boldsymbol{\phi}_{k-1}^{-1} \mathbf{J}_k^T \tag{6.6.27}$$

$$\mathbf{C}_k \rightarrow -\mathbf{Q}_{k-1} \boldsymbol{\phi}_{k-1}^{-1} \mathbf{J}_k^T \tag{6.6.28}$$

where $\rightarrow$ means "is replaced by." After the indicated replacements are made in the recursive equations, the result is a relatively complicated set of equations that involve, among other things, the inverse of $\boldsymbol{\phi}_{k-1}$. This is a computation that is not required in the usual recursive equations, and it can be eliminated with appropriate algebraic substitutions. The key step is to eliminate $\mathbf{Q}_{k-1}$, by noting that

$$\mathbf{Q}_{k-1} = \mathbf{P}_k^- - \boldsymbol{\phi}_{k-1} \mathbf{P}_{k-1} \boldsymbol{\phi}_{k-1}^T \tag{6.6.29}$$

and that the inverse of the transpose is the transpose of the inverse. The final resulting recursive equations for the delayed-state measurement situation can then be written in the form:

Estimate update:

$$\hat{\mathbf{x}}_k = \hat{\mathbf{x}}_k^- + \mathbf{K}_k \left(\mathbf{z}_k - \hat{\mathbf{z}}_k^- \right) \tag{6.6.30}$$

where

$$\hat{\mathbf{z}}_k^- = \mathbf{H}_k \hat{\mathbf{x}}_k^- + \mathbf{J}_k \hat{\mathbf{x}}_{k-1} \tag{6.6.31}$$

Gain:

$$\begin{aligned}\mathbf{K}_k = \left[\mathbf{P}_k^- \mathbf{H}_k^T + \boldsymbol{\phi}_{k-1} \mathbf{P}_{k-1} \mathbf{J}_k^T \right] \Big[\mathbf{H}_k \mathbf{P}_k^- \mathbf{H}_k^T + \mathbf{R}_k + \mathbf{J}_k \mathbf{P}_{k-1} \boldsymbol{\phi}_{k-1}^T \mathbf{H}_k^T \\ + \mathbf{H}_k \boldsymbol{\phi}_{k-1} \mathbf{P}_{k-1} \mathbf{J}_k^T + \mathbf{J}_k \mathbf{P}_{k-1} \mathbf{J}_k^T \Big]^{-1}\end{aligned} \tag{6.6.32}$$

Error covariance update:

$$\mathbf{P}_k = \mathbf{P}_k^- - \mathbf{K}_k \mathbf{L}_k \mathbf{K}_k^T \tag{6.6.33}$$

where

$$\mathbf{L}_k = \mathbf{H}_k \mathbf{P}_k^- \mathbf{H}_k^T + \mathbf{R}_k + \mathbf{J}_k \mathbf{P}_{k-1} \boldsymbol{\phi}_{k-1}^T \mathbf{H}_k^T + \mathbf{H}_k \boldsymbol{\phi}_{k-1} \mathbf{P}_{k-1} \mathbf{J}_k^T + \mathbf{J}_k \mathbf{P}_{k-1} \mathbf{J}_k^T \tag{6.6.34}$$

Projection:

$$\hat{\mathbf{x}}_{k+1}^- = \boldsymbol{\phi}_k \hat{\mathbf{x}}_k \tag{6.6.35}$$

$$\mathbf{P}_{k+1}^- = \boldsymbol{\phi}_k \mathbf{P}_k \boldsymbol{\phi}_k^T + \mathbf{Q}_k \tag{6.6.36}$$

Equations (6.6.30) through (6.6.36) comprise the complete set of recursive equations that must be implemented for the exact (i.e., optimal) solution for the delayed-state measurement problem.* Note that the general form of the equations is

* It is of interest to note that Eqs. (6.6.30) through (6.6.36) can also be derived by a completely different method. See Section 9.4 of (15).

the same as for the usual Kalman filter equations. It is just that there are a few additional terms that have to be calculated in the gain and **P**-update expressions. Thus, the extra effort in programming the exact equations is quite modest.

6.7 DECENTRALIZED KALMAN FILTERING

In our previous discussions of Kalman filtering, we always considered all of the measurements being input directly into a single filter. This mode of operation is usually referred to as a *centralized* or *global* Kalman filter. Before we look at alternatives, it should be recognized that the centralized filter yields optimal estimates. We cannot expect to find any other filter that will produce any better MMSE (minimum mean-square error) estimates, subject, of course, to the usual assumptions of linear dynamics and measurement connections, and the validity of the state models that describe the various random processes. Occasionally, though, there are applications where, for one reason or another, it is desirable to divide the global filtering job among a bank of subfilters, each of which is operating on a separate subset of the total measurement suite. This is shown in Fig. 6.10 where the separate filters are denoted as Local KF1, and Local KF2, etc. Note that the local filters are all operating autonomously. We now wish to develop a way of combining the results of the local filters such as to obtain a better estimate than would be obtained with any of the individual filters—"better" here meaning as close to the optimal as possible.

We begin with four basic assumptions:

1. The state vector **x** is the same for all local filters and the master (fusion) filter.
2. There is no information sharing among the local filters, and there is no feedback from the master filter back to the local filters.
3. The measurement errors in $z_1, z_2, \ldots z_N$ (see Fig. 6.10) are mutually uncorrelated. Thus the global **R** matrix is block diagonal.
4. None of the measurements $z_1, z_2, \ldots z_N$ are fed to the master filter directly.

We note in passing that there are no constraints on the dimensionality of the $z_1, z_2, \ldots z_N$ measurement vectors.

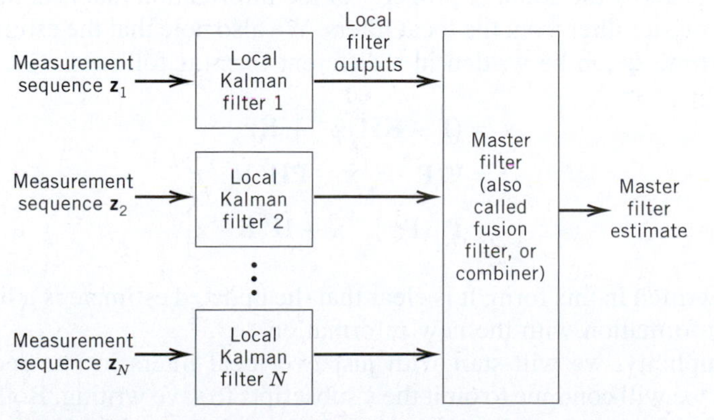

Figure 6.10 Decentralized filter—No feedback.

Basic Decentralized Filter Development

To facilitate the development, we use the information form of the Kalman filter for the error covariance update, but retain the usual equations for state estimate update and the state and covariance projection. This mixture of the key equations is repeated for convenience (see Sections 5.1 and 5.2).

1. Information matrix update:

$$\mathbf{P}_k^{-1} = \left(\mathbf{P}_k^-\right)^{-1} + \mathbf{H}_k^T \mathbf{R}_k^{-1} \mathbf{H}_k \tag{6.7.1}$$

2. Gain computation:

$$\mathbf{K}_k = \mathbf{P}_k \mathbf{H}_k^T \mathbf{R}_k^{-1} \tag{6.7.2}$$

3. Estimate update:

$$\hat{\mathbf{x}}_k = \hat{\mathbf{x}}_k^- + \mathbf{K}_k\left(\mathbf{z}_k - \mathbf{H}_k \hat{\mathbf{x}}_k^-\right) \tag{6.7.3}$$

4. Project ahead to next step:

$$\hat{\mathbf{x}}_{k+1}^- = \boldsymbol{\phi}_k \hat{\mathbf{x}}_k \tag{6.7.4}$$

$$\mathbf{P}_{k+1}^- = \boldsymbol{\phi}_k \mathbf{P}_k \boldsymbol{\phi}_k^T + \mathbf{Q}_k \tag{6.7.5}$$

Recall that $\mathbf{P}^{-1}$ is called the *information* matrix. In terms of information, Eq. (6.7.1) says that the updated information is equal to the prior information plus the additional information obtained from the measurement at time t_k. Furthermore, if $\mathbf{R}_k$ is block diagonal, the total "added" information can be divided into separate components, each representing the contribution from the respective measurement blocks. That is, we have (omitting the k subscripts for convenience)

$$\mathbf{H}^T\mathbf{R}^{-1}\mathbf{H} = \mathbf{H}_1^T\mathbf{R}_1^{-1}\mathbf{H}_1 + \mathbf{H}_2^T\mathbf{R}_2^{-1} + \cdots \mathbf{H}_N^T\mathbf{R}_N^{-1}\mathbf{H}_N \tag{6.7.6}$$

Note especially the additive property of the information that is being accumulated by the master filter from the local filters. We also note that the estimate update equation at time t_k can be written in a different form as follows:

$$\begin{aligned} \hat{\mathbf{x}} &= (\mathbf{I} - \mathbf{K}\mathbf{H})\hat{\mathbf{x}}^- + \mathbf{K}\mathbf{z} \\ &= \mathbf{P}(\mathbf{P}^-)^{-1}\hat{\mathbf{x}} + \mathbf{P}\mathbf{H}^T\mathbf{R}^{-1}\mathbf{z} \\ &= \mathbf{P}\left[(\mathbf{P}^-)^{-1}\hat{\mathbf{x}} + \mathbf{H}^T\mathbf{R}^{-1}\mathbf{z}\right] \end{aligned} \tag{6.7.7}$$

When written in this form, it is clear that the updated estimate is a linear blend of the old information with the new information.

For simplicity, we will start with just two local filters in our decentralized system, and we will continue to omit the k subscripts to save writing. Both filters are assumed to implement the full-order state vector, and at step k both are assumed to have available their respective prior estimates $\mathbf{m}_1$ and $\mathbf{m}_2$ and their associated error covariances $\mathbf{M}_1$ and $\mathbf{M}_2$. For Gaussian processes, $\mathbf{m}_1$ and $\mathbf{m}_2$ will be the means of $\mathbf{x}$ conditioned on their respective measurement streams up to, but not including, time t_t. The measurements presented to filters 1 and 2 at time t_k are $\mathbf{z}_1$ and $\mathbf{z}_2$, and they have the usual relationships to $\mathbf{x}$:

$$\mathbf{z}_1 = \mathbf{H}_1\mathbf{x} + \mathbf{v}_1 \tag{6.7.8}$$

$$\mathbf{z}_2 = \mathbf{H}_2\mathbf{x} + \mathbf{v}_2 \tag{6.7.9}$$

where $\mathbf{v}_1$ and $\mathbf{v}_2$ are zero mean random variables with covariances $\mathbf{R}_1$ and $\mathbf{R}_2$. The state $\mathbf{x}$ and noises $\mathbf{v}_1$ and $\mathbf{v}_2$ are assumed to be mutually uncorrelated as usual.

If we assume now that local filters 1 and 2 do not have access to each other's measurements, the filters will form their respective error covariances and estimates according to Eqs. (6.7.1) and (6.7.7).

Local filter 1:

$$\mathbf{P}_1^{-1} = \mathbf{M}_1^{-1} + \mathbf{H}_1^T\mathbf{R}_1^{-1}\mathbf{H}_1 \tag{6.7.10}$$

$$\hat{\mathbf{x}}_1 = \mathbf{P}_1\left(\mathbf{M}_1^{-1}\mathbf{m}_1 + \mathbf{H}_1^T\mathbf{R}_1^{-1}\mathbf{z}_1\right) \tag{6.7.11}$$

Local filter 2:

$$\mathbf{P}_2^{-1} = \mathbf{M}_2^{-1} + \mathbf{H}_2^T\mathbf{R}_2^{-1}\mathbf{H}_2 \tag{6.7.12}$$

$$\hat{\mathbf{x}}_2 = \mathbf{P}_2\left(\mathbf{M}_2^{-1}\mathbf{m}_2 + \mathbf{H}_2^T\mathbf{R}_2^{-1}\mathbf{z}_2\right) \tag{6.7.13}$$

Note that the local estimates will be optimal, conditioned on their *respective* measurement streams, but not with respect to the combined measurements. (Remember, the filters are operating autonomously.)

Now consider the master filter. It is looking for an optimal global estimate of $\mathbf{x}$ conditioned on both measurement streams 1 and 2. Let

$\mathbf{m}$ = optimal estimate of $\mathbf{x}$ conditioned on both measurement streams up to but not including t_k

$\mathbf{M}$ = covariance matrix associated with $\mathbf{m}$

The optimal global estimate and associated error covariance are then

$$\begin{aligned}\mathbf{P}^{-1} &= \left[\mathbf{H}_1^T\mathbf{H}_2^T\right]\begin{bmatrix}\mathbf{R}_1^{-1} & \mathbf{0}\\ \mathbf{0} & \mathbf{R}_2^{-1}\end{bmatrix}\begin{bmatrix}\mathbf{H}_1\\ \mathbf{H}_2\end{bmatrix} + \mathbf{M}^{-1}\\ &= \mathbf{M}^{-1} + \mathbf{H}_1^T\mathbf{R}_1^{-1}\mathbf{H}_1 + \mathbf{H}_2^T\mathbf{R}_2^{-1}\mathbf{H}_2\end{aligned} \tag{6.7.14}$$

$$\hat{\mathbf{x}} = \mathbf{P}\left(\mathbf{M}^{-1}\mathbf{m} + \mathbf{H}_1^T\mathbf{R}_1^{-1}\mathbf{z}_1 + \mathbf{H}_2^T\mathbf{R}_2^{-1}\mathbf{z}_2\right) \tag{6.7.15}$$

However, the master filter does not have direct access to $\mathbf{z}_1$ and $\mathbf{z}_2$, so we will rewrite Eqs. (6.7.14) and (6.7.15) in terms of the local filter's computed estimates and covariances. The result is

$$\mathbf{P}^{-1} = \left(\mathbf{P}_1^{-1} - \mathbf{M}_1^{-1}\right) + \left(\mathbf{P}_2^{-1} - \mathbf{M}_2^{-1}\right) + \mathbf{M}^{-1} \tag{6.7.16}$$

$$\hat{\mathbf{x}} = \mathbf{P}\left[\left(\mathbf{P}_1^{-1}\hat{\mathbf{x}}_1 - \mathbf{M}_1^{-1}\mathbf{m}_1\right) + \left(\mathbf{P}_2^{-1}\hat{\mathbf{x}}_2 - \mathbf{M}_2^{-1}\mathbf{m}_2\right) + \mathbf{M}^{-1}\mathbf{m}\right] \tag{6.7.17}$$

It can now be seen that the local filters can pass their respective $\hat{\mathbf{x}}_i$, $\mathbf{P}_i^{-1}$, $\mathbf{m}_i$, $\mathbf{M}_i^{-1}(i = 1, 2)$ on to the master filter, which, in turn, can then compute its global estimate. The local filters can, of course, do their own local projections and then repeat the cycle at step $k+1$. Likewise, the master filter can project its global estimate and get a new **m** and **M** for the next step. Thus, we see that this architecture permits complete autonomy of the local filters, and it yields local optimality with respect to the respective measurement streams. The system also achieves global optimality in the master filter. Thus, with this system, we have the "best of both worlds." We maintain the independence of the local filters; and at the same time, we also have the global optimal solution.

6.8 DIFFICULTY WITH HARD-BANDLIMITED PROCESSES

It should be apparent at this point in our filtering discussion that we are severely limited in the type of random processes that can be estimated with Kalman filtering. The restriction is imposed right at the start with the assumption that x must satisfy the difference equation

$$\mathbf{x}_{k+1} = \boldsymbol{\phi}_k \mathbf{x}_k + \mathbf{w}_k \tag{6.8.1}$$

If the continuous $x(t)$ has a spectral function that is rational in form (i.e., the PSD is represented by a ratio of polynomials in ω^2), then we can usually find a sampled form of $\mathrm{x}(t)$ that will fit the form required by Eq. (6.8.1). There are, though, other random process models that are not rational, but yet they fit certain physical situations reasonably well. One example is hard-bandlimited noise. This appears frequently in communications applications, usually as a special case of bandlimited white noise. This process is discussed briefly in Chapter 3, Section 3.4, but is defies estimation using Kalman filtering methods. However, do not despair. Some of the ideas of Wiener filtering (20) can sometimes be brought to bear on bandlimited noise problems, and this will now be explored further.

The Wiener approach to least-squares filtering is basically a weighting function approach. When viewed this way, the problem always reduces to: How should the past history of the input be weighted in order to yield the best estimate of the variable of interest (i.e., the signal)? We will only consider the discrete case here. We will assume that the input is a sequence of noisy measurements, $z_1, z_2, \ldots, z_k$, much the same as we have in Kalman filtering, and we will now proceed to illustrate the Wiener approach with a specific example.

EXAMPLE 6.5 BANDLIMITED WHITE NOISE

Consider the following scenario in the content of least-squares filtering. Both the signal and the noise are bandlimited white noises, but they have significantly different bandwidths. The signal has mostly low-frequency components, whereas the noise is more high frequency in character. Denote the signal as x and the noise as n. The measurement is an additive combination of x and n, i.e.,

$$\begin{aligned} z_1 &= x_1 + n_1 \\ z_2 &= x_2 + n_2 \\ &\vdots \\ z_k &= x_k + n_k \end{aligned} \tag{6.8.2}$$

where k is the "present" sample time. There will be a total of k measurement samples being considered, with z_1 being the "oldest" sample in the weighted sum. We are seeking the steady-state least-squares estimate, and the total number of samples used in the estimate will be adjusted as needed.

The key parameters are as follows:

Signal (Bandlimited White)

$\sigma_x^2 = 16.0\,m^2$
$W_x = 1.0$ Hz (bandwidth)
$PSD_x = 8\,m^2/Hz$ (power spectral density)

Noise (Bandlimited White)

$\sigma_n^2 = 2.0\,m^2$
$W_n = 10.0$ Hz (bandwidth)
$PSD_n = 0.1\,m^2/Hz$ (power spectral density)

Sampling Parameters

$\Delta t = 1/2W_n = 0.05$ s (sampling interval)
Sampling rate = 20 samples/s (Nyquist rate for noise process)
Total number of samples = 50

We now proceed with the minimization. We arbitrarily say that the estimate at t_k is to be the weighted sum

$$\hat{x}_k = w_1 z_1 + w_2 z_2 + \cdots + w_k z_k \tag{6.8.3}$$

Thus, the error is

$$\begin{aligned} e_k &= x_k - \hat{x}_k \\ &= x_k - (w_1 z_1 + w_2 z_2 + \cdots + w_k z_k) \end{aligned} \tag{6.8.4}$$

and the mean-square error is

$$E\left[e_k^2\right] = E[x_k - (w_1 z_1 + w_2 z_2 + \cdots + w_k z_k)]^2 \tag{6.8.5}$$

We wish to minimize $E\left[e_k^2\right]$ with respect to the weight factors $w_1, w_2, \cdots w_k$. Thus, using ordinary differential calculus, we can differentiate Eq. (6.8.5) accordingly and set the respective derivatives equal to zero. Omitting the routine algebra, this leads to the following set of linear equations in the weight factors:

$$\begin{bmatrix} E(z_1^2) & E(z_1 z_2) & \cdots & \\ E(z_2 z_1) & & & \\ \vdots & & \ddots & \\ E(z_k z_1) & & & E(z_k^2) \end{bmatrix} \begin{bmatrix} w_1 \\ w_2 \\ \vdots \\ w_k \end{bmatrix} = \begin{bmatrix} E(z_1 x_k) \\ E(z_2 x_k) \\ \vdots \\ E(z_k x_k) \end{bmatrix} \tag{6.8.6}$$

Or

$$\mathbf{Tw} = \mathbf{f} \tag{6.8.7}$$

Equation (6.8.7) is easily solved for **w** once the parameters are specified, and they are known from the correlation structure assumed for signal and noise, x and n. Recall that x and n are mutually uncorrelated, and the respective samples of n are mutually uncorrelated among themselves because of the Nyquist sampling rate. Thus, the terms in **T** are given by

$$\begin{aligned} E(z_1^2) &= E(z_2^2) = \cdots = (\sigma_x^2 + \sigma_n^2) \\ E(z_1 z_2) &= E(x_1 x_2) = R_x(\Delta t) \\ E(z_1 z_3) &= E(x_1 x_3) = R_k(2\Delta t) \\ &\vdots \\ E(z_1 z_k) &= E(x_1 x_k) = R_x((k-1)\Delta t) \end{aligned} \tag{6.8.8}$$

and the terms in **f** are specified by

$$\begin{aligned} E(z_1 x_k) &= R_x((k-1)\Delta t) \\ E(z_2 x_k) &= R_x((k-2)\Delta t) \\ &\vdots \\ E(z_k x_k) &= R_x((k-k)\Delta t) = R_x(0) \end{aligned} \tag{6.8.9}$$

and R_x is the autocorrelation function for the signal x. [Recall that $R_x(\tau) = \mathfrak{F}^{-1}$ (PSD$_x$). See Section 2.7.]

The square matrix **T** is especially easy to program in MATLAB because it works out to be a Toeplitz matrix. (A Toeplitz matrix is real and symmetric and has further special symmetry: All terms on the major diagonal are equal; then further, all terms on each of the respective sub-diagonals below and above the major diagonal are also equal. Thus, an $n \times n$ Toeplitz matrix may be completely defined by specifying all the terms in the first column vector of the matrix.)

Equation (6.8.7) was solved using MATLAB with the number of measurements set at 50 for this example, and a plot of the calculated weights is shown in Fig. 6.11.

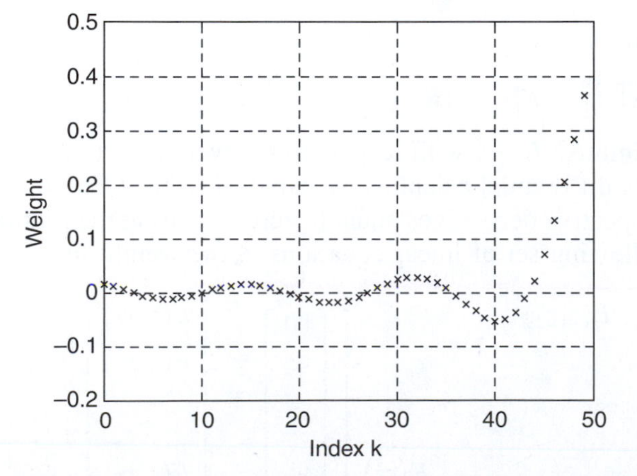

Figure 6.11 Plot of measurement weights for Example 6.5.

Recall that, in the numbering scheme used her, the beginning weight on the right side of the plot is the weight given to the current measurement. Then, moving toward the left end of the plot, we have the weights given to the respective "older" measurements. As might be expected, the weights oscillate and diminish as we proceed further into the past measurements. The oscillations are due to the sinc autocorrelation functions in Eq. (6.8.7).

If this batch scheme were to be programmed in real time, the weights would be held fixed, and they would be applied to the measurement stream according to age (relative to the "present") on a running time basis. This would be easy to program. The steady-state mean-square error can also be calculated without undue difficulty, but this will not be done here. It will be left as an exercise (see Problem 6.8).

6.9 THE RECURSIVE BAYESIAN FILTER

In the Conditional Density Viewpoint given in Section 4.7, it was shown that the Kalman filter's estimate that achieves minimum mean-square error can be represented by Eq. (4.7.3). In the general case, this estimate comes from a conditional probability density function of the state $\mathbf{x}_k$ conditioned on the set of measurements $\mathbf{z}_k^*$ that include all measurements up to and including $\mathbf{z}_k$. We will now proceed to develop the corresponding recursive equations while keeping the conditions general, i.e., without limiting the problem to Gaussian and linear assumptions.

Thus, the measurement model will simply be

$$\mathbf{z}_k = \mathbf{H}(\mathbf{x}_k, \mathbf{v}_k) \tag{6.9.1}$$

with a corresponding probabilistic model given by the density function $p(\mathbf{z}_k | \mathbf{x}_k)$.

The measurement information are all contained in the measurement set

$$\mathbf{z}_k^* = (\mathbf{z}_0, \mathbf{z}_1, \ldots, \mathbf{z}_k)$$

Similarly, we write the system process model as a nonlinear function but, without loss of generality, include the process noise as an additive term:

$$\mathbf{x}_k = \boldsymbol{\phi}(\mathbf{x}_{k-1}) + \mathbf{w}_k \tag{6.9.2}$$

Its probabilistic model is represented by the density function $p(\mathbf{x}_k | \mathbf{x}_{k-1})$.

Measurement Update

We start with the measurement update by seeking a relationship between the conditional density $p(\mathbf{x}_k | \mathbf{z}_k^*)$ based on the measurement set $\mathbf{z}_k^*$ and a similar conditional density $p(\mathbf{x}_k | \mathbf{z}_{k-1}^*)$ based on the measurement set $\mathbf{z}_{k-1}^*$. The updated conditional density function is given by (15)

$$\begin{aligned} p(\mathbf{x}_k|\mathbf{z}_k^*) &= p(\mathbf{x}_k|\mathbf{z}_k, \mathbf{z}_0, \ldots, \mathbf{z}_{k-1}) = p(\mathbf{x}_k|\mathbf{z}_k, \mathbf{z}_{k-1}^*) \\ &= \frac{p(\mathbf{z}_k|\mathbf{x}_k, \mathbf{z}_{k-1}^*)p(\mathbf{x}_k|\mathbf{z}_{k-1}^*)}{p(\mathbf{z}_k|\mathbf{z}_{k-1}^*)} \\ &= \frac{p(\mathbf{z}_k|\mathbf{x}_k)p(\mathbf{x}_k|\mathbf{z}_{k-1}^*)}{p(\mathbf{z}_k|\mathbf{z}_{k-1}^*)} \end{aligned} \tag{6.9.3}$$

(One of two fundamental Markov properties provides that $p(\mathbf{z}_k|\mathbf{x}_k, \mathbf{z}_{k-1}^*) = p(\mathbf{z}_k|\mathbf{x}_k)$ (16).)

As for the prediction, we seek a relationship between the conditional density of the future state $p(\mathbf{x}_{k+1}|\mathbf{z}_k^*)$ and that of the present state $p(\mathbf{x}_k|\mathbf{z}_k^*)$, both conditioned on the same measurement set $\mathbf{z}_k^*$:

$$\begin{aligned} p(\mathbf{x}_{k+1}|\mathbf{z}_k^*) &= \int p(\mathbf{x}_k, \mathbf{x}_{k+1}|\mathbf{z}_k^*)d\mathbf{x}_k \\ &= \int p(\mathbf{x}_{k+1}|\mathbf{x}_k, \mathbf{z}_k^*)p(\mathbf{x}_k|\mathbf{z}_k^*)d\mathbf{x}_k \\ &= \int p(\mathbf{x}_{k+1}|\mathbf{x}_k)p(\mathbf{x}_k|\mathbf{z}_k^*)d\mathbf{x}_k \end{aligned} \tag{6.9.4}$$

(The second of two fundamental Markov properties provides that $p(\mathbf{x}_{k+1}|\mathbf{x}_k, \mathbf{z}_k^*) = p(\mathbf{x}_{k+1}|\mathbf{x}_k)$ (16).)

We can rewrite the update equation of (6.9.3) as:

$$p(\mathbf{x}_k|\mathbf{z}_k^*) = \kappa_k p(\mathbf{z}_k|\mathbf{x}_k)p(\mathbf{x}_k|\mathbf{z}_{k-1}^*) \tag{6.9.5}$$

where, as described in words, the conditional density of the state $\mathbf{x}_k$ given the measurement set $\mathbf{z}_k^*$ is related to the conditional density of the state $\mathbf{x}_k$ given the measurement set $\mathbf{z}_{k-1}^*$ as a product with the conditional density $p(\mathbf{z}_k|\mathbf{x}_k)$ that describes the measurement model of Eq. 6.9.1 and some multiplier κ_k. Related to the Kalman filter language we have already established, $p(\mathbf{x}_k|\mathbf{z}_k^*)$ is the *a posteriori* conditional density, while $p(\mathbf{x}_k|\mathbf{z}_{k-1}^*)$ is the *a priori* conditional density.

State Prediction

The prediction equation extended from Eq. (6.9.4) can be rewritten as (17):

$$p(\mathbf{x}_{k+1}|\mathbf{z}_k^*) = \int p(\mathbf{x}_{k+1}|\mathbf{x}_k)p(\mathbf{x}_k|\mathbf{z}_k^*)d\mathbf{x}_k \tag{6.9.6}$$

where the conditional density of the future state $\mathbf{x}_{k+1}$ is related to the conditional density of the present state $\mathbf{x}_k$, both given the same measurement set $\mathbf{z}_k^*$, through a product with the conditional density $p(\mathbf{x}_{k+1}|\mathbf{x}_k)$ that describes the system process model of Eq. 6.9.2, and an integral over changes in the state $\mathbf{x}_k$.

Equations (6.9.5) and (6.9.6), along with an initial condition $p(\mathbf{x}_0)$, therefore make up what is known as a recursive Bayesian filter, which is a generalized formulation of the Kalman filter. The recursive Bayesian filter, for good reason, is not particularly useful in its general form because real solutions to Eqs. (6.9.5) and (6.9.6) are hard to come by analytically. Only with the assumption of Gaussian probabilistic models and the imposition of linearity to both the process and measurement models did we end up with a very useful result in the Kalman filter, a solution that has been used to tackle a virtually endless variety of real-world problems.

The Gaussian Approximation

Let us explore some simplifications to be had when the probability density in Eq. (6.9.5) can be assumed to be Gaussian. With this, we only need to represent it with a mean and covariance, the two terms that are the essential parameters of the Kalman filter:

$$E\left(\mathbf{x}_k|\mathbf{z}_k^*\right) = \hat{\mathbf{x}}_k \tag{6.9.7}$$

$$Cov\left(\mathbf{x}_k|\mathbf{z}_k^*\right) = \mathbf{P}_k \tag{6.9.8}$$

We rewrite Eq. (6.9.3) by converting the conditional density on the right-hand side of the equation into a joint probability density:

$$p\left(\mathbf{x}_k|\mathbf{z}_k^*\right) = \frac{p\left(\mathbf{z}_k, \mathbf{x}_k|\mathbf{z}_{k-1}^*\right)}{p\left(\mathbf{z}_k|\mathbf{z}_{k-1}^*\right)} \tag{6.9.9}$$

With the Gaussian assumption just invoked, we can also rewrite this as a relationship of Gaussian probability density functions (and, for convenience, drop the time index k as this is implied presently):

$$\mathcal{N}(\hat{\mathbf{x}}, \mathbf{P}_{\tilde{\mathbf{x}}\tilde{\mathbf{x}}}) = \frac{\mathcal{N}\left(\begin{bmatrix} \hat{\mathbf{x}}^- \\ \hat{\mathbf{z}}^- \end{bmatrix}, \begin{bmatrix} \mathbf{P}_{\tilde{\mathbf{x}}\tilde{\mathbf{x}}}^- & \mathbf{P}_{\tilde{\mathbf{x}}\tilde{\mathbf{z}}}^- \\ \mathbf{P}_{\tilde{\mathbf{z}}\tilde{\mathbf{x}}}^- & \mathbf{P}_{\tilde{\mathbf{z}}\tilde{\mathbf{z}}}^- \end{bmatrix}\right)}{\mathcal{N}\left(\hat{\mathbf{z}}^-, \mathbf{P}_{\tilde{\mathbf{z}}\tilde{\mathbf{z}}}^-\right)} \tag{6.9.10}$$

where $\mathbf{P}_{\tilde{\mathbf{x}}\tilde{\mathbf{x}}}^-$ is the *a priori* error covariance matrix (seen before as $\mathbf{P}_k^-$) associated with the *a priori* state estimate $\hat{\mathbf{x}}^-$, $\mathbf{P}_{\tilde{\mathbf{z}}\tilde{\mathbf{z}}}^-$ is the measurement residual covariance matrix associated with the predicted measurement $\hat{\mathbf{z}}^-$, and $\mathbf{P}_{\tilde{\mathbf{x}}\tilde{\mathbf{z}}}^-$ is the cross-covariance between $\hat{\mathbf{x}}^-$ and $\hat{\mathbf{z}}^-$. On the left-hand side of Eq. (6.9.10), $\hat{\mathbf{x}}$ is the *a posteriori* state estimate with an error covariance of $\mathbf{P}_{\tilde{\mathbf{x}}\tilde{\mathbf{x}}}$ also denoted elsewhere as $\mathbf{P}_k$.

Note that Eq. (6.9.10) essentially relates the *a priori* state estimate $\hat{\mathbf{x}}^-$ and the associated predicted measurement $\hat{\mathbf{z}}^-$ on the right-hand side of the equation to the *a posteriori* state estimate $\hat{\mathbf{x}}$ on the left-hand side of the equation, via the relevant covariance matrices. But to extract the direct relationships, we equate the essential part of the exponential terms in the Gaussian densities on both sides of Eq. (6.9.10):

$$(\tilde{\mathbf{x}})^T [\mathbf{P}_{\tilde{\mathbf{x}}\tilde{\mathbf{x}}}]^{-1} \tilde{\mathbf{x}} = \begin{bmatrix} \tilde{\mathbf{x}}^- \\ \tilde{\mathbf{z}}^- \end{bmatrix}^T \begin{bmatrix} \mathbf{P}_{\tilde{\mathbf{x}}\tilde{\mathbf{x}}}^- & \mathbf{P}_{\tilde{\mathbf{x}}\tilde{\mathbf{z}}}^- \\ \mathbf{P}_{\tilde{\mathbf{z}}\tilde{\mathbf{x}}}^- & \mathbf{P}_{\tilde{\mathbf{z}}\tilde{\mathbf{z}}}^- \end{bmatrix}^{-1} \begin{bmatrix} \tilde{\mathbf{x}}^- \\ \tilde{\mathbf{z}}^- \end{bmatrix} - (\tilde{\mathbf{z}}^-)^T [\mathbf{P}_{\tilde{\mathbf{z}}\tilde{\mathbf{z}}}^-]^{-1} \tilde{\mathbf{z}}^- \tag{6.9.11}$$

We deconstruct the equation by first addressing the joint covariance matrix using the following matrix inversion lemma:

$$\begin{bmatrix} A & U \\ V & C \end{bmatrix}^{-1} = \begin{bmatrix} [A - UC^{-1}V]^{-1} & -[A - UC^{-1}V]^{-1}UC^{-1} \\ -C^{-1}V[A - UC^{-1}V]^{-1} & C^{-1}V[A - UC^{-1}V]^{-1}UC^{-1} + C^{-1} \end{bmatrix}$$
$$= \begin{bmatrix} B^{-1} & -B^{-1}UC^{-1} \\ -C^{-1}VB^{-1} & C^{-1}VB^{-1}UC^{-1} + C^{-1} \end{bmatrix} \tag{6.9.12}$$

where $B = [A - UC^{-1}V]$

Therefore,

$$\begin{bmatrix} \mathbf{P}_{\tilde{\mathbf{x}}\tilde{\mathbf{x}}}^{-} & \mathbf{P}_{\tilde{\mathbf{x}}\tilde{\mathbf{z}}}^{-} \\ \mathbf{P}_{\tilde{\mathbf{z}}\tilde{\mathbf{x}}}^{-} & \mathbf{P}_{\tilde{\mathbf{z}}\tilde{\mathbf{z}}}^{-} \end{bmatrix}^{-1} = \begin{bmatrix} \mathbf{B}^{-1} & -\mathbf{B}^{-1}\mathbf{P}_{\tilde{\mathbf{x}}\tilde{\mathbf{z}}}^{-}\left(\mathbf{P}_{\tilde{\mathbf{z}}\tilde{\mathbf{z}}}^{-}\right)^{-1} \\ -\left(\mathbf{P}_{\tilde{\mathbf{z}}\tilde{\mathbf{z}}}^{-}\right)^{-1}\mathbf{P}_{\tilde{\mathbf{z}}\tilde{\mathbf{x}}}^{-}\mathbf{B}^{-1} & \left(\mathbf{P}_{\tilde{\mathbf{z}}\tilde{\mathbf{z}}}^{-}\right)^{-1}\mathbf{P}_{\tilde{\mathbf{z}}\tilde{\mathbf{x}}}^{-}\mathbf{B}^{-1}\mathbf{P}_{\tilde{\mathbf{x}}\tilde{\mathbf{z}}}^{-}\left(\mathbf{P}_{\tilde{\mathbf{z}}\tilde{\mathbf{z}}}^{-}\right)^{-1} + \left(\mathbf{P}_{\tilde{\mathbf{z}}\tilde{\mathbf{z}}}^{-}\right)^{-1} \end{bmatrix} \tag{6.9.13}$$

where we define $\mathbf{B} = \left[\mathbf{P}_{\tilde{\mathbf{x}}\tilde{\mathbf{x}}}^{-} - \mathbf{P}_{\tilde{\mathbf{x}}\tilde{\mathbf{z}}}^{-}\left(\mathbf{P}_{\tilde{\mathbf{z}}\tilde{\mathbf{z}}}^{-}\right)^{-1}\mathbf{P}_{\tilde{\mathbf{z}}\tilde{\mathbf{x}}}^{-}\right]$.

Now, to extend the right-hand side of Eq. (6.9.11),

$$\begin{bmatrix} \tilde{\mathbf{x}}^{-} \\ \tilde{\mathbf{z}}^{-} \end{bmatrix}^{T} \begin{bmatrix} \mathbf{P}_{\tilde{\mathbf{x}}\tilde{\mathbf{x}}}^{-} & \mathbf{P}_{\tilde{\mathbf{x}}\tilde{\mathbf{z}}}^{-} \\ \mathbf{P}_{\tilde{\mathbf{z}}\tilde{\mathbf{x}}}^{-} & \mathbf{P}_{\tilde{\mathbf{z}}\tilde{\mathbf{z}}}^{-} \end{bmatrix}^{-1} \begin{bmatrix} \tilde{\mathbf{x}}^{-} \\ \tilde{\mathbf{z}}^{-} \end{bmatrix} - (\tilde{\mathbf{z}}^{-})^{T}\left[\mathbf{P}_{\tilde{\mathbf{z}}\tilde{\mathbf{z}}}^{-}\right]^{-1}\tilde{\mathbf{z}}^{-}$$
$$= (\tilde{\mathbf{x}}^{-})^{T}\mathbf{B}^{-1}\tilde{\mathbf{x}}^{-} - (\tilde{\mathbf{x}}^{-})^{T}\mathbf{B}^{-1}\mathbf{P}_{\tilde{\mathbf{x}}\tilde{\mathbf{z}}}^{-}\left(\mathbf{P}_{\tilde{\mathbf{z}}\tilde{\mathbf{z}}}^{-}\right)^{-1}\tilde{\mathbf{z}}^{-} - (\tilde{\mathbf{z}}^{-})^{T}\left(\mathbf{P}_{\tilde{\mathbf{z}}\tilde{\mathbf{z}}}^{-}\right)^{-1}\mathbf{P}_{\tilde{\mathbf{z}}\tilde{\mathbf{x}}}^{-}\mathbf{B}^{-1}\tilde{\mathbf{x}}^{-}$$
$$+ (\tilde{\mathbf{z}}^{-})^{T}\left(\mathbf{P}_{\tilde{\mathbf{z}}\tilde{\mathbf{z}}}^{-}\right)^{-1}\mathbf{P}_{\tilde{\mathbf{z}}\tilde{\mathbf{x}}}^{-}\mathbf{B}^{-1}\mathbf{P}_{\tilde{\mathbf{x}}\tilde{\mathbf{z}}}^{-}\left(\mathbf{P}_{\tilde{\mathbf{z}}\tilde{\mathbf{z}}}^{-}\right)^{-1}\tilde{\mathbf{z}}^{-}$$
$$= \left[\tilde{\mathbf{x}}^{-} - \mathbf{P}_{\tilde{\mathbf{x}}\tilde{\mathbf{z}}}^{-}\left(\mathbf{P}_{\tilde{\mathbf{z}}\tilde{\mathbf{z}}}^{-}\right)^{-1}\tilde{\mathbf{z}}^{-}\right]^{T}\mathbf{B}^{-1}\left[\tilde{\mathbf{x}}^{-} - \mathbf{P}_{\tilde{\mathbf{x}}\tilde{\mathbf{z}}}^{-}\left(\mathbf{P}_{\tilde{\mathbf{z}}\tilde{\mathbf{z}}}^{-}\right)^{-1}\tilde{\mathbf{z}}^{-}\right] \tag{6.9.14}$$

If we then equate this with the left-hand side of Eq. (6.9.11), we get the following relationship for the *a posteriori* state estimate:

$$\tilde{\mathbf{x}} = \tilde{\mathbf{x}}^{-} - \mathbf{P}_{\tilde{\mathbf{x}}\tilde{\mathbf{z}}}^{-}\left(\mathbf{P}_{\tilde{\mathbf{z}}\tilde{\mathbf{z}}}^{-}\right)^{-1}\tilde{\mathbf{z}}^{-}$$
$$\mathbf{x} - \hat{\mathbf{x}} = \mathbf{x} - \hat{\mathbf{x}}^{-} - \mathbf{P}_{\tilde{\mathbf{x}}\tilde{\mathbf{z}}}^{-}\left(\mathbf{P}_{\tilde{\mathbf{z}}\tilde{\mathbf{z}}}^{-}\right)^{-1}\tilde{\mathbf{z}}^{-}$$

Now reapplying the time index k, it should be evident that, lo and behold, this is the state estimate update equation of the Kalman filter:

$$\hat{\mathbf{x}}_k = \hat{\mathbf{x}}_k^{-} + \underbrace{\left(\mathbf{P}_{\tilde{\mathbf{x}}\tilde{\mathbf{z}}}^{-}\right)_k\left(\mathbf{P}_{\tilde{\mathbf{z}}\tilde{\mathbf{z}}}^{-}\right)_k^{-1}}_{\mathbf{K}_k}\tilde{\mathbf{z}}_k^{-} \tag{6.9.15}$$

where $\left(\mathbf{P}_{\tilde{\mathbf{x}}\tilde{\mathbf{z}}}^{-}\right)_k\left(\mathbf{P}_{\tilde{\mathbf{z}}\tilde{\mathbf{z}}}^{-}\right)_k^{-1}$ is the Kalman gain $\mathbf{K}_k$.

The error covariance of the *a posteriori* Gaussian density function found in its exponent, given by (Eq. 6.9.14), is $\mathbf{B}$:

$$
\begin{aligned}
(\mathbf{P}_{\tilde{\mathbf{x}}\tilde{\mathbf{x}}})_k = \mathbf{B}_k &= \left[(\mathbf{P}^-_{\tilde{\mathbf{x}}\tilde{\mathbf{x}}})_k - (\mathbf{P}^-_{\tilde{\mathbf{x}}\tilde{\mathbf{z}}})_k(\mathbf{P}^-_{\tilde{\mathbf{z}}\tilde{\mathbf{z}}})_k^{-1}(\mathbf{P}^-_{\tilde{\mathbf{z}}\tilde{\mathbf{x}}})_k\right] \\
&= (\mathbf{P}^-_{\tilde{\mathbf{x}}\tilde{\mathbf{x}}})_k - (\mathbf{P}^-_{\tilde{\mathbf{x}}\tilde{\mathbf{z}}})_k(\mathbf{P}^-_{\tilde{\mathbf{z}}\tilde{\mathbf{z}}})_k^{-1}(\mathbf{P}^-_{\tilde{\mathbf{z}}\tilde{\mathbf{z}}})_k(\mathbf{P}^-_{\tilde{\mathbf{z}}\tilde{\mathbf{z}}})_k^{-1}(\mathbf{P}^-_{\tilde{\mathbf{x}}\tilde{\mathbf{z}}})_k^T \\
&= (\mathbf{P}^-_{\tilde{\mathbf{x}}\tilde{\mathbf{x}}})_k - \underbrace{\left[(\mathbf{P}^-_{\tilde{\mathbf{x}}\tilde{\mathbf{z}}})_k(\mathbf{P}^-_{\tilde{\mathbf{z}}\tilde{\mathbf{z}}})_k^{-1}\right]}_{\mathbf{K}_k}(\mathbf{P}^-_{\tilde{\mathbf{z}}\tilde{\mathbf{z}}})_k\underbrace{\left[(\mathbf{P}^-_{\tilde{\mathbf{x}}\tilde{\mathbf{z}}})_k(\mathbf{P}^-_{\tilde{\mathbf{z}}\tilde{\mathbf{z}}})_k^{-1}\right]^T}_{\mathbf{K}_k}
\end{aligned}
\tag{6.9.16}
$$

To recall how we got here, this section first introduced the generalized estimator known as the Bayesian filter. With that representation using general probability distribution or density functions, its analytical solution is very difficult to obtain. However, by invoking a Gaussian assumption to the random processes, the recursive Bayesian filter was simplified to a form that we can recognize as the Kalman filter, at least for the update equations:

$$\text{Kalman gain:}\quad \mathbf{K}_k = (\mathbf{P}^-_{\tilde{\mathbf{x}}\tilde{\mathbf{z}}})_k(\mathbf{P}^-_{\tilde{\mathbf{z}}\tilde{\mathbf{z}}})_k^{-1} \tag{6.9.17}$$

$$\text{State update:}\quad \hat{\mathbf{x}}_k = \hat{\mathbf{x}}_k^- + \mathbf{K}_k\tilde{\mathbf{z}}_k^- \tag{6.9.18}$$

$$\text{Error covariance update:}\quad (\mathbf{P}_{\tilde{\mathbf{x}}\tilde{\mathbf{x}}})_k = (\mathbf{P}^-_{\tilde{\mathbf{x}}\tilde{\mathbf{x}}})_k - \mathbf{K}_k(\mathbf{P}^-_{\tilde{\mathbf{z}}\tilde{\mathbf{z}}})_k\mathbf{K}_k^T \tag{6.9.19}$$

For the prediction equation, the projected state estimate is the expected value of the probability density given by Eq. (6.9.6):

$$
\begin{aligned}
\hat{\mathbf{x}}_{k+1}^- &= E\left[\mathbf{x}_{k+1}|\mathbf{z}_k^*\right] = E\left[(\Phi(\mathbf{x}_k) + \mathbf{w}_k)|\mathbf{z}_k^*\right] = E\left[\Phi(\mathbf{x}_k)|\mathbf{z}_k^*\right] \\
\hat{\mathbf{x}}_{k+1}^- &= \int \Phi(\mathbf{x}_k)p\left(\mathbf{x}_k|\mathbf{z}_k^*\right)d\mathbf{x}_k
\end{aligned}
$$

where

$$p\left(\mathbf{x}_k|\mathbf{z}_k^*\right) \sim \mathcal{N}(\hat{\mathbf{x}}_k, \mathbf{P}_{\tilde{\mathbf{x}}\tilde{\mathbf{x}}})$$

The error covariance projection to the next step ahead is given by (18):

$$
\begin{aligned}
(\mathbf{P}^-_{\tilde{\mathbf{x}}\tilde{\mathbf{x}}})_{k+1} &= E\left[\left(\mathbf{x}_{k+1} - \hat{\mathbf{x}}_{k+1}^-\right)\left(\mathbf{x}_{k+1} - \hat{\mathbf{x}}_{k+1}^-\right)^T|\mathbf{z}_k^*\right] \\
&= E\left[\mathbf{x}_{k+1}\mathbf{x}_{k+1}{}^T|\mathbf{z}_k^*\right] - E\left[\mathbf{x}_{k+1}|\mathbf{z}_k^*\right]\left(\hat{\mathbf{x}}_{k+1}^-\right)^T \\
&\quad -\hat{\mathbf{x}}_{k+1}^- E\left[\mathbf{x}_{k+1}^T|\mathbf{z}_k^*\right] + \hat{\mathbf{x}}_{k+1}^-\left(\hat{\mathbf{x}}_{k+1}^-\right)^T \\
&= \int \Phi(\mathbf{x}_k)\Phi(\mathbf{x}_k)^T p(\mathbf{x}_k|\mathbf{z}_k^*)d\mathbf{x}_k - \hat{\mathbf{x}}_{k+1}^-\left(\hat{\mathbf{x}}_{k+1}^-\right)^T + \mathbf{Q}_k
\end{aligned}
\tag{6.9.20}
$$

At this stage, we have still not invoked any linear assumptions. When we do, Eqs. (6.9.17)–(6.9.19) can then be further reduced ultimately to the linear Kalman filter equations given by Eqs. (4.2.17), (4.2.8), and (4.2.21). Under such

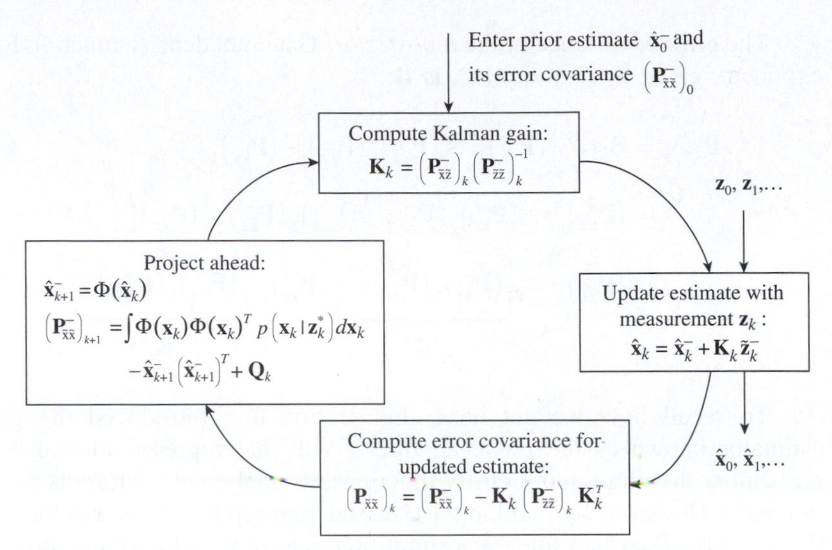

Figure 6.12 The recursive Bayesian filter with Gaussian approximation.

circumstances, the following correspondences apply:

$$\left(\mathbf{P}_{\tilde{\mathbf{x}}\tilde{\mathbf{x}}}^-\right)_k \equiv \mathbf{P}_k^-$$
$$\left(\mathbf{P}_{\tilde{\mathbf{x}}\tilde{\mathbf{z}}}^-\right)_k \equiv \mathbf{P}_k^- \mathbf{H}_k^T$$
$$\left(\mathbf{P}_{\tilde{\mathbf{z}}\tilde{\mathbf{z}}}^-\right)_k \equiv \left(\mathbf{H}_k \mathbf{P}_k^- \mathbf{H}_k^T + \mathbf{R}_k\right)$$
$$\left(\mathbf{P}_{\tilde{\mathbf{x}}\tilde{\mathbf{x}}}\right)_k \equiv \mathbf{P}_k$$

The recursive loop for the Bayesian filter is summarized in Fig. 6.12. In problems where nonlinearity is a particularly severe condition, there has been a growing body of work that leans on this generalized form of the recursive Bayesian filter we have presented in this section, as well as the more accessible form where an approximation of Gaussian statistics was invoked while still avoiding linear assumptions. We defer further discussion on nonlinear filter problems until Chapter 7.

PROBLEMS

6.1 Consider the following elementary statistical estimation problem. Suppose we have two independent measurements of a random variable x, and we know nothing about x other than it is Gaussian. We will call the two measurement z_1 and z_2; and their errors e_1 and e_2 are zero-mean Gaussian, independent, and with rms values σ_1 and σ_2. Our objective is to combine the z_1 and z_2 measurements such as to obtain a minimum-mean-square-error estimate of x, which we will call $\hat{x}$ (without a subscript for brevity). We know from the discussion in Section 4.7 that $\hat{x}$ will be a linear combination of z_1 and z_2, and we demand that $\hat{x}$ be unbiased. Therefore, we can write $\hat{x}$ as

$$\hat{x} = k_1 z_1 + k_2 z_2 \tag{P6.1.1}$$

where

$$k_1 + k_2 = 1 \tag{P6.1.2}$$

Or

$$\hat{x} = k_1 z_1 + (1 - k_1) z_2 \tag{P6.1.3}$$

Also, we know that $z_1 = x + e_1$ and $z_2 = x + e_2$, so the error in the combined estimate of x is

$$e = k_1(e_1 - e_2) + e_2 \tag{P6.1.4}$$

and

$$E(e^2) = E\left\{[k_1(e_1 - e_2) + e_2]^2\right\} \tag{P6.1.5}$$

We wish to minimize $E(e^2)$, so this reduces to a simple problem in differential calculus.

Show that k_1 and k_2 are given by

$$k_1 = \frac{\sigma_2^2}{\sigma_1^2 + \sigma_2^2}, \text{ and } k_2 = \frac{\sigma_1^2}{\sigma_1^2 + \sigma_2^2} \tag{P6.1.6}$$

and that the variance of the combined estimation error is

$$E(e^2) = \frac{\sigma_1^2 \sigma_2^2}{\sigma_1^2 + \sigma_2^2} \tag{P6.1.7}$$

The variance expression in Eq. (P6.1.7) is sometimes referred to as the "product over sum" formula. **Remember:** Knowing nothing about the sigma of x is an important condition in deriving the product-over-sum formula. If we were to assume that the sigma of x was known and did not put a constraint between k_1 and k_2, then the result would be entirely different.

6.2 In 1969, D.C. Fraser and J.E. Potter presented a novel solution to the fixed-interval smoothing problem (19). Their approach was to filter the measurement data from both ends to the interior point of interest and then combine the two filter estimates to obtain the smoothed estimate. Application of this forward/backward filtering method is fairly straightforward for the continuous smoother, but it is considerably more complicated for the corresponding discrete problem. Therefore, in this exercise we will look at a simpler method which is approximate, but it is still useful in certain special scenarios.

First of all, the random process x in question must be stationary, and the measurement process must be uniform throughout the fixed interval of interest. Secondly, we will assume that the x process variance is large relative to the measurement error variance. This is equivalent to saying that we know very little about the amplitude of the x process. This assumption will enable us to use the product-over-sum formula in blending the forward and backward estimates together (see Problem 6.1).

Now consider the same fixed-interval smoothing scenario in Example 6.1, except that the sigma of the Markov process is specified to be 30 m rather than 1 m. This makes the x process variance roughly three orders of magnitude greater than the measurement noise variance. This being the case, we should be able to use the simple product-over-sum formula in combining the forward and backward estimates. To be more specific, let us say that we are especially interested in the smoothed estimate at $k = 25$. This is the exact midpoint of the 51-point fixed interval. Now proceed as follows:

(a) First, program and run the RTS covariance solution with the x process sigma set at 30 m. The other parameters are the same as in Example 6.1. Save both the *a priori* and *a posteriori* filter P's for future reference. Also save the RTS smoother Ps. (Note we will only be doing covariance analysis here, so there is no need to generate Monte Carlo measurements. Also, note that the RTS algorithm is "tried-and-true," and it is exact—no approximations.)

(b) Now consider the forward-filter part of the forward/backward smoother method. This is to be the usual Kalman filter run from $k = 0$ to $k = 25$, 26 recursive steps in total. Of course, this is just a subset of the data saved from the RTS run. All we need to do here is identify $P_{25|25}$ and save it for use later with the corresponding result from the backward filter run.

(c) Next, consider the backward filter going from $k = 50$ back to $k = 26$. The filter parameters and recursive procedure here are the same as for the forward filter, except at the end points. (After all, the correlation structure is the same going in either direction timewise.) Note that the "initial" P^- should be set at an extremely large value in order that the filter will ignore any *a priori* estimate on the first backward step. (It would be redundant to use this information in both forward and backward filters.) Also, note that we stop this filter at $k = 26$. We then project the estimate (conceptually) and its error covariance back one more step without benefit of an update at $k = 25$. In this way, we avoid using the z_{25} measurement twice (conceptually) which would be redundant. Thus, the error covariance that we save from the backward filter is $P(25|$"measurement stream from $k = 50$ to $k = 26$"), which we will identify as $P(25|50 \rightarrow 26)$.

(d) We now have two error covariances: the forward one accounting for the measurements $z_0, z_1, \ldots z_{25}$, and the backward one accounting for $z_{50}, z_{49}, \ldots z_{26}$. We then combine them for the final result

$$P(25|50) = \frac{P(25|25) \cdot P(25|50 \rightarrow 26)}{P(25|25) + P(25|50 \rightarrow 26)}$$

(e) Carry out the procedure described in (a) through (d), and then compare the final result with the corresponding result using the RTS algorithm. You should find that the two error covariances match within a fraction of a percent.

6.3 In Problem 6.2 the parameters were such that the smoother covariance plot was nearly constant within most of the interior portion of the 51-point interval. This indicates that the initial conditions imposed at either end had very little influence on the smoother results at the midpoint of the interval. Therefore, in the interest of assessing the effect of the "initial" conditions on the backward filter at $k=50$, compute the smoother error covariance $P(49|50)$ using the forward/backward filter method, and compare the result with the corresponding result obtained with the RTS algorithm. Note that the filter parameters are to be the same here as in Problem 6.2. Note also that the RTS results worked out in Problem 6.1 still apply here, and they may be considered as "truth," not approximations.

6.4 A second-order Gauss-Markov Kalman filter was described in Example 4.4, Section 4.5, and a plot of the filter rms error is shown in Fig. 4.7. Using the same numerical values as in Example 4.4, do a 50-second fixed-interval smoother covariance run for this stochastic setting, and compare the smoother rms error with the corresponding filter error. Are the smoother results significantly better than the filter results; and if so, over what region of the fixed-interval is this true?

6.5 The Magill adaptive Kalman filter can be effectively used to handle a random process model with unknown tuning parameters. Suppose we are given an integrated random walk process model described as

$$\begin{bmatrix} x_1 \\ x_2 \end{bmatrix}_{k+1} = \begin{bmatrix} 1 & \Delta t \\ 0 & 1 \end{bmatrix} \begin{bmatrix} x_1 \\ x_2 \end{bmatrix}_k + \begin{bmatrix} w_1 \\ w_2 \end{bmatrix}_k \qquad \mathbf{Q}_k = E\mathbf{w}_k\mathbf{w}_k^T = \begin{bmatrix} \frac{1}{3}S\Delta t^3 & \frac{1}{2}S\Delta t^2 \\ \frac{1}{2}S\Delta t^2 & S\Delta t \end{bmatrix} \tag{P6.5.1}$$

and its corresponding measurement model described as

$$z_k = [1 \quad 0] \begin{bmatrix} x_1 \\ x_2 \end{bmatrix}_k + v_k \qquad \mathbf{R}_k = E\mathbf{v}_k\mathbf{v}_k^T \tag{P6.5.2}$$

We do know that the measurement noise variance $\mathbf{R}=1$ unit, but that the process noise covariance $\mathbf{Q}$ has an unknown spectral amplitude S that can take on a value of either 0.01 or 1 unit.

(a) Let $\Delta t=1$ second. Arbitrarily starting the state vector at $\mathbf{x}_0=[0 \quad 0]^T$, generate a 20-second sequence of the dynamic process along with a 20-second record of noisy measurements. For the true process, let $S=1$ unit.

(b) Construct a Magill adaptive Kalman filter that consists of two models, both with the same structure given above, except differing in the value of S that defines each $\mathbf{Q}$ matrix. Initialize both models with the same initial conditions:

$$\begin{bmatrix} \hat{x}_1 \\ \hat{x}_2 \end{bmatrix}_0 = \begin{bmatrix} 0 \\ 0 \end{bmatrix} \qquad \mathbf{P}_0^- = \begin{bmatrix} 10^4 & 0 \\ 0 & 10^4 \end{bmatrix}$$

(c) Process the 20-second measurement sequence that had been generated before and plot the composite output of the Magill adaptive Kalman filter.

(d) Repeat Parts (a), (b), and (c), except using $S=0.01$ unit in the true process. When comparing the results for $S=0.01$ and $S=1$, note that the resolution of the correct model, i.e., the speed of convergence of the probability, is noticeably different between the two cases. What might be a qualitative explanation for this difference?

6.6 The Magill adaptive Kalman filter is sometimes used to differentiate between hypotheses that have different discrete bias values of a state, even though their random models are identical. Consider the following two-state process model that combines a random walk and a random bias.

Process model:

$$\begin{bmatrix} x_1 \\ x_2 \end{bmatrix}_{k+1} = \begin{bmatrix} 1 & 0 \\ 0 & 1 \end{bmatrix} \begin{bmatrix} x_1 \\ x_2 \end{bmatrix}_k + \begin{bmatrix} w \\ 0 \end{bmatrix}_k \tag{P6.6.1}$$

At t = 0,

$$\begin{bmatrix} x_1 \\ x_2 \end{bmatrix}_0 = \begin{bmatrix} \mathcal{N}(0,100) \\ 4 \end{bmatrix}_k + \begin{bmatrix} w \\ 0 \end{bmatrix}_k \tag{P6.6.2}$$

Measurement model:

$$z_k = \begin{bmatrix} \cos\left(\dfrac{\pi k}{100}\right) & 1 \end{bmatrix} \begin{bmatrix} x_1 \\ x_2 \end{bmatrix}_k + v_k \tag{P6.6.3}$$

Instead of putting together a parallel bank of two-state Kalman filters for the Magill structure, the measurement model can be rewritten as

$$z_k - \alpha = \cos\left(\frac{\pi k}{100}\right) x_k + v_k \tag{P6.6.4}$$

where $x = x_1$ is the sole remaining state, and $\alpha = x_2$ is the hypothesized random bias. With this the different elemental models in the Magill filter bank are all 1-state processes.

(a) Generate a sequence of noisy measurements using the following parameters: $Ew_k^2 = 1, \quad Ev_k^2 = 1, \quad x_0 = \mathcal{N}(0,1)$

(b) Construct a Magill Kalman filter where each element is a one-state Kalman filter describing the random walk process given above, and consisting of five elements over the range of $\alpha = 1, \ldots, 5$

(c) Plot the conditional probabilities associated with each of the five hypotheses. Does the correct hypothesis win out after 100 steps?

(d) Replace the measurement connection parameter in Eq. (P6.6.4) from the time-varying cosine function to a fixed constant of 1 instead, and rerun Parts (b) and (c). Explain the outcome of your results.

6.7 Consider the following vehicle positioning scenario. We have two position sensors that operate independently to provide a user with two separate measures of horizontal position. The internal workings of the individual sensors are not made available to the user. Only the measured x and y positions and their respective estimated accuracies are reported out to the user at periodic intervals (i.e., the raw measurements are not made available to the user). It is desired to merge the two sensor outputs together in such a way as to obtain the best possible estimate of vehicle position in an rms sense. Make reasonable assumptions for vehicle motion and instrument accuracies, and demonstrate that the decentralized Kalman filter described in Section 6.7 will yield the same results as a hypothetical optimal centralized filter, which is not possible in the described restrictive circumstances. One set of reasonable assumptions for this demonstration would be as follows:

(a) The motions in both the x and y directions are random walk and independent. The Q for a Δt interval of 1 second is $0.01\,\text{m}^2$.

(b) The initial uncertainty in the x and y position estimates is 1 m rms in each direction, and the assumed initial position estimate is $[0, 0]^T$ for the local as well as the global filter.

(c) Let the true vehicle position be $[1, 1]^T$ in the x-y coordinate frame at $t = 0$.

(d) The measurement errors for the local Kalman filters are white and $1\,\text{m}^2$ for Sensor No. 1 and $4\,\text{m}^2$ for Sensor No. 2.

6.8 In Example 6.5, Section 6.5, we illustrated how measurement weight factors can be computed to yield a minimum mean square estimation error on a batch basis. Now complete this example by computing the mean square error associated with the resulting estimate.

REFERENCES CITED IN CHAPTER 6

1. J.S. Meditch, *Stochastic Optimal Linear Estimation and Control*, New York: McGraw-Hill, 1969.
2. M.S. Grewal and A.P. Andrews, *Kalman Filtering: Theory and Practice Using MATLAB*, New York: Wiley, 2008.
3. H.E. Rauch, "Solutions to the Linear Smoothing Problem," *IEEE Trans. Auto. Control*, AC-8: 371 (1963).
4. H.E. Rauch, F. Tung, and C.T. Striebel, "Maximum Likelihood Estimates of Linear Dynamic Systems," *AIAA J.*, 3: 1445 (1965).
5. D.T. Magill, "Optimal Adaptive Estimation of Sampled Stochastic Processes," *IEEE Transactions Automatic Control*, AC-10(4): 434–439 (Oct. 1965).
6. G.L. Mealy and W. Tang, "Application of Multiple Model Estimation to a Recursive Terrain Height Correlation System," *IEEE Trans. Automatic Control*, AC-28: 323–331 (March 1983).
7. A.A. Girgis and R.G. Brown, "Adaptive Kalman Filtering in Computer Relaying: Fault Classification Using Voltage Models," *IEEE Trans. Power Apparatus Systs.*, PAS-104(5): 1168–1177 (May 1985).
8. R.G. Brown, "A New Look at the Magill Adaptive Filter as a Practical Means of Multiple Hypothesis Testing," *IEEE Trans. Circuits Systs.*, CAS-30: 765–768 (Oct. 1983).
9. R.G. Brown and P.Y.C. Hwang, "A Kalman Filter Approach to Precision GPS Geodesy," *Navigation, J. Inst. Navigation*, 30(4): 338–349 (Winter 1983-84).
10. H.E. Rauch, "Autonomous Control Reconfiguration," *IEEE Control Systems*, 15(6): 37–48 (Dec. 1995).
11. W.D. Blair and Y. Bar-Shalom, "Tracking Maneuvering Targets with Multiple Sensors: Does More Data Always Mean Better Estimates?", *IEEE Trans. Aerospace Electronic Systs.*, 32(1): 450–456 (Jan. 1996).
12. M.J. Caputi, "A Necessary Condition for Effective Performance of the Multiple Model Adaptive Estimator," *IEEE Trans. Aerospace Electronic Systs.*, 31(3): 1132–1138 (July 1995).
13. R.G. Brown and P.Y.C. Hwang, "Application of GPS to Geodesy: A Combination Problem in Estimation and Large-Scale Multiple Hypothesis Testing," *Proc. IEEE National Telesystems Conf. 1983*, San Francisco, CA, Nov. 1983, pp. 104–111.

14. R.G. Brown and G.L. Hartman, "Kalman Filter with Delayed States as Observables," *Proceedings of the National Electronics Conference*, Chicago, IL, 1968.
15. L. Levy, *Applied Kalman Filtering*, Course 447, Navtech Seminars, 2007.
16. J.V. Candy, *Bayesian Signal Processing: Classical, Modern and Particle Filtering Methods*, New York: Wiley, 2009.
17. W. Koch, "On Bayesian Tracking and Data Fusion: A Tutorial Introduction with Examples," *IEEE Aerospace and Electronic Systems Magazine*, 25(7): 29–51 (July 2010).
18. A.J. Haug, "A Tutorial on Bayesian Estimation and Tracking Techniques Applicable to Nonlinear and Non-Gaussian Processes," *MITRE Technical Report*, Jan. 2005.
19. D.C. Fraserand J.E. Potter, "The Optimum Linear Smoother as a Combination of Two Optimum Linear Filters," *IEEE Trans. Auto. Control*, AC-14(4): 387 (Aug. 1969).
20. R.G. Brown and P.Y.C. Hwang, *Introduction to Random Signals and Applied Kalman Filtering*, 3rd Edition, New York: Wiley, 1997, Chapter 8.
21. U.B. Desai, H.L. Weinert, and G.J. Yusypchuk, "Discrete-Time Complementary Models and Smoothing Algorithms: The Correlated Noise Case," *IEEE Trans. Auto. Control*, AC-28(4):536–539 (Apr. 1983).

7

Linearization, Nonlinear Filtering, and Sampling Bayesian Filters

Many of the early applications of Kalman filtering were in navigation where the measurements were nonlinear. Thus, linearization has been an important consideration in applied Kalman filtering right from the start, and it continues to be so to this very day. Linearization is the main topic of Sections 7.1 and 7.2, and it is the authors' recommendation that the student just new to Kalman filtering should begin with these sections, because they are very basic. In particular, the extended Kalman filter that is discussed in Section 7.2 was one of the early means of coping with nonlinear measurements, and it is still the method of choice in many applications. However, there are some newer extensions of basic Kalman filtering that have been introduced in recent years that are also important. Three of these, namely the ensemble, unscented, and particle filters have been selected for discussion here in Sections 7.4, 7.5, and 7.6. Research in Kalman filtering is still quite active, so it is reasonable to expect to see further extensions and variations on the basic filter in the years ahead.

7.1 LINEARIZATION

Some of the most successful applications of Kalman filtering have been in situations with nonlinear dynamics and/or nonlinear measurement relationships. We now examine two basic ways of linearizing the problem. One is to linearize about some nominal trajectory in state space that does not depend on the measurement data. The resulting filter is usually referred to as simply a *linearized Kalman filter*. The other method is to linearize about a trajectory that is continually updated with the state estimates resulting from the measurements. When this is done, the filter is called an *extended Kalman filter*. A brief discussion of each will now be presented.

Linearized Kalman Filter

We begin by assuming the process to be estimated and the associated measurement relationship may be written in the form

$$\dot{\mathbf{x}} = \mathbf{f}(\mathbf{x}, \mathbf{u}_d, t) + \mathbf{u}(t) \tag{7.1.1}$$

$$\mathbf{z} = \mathbf{h}(\mathbf{x}, t) + \mathbf{v}(t) \tag{7.1.2}$$

where **f** and **h** are known functions, $\mathbf{u}_d$ is a deterministic forcing function, and **u** and **v** are white noise processes with zero crosscorrelation as before. Note that nonlinearity may enter into the problem either in the dynamics of the process or in the measurement relationship. Also, note that the forms of Eqs. (7.1.1) and (7.1.2) are somewhat restrictive in that **u** and **v** are assumed to be separate additive terms and are not included with the **f** and **h** terms. However, to do otherwise complicates the problem considerably, and thus, we will stay with these restrictive forms.

Let us now assume that an approximate trajectory $\mathbf{x}^*(t)$ may be determined by some means. This will be referred to as the nominal or reference trajectory, and it is illustrated along with the actual trajectory in Fig. 7.1. The actual trajectory $\mathbf{x}(t)$ may then be written as

$$\mathbf{x}(t) = \mathbf{x}^*(t) + \Delta\mathbf{x}(t) \tag{7.1.3}$$

Equations (7.1.1) and (7.1.2) then become

$$\dot{\mathbf{x}}^* + \Delta\dot{\mathbf{x}} = \mathbf{f}(\mathbf{x}^* + \Delta\mathbf{x}, \mathbf{u}_d, t) + \mathbf{u}(t) \tag{7.1.4}$$

$$\mathbf{z} = \mathbf{h}(\mathbf{x}^* + \Delta\mathbf{x}, t) + \mathbf{v}(t) \tag{7.1.5}$$

We now assume $\Delta\mathbf{x}$ is small and approximate the **f** and **h** functions with Taylor's series expansions, retaining only first-order terms. The result is

$$\dot{\mathbf{x}}^* + \Delta\dot{\mathbf{x}} \approx \mathbf{f}(\mathbf{x}^*, \mathbf{u}_d, t) + \left[\frac{\partial \mathbf{f}}{\partial \mathbf{x}}\right]_{\mathbf{x}=\mathbf{x}^*} \cdot \Delta\mathbf{x} + \mathbf{u}(t) \tag{7.1.6}$$

$$\mathbf{z} \approx \mathbf{h}(\mathbf{x}^*, t) + \left[\frac{\partial \mathbf{h}}{\partial \mathbf{x}}\right]_{\mathbf{x}=\mathbf{x}^*} \cdot \Delta\mathbf{x} + \mathbf{v}(t) \tag{7.1.7}$$

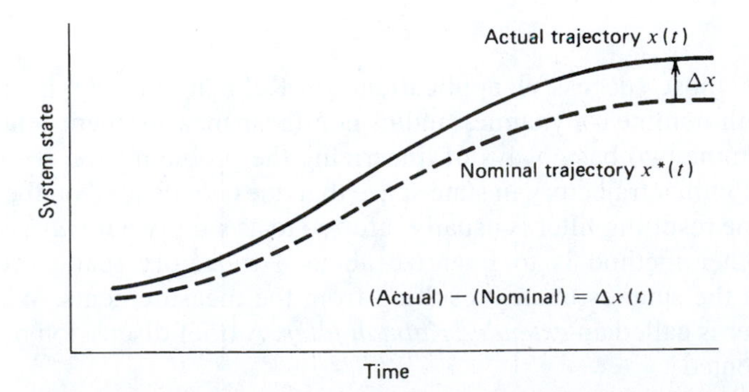

Figure 7.1 Nominal and actual trajectories for a linearized Kalman filter.

where

$$\frac{\partial \mathbf{f}}{\partial \mathbf{x}} = \begin{bmatrix} \frac{\partial f_1}{\partial x_1} & \frac{\partial f_1}{\partial x_2} & \cdots \\ \frac{\partial f_2}{\partial x_1} & \frac{\partial f_2}{\partial x_2} & \cdots \\ & \vdots & \end{bmatrix}; \quad \frac{\partial \mathbf{h}}{\partial \mathbf{x}} = \begin{bmatrix} \frac{\partial h_1}{\partial x_1} & \frac{\partial h_1}{\partial x_2} & \cdots \\ \frac{\partial h_2}{\partial x_1} & \frac{\partial h_2}{\partial x_2} & \cdots \\ & \vdots & \end{bmatrix} \tag{7.1.8}$$

It is customary to choose the nominal trajectory $\mathbf{x}^*(t)$ to satisfy the deterministic differential equation

$$\dot{\mathbf{x}}^* = \mathbf{f}(\mathbf{x}^*, \mathbf{u}_d, t) \tag{7.1.9}$$

Substituting this into (7.1.7) then leads to the linearized model

$$\Delta\dot{\mathbf{x}} = \left[\frac{\partial \mathbf{f}}{\partial \mathbf{x}}\right]_{\mathbf{x}=\mathbf{x}^*} \cdot \Delta\mathbf{x} + \mathbf{u}(t) \quad \text{(linearized dynamics)} \tag{7.1.10}$$

$$[\mathbf{z} - \mathbf{h}(\mathbf{x}^*, t)] = \left[\frac{\partial \mathbf{h}}{\partial \mathbf{x}}\right]_{\mathbf{x}=\mathbf{x}^*} \cdot \Delta\mathbf{x} + \mathbf{v}(t) \quad \text{(linearized measurement equation)} \tag{7.1.11}$$

Note that the "measurement" in the linear model is the actual measurement less that predicted by the nominal trajectory in the absence of noise. Also the equivalent $\mathbf{F}$ and $\mathbf{H}$ matrices are obtained by evaluating the partial derivative matrices (Eq. 7.1.8) along the *nominal* trajectory. We will now look at two examples that illustrate the linearization procedure. In the first example the nonlinearity appears only in the measurement relationship, so it is relatively simple. In the second, nonlinearity occurs in both the measurement and process dynamics, so it is somewhat more involved than the first.

EXAMPLE 7.1

In many electronic navigation systems the basic observable is a noisy measurement of range (distance) from the vehicle to a known location. One such system that has enjoyed wide use in aviation is distance-measuring equipment (DME)* (1). We do not need to go into detail here as to how the equipment works. It suffices to say that the airborne equipment transmits a pulse that is returned by the ground station, and then the aircraft equipment interprets the transit time in terms of distance. In our example we will simplify the geometric situation by assuming that the aircraft and the two DME stations are all in a horizontal plane as shown in Fig. 7.2 (slant range ≈ horizontal range). The coordinates of the two DME stations are assumed to be known, and the aircraft coordinates are unknown and to be estimated.

* It is of interest to note that the DME land-based system has survived into the satellite navigation age. It is the U. S. government's current policy to keep the DME system operational within the continental U.S. as a backup to GPS.

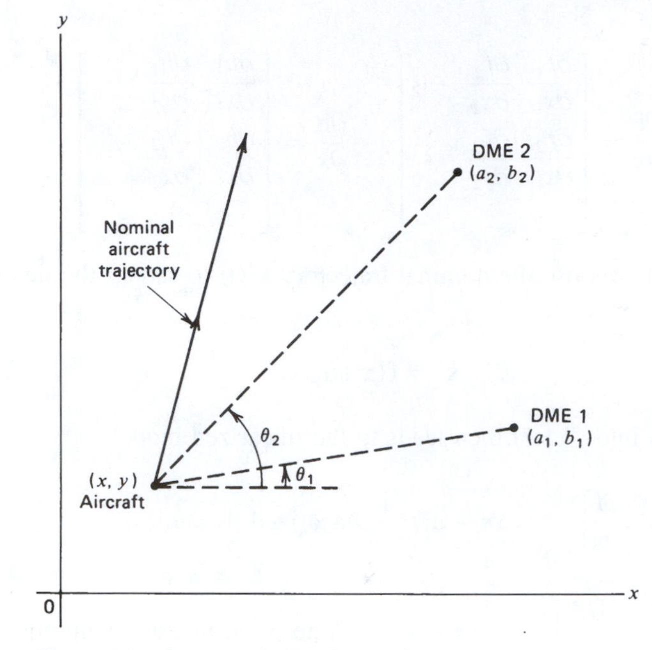

Figure 7.2 Geometry for DME example.

We will look at the aircraft dynamics first. This, in turn, will determine the process state model. To keep things as simple as possible, we will assume a nominal straight-and-level flight condition with constant velocity. The true trajectory will be assumed to be the nominal one plus small perturbations due to random horizontal accelerations, which will be assumed to be white. This leads to random walk in velocity and integrated random walk in position. This is probably unrealistic for long time spans because of the control applied by the pilot (or autopilot). However, this would be a reasonable model for short intervals of time. The basic differential equations of motion in the x and y directions are then

$$\begin{aligned} \ddot{x} &= \underbrace{\begin{matrix} 0 \\ 0 \end{matrix}}_{\substack{\text{Deterministic} \\ \text{forcing} \\ \text{function}}} + \underbrace{\begin{matrix} u_x \\ u_y \end{matrix}}_{\substack{\text{Random} \\ \text{forcing} \\ \text{function}}} \\ \ddot{y} &= \end{aligned} \tag{7.1.12}$$

The dynamical equations are seen to be linear in this case, so the differential equations for the incremental quantities are the same as for the total x and y, that is,

$$\begin{aligned} \Delta\ddot{x} &= u_x \\ \Delta\ddot{y} &= u_y \end{aligned} \tag{7.1.13}$$

We now define filter state variables in terms of the incremental positions and velocities:

$$\begin{aligned} x_1 &= \Delta x, \qquad & x_2 &= \Delta\dot{x} \\ x_3 &= \Delta y, \qquad & x_4 &= \Delta\dot{y} \end{aligned} \tag{7.1.14}$$

The state equations are then

$$\begin{bmatrix} \dot{x}_1 \\ \dot{x}_2 \\ \dot{x}_3 \\ \dot{x}_4 \end{bmatrix} = \begin{bmatrix} 0 & 1 & 0 & 0 \\ 0 & 0 & 0 & 0 \\ 0 & 0 & 0 & 1 \\ 0 & 0 & 0 & 0 \end{bmatrix} \begin{bmatrix} x_1 \\ x_2 \\ x_3 \\ x_4 \end{bmatrix} + \begin{bmatrix} 0 \\ u_x \\ 0 \\ u_y \end{bmatrix} \tag{7.1.15}$$

The state variables are driven by the white noise processes u_x and u_y, so we are assured that the corresponding discrete equations will be in the appropriate form for a Kalman filter.

We now turn to the measurement relationships. We will assume that we have two simultaneous range measurements, one to DME_1 and the other to DME_2. The two measurement equations in terms of the total x and y are then

$$\begin{aligned} z_1 &= \sqrt{(x-a_1)^2 + (y-b_1)^2} + v_1 \\ z_2 &= \sqrt{(x-a_2)^2 + (y-b_2)^2} + v_2 \end{aligned} \tag{7.1.16}$$

where v_1 and v_2 are additive white measurement noises. We see immediately that the connection between the observables (z_1 and z_2) and the quantities to be estimated (x and y) is nonlinear. Thus, linearization about the nominal trajectory is in order. We assume that an approximate nominal position is known at the time of the measurement, and that the locations of the two DME stations are known exactly. We now need to form the $\partial\mathbf{h}/\partial\mathbf{x}$ matrix as specified by Eq. (7.1.8). [We note a small notational problem here. The variables x_1, x_2, x_3, and x_4 are used in Eq. (7.1.8) to indicate total state variables, and then the same symbols are used again to indicate incremental state variables as defined by Eqs. (7.1.14). However, the meanings of the symbols are never mixed in any one set of equations, so this should not lead to confusion.] We now note that the x and y position variables are the first and third elements of the state vector. Thus, evaluation of the partial derivatives indicated in Eq. (7.1.8) leads to

$$\frac{\partial\mathbf{h}}{\partial\mathbf{x}} = \begin{bmatrix} \dfrac{(x_1-a_1)}{\sqrt{(x_1-a_1)^2+(x_3-b_1)^2}} & 0 & \dfrac{(x_3-b_1)}{\sqrt{(x_1-a_1)^2+(x_3-b_1)^2}} & 0 \\ \dfrac{(x_1-a_2)}{\sqrt{(x_1-a_2)^2+(x_3-b_2)^2}} & 0 & \dfrac{(x_3-b_2)}{\sqrt{(x_1-a_2)^2+(x_3-b_2)^2}} & 0 \end{bmatrix} \tag{7.1.17}$$

or

$$\frac{\partial\mathbf{h}}{\partial\mathbf{x}} = \begin{bmatrix} -\cos\theta_1 & 0 & -\sin\theta_1 & 0 \\ -\cos\theta_2 & 0 & -\sin\theta_2 & 0 \end{bmatrix} \tag{7.1.18}$$

Finally, we note that Eq. (7.1.18) can be generalized even further, since the sine and cosine terms are actually direction cosines between the x and y axes and the respective lines of sight to the two DME stations. Therefore, we will write the linearized $\mathbf{H}$ matrix in its final form as

$$\mathbf{H} = \left.\frac{\partial\mathbf{h}}{\partial\mathbf{x}}\right|_{\mathbf{x}=\mathbf{x}^*} = \begin{bmatrix} -\cos\theta_{x1} & 0 & -\cos\theta_{y1} & 0 \\ -\cos\theta_{x2} & 0 & -\cos\theta_{y2} & 0 \end{bmatrix} \tag{7.1.19}$$

where the subscripts on θ indicate the respective axes and lines of sight to the DME stations. Note that $\mathbf{H}$ is evaluated at a point on the *nominal* trajectory. (The true trajectory is not known to the filter.) The nominal aircraft position will change with each step of the recursive process, so the terms of $\mathbf{H}$ are time-variable and must be recomputed with each recursive step. Also, recall from Eq. (7.1.11) that the measurement presented to the linearized filter is the total $\mathbf{z}$ minus the predicted $\mathbf{z}$ based on the nominal position $\mathbf{x}^*$.

Strictly speaking, the linearized filter is always estimating incremental quantities, and then the total quantity is reconstructed by adding the incremental estimate to the nominal part. However, we will see later that when it comes to the actual mechanics of handling the arithmetic on the computer, we can avoid working with incremental quantities if we choose to do so. This is discussed further in the section on the extended Kalman filter. We will now proceed to a second linearization example, where the process dynamics as well as the measurement relationship has to be linearized.

EXAMPLE 7.2

This example is taken from Sorenson (2) and is a classic example of linearization of a nonlinear problem. Consider a near earth space vehicle in a nearly circular orbit. It is desired to estimate the vehicle's position and velocity on the basis of a sequence of angular measurements made with a horizon sensor. With reference to Fig. 7.3, the horizon sensor is capable of measuring:

1. The angle γ between the earth's horizon and the local vertical.
2. The angle α between the local vertical and a known reference line (say, to a celestial object).

In the interest of simplicity, we assume all motion and measurements to be within a plane as shown in Fig. 7.3. Thus, the motion of the vehicle can be described with the usual polar coordinates r and θ.

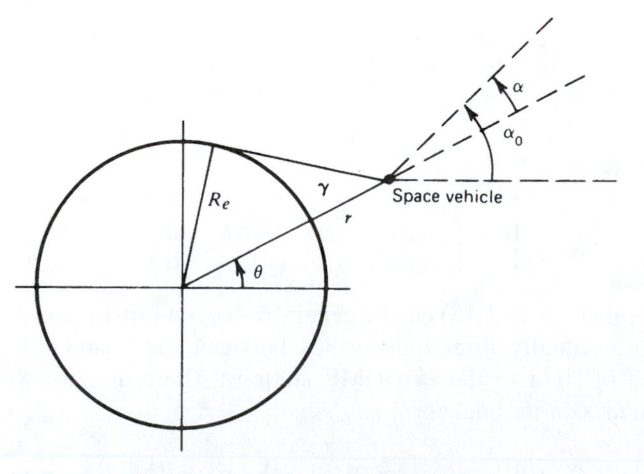

Figure 7.3 Coordinates for space vehicle example.

The equations of motion for the space vehicle may be obtained from either Newtonian or Lagrangian mechanics. They are (see Section 2.10, ref. 3):

$$\ddot{r} - r\dot{\theta}^2 + \frac{K}{r^2} = u_r(t) \tag{7.1.20}$$

$$r\ddot{\theta} + 2\dot{r}\dot{\theta} = u_\theta(t) \tag{7.1.21}$$

where K is a constant proportional to the universal gravitational constant, and u_r and u_θ are small random forcing functions in the r and θ directions (due mainly to gravitational anomalies unaccounted for in the K/r^2 term). It can be seen that the constant K must be equal to gR_e^2 if the gravitational forcing function is to match the earth's gravity constant g at the surface. The random forcing functions u_r and u_θ will be assumed to be white. We will look at the linearized process dynamics first and then consider the nonlinear measurement situation later.

The equations of motion, Eqs. (7.1.20) and (7.1.21), are clearly nonlinear so we must linearize the dynamics if we are to apply Kalman filter methods. We have assumed that random forcing functions u_r and u_θ are small, so the corresponding perturbations from a circular orbit will also be small. By direct substitution into Eqs. (7.1.20) and (7.1.21), it can be verified that

$$r^* = R_0 \quad \text{(a constant radius)} \tag{7.1.22}$$

$$\theta^* = \omega_0 t \quad \left(\omega_0 = \sqrt{\frac{K}{R_0^3}}\right) \tag{7.1.23}$$

will satisfy the differential equations. Thus, this will be the reference trajectory that we linearize about.

We note that we have two second-order differential equations describing the dynamics. Therefore, we must have four state variables in our state model. We choose the usual phase variables as state variables as follows:

$$\begin{aligned} x_1 &= r, & x_2 &= \dot{r} \\ x_3 &= \theta, & x_4 &= \dot{\theta} \end{aligned} \tag{7.1.24}$$

The nonlinear state equations are then

$$\begin{aligned} \dot{x}_1 &= x_2 \\ \dot{x}_2 &= x_1 x_4^2 - \frac{K}{x_1^2} + u_r(t) \\ \dot{x}_3 &= x_4 \\ \dot{x}_4 &= -\frac{2x_2 x_4}{x_1} + \frac{u_\theta(t)}{x_1} \end{aligned} \tag{7.1.25}$$

We must now form the $\partial\mathbf{f}/\partial\mathbf{x}$ matrix indicated in Eq. (7.1.8) to get the linearized **F** matrix.

$$\frac{\partial\mathbf{f}}{\partial\mathbf{x}} = \begin{bmatrix} 0 & 1 & 0 & 0 \\ \left(x_4^2 + \dfrac{2K}{x_1^3}\right) & 0 & 0 & 2x_1x_4 \\ 0 & 0 & 0 & 1 \\ \dfrac{2x_2x_4}{x_1^2} & \dfrac{-2x_4}{x_1} & 0 & \dfrac{-2x_2}{x_1} \end{bmatrix} = \begin{bmatrix} 0 & 1 & 0 & 0 \\ \left(\dot{\theta}^2 + \dfrac{2K}{r^3}\right) & 0 & 0 & 2r\dot{\theta} \\ 0 & 0 & 0 & 1 \\ \dfrac{2\dot{r}\dot{\theta}}{r^2} & \dfrac{-2\dot{\theta}}{r} & 0 & \dfrac{-2\dot{r}}{r} \end{bmatrix} \tag{7.1.26}$$

Next, we evaluate $\partial\mathbf{f}/\partial\mathbf{x}$ along the reference trajectory.

$$\left.\frac{\partial\mathbf{f}}{\partial\mathbf{x}}\right|_{\substack{r=R_0 \\ \theta=\omega_0 t}} = \begin{bmatrix} 0 & 1 & 0 & 0 \\ 3\omega_0^2 & 0 & 0 & 2R_0\omega_0 \\ 0 & 0 & 0 & 1 \\ 0 & \dfrac{-2\omega_0}{R_0} & 0 & 0 \end{bmatrix} \tag{7.1.27}$$

Equation (7.1.27) then defines the **F** matrix that characterizes the linearized dynamics. Note that in the linear equations, Δr, $\Delta\dot{r}$, $\Delta\theta$, and $\Delta\dot{\theta}$ become the four state variables.

We now turn to the measurement model. The idealized (no noise) relationships are given by

$$\begin{bmatrix} z_1 \\ z_2 \end{bmatrix} = \begin{bmatrix} \gamma \\ \alpha \end{bmatrix} = \begin{bmatrix} \sin^{-1}\left(\dfrac{R_e}{r}\right) \\ \alpha_0 - \theta \end{bmatrix} \tag{7.1.28}$$

We next replace r with x_1 and θ with x_3, and then perform the partial derivatives indicated by Eq. (7.1.8). The result is

$$\left[\frac{\partial\mathbf{h}}{\partial\mathbf{x}}\right] = \begin{bmatrix} -\dfrac{R_e}{r\sqrt{r^2 - R_e^2}} & 0 & 0 & 0 \\ 0 & 0 & -1 & 0 \end{bmatrix} \tag{7.1.29}$$

Finally, we evaluate $\partial\mathbf{h}/\partial\mathbf{x}$ along the reference trajectory

$$\left.\frac{\partial\mathbf{h}}{\partial\mathbf{x}}\right|_{\substack{r=R_0 \\ \theta=\omega_0 t}} = \begin{bmatrix} -\dfrac{R_e}{R_0\sqrt{R_0^2 - R_e^2}} & 0 & 0 & 0 \\ 0 & 0 & -1 & 0 \end{bmatrix} \tag{7.1.30}$$

This then becomes the linearized **H** matrix of the Kalman filter. The linearized model is now complete with the determination of the **F** and **H** matrices. Before we leave this example, though, it is worth noting that the forcing function $u_\theta(t)$ must be scaled by $1/R_0$ in the linear model because of the $1/x_1$ factor in Eq. (7.1.25).

7.2 THE EXTENDED KALMAN FILTER

The extended Kalman filter is similar to a linearized Kalman filter except that the linearization takes place about the filter's estimated trajectory, as shown in Fig. 7.4, rather than a precompiled nominal trajectory. That is, the partial derivatives of Eq. (7.1.8) are evaluated along a trajectory that has been updated with the filter's estimates; these, in turn, depend on the measurements, so the filter gain sequence will depend on the sample measurement sequence realized on a particular run of the experiment. Thus, the gain sequence is not predetermined by the process model assumptions as in the usual Kalman filter.

A general analysis of the extended Kalman filter is difficult because of the feedback of the measurement sequence into the process model. However, qualitatively it would seem to make sense to update the trajectory that is used for the linearization—after all, why use the old trajectory when a better one is available? The flaw in this argument is this: The "better" trajectory is only better in a *statistical* sense. There is a chance (and maybe a good one) that the updated trajectory will be poorer than the nominal one. In that event, the estimates may be poorer; this, in turn, leads to further error in the trajectory, which causes further errors in the estimates, and so forth and so forth, leading to eventual divergence of the filter. The net result is that the extended Kalman filter is a somewhat riskier filter than the regular linearized filter, especially in situations where the initial uncertainty and measurement errors are large. It may be better *on the average* than the linearized filter, but it is also more likely to diverge in unusual situations.

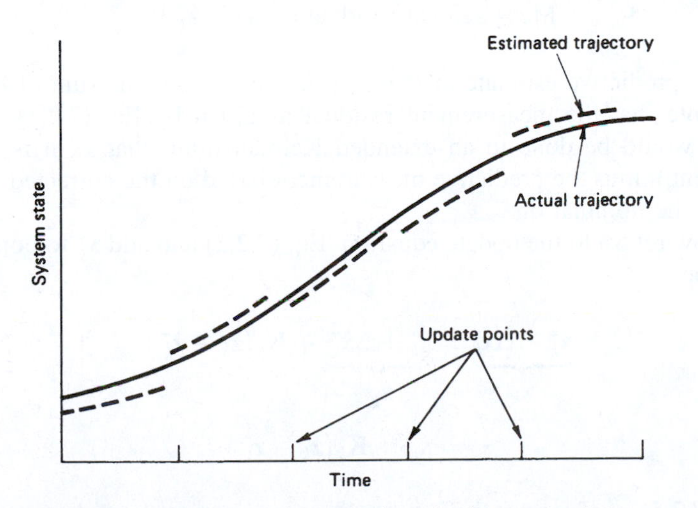

Figure 7.4 Reference and actual trajectories for an extended Kalman filter.

Both the regular linearized Kalman filter and the extended Kalman filter have been used in a variety of applications. Each has its advantages and disadvantages, and no general statement can be made as to which is best because it depends on the particular situation at hand. Aided inertial navigation systems serve as good examples of both methods of linearization, and this is discussed further in Chapters 8 and 9.

Keeping Track of Total Estimates in an Extended Kalman Filter

It should be remembered that the basic state variables in a linearized Kalman filter are incremental quantities, and not the total quantities such as position, velocity, and so forth. However, in an extended Kalman filter it is usually more convenient to keep track of the total estimates rather than the incremental ones, so we will now proceed to show how this is done and why it is valid to do so.

We begin with the basic linearized measurement equation, Eq. (7.1.11)

$$\mathbf{z} - \mathbf{h}(\mathbf{x}^*) = \mathbf{H}\Delta\mathbf{x} + \mathbf{v} \tag{7.2.1}$$

Note that when working with incremental state variables, the measurement presented to the Kalman filter is $[\mathbf{z} - \mathbf{h}(\mathbf{x}^*)]$ rather than the total measurement $\mathbf{z}$. Next, consider the incremental estimate update equation at time t_k

$$\Delta\hat{\mathbf{x}}_k = \Delta\hat{\mathbf{x}}_k^- + \mathbf{K}_k[\underbrace{\mathbf{z}_k - \mathbf{h}(\mathbf{x}_k^*)}_{\text{Inc. meas.}} - \mathbf{H}_k\Delta\hat{\mathbf{x}}_k^-] \tag{7.2.2}$$

Now, in forming the measurement residual in Eq. (7.2.2), suppose we associate the $\mathbf{h}(\mathbf{x}_k^*)$ term with $\mathbf{H}_k\Delta\hat{\mathbf{x}}_k^-$ rather than $\mathbf{z}_k$. This measurement residual can then be written as

$$\text{Measurement residual} = (\mathbf{z}_k - \hat{\mathbf{z}}_k^-) \tag{7.2.3}$$

because the predictive estimate of the measurement is just the sum of $\mathbf{h}(\mathbf{x}_k^*)$ and $\mathbf{H}_k\Delta\hat{\mathbf{x}}_k^-$. Note that the measurement residual as given by Eq. (7.2.3) is formed exactly as would be done in an extended Kalman filter, that is, it is the noisy measurement minus the predictive measurement based on the corrected trajectory rather than the nominal one.

We now return to the update equation, Eq. (7.2.2) and add $\mathbf{x}_k^*$ to both sides of the equation:

$$\underbrace{\mathbf{x}_k^* + \Delta\hat{\mathbf{x}}_k}_{\hat{\mathbf{x}}_k} = \underbrace{\mathbf{x}_k^* + \Delta\hat{\mathbf{x}}_k^-}_{\hat{\mathbf{x}}_k^-} + \mathbf{K}_k(\mathbf{z}_k - \hat{\mathbf{z}}_k^-) \tag{7.2.4}$$

$$\hat{\mathbf{x}}_k = \hat{\mathbf{x}}_k^- + \mathbf{K}_k(\mathbf{z}_k - \hat{\mathbf{z}}_k^-) \tag{7.2.5}$$

Equation (7.2.5) is, of course, the familiar linear estimate update equation written in terms of *total* rather than incremental quantities. It simply says that we correct the *a priori* estimate by adding the measurement residual appropriately weighted by the Kalman gain $\mathbf{K}_k$. Note that after the update is made in the extended Kalman filter, the incremental $\Delta\hat{\mathbf{x}}_k$ is reduced to zero. Its projection to the next step is then trivial. The only nontrivial projection is to project $\hat{\mathbf{x}}_k$ (which has become the reference $\mathbf{x}$ at t_k) to $\hat{\mathbf{x}}_{k+1}^-$. This must be done through the nonlinear dynamics as dictated by Eq. (7.1.1). That is,

$$\hat{\mathbf{x}}_{k+1}^- = \left\{ \begin{array}{l} \text{Solution of the nonlinear differential equation} \\ \dot{\mathbf{x}} = \mathbf{f}(\mathbf{x}, \mathbf{u}_d, t) \text{ at } t = t_{k+1}, \text{ subject to the} \\ \text{initial conditon } \mathbf{x} = \hat{\mathbf{x}}_k \text{ at } t_k \end{array} \right\} \tag{7.2.6}$$

Note that the additive white noise forcing function $\mathbf{u}(t)$ is zero in the projection step, but the deterministic $\mathbf{u}_d$ is included in the $\mathbf{f}$ function. Once $\hat{\mathbf{x}}_{k+1}^-$ is determined, the predictive measurement $\hat{\mathbf{z}}_{k+1}^-$ can be formed as $\mathbf{h}(\hat{\mathbf{x}}_{k+1}^-)$, and the measurement residual at t_{k+1} is formed as the difference $(\mathbf{z}_{k+1} - \hat{\mathbf{z}}_{k+1}^-)$. The filter is then ready to go through another recursive loop.

For completeness, we repeat the familiar error covariance update and projection equations:

$$\mathbf{P}_k = (\mathbf{I} - \mathbf{K}_k\mathbf{H}_k)\mathbf{P}_k^- \tag{7.2.7}$$

$$\mathbf{P}_{k+1}^- = \boldsymbol{\phi}_k\mathbf{P}_k\boldsymbol{\phi}_k^T + \mathbf{Q}_k \tag{7.2.8}$$

where $\boldsymbol{\phi}_k$, $\mathbf{H}_k$, and $\mathbf{Q}_k$ come from the linearized model. Equations (7.2.7) and (7.2.8) and the gain equation (which is the same as in the linear Kalman filter) should serve as a reminder that the extended Kalman filter is still working in the world of linear dynamics, even though it keeps track of total estimates rather than incremental ones.

Getting the Extended Kalman Filter Started

It was mentioned previously that the extended Kalman filter can diverge if the reference about which the linearization takes place is poor. The most common situation of this type occurs at the initial starting point of the recursive process. Frequently, the *a priori* information about the true state of the system is poor. This causes a large error in $\hat{\mathbf{x}}_0^-$ and forces $\mathbf{P}_0^-$ to be large. Thus, two problems can arise in getting the extended filter started:

1. A very large $\mathbf{P}_0^-$ combined with low-noise measurements at the first step will cause the $\mathbf{P}$ matrix to "jump" from a very large value to a small value in one step. In principle, this is permissible. However, this can lead to numerical problems. A non-positive-definite $\mathbf{P}$ matrix at any point in the recursive process usually leads to divergence.
2. The initial $\hat{\mathbf{x}}_0^-$ is presumably the best estimate of $\mathbf{x}$ prior to receiving any measurement information, and thus, it is used as the reference for linearization. If the error in $\hat{\mathbf{x}}_0^-$ is large, the first-order approximation used in the linearization will be poor, and divergence may occur, even with perfect arithmetic.

With respect to problem 1, the filter designer should be especially careful to use all the usual numerical precautions to preserve the symmetry and positive definiteness of the **P** matrix on the first step. In some cases, simply using the symmetric form of the **P**-update equation is sufficient to ward off divergence. This form, Eq. (4.2.18), is repeated here for convenience (sometimes called the Joseph form)

$$\mathbf{P}_k = (\mathbf{I} - \mathbf{K}_k\mathbf{H}_k)\mathbf{P}_k^-(\mathbf{I} - \mathbf{K}_k\mathbf{H}_k)^T + \mathbf{K}_k\mathbf{R}_k\mathbf{K}_k^T \tag{7.2.9}$$

Another way of mitigating the numerical problem is to let $\mathbf{P}_0^-$ be considerably smaller than would normally be dictated by the true a priori uncertainty in $\hat{\mathbf{x}}_0^-$. This will cause suboptimal operation for the first few steps, but this is better than divergence! A similar result can be accomplished by letting $\mathbf{R}_k$ be abnormally large for the first few steps. There is no one single cure for all numerical problems. Each case must be considered on its own merits.

Problem 2 is more subtle than problem 1. Even with perfect arithmetic, poor linearization can cause a poor $\hat{\mathbf{x}}_0^-$ to be updated into an even poorer *a posteriori* estimate, which in turn gets projected on ahead, and so forth. Various "fixes" have been suggested for the poor-linearization problem, and it is difficult to generalize about them (4–7). All are *ad hoc* procedures. This should come as no surprise, because the extended Kalman filter is, itself, an *ad hoc* procedure. One remedy that works quite well when the information contained in $\mathbf{z}_0$ is sufficient to determine $\mathbf{x}$ algebraically is to use $\mathbf{z}_0$ to solve for $\mathbf{x}$, just as if there were no measurement error. This is usually done with some tried-and-true numerical algorithm such as the Newton-Raphson method of solving algebraic equations. It is hoped this will yield a better estimate of $\mathbf{x}$ than the original coarse $\hat{\mathbf{x}}_0^-$. The filter can then be linearized about the new estimate (and a smaller $\mathbf{P}_0^-$ than the original $\mathbf{P}_0^-$ can be used), and the filter is then run as usual beginning with $\mathbf{z}_0$ and with proper accounting for the measurement noise. Another *ad hoc* procedure that has been used is to let the filter itself iterate on the estimate at the first step. The procedure is fairly obvious. The linearized filter parameters that depend on the reference $\mathbf{x}$ are simply relinearized with each iteration until convergence is reached within some predetermined tolerance. $\mathbf{P}_0^-$ may be held fixed during the iteration, but this need not be the case. Also, if $\mathbf{x}$ is not observable on the basis of just one measurement, iteration may also have to be carried out at a few subsequent steps in order to converge on good estimates of all the elements of $\mathbf{x}$. There is no guarantee that iteration will work in all cases, but it is worth trying.

Before leaving the subject of getting the filter started, it should be noted that neither the algebraic solution nor the iteration remedies just mentioned play any role in the basic "filtering" process. Their sole purpose is simply to provide a good reference for linearization, so that the linearized Kalman filter can do its job of optimal estimation.

7.3 "BEYOND THE KALMAN FILTER"

Even though the linearized and extended forms of the Kalman filter have served the navigation community very well for the past half century, a healthy amount of research on more general forms of nonlinear filtering has steadily persisted alongside for quite sometime now. More recently, there has even been a considerable upsurge of interest in nonlinear and non-Gaussian filtering. While some of these results can be quite interesting in their own right, our treatment of this subject here can only be brief but one at a tutorial pace.

When dealing with a nonlinear function, either in the process or the measurement model or both, the results of the linearized or extended Kalman filter strays from the theoretical optima and, depending on the severity of the nonlinearity, can lead to misleading performance or, worse, divergence. With its linear assumption, the Kalman filter is an elegant, almost minimalist, recursive algorithm. Formulations of Kalman filters that accommodate second- or higher-order relationships in the nonlinear (process or measurement) functions do exist (8, 9), but they naturally involve solutions with extra complexity. These belong in the same class as the extended Kalman filter where the nonlinear function is being approximated, except with higher degrees of complexity.

Another class of methods that has proven even more popular of late involves the random *sampling* of the *a posteriori* conditional density function (see Section 6.9) whose expected value would represent the optimum state estimate, in a minimum mean square error sense. Of course, this class of methods is also suboptimal because it is an approximation as well, albeit of the density function instead of an approximation of the nonlinear function(s) in the model that is what the Extended Kalman filter uses. In the later sections of this chapter, we will be delving into several of the nonlinear methods from this sampling class. At first glance, they may appear daunting due to their use of a whole new and unfamiliar approach to Kalman filtering but we shall first broadly outline how they are related to each other. The first two to be discussed, the Ensemble Kalman filter and the Unscented Kalman filter, both invoke the Gaussian approximation of the Bayesian filter. In contrast, the more general Particle filter does not do so. Of the two Gaussian approximation methods to be discussed next, the Ensemble Kalman filter is a form of a Monte Carlo-type filter and depends on random sampling of the probability density function. The Unscented Kalman filter, on the other hand, uses a deterministic sampling scheme.

We begin by revisiting the recursive Bayesian filter that was first introduced back in Chapter 6, in particular, the form derived under the Gaussian approximation. Even with the Gaussian assumption, we still face the non-trivial task of deriving the associated covariance matrices to obtain an analytical solution. Rather, the idea behind the sampling methods is to approximate these parameters via Monte Carlo sampling. From the recursive loop of Fig. 6.12, we shall insert Gaussian random number generators into the loop at two junctures, one before the projection computations and one after thus resulting in Fig. 7.5. This is the basic Monte Carlo Kalman filter (10) where the associated gains and error covariances are derived, not analytically, but statistically via the random samples generated.

One notable deviation from our usual flow diagram here, and for subsequent cases in this section, is that the initial conditions feed into the loop just before projection step to the next cycle as opposed to at the measurement update step, as have been seen before in earlier chapters. This convention appears to have been widely adopted in the nonlinear filtering literature so we will follow it here as well. The Monte Carlo Kalman filter provides a good intuitive stepping stone but we will not dwell on it any further and move on next to another similar and more popular sampling method called the Ensemble Kalman filter in the next section.

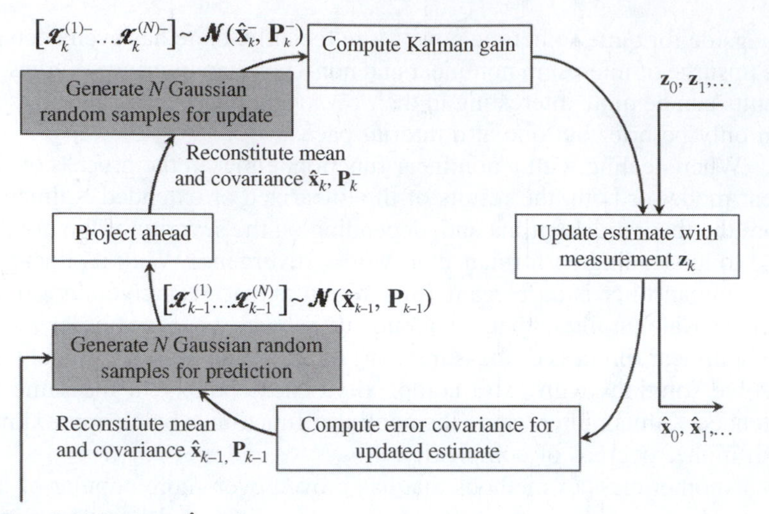

Figure 7.5 The Monte Carlo Kalman filter.

7.4 THE ENSEMBLE KALMAN FILTER

The Ensemble Kalman filter method grew out of research activities from the world of geophysical sciences and is finding use in large-scale systems such as those found in dynamic meteorological forecasting and ocean systems (11,12). In place of inserting the two Gaussian random number generators shown in Fig. 7.5, we modify the scheme utilize a Gaussian random number generator at the very start to generate random samples representing the initial state estimate with its associated error covariance and then two for the additive process and measurement noise sequences encountered in the recursive loop. Fig. 7.6 shows the modifications (gray blocks) for the Ensemble Kalman filter.

After generating $[\mathscr{X}_k^{(1)-} \ldots \mathscr{X}_k^{(N)-}]$, we compute the following:

$$\hat{\mathbf{z}}_k^- = \frac{1}{N}\sum_{i=1}^{N} \mathbf{h}\left(\mathscr{X}_k^{(i)-}\right) \tag{7.4.1}$$

$$\left(\mathbf{P}_{\tilde{\mathbf{z}}\tilde{\mathbf{z}}}^-\right)_k = \left\{\frac{1}{N}\sum_{i=1}^{N} \mathbf{h}\left(\mathscr{X}_k^{(i)-}\right)\left[\mathbf{h}\left(\mathscr{X}_k^{(i)-}\right)\right]^T\right\} - \hat{\mathbf{z}}_k^-\left(\hat{\mathbf{z}}_k^-\right)^T + \mathbf{R}_k \tag{7.4.2}$$

$$\left(\mathbf{P}_{\tilde{\mathbf{x}}\tilde{\mathbf{z}}}^-\right)_k = \frac{1}{N}\sum_{i=1}^{N} \left[\mathscr{X}_k^{(i)-} - \hat{\mathbf{x}}_k^-\right]\left[\mathbf{h}\left(\mathscr{X}_k^{(i)-}\right) - \hat{\mathbf{z}}_k^-\right]^T \tag{7.4.3}$$

Then, the Kalman gain:

$$\mathbf{K}_k = \left(\mathbf{P}_{\tilde{\mathbf{x}}\tilde{\mathbf{z}}}^-\right)_k \left(\mathbf{P}_{\tilde{\mathbf{z}}\tilde{\mathbf{z}}}^-\right)_k^{-1} \tag{7.4.4}$$

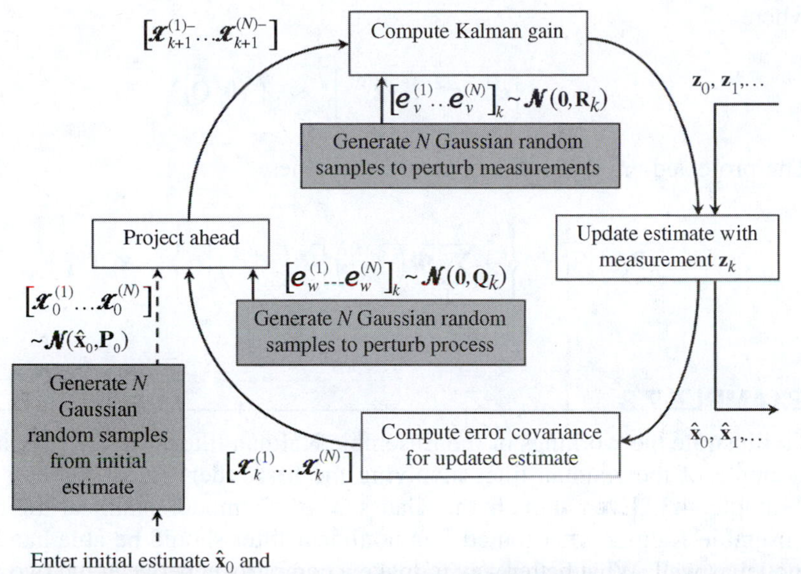

Figure 7.6 The Ensemble Kalman filter.

The state estimate update equation is given by:

$$\mathscr{X}_k^{(i)} = \mathscr{X}_k^{(i)-} + \mathbf{K}_k\left(\mathbf{z}_k + \boldsymbol{\varepsilon}_v^{(i)} - \hat{\mathbf{z}}_k^-\right) \tag{7.4.5}$$

where

$$\left[\boldsymbol{\varepsilon}_v^{(1)}, \boldsymbol{\varepsilon}_v^{(2)} \ldots \boldsymbol{\varepsilon}_v^{(N)}\right]_k \sim \mathcal{N}(\mathbf{0}, \mathbf{R}_k)$$

At the stage of the processing cycle, if we wish to extract an updated state estimate, we would simply reconstruct it as a mean of the sampled distribution:

$$\hat{\mathbf{x}}_k = \frac{1}{N}\sum_{i=1}^{N} \mathscr{X}_k^{(i)} \tag{7.4.6}$$

And, to avoid the explicit use of any linearized measurement connection, $\mathbf{H}$, we revert to an equivalent form of Eq. () for the error covariance update:

$$(\mathbf{P}_{\tilde{\mathbf{x}}\tilde{\mathbf{x}}})_k = \left(\mathbf{P}_{\tilde{\mathbf{x}}\tilde{\mathbf{x}}}^-\right)_k - \mathbf{K}_k\left(\mathbf{P}_{\tilde{\mathbf{z}}\tilde{\mathbf{z}}}^-\right)_k \mathbf{K}_k^T \tag{7.4.7}$$

The projection of the samples to the next step is made individually but also with random perturbations for each sample:

$$\hat{\mathbf{x}}_{k+1}^- = \frac{1}{N}\sum_{i=1}^{N}\left[\boldsymbol{\Phi}\left(\mathscr{X}_k^{(i)}\right) + \left(\boldsymbol{\varepsilon}_w^{(i)}\right)_k\right] \tag{7.4.8}$$

where

$$\left[\boldsymbol{\varepsilon}_w^{(1)}, \boldsymbol{\varepsilon}_w^{(2)} \ldots \boldsymbol{\varepsilon}_w^{(N)}\right]_k \sim \mathcal{N}(\mathbf{0}, \mathbf{Q}_k)$$

The projected error covariance matrix becomes:

$$\left(\mathbf{P}_{\hat{\mathbf{x}}\hat{\mathbf{x}}}^{-}\right)_{k+1} = \left\{\frac{1}{N}\sum_{i=1}^{N} \boldsymbol{\Phi}\left(\mathscr{X}_k^{(i)}\right)\left[\boldsymbol{\Phi}\left(\mathscr{X}_k^{(i)}\right)\right]^T\right\} - \hat{\mathbf{x}}_k^{-}\left(\hat{\mathbf{x}}_k^{-}\right)^T \qquad (7.4.9)$$

EXAMPLE 7.3

To illustrate the workings of the Ensemble Kalman filter, we will revisit an earlier example of the Kalman filter involving the first-order Gauss–Markov process in Example 4.3. Even though the Gauss–Markov model is linear in nature, the Ensemble Kalman filter touted as a nonlinear filter should be able handle a linear model as well. What better way to make a comparison between the two so that one can get a better understanding of their associated similarities and differences.

If we implement the Ensemble Kalman filter as outlined by Eqs. (7.4.1)–(7.4.9) and simply let the measurement equation be

$$\mathbf{h}\left(\mathscr{X}_k^{(i)-}\right) = H_k \mathscr{X}_k^{(i)-} \quad \text{where } H_k = 1 \text{ (see Eq. 4.5.6)},$$

and the state transition be

$$\boldsymbol{\Phi}\left(\mathscr{X}_k^{(i)}\right) = \phi_k \mathscr{X}_k^{(i)} \quad \text{where } \phi_k = e^{-0.1}$$

while adopting all the remaining parameters from Example 4.3, we should expect to get state estimates from the Ensemble Kalman filter that are nearly the same as from the results of Example 4.3, when working off the same true first-order Gauss–Markov process. Fig. 7.7 shows a comparison between the results from the two

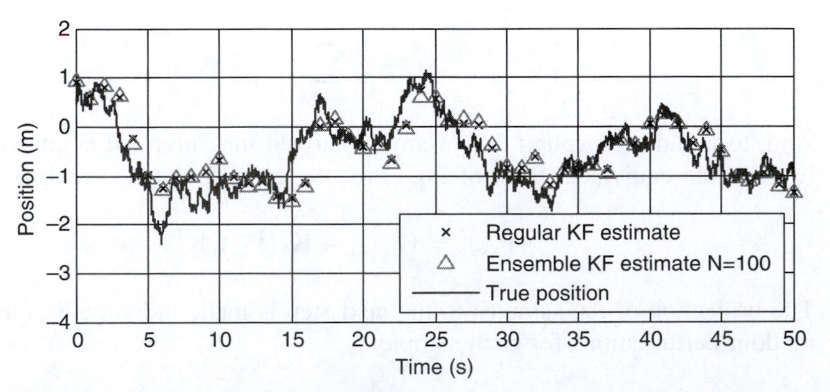

Figure 7.7 Comparing state estimates from regular Kalman filter versus Ensemble Kalman filter (for N = 100 samples) processing identical measurements from same first-order Gauss-Markov process.

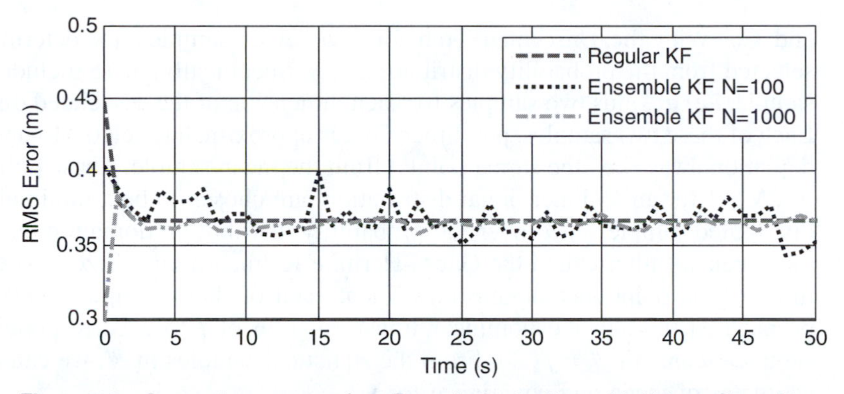

Figure 7.8 Comparing rms error from **P** matrix, for the Ensemble Kalman filter with different sampling sizes and for linear Kalman filter.

filters, when the Ensemble Kalman filter uses $N = 100$ samples. The rms error from the **P** matrix, shown in Fig. 7.8, reflects the random aspect of the sampling method used in the Monte Carlo Kalman filter, unlike the analytical nature of the same result from Example 4.3 as depicted in Fig. 4.5. It compares the cases where $N = 100$ and 1,000 samples to the analytical result of the regular Kalman filter.

Clearly, a linear example was chosen to demonstrate a method that is capable of handling nonlinear models. Nevertheless, what is important is that the comparison of results helps validate the method. More importantly, the example was intended for readers to make the connection between the newer method to the more familiar Kalman filter seen before.

■

There are many other variations that are similar to the Ensemble Kalman filter but we leave the reader to explore these, bearing in mind that this remains a very dynamic field of study.

7.5 THE UNSCENTED KALMAN FILTER

Now that we have a better understanding of what a sampling method is all about from the preceding discussion, we will take a look next at a popular Gaussian sampling method called the Unscented Kalman filter. Rather than picking out a bunch of random samples that represent the state estimate's conditional density, a deterministic choice of sampling points, usually called *sigma points*. These samples have something to do with the sigma of a distribution although they are not necessarily at exactly "one sigma." The name of this filter solution takes after the *Unscented Transform*, a method for calculating the statistics of a random variable that has been subject to a nonlinear transformation (13,14). This method provides estimates of the mean and covariance of the random variable based on discrete samples projected exactly through the associated nonlinear transform.

Consider a nonlinear function $f(\cdot)$ that propagates a random variable $\mathbf{x}$ to result in another random variable $\mathbf{y} = f(\mathbf{x})$. Given relevant parameters of $\mathbf{x}$ (mean of $\boldsymbol{\mu}_x$ and covariance of $\mathbf{C}_x$), we seek to determine the mean and covariance of $\mathbf{y}$, i.e., $\boldsymbol{\mu}_y$

and $\mathbf{C}_y$. With the Unscented Transform, a set of samples are deterministically selected from the probability distribution of $\mathbf{x}$. Specifically, these include the mean sample, i.e. $\boldsymbol{\mu}_x$, plus two samples for each dimension of the associated distribution. Each of these two samples, per dimension, is approximately related to one standard deviation away (i.e., the sigma points) from the mean sample. These points, a total of $2N+1$ for an N-dimensional distribution, are chosen to be a minimal yet well-distributed sample set $\mathcal{X}$ over the probability distribution domain of $\mathbf{x}$. (An even more general filter called the Gauss-Hermite Kalman filter utilizes more than just three sigma points per dimension.) Then, each of these sample points in $\mathcal{X}$ are propagated through the nonlinear transformation of f to a corresponding set of samples points in $\mathcal{Y} = f(\mathcal{X})$. From the statistical samples in $\mathcal{Y}$, we can derive the estimates of mean and covariance for $\mathbf{y}$.

The entire procedure can be summarized as follows (15):

(a) Selecting the unscented samples:

$$\mathcal{X}_i = \begin{cases} \boldsymbol{\mu}_x & i = 0 \\ \boldsymbol{\mu}_x + \sqrt{(N+\lambda)\mathbf{C}_x} & i = 1, \ldots N \\ \boldsymbol{\mu}_x - \sqrt{(N+\lambda)\mathbf{C}_x} & i = N+1, \ldots 2N \end{cases} \tag{7.5.1}$$

where

$\lambda = \alpha^2 (N + \kappa) - N$

$N =$ dimension of state $\mathbf{x}$

$\alpha =$ determines spread of the unscented samples about the mean sample

$\beta =$ dependent on knowledge of distribution (for Gaussian, $\beta = 2$)

$\kappa =$ scaling factor, usually equal to 3-N

(b) Transforming the unscented samples: $\mathcal{Y} = f(\mathcal{X})$

(c)

$$\text{Define weights}: \begin{cases} \omega_0^{\mu} = \dfrac{\lambda}{\lambda + N} \\ \omega_{0<i\leq 2N}^{\mu} = \dfrac{1}{2(\lambda + N)} \\ \omega_0^{C} = \omega_0^{\mu} + 1 - \alpha^2 + \beta \\ \omega_{0<i\leq 2N}^{C} = \dfrac{1}{2(\lambda + N)} \end{cases} \tag{7.5.2}$$

(d)

$$\hat{\boldsymbol{\mu}}_y = \sum_{i=0}^{2N} \omega_i^{\mu} \mathcal{Y}_i \tag{7.5.3}$$

and

$$\mathbf{C}_y = \sum_{i=0}^{2N} \omega_i^{C} (\mathcal{Y}_i - \hat{\boldsymbol{\mu}}_y)(\mathcal{Y}_i - \hat{\boldsymbol{\mu}}_y)^T \tag{7.5.4}$$

The reconstitution of the estimates in Step (d) based on the unscented samples derived in Step (b) require the derivation of weights (Eq. 7.5.2) outlined by Step (c). The choice of tuning parameters that make up the weights is not an exact one but nominal values can be found for use from various reference sources (15).

EXAMPLE 7.4

Consider a one-dimensional random variable x that is subject to a nonlinear transformation with the following function:

$y = 0.1x\cos(0.01\,x^2)$, where x is a Gaussian random variable: $x \sim \mathcal{N}\left(48.0, 0.75^2\right)$.

In a comparison of the statistics derived for the nonlinear function output y, we can derive a mean and standard deviation (sigma) from the true probability density function, from the linearized method, and from the Unscented Transform, all shown in Fig. 7.9.

For the linearized method, we would simply approximate the nonlinear function in the form of a Taylor Series to the first order, ignoring higher order terms:

$$y = f(x) \cong f(x_0) + \left.\frac{df}{dx}\right|_{x=x_0}$$

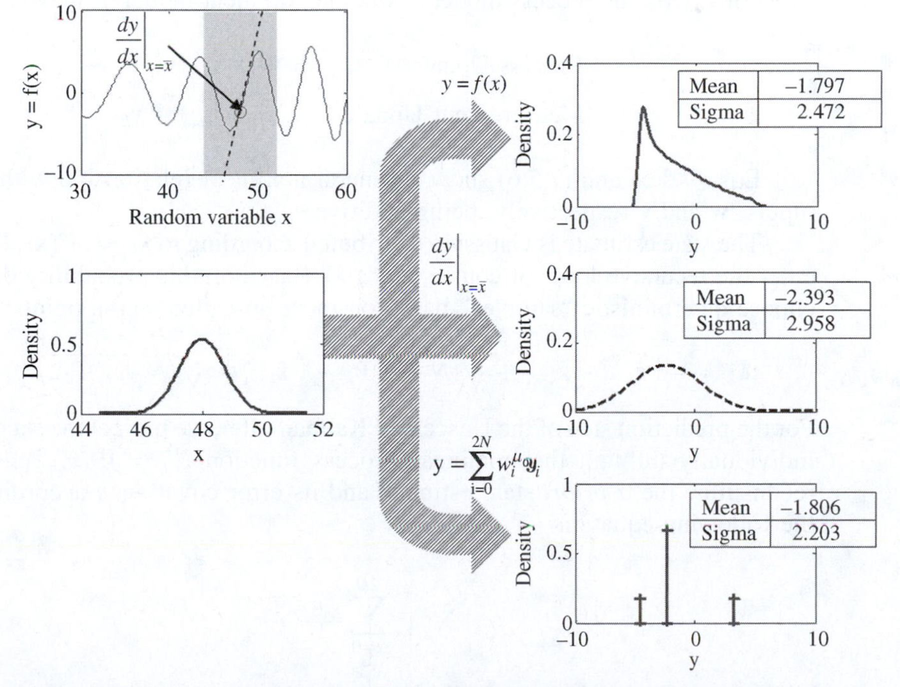

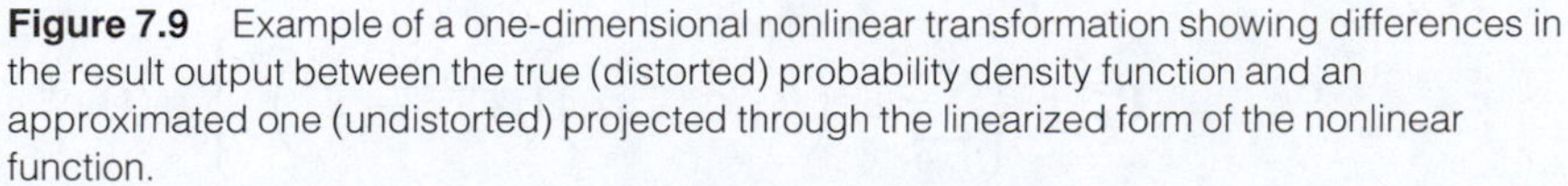

Figure 7.9 Example of a one-dimensional nonlinear transformation showing differences in the result output between the true (distorted) probability density function and an approximated one (undistorted) projected through the linearized form of the nonlinear function.

The two terms in the approximation, the nonlinear function f and its derivative, are evaluated at the mean of the input density function, i.e., $x_0 = 48.0$.

$$f(48.0) = -2.393; \quad \left.\frac{dy}{dx}\right|_{x=48.0} = 3.945$$

This derivative serves as the multiplier for any variation in x that results in the same for y:

$$\sigma_y \approx \sigma_x \cdot \left.\frac{dy}{dx}\right|_{x=48} = 0.75 \cdot 3.945 = 2.958$$

The Unscented Transform estimates were obtained with Eqs. (7.5.1) to (7.5.4) using the following parameters: $N=1$; $\alpha=1$; $\beta=2$; $\kappa=3\text{-}N$

As the Example 7.4 shows, the Unscented Transform can do a slightly better job of estimating the mean and sigma of a random variable that has been subject to a nonlinear transformation over a linearized approach. Note that the Unscented Transform really only involves selecting the mean and "sigma point" (from the covariance matrix) samples prior to the transformation, transforming them through the corresponding nonlinear function and then estimating the mean and the covariance of the resulting transformation. Clearly, its application to the Kalman filter will concern strictly the process model or the measurement model or maybe both:

$$\text{Process Update:} \quad \mathbf{x}_{k+1} = \mathbf{\Phi}(\mathbf{x}_k) + \mathbf{w}_k \tag{7.5.5}$$

$$\text{Measurement Update:} \quad \mathbf{z}_k = \mathbf{h}(\mathbf{x}_k) + \mathbf{v}_k \tag{7.5.6}$$

Eqs. (7.5.5) and (7.5.6) show a general nonlinear relationship with the noise inputs, **w** and **v** respectively, being additive.

The state estimate is Gaussian distributed according to $\hat{\mathbf{x}}_k \sim \mathcal{N}(\mathbf{x}_k, \mathbf{P}_k)$ and we enter the recursive loop, of course, at $k=0$. Based on this probability density, we choose deterministic "samples" based on those so-called sigma points:

$$\mathcal{X}_{k-1} = \left[\hat{\mathbf{x}}_{k-1} \quad \hat{\mathbf{x}}_{k-1} + \sqrt{(N+\lambda)\mathbf{P}_{k-1}} \quad \hat{\mathbf{x}}_{k-1} - \sqrt{(N+\lambda)\mathbf{P}_{k-1}}\right] \tag{7.5.7}$$

For the prediction step of the Unscented Kalman filter, we project the set of samples individually through the nonlinear process function $\mathcal{X}_k^- = \mathbf{\Phi}(\mathcal{X}_{k-1})$, and then reconstitute the *a priori* state estimate and its error covariance accordingly, with the following equation:

$$\hat{\mathbf{x}}_k^- = \sum_{i=0}^{2N} \omega_i^{\mu} \mathcal{X}_k^{(i)-} \tag{7.5.8}$$

$$\mathbf{P}_k^- = \left\{ \sum_{i=0}^{2N} \omega_i^C \left(\mathcal{X}_k^{(i)-} - \hat{\mathbf{x}}_k^-\right)\left(\mathcal{X}_k^{(i)-} - \hat{\mathbf{x}}_k^-\right)^T \right\} + \mathbf{Q}_k \tag{7.5.9}$$

We are now ready to re-sample new sigma points for the remainder of the recursive cycle.

$$\mathscr{X}_k^- = \left[\hat{\mathbf{x}}_k^- \quad \hat{\mathbf{x}}_k^- + \sqrt{(N+\lambda)\mathbf{P}_k^-} \quad \hat{\mathbf{x}}_k^- - \sqrt{(N+\lambda)\mathbf{P}_k^-}\right] \tag{7.5.10}$$

To prepare the measurement update step of the Unscented Kalman filter, we project the sigma point samples through the nonlinear measurement function and evaluate the associated covariances needed to compute the gain:

$$\mathscr{Z}_k^- = \mathbf{h}\left(\mathscr{X}_k^{(i)-}\right) \tag{7.5.11}$$

$$\hat{\mathbf{z}}_k^- = \sum_{i=0}^{2N} \omega_i^{\mu} \mathscr{Z}_k^{(i)-} \tag{7.5.12}$$

$$\left(\mathbf{P}_{\tilde{\mathbf{z}}\tilde{\mathbf{z}}}^-\right)_k = \left\{\sum_{i=0}^{2N} \omega_i^C \left(\mathscr{Z}_k^{(i)-} - \hat{\mathbf{z}}_k^-\right)\left(\mathscr{Z}_k^{(i)-} - \hat{\mathbf{z}}_k^-\right)^T\right\} + \mathbf{R}_k \tag{7.5.13}$$

$$\left(\mathbf{P}_{\tilde{\mathbf{x}}\tilde{\mathbf{z}}}^-\right)_k = \sum_{i=0}^{2N} \omega_i^C \left(\mathscr{X}_k^{(i)-} - \hat{\mathbf{x}}_k^-\right)\left(\mathscr{Z}_k^{(i)-} - \hat{\mathbf{z}}_k^-\right)^T \tag{7.5.14}$$

$$\mathbf{K}_k = \left(\mathbf{P}_{\tilde{\mathbf{x}}\tilde{\mathbf{z}}}^-\right)_k \left(\mathbf{P}_{\tilde{\mathbf{z}}\tilde{\mathbf{z}}}^-\right)_k^{-1} \tag{7.5.15}$$

The Kalman gain $\mathbf{K}_k$ is used to update the state estimate in the usual way:

$$\hat{\mathbf{x}}_k = \hat{\mathbf{x}}_k^- + \mathbf{K}_k\left(\mathbf{z}_k - \hat{\mathbf{z}}_k^-\right) = \hat{\mathbf{x}}_k^- + \mathbf{K}_k \tilde{\mathbf{z}}_k^- \tag{7.5.16}$$

For the error covariance update, as was from the Bayesian filter, we use the following form to avoid a need to linearize the measurement connection matrix $\mathbf{H}$:

$$\mathbf{P}_k = \mathbf{P}_k^- - \mathbf{K}_k\left(\mathbf{P}_{\tilde{\mathbf{z}}\tilde{\mathbf{z}}}^-\right)_k \mathbf{K}_k^T \tag{7.5.17}$$

At this point, the recursive cycle then repeats itself by returning back to Eq. (7.5.7).

To close this sub-section, there are other variants of the Unscented Kalman filter that have been proposed and used. We can only refer the reader to other references for notes on these (16,17).

7.6 THE PARTICLE FILTER

With its random samples being treated as "particles," some would consider the Ensemble Kalman filter as belonging to the family of Particle filters. In most instances, however, the Particle filter referred to in the contemporary literature takes one step further in generalization to detach itself from the Gaussian assumption. In that regard, Particle filters have become very popular when there is a need to handle non-Gaussian problems, not just nonlinear ones, and particularly multimodal density functions and not just unimodal ones. Whole books have been devoted to the

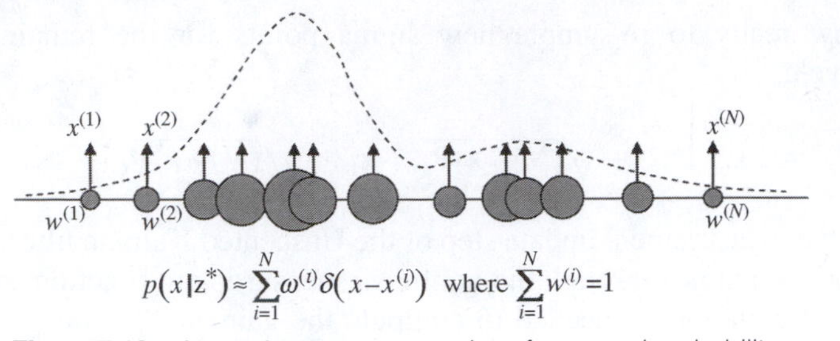

Figure 7.10 Approximate representation of a general probability density function using appropriately-weighted random samples.

different variations of the Particle filter so this is an expansive topic in and of itself (18,19). Our treatment of this subject here cannot be much beyond an introductory tutorial treatment of one commonly-used variation with an accompanying example. For those attempting to learn about the rudiments of the Particle filter for the first time, the terminology can be rather confusing so we shall pay careful attention to clarifying seemingly ambiguous terms.

Being more like the Ensemble Kalman filter than the Unscented Kalman filter, the Particle filter goes to great lengths at approximating the relevant probability distributions involved in the estimation process except in an even more elaborate manner. The utility of a Particle filter lies in its ability to approximate a continuous probability density function with discrete weights and at sample points that are generally unevenly spaced, rather as the very *particles* being associated with the filter name.

Fig. 7.10 shows an example of how a continuous probability density function can be approximated by a set of such weights. If the samples were in fact uniformly spaced, then one might be inclined to think that the weights have some direct relationship to the value of the density function at the location of the samples. However, it is incorrect to think that the weights and density function have such a direct relationship only because the samples are not uniformly spaced, in general. Therefore, the "weights" of the particles, which account for both the value of the density function and the sparseness or compactness of the other particles surrounding it, are often depicted, here in Fig. 7.10 and elsewhere, in terms of relative size of circles.

In a Particle filter, this discrete approximation is used to represent the *a posteriori* conditional density:

$$p\left(\mathbf{x}_k | \mathbf{z}_k^*\right) \approx \sum_{i=1}^{N} \omega_k^{(i)} \delta\left(\mathbf{x}_k - \mathbf{x}_k^{(i)}\right) \tag{7.6.1}$$

where $\delta(\cdot)$ is the Dirac delta function.

Recall from Section 6.9 on the recursive Bayesian filter that we can derive from this conditional density, among other things, an expected value that represents the optimal estimate of a Kalman filter.

However, unlike a Monte Carlo-type Ensemble Kalman filter that assumes a Gaussian approximation in which we are able to count on drawing random samples from an easily-accessible distribution such as the Gaussian, the Particle filter must deal with a dynamic set of random samples that may represent just about any distribution and most likely one that we cannot easily draw random samples from, or worse, one that we may not be able to even describe analytically!

This leads to another important concept we use to construct the Particle filter as something called *Importance Sampling*. This concept allows us to work around the difficulty of not always being able to directly generate random samples where we need to do so. Rather, for a general density function, we attempt to do the same thing but indirectly via what is called a *proposal* function that we are assured of generating such samples (e.g., Gaussian, uniform probability density function). This proposal function is known as the *Importance Density* and so the process of drawing samples from a general density function is known accordingly as *Importance Sampling*. Used in the Particle filter, its recursive nature results in an algorithm to determine the weight sequence w_k, called *Sequential Important Sampling* or *SIS*:

$$\omega_k^i \propto \omega_{k-1}^i \frac{p\left(\mathbf{z}_k|\mathbf{x}_k^{(i)}\right)p\left(\mathbf{x}_k^{(i)}|\mathbf{x}_{k-1}^{(i)}\right)}{q\left(\mathbf{x}_k^{(i)}|\mathbf{x}_{k-1}^{(i)},\mathbf{z}_k\right)} \tag{7.6.2}$$

where

$p\left(\mathbf{z}_k|\mathbf{x}_k^{(i)}\right)$ is known as the *likelihood* function;

$p\left(\mathbf{x}_k^{(i)}|\mathbf{x}_{k-1}^{(i)}\right)$ is known as the *transition prior*; and

$q\left(\mathbf{x}_k^{(i)}|\mathbf{x}_{k-1}^{(i)},\mathbf{z}_k\right)$ is the proposal *Importance Density* that is to be sampled from.

When it comes to choosing the density function q, one suboptimal choice that has been favored is to choose $q\left(\mathbf{x}_k^{(i)}|\mathbf{x}_{k-1}^{(i)},\mathbf{z}_k\right)$ to equal $p\left(\mathbf{x}_k^{(i)}|\mathbf{x}_{k-1}^{(i)}\right)$, the density function that describes the state projection from one time step to the next, i.e., the transition prior. In that particular case, $\omega_k^i \propto \omega_{k-1}^i p\left(\mathbf{z}_k|\mathbf{x}_k^{(i)}\right)$. This is sometimes known as the "bootstrap filter." If we use such a scheme and also a special condition that the weights are resampled every cycle such that they turn into uniformly-distributed weights, we end up with the rather concise expression of $\omega_k^i = p\left(\mathbf{z}_k|\mathbf{x}_k^{(i)}\right)$. (Note that the weights are normalized so that they sum to one.) We shall explore this notion of resampling next.

There are many variants of the Particle filter but one that uses the above choice is a relatively popular scheme called the *Sequential Importance Resampling*† or *SIR* Particle filter. A detrimental effect commonly associated with the SIS Particle filter is known as the "degeneracy phenomenon" where the normalized weights tend to concentrate into one particle, after a certain number of recursive steps, leaving all other particles to be essentially degenerate. An example of this can be seen in Fig. 7.11, which depicts what might happen to 20 particles without resampling after only two cycles of a particular sample realization of a given problem. Even though the particles are still spread around in their locations (x), the weight of one particular particle has become dominant (Particle 10 highlighted with the square) along with another one slight less so (Particle 19 highlighted with the circle). Both Particle 10

† The acronym "SIR" is sometimes found in the literature to be "Sampling Importance Resampling" which is somewhat confounding. It is related to the SIS Particle filter, where the "SIS" stands for "Sequential Importance Sampling," so we follow Wikipedia's use of "Sequential Importance Resampling" as the most cogent alternative.

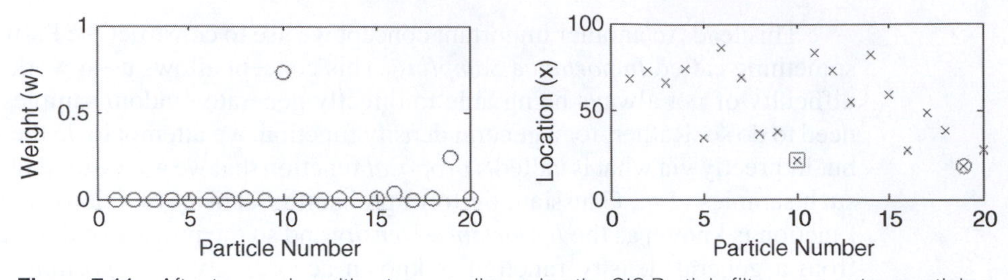

Figure 7.11 After two cycles without resampling using the SIS Particle filter, we see two particle weights dominate over the others.

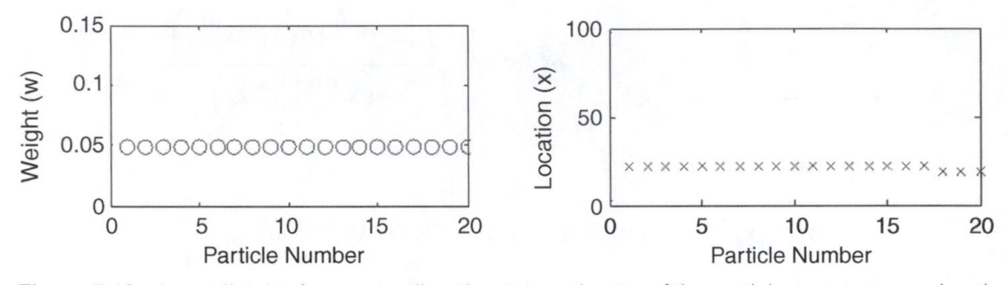

Figure 7.12 Immediately after *resampling*, the state estimates of the particles are remapped and their weights rebalanced, usually equalized by a uniform density function.

and Particle 19 are strongly weighted because they carry values that are near the mean of the conditional density function (i.e., the true value). After resampling, the particles are remapped with more meaningful weights, in most cases uniformly distributed (Fig. 7.12).

Made once every recursive cycle, usually, the process of *resampling* generally uses the cumulative profile of the discrete weights for the remapping procedure. Fig. 7.13 shows how a new drawing for all particles from a uniform distribution will end up favoring usurping the values from the very few dominant particles from before.

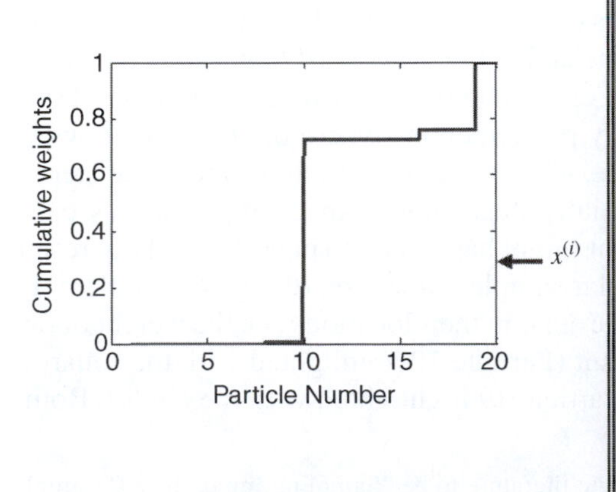

In *Resampling*, each particle is remapped to redistribute its weight. If, for Particle i, we draw a random number $x^{(i)}$ from a uniform distribution over (0, 1), we are most likely to map the new Particle i to Particle 10 from before resampling, thereby picking up the location value of 22.02. Or another less likely choice might be the previous Particle 19 with its location value of 19.11. The new weight of Particle i would be set equal to all other particles due to our drawing from a uniform distribution.

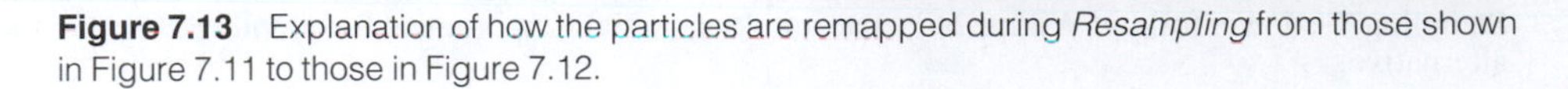

Figure 7.13 Explanation of how the particles are remapped during *Resampling* from those shown in Figure 7.11 to those in Figure 7.12.

Since the SIR filter was first introduced in 1993 (20), many other variants have been spawned. One such class of variants include the use of the Extended Kalman filter or Unscented Kalman filter with local linearization (for individual particles) to generate the Importance Density needed in Eq. (7.6.2). Although the subject of the Particle filter has, in some ways, veered from the original topic of this book, we shall take one last look at a Particle filter that actually uses a parallel set of Kalman filters for its inner workings. We do so in the form of an example.

EXAMPLE 7.5

In terrain-referenced navigation or TRN, the position of an aircraft is determined by comparing a series of height measurements against an accurate database of terrain heights above a reference frame. In its simplest form, measurements from a radar altimeter provide information of aircraft height above the terrain while measurements from a baro-altimeter provide information about the aircraft altitude above the reference frame. Flat terrain presents the worst conditions for solving position while rugged terrain with distinct height variations are the best to work with (21).

Our example deals with a one-dimensional case so we only need to deal with a single position state. We will also process only the radar altimeter measurement and assume the baro-altimeter measurement is perfect. We can describe the process and measurement models according to the following:

$$\mathbf{x}_{k+1} = \mathbf{x}_k + \mathbf{u}_k + \mathbf{w}_k \tag{7.6.3}$$

$$\mathbf{z}_k = \mathbf{h}(\mathbf{x}_k) + \mathbf{v}_k \tag{7.6.4}$$

Our process model here is entirely linear, but the measurement model is nonlinear with additive measurement noise:

$$\underbrace{h_{radalt} - h_{baro}}_{\mathbf{z}_k} = h_{terrain}(x_k) + v_k \tag{7.6.5}$$

Many descriptions of the various Particle filter implementations in the published literature are often generalized, almost overly so. We will dissect a particular implementation of a Particle filter where the Importance Density is derived from

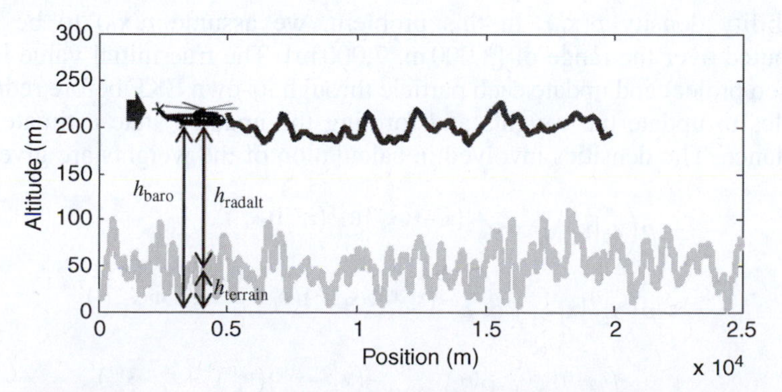

Figure 7.14 Terrain-referenced navigation example illustrating a Particle filter implementation to address a nonlinear measurement situation.

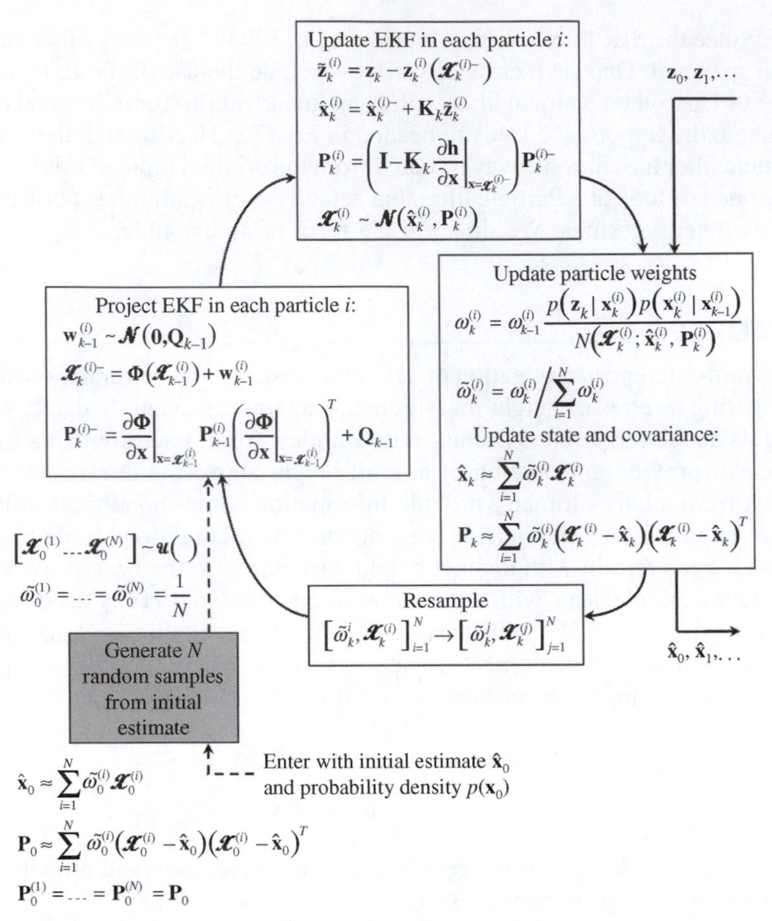

Figure 7.15 Particle filter using Extended Kalman filters for Local Linearization.

local linearization made independently for each particle with the use of an Extended Kalman filter. The algorithm used is shown in Fig. 7.15.

We enter the loop shown in Fig. 7.15 with an initial estimate of $\mathbf{x}_0$ and its probability density $p(\mathbf{x}_0)$. In this problem, we assume $p(\mathbf{x}_0)$ to be uniformly distributed over the range of [3,000 m, 7,000 m]. The true initial value is 5,100 m. We then project and update each particle through its own EKF before redrawing the particles to update the weights and forming the updated state estimate and error covariance. The densities involved in calculation of the weights are given by:

$$p\left(\mathbf{z}_k \middle| \mathbf{x}_k^{(i)}\right) \propto e^{-\frac{1}{2}\left(\mathbf{z}_k - \mathbf{H}_k \hat{\mathbf{x}}_k^{-}\right) \mathbf{R}^{-1} \left(\mathbf{z}_k - \mathbf{H}_k \hat{\mathbf{x}}_k^{-}\right)^T}$$

$$p\left(\mathbf{x}_k^{(i)} \middle| \mathbf{x}_{k-1}^{(i)}\right) \propto e^{-\frac{1}{2}\left(\mathbf{x}_k^{(i)} - \mathbf{\Phi}(\hat{\mathbf{x}}_{k-1})\right) \left(\mathbf{P}_k^{(i)-}\right)^{-1} \left(\mathbf{x}_k^{(i)} - \mathbf{\Phi}(\hat{\mathbf{x}}_{k-1})\right)^T}$$

$$\mathcal{N}\left(\mathcal{X}_k^{(i)}; \hat{\mathbf{x}}_k^{(i)}, \mathbf{P}_k^{(i)}\right) \propto e^{-\frac{1}{2}\left(\mathcal{X}_k^{(i)} - \hat{\mathbf{x}}_k^{(i)}\right) \left(\mathbf{P}_k^{(i)}\right)^{-1} \left(\mathcal{X}_k^{(i)} - \hat{\mathbf{x}}_k^{(i)}\right)^T}$$

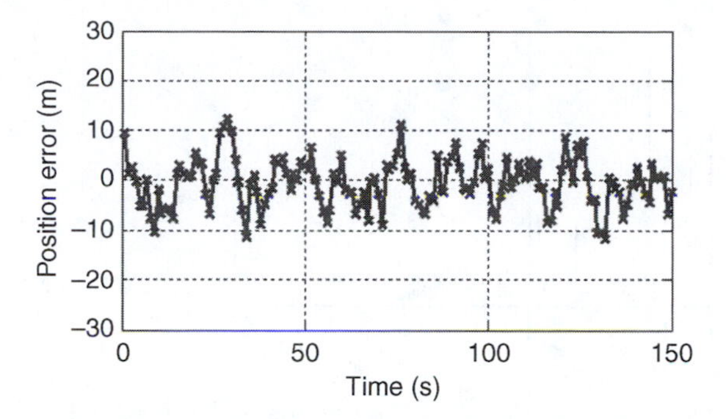

Figure 7.16 Estimation error of the Particle filter using EKF for local linearization.

The weights are resampled and then the entire cycle is repeated for subsequent steps. The profile of the estimation error for this example is shown in Figure 7.16.

The previous example is simply one out of a myriad of Particle filter variants. However, its use of the EKF for local linearization of the particles warrants its introduction here. There is also another type of Particle filter, very similar to one we have just examined, that uses an Unscented Kalman filter instead of the EKF. The Unscented Particle filter is also popular choice in being able to handle the local linearization approximation slightly better than an EKF at the expense of even more algorithmic complexity (22). Few problems contain nonlinear functions in their entirety for both process and measurement models. Many may contain a mixture such that the state vector can be partitioned into nonlinear and linear states. For such situations, there is a popular variant called the Rao-Blackwellized (or Marginalized) Particle Filter to exploit what can be a substantial reduction in computational burden (23).

Clearly, the many different variants of Particle filters all have their strengths and weaknesses, whether in numerical terms or in informational terms. However, one has to be a little circumspect when evaluating the usefulness of any form of particle filter depending on the problem that is being solved for. One can always contrive a problem to maximize the strengths of the solution but the realism of that particular problem should be fairly weighed against the extra computational burden undertaken. This subject is still very dynamic today even though there are some schemes that work better than others for certain problems, and the potential for even more innovative variations remains to be uncovered. We shall leave this topic at this point in time by saying that the end of the story of the Particle filter is far from being settled!

PROBLEMS

7.1 Consider a two-dimensional problem where the position of an observer is moving in an oscillatory pattern, and it is to be determined based on ranging measurements from two known references A and B. See Fig. P7.1.

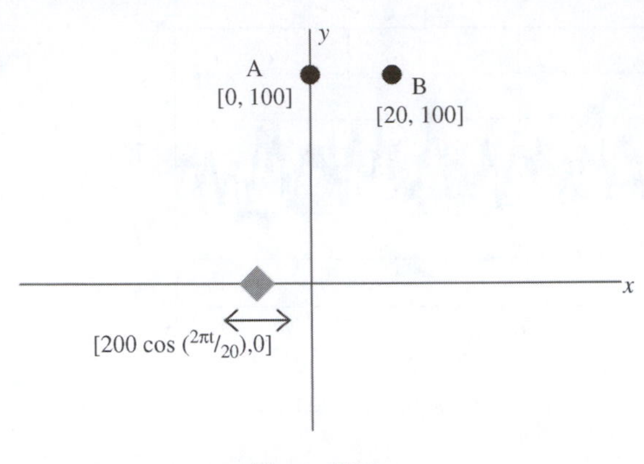

Figure P7.1

Given that the motion occurs entirely along the x-axis, let us assume a two-state position-velocity model for the process dynamics along the x-dimension. Since the motion is somewhat deterministic where the maximum acceleration is just slightly less than $20\,\text{m/s}^2$, we can assume a spectral amplitude S for the position-velocity driving process noise of $400\,(\text{m/s}^2)^2/\text{Hz}$.

The measurement model will be nonlinear and must be linearized to a nominal position that is approximately near the true position. Assume that the measurement noise associated with ranging to each reference station to have a variance of $(4\,\text{m})^2$. Choose appropriate initial uncertainties for the position and velocity states.

Write out the parameters of this two-state Kalman filter for $\boldsymbol{\phi}$, $\mathbf{Q}$, $\mathbf{H}$, $\mathbf{R}$, and $\mathbf{P}_0^-$.

(a) Generate noisy ranging measurements from the two references to the true position of the moving observer for 100 steps. The step size is 1 second.

(b) To formulate an Extended Kalman Filter (EKF), a nominal position must be assumed for linearization to generate the time-varying **H** matrix. Without any external help, this nominal position is determined by the best prediction solution of the filter. Recall that the EKF works off the difference between the noisy range measurements and the predicted range measurements (based on the nominal position). At each step, what the EKF estimates then is the difference between the total position and the nominal position. This estimate gets recombined with the nominal position into a best estimate of the total position, which then gets projected ahead to form the next nominal position. At the start of the next cycle, the *a priori* state estimate for the position error starts out at zero again. Run the filter over 100 steps and plot out the position error. Note the characteristic of the error profile.

(c) The non-trivial oscillations seen in the error profile is due to the severe nonlinearity of the measurement model. Find a way to run the EKF such that this effect due to the nonlinear measurement situation is drastically reduced. (*Hint*: The degree of approximation in the linearization of the nonlinear measurement model is dependent on the choice of the nominal position used for the point of linearization. Choose a better nominal position after a given processing cycle to do that same cycle over with.) Run your improved filter over 100 steps and plot out the position error.

7.2 Consider a simple one-dimensional trajectory determination problem as follows. A small object is launched vertically in the earth's atmosphere. The initial thrust exists for a very short time, and the object "free falls" ballistically for essentially all of its straight-up, straight-down trajectory. Let y be measured in the up direction, and assume that the nominal trajectory is governed by the following dynamical equation:

$$m\ddot{y} = -mg - D\dot{y}|\dot{y}|$$

where

$m =$ .05 kg (mass of object)
$g =$ 9.087 m/s^2 (acceleration of gravity)
$D = 1.4 \times 10^{-4}$ n/(m/s)^2 (drag coefficient)

The drag coefficient will be assumed to be constant for this relatively short trajectory, and note that drag force is proportional to (velocity)^2, which makes the differential equation nonlinear. Let the initial conditions for the nominal trajectory be as follows:

$$y(0) = 0$$
$$\dot{y}(0) = 85 \text{ m/s}$$

The body is tracked and noisy position measurements are obtained at intervals of .1 sec. The measurement error variance is .25 m^2. The actual trajectory will differ from the nominal one primarily because of uncertainty in the initial velocity. Assume that the initial position is known perfectly but that initial velocity is best modeled as a normal random variable described by $\mathcal{N}$ (85 m/sec, 1 m^2/sec^2). Work out the linearized discrete Kalman filter model for the up portion of the trajectory.
(*Hint:* An analytical solution for the nominal trajectory may be obtained by considering the differential equation as a first-order equation in velocity. Note $|\dot{y}| = \dot{y}$ during the up portion of the trajectory. Since variables are separable in the velocity equation, it can be integrated. The velocity can then be integrated to obtain position.)

7.3 (a) At the kth step of the usual nonadaptive Kalman filter, the measurement residual is $(\mathbf{z}_k - \mathbf{H}_k\hat{\mathbf{x}}_k^-)$. Let z_k be scalar and show that the expectation of the squared residual is minimized if $\hat{\mathbf{x}}_k^-$ is the optimal *a priori* estimate of $\mathbf{x}_k$, that is, the one normally computed in the projection step of the Kalman filter loop.
(*Hint:* Use the measurement relationship $z_k = \mathbf{H}_k\mathbf{x}_k + v_k$ and note that v_k and the a priori estimation error have zero crosscorrelation. Also note that the a priori estimate $\hat{\mathbf{x}}_k^-$, optimal or otherwise, can only depend on the measurement sequence up through z_{k-1} and not z_k.)

(b) Show that the time sequence of residuals $\left(z_k - \mathbf{H}_k\hat{\mathbf{x}}_k^-\right)$, $\left(z_{k+1} - \mathbf{H}_{k+1}\hat{\mathbf{x}}_{k+1}^-\right), \ldots$, is a white sequence if the filter is optimal. As a matter of terminology, this sequence is known as an *innovations* sequence. See Section 5.3.

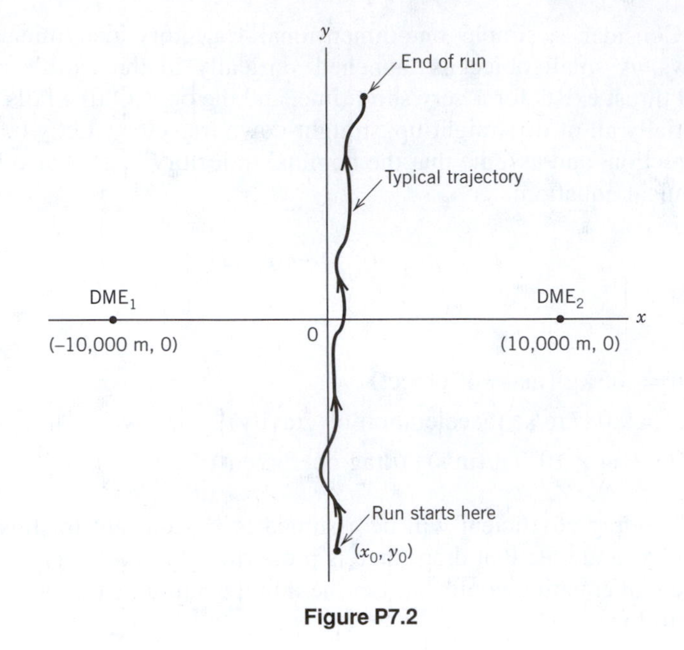

Figure P7.2

7.4 This problem is a variation on the DME example given in Section 7.1 (Example 7.1 with a simplification in the model of the aircraft dynamics). Suppose that the two DME stations are located on the x-axis as shown in the accompanying figure, and further suppose that the aircraft follows an approximate path from south to north as shown. The aircraft has a nominal velocity of 100 m/s in a northerly direction, but there is random motion superimposed on this in both the x and y directions. The flight duration (for our purposes) is 200 s, and the initial coordinates at $t = 0$ are properly described as normal random variables as follows:

$$x_0 \sim \mathcal{N}(0, 2{,}000\,\text{m}^2)$$

$$y_0 \sim \mathcal{N}(-10{,}000\,\text{m},\ 2{,}000\,\text{m}^2)$$

The aircraft kinematics are described by the following equations:

$$x_{k+1} = x_k + w_{1k}, \quad k = 0, 1, 2, \ldots, 200$$
$$y_{k+1} = y_k + w_{2k} + 100\,\Delta t, \quad k = 0, 1, 2, \ldots, 200$$

where w_{1k} and w_{2k} are independent white Gaussian sequences described by

$$w_{1k} \sim \mathcal{N}(0, 400\,\text{m}^2)$$
$$w_{2k} \sim \mathcal{N}(0, 400\,\text{m}^2)$$

The sampling interval Δt is 1 s. The aircraft motion will be recognized as simple random walk in the x-coordinate, and random walk superimposed on linear motion for the y-coordinate.

The aircraft obtains simultaneous discrete range measurements at 1 s intervals on both DME stations, and the measurement errors consist of a superposition of Markov and white components. Thus, we have (for a typical range measurement)

$$\begin{bmatrix} \text{Total} \\ \text{noisy} \\ \text{measurement} \end{bmatrix} = \begin{bmatrix} \text{true} \\ \text{total} \\ \text{range} \end{bmatrix} + \begin{bmatrix} \text{Markov} \\ \text{error} \\ \text{component} \end{bmatrix} + \begin{bmatrix} \text{white} \\ \text{error} \\ \text{component} \end{bmatrix}$$

We have two DME stations, so the measurement vector is a two-tuple at each sample point beginning at $k=0$ and continuing until $k=200$. The Markov errors for each of the DME stations are independent Gaussian random processes, which are described by the autocorrelation function

$$R_m(\tau) = 900e^{-0.01|\tau|}\ \text{m}^2$$

The white measurement errors for both stations have a variance of $225\ \text{m}^2$.

(a) Work out the linearized Kalman filter parameters for this situation. The linearization is to be done about the nominal linear-motion trajectory exactly along the y-axis. That is, the resultant filter is to be an "ordinary" linearized Kalman filter and not an extended Kalman filter.

(b) After the key filter parameters have been worked out, run a covariance analysis for $k=0, 1, 2, \ldots, 200$, and plot the estimation error variances for both the x and y position coordinates.

(c) You should see a pronounced peak in the y-coordinate error curve as the aircraft goes through (or near) the origin. This simply reflects a bad geometric situation when the two DME stations are 180 degrees apart relative to the aircraft. One might think that the estimation error variance should go to infinity at exactly $k=100$. Explain qualitatively why this is not true.

7.5 Recall from Chapter 1 that the linear transformation of a Gaussian distribution results in another Gaussian distribution. We will demonstrate this here by Monte Carlo sampling. The linear transformation equation we will use in this problem is the Kalman filter state and covariance projection equations:

True process of a random sample $\mathbf{x}_k$:

$$\mathbf{x}_{k+1} = \boldsymbol{\phi}_k \mathbf{x}_k + \mathbf{w}_k \tag{P7.5.1}$$

Linear transformation of a Gaussian distribution:

$$\begin{aligned} \hat{\mathbf{x}}_{k+1}^- &= \boldsymbol{\phi}_k \hat{\mathbf{x}}_k \\ \mathbf{P}_{k+1}^- &= \boldsymbol{\phi}_k \mathbf{P}_k \boldsymbol{\phi}_k^T + \mathbf{Q}_k \end{aligned} \tag{P7.5.2}$$

To begin with, we are given the mean and covariance of the Gaussian distribution,

$$\hat{\mathbf{x}}_k = \begin{bmatrix} -16.5 \\ -6.6 \\ 2.9 \end{bmatrix} \quad \mathbf{P}_k = \begin{bmatrix} 3.2 & -1.2 & 0.2 \\ -1.2 & 4.5 & 0.02 \\ 0.2 & 0.02 & 1.2 \end{bmatrix} \tag{P7.5.3}$$

The transformation parameters given are

$$\boldsymbol{\phi}_k = \begin{bmatrix} 1 & 1 & 0 \\ 0 & 1 & 0 \\ 0 & 0 & 0.98 \end{bmatrix} \quad \mathbf{Q}_k = \begin{bmatrix} 1/3 & 1/2 & 0 \\ 1/2 & 1 & 0 \\ 0 & 0 & 0.01 \end{bmatrix} \tag{P7.5.4}$$

(a) With the distribution parameters given by Eq. (P7.5.3), compute the mean and covariance of the random variable after the linear transformation given by Eq. (P7.5.2).

(b) Alternatively, we can determine approximate values for this by way of a Monte Carlo experiment. To begin, generate a set of 10,000 samples of the distribution (a process called "*drawing*") whose parameters are given by Eq. (P7.5.3).

(c) Project these 10,000 samples with the transformation equation of Eq. (P7.5.1) to obtain a resulting set of 10,000 samples. Evaluate the mean and covariance of the transformed set. Note that Eq. (P7.5.1) also involves another random variable **w**. This is a zero-mean random vector that must be generated separately from its distribution, as specified by the covariance matrix **Q**.

(d) Compare your results from Part (c) with the theoretical results of Part (a). Is the error within 10% of the magnitude of the values?

(This Monte Carlo exercise is continued in a nonlinear setting in Problem 7.6.)

7.6 Let the measurement model be the nonlinear function given by:

$$\begin{bmatrix} z_1 \\ z_2 \end{bmatrix} = \begin{bmatrix} x_1 \cos(x_2) \\ x_1 \sin(x_2) \end{bmatrix} + \begin{bmatrix} v_1 \\ v_2 \end{bmatrix} \tag{P7.6.1}$$

Let the state estimates and associated covariance be

$$\begin{bmatrix} \hat{x}_1 \\ \hat{x}_2 \end{bmatrix} = \begin{bmatrix} 100\,\text{units} \\ 0.3\,\text{radians} \end{bmatrix} \quad \mathbf{P} = E\left\{ \begin{bmatrix} \hat{x}_1 \\ \hat{x}_2 \end{bmatrix} [\hat{x}_1 \;\; \hat{x}_2] \right\} = \begin{bmatrix} 45 & 0.2 \\ 0.2 & 0.001 \end{bmatrix}$$

(a) Assuming that the state estimates have a bivariate Gaussian probability distribution with mean and covariance specified above, use a Monte Carlo method with 10,000 samples to compute the measurement residual covariance (Eq. 7.4.2) and the Kalman gain (given by Eq. 7.4.4).

(b) Repeat Part (a) with the Unscented Transform (Eqs. 7.5.13 and 7.5.15). Choose the relevant parameters for the transform to be the same as those used in Example 7.4, except $N = 2$ (for the two-state problem at hand). The other parameters are $\alpha = 1$, $\beta = 2$, $\kappa = 3\text{-}N$.

(c) Finally, linearize the measurement equation of Eq. (P7.6.1) and compute the measurement residual covariance and Kalman gain in the usual way. How does this compare to the results of Parts (a) and (b)?

7.7 The resampling process described for the particle filter in Section 7.6 shows an example of a sample realization in Fig. 7.11 (before resampling) and Fig 7.12 (after resampling). The table below lists the 20 particle samples of weights and values as depicted in Fig. 7.11.

N	Weight	Value
1	0.0000	68.0845
2	0.0000	73.5830
3	0.0000	65.8939
4	0.0000	74.2335
5	0.0000	34.3776
6	0.0000	86.4921
7	0.0000	69.3185
8	0.0000	36.5419
9	0.0000	37.5836
10	0.7266	22.0238
11	0.0000	83.5249
12	0.0000	73.3241
13	0.0000	55.4148
14	0.0000	82.0355
15	0.0000	59.0248
16	0.0338	27.6553
17	0.0000	49.1076
18	0.0000	38.9061
19	0.2396	19.1080
20	0.0000	27.5404

Write a program function that will perform a resampling of these particles using the idea of remapping shown in Fig. 7.13. Since the resampling process involves drawing random numbers, your results might differ slightly from the outcome shown in Fig. 7.12, but they should nevertheless turn out to be very similar.

7.8 This problem re-creates the terrain-referenced navigation example given by Example 7.5. Recall that this is a one-dimensional problem where the state to be estimated is horizontal position. The measurement is related to the height of the rotorcraft over the known undulating terrain.

(a) First, we need to generate a representative terrain profile. Use a second-order Gauss-Markov model with $\sigma = 20\,\text{m}$ and $\omega = 0.01\,\text{rad/m}$. Note that the independent variable in this random process model is now space instead of time, hence, the units of rad/m for ω. Generate profile with a discrete-space interval of 1m over the range from 0 to 25,000 m.

(b) Next, generate the rotorcraft's motion profile. Use a second-order Gauss-Markov model for the horizontal and another for the vertical axis to generate a bounded random process model. Use a sampling interval of $\Delta t = 1$s, and the parameters $\sigma = 10$ m and $\omega = 0.51$ rad/s. Then add this random process to a nominal motion that moves horizontally at 100 m/s (in the positive x-direction) but remains at a fixed altitude, starting at coordinates (5,000 m, 200 m). Generate the profile for a 150-s duration. At each sampling time, compute the radar altimeter measurement by differencing the appropriate terrain height from the rotorcraft altitude. The terrain height required will, in general, need to be computed by linear interpolation between samples generated in Part (a). To each radar altimeter measurement, add a measurement noise random sample from a Gaussian distribution with zero mean and a sigma of 1 meter.

(c) When processing the radar altimeter measurements generated in Part (b), the true altitude of the rotorcraft should be known. Assume that this information comes from a perfect barometric altitude measurement. The difference between this "perfect" baro-altitude measurement and the radar altimeter measurement feeds the nonlinear measurement model described by Eq. (7.6.5). Design a one-state particle filter based on the flowchart of Fig. 7.15 using N = 100 particles. Assume that, in the processing, the 100 m/s true nominal motion is nearly fully compensated for with a predicted nominal motion of 101 m/s. For the partial derivative associated with the local linearization of each particle, use the finite-difference approximation:

$$\frac{\partial \mathbf{h}}{\partial \mathbf{x}} = \frac{h(x-\delta) - h(x+\delta)}{2\delta}$$

(d) Process the measurements generated in Part (b) with the design of Part (c) to generate the solution estimates of x. The initial estimates of the set of particle filters are uniformly distributed over a range of [−50 m, 50 m] centered at 5,000 m, the true horizontal position at $t = 0$. Compute the horizontal position error by differencing the solution from true value of x, for the 150-s sequence. Plot this horizontal position error over time.

REFERENCES CITED IN CHAPTER 7

1. M. Kayton and W.R. Fried (eds.), *Avionics Navigation Systems,* 2nd Edition, New York: Wiley, 1997, p. 375.
2. H.W. Sorenson,"Kalman Filtering Techniques," C.T. Leondes (ed.), *Advances in Control Systems*, Vol. 3, New York: Academic Press, 1966, pp. 219–289.
3. R.G. Brown and J.W. Nilsson, *Introduction to Linear Systems Analysis*, New York: Wiley, 1962.
4. M.S. Grewal and A.P. Andrews, *Kalman Filtering Theory and Practice*, Englewood Cliffs, NJ: Prentice-Hall, 1993 (see Section 4.2 and Chapter 6).
5. A. Gelb (ed.), *Applied Optimal Estimation*, Cambridge, MA: MIT Press, 1974.

6. C.K. Chui, G. Chen,and H.C. Chui, "Modified Extended Kalman Filtering and a Real-Time Parallel Algorithm for System Parameter Identification," *IEEE Trans. Automatic Control*, 35(1): 100–104 (Jan. 1990).
7. S.T. Park and J.G. Lee, "Comments on "Modified Extended Kalman Filtering and a Real-Time Parallel Algorithm for System Parameter Identification," *IEEE Trans. Automatic Control*, 40(9): 1661–1662 (Sept. 1995).
8. V. E. Benes, "Exact Finite-Dimensional Filters with Certain Diffusion Nonlinear Drift," *Stochastics*, vol. 5, pp. 65–92, 1981.
9. F.E. Daum, "Beyond Kalman Filters: Practical Design of Nonlinear Filters," *Proceedings of SPIE*, vol. 2561, pp. 252–262, 1995.
10. A.J. Haug, "A Tutorial on Bayesian Estimation and Tracking Techniques Applicable to Nonlinear and Non-Gaussian Processes," *MITRE Technical Report*, Jan. 2005.
11. G. Evensen, "Sequential data assimilation with nonlinear quasi-geostrophic model using Monte Carlo methods for forecast error statistics," *Journal of Geophysical Research*, Vol. 99, pp, 10143–10162, 1994.
12. S. Lakshmivarahan and D.J. Stensrud, "Ensemble Kalman Filter: Application to Meteorological Data Assimilation," *IEEE Control Systems Magazine*, June 2009, pp. 34–46.
13. S.J. Julier, J.K. Uhlmann, and H. F. Durrant-Whyte, "A new approach for filtering nonlinear systems," *Proceedings of the American Control Conference*, 1995, pp. 1628–1632.
14. S. J. Julier and J.K. Uhlmann, "A new extension of the Kalman filter to nonlinear systems," *Proceedings of AeroSense: The 11th International Symposium on Aerospace/ Defence Sensing, Simulation and Controls*, Orlando, Florida, 1997.
15. E.A. Wan and R. van der Merwe,"The Unscented Kalman Filter," in *Kalman Filtering and Neural Networks*, Chapter 7, S. Haykin (Ed.), New York: Wiley, 2001.
16. D. Tenne and T. Singh, "The higher order unscented filter," *Proceedings of the American Control Conference*, 2002, pp. 4555–4559.
17. R. Merwe and E. Wan, "The square-root unscented Kalman filter for state and parameter-estimation," *Proceedings of the IEEE Int. Conf. Acoustics, Speech and Signal Processing (ICASSP)*, vol. 6, 2001, pp. 3461–3464.
18. B. Ristic, S. Arulampalam, and N.J. Gordon, *Beyond the Kalman Filter: Particle Filters for Tracking Applications*, Artech House, 2004.
19. A. Doucet, N. deFreitas, and N. Gordon, Eds., *Sequential Monte Carlo Methods in Practice*, Springer-Verlag, New York, 2001.
20. N.J. Gordon, D.J. Salmond, and A.F.M. Smith, "Novel approach to nonlinear/non-Gaussian Bayesian state estimation," *IEE Proceedings-F*, vol. 140, no. 2, pp. 107–113, April 1993.
21. F. Gustafsson, "Particle Filter Theory and Practice with Positioning Applications," *IEEE Aerospace & Electronic Systems Magazine*, 25: 53–82 (July 2010).
22. R. van der Merwe, A. Doucet, N. deFreitas, and E. Wan, "The Unscented Particle Filter," Technical Report CUED/F-INTENG/TR 380, Cambridge University Engineering Department, August 2000.
23. K. Murphy and S. Russell,"Rao-Blackwellised Particle Filtering for Dynamic Bayesian Networks," in *Sequential Monte Methods in Practice*, A. Doucet, J.F.G. de Freitas, and N.J. Gordon (Eds.), Springer-Verlag, New York, 2001.

8

The "Go-Free" Concept, Complementary Filter, and Aided Inertial Examples

In the usual form of Kalman filtering we are forced to make assumptions about the probabilistic character of the random variables under consideration. However, in some applications we would like to avoid such demanding assumptions, to some extent at least. There is a methodology for accomplishing this, and it will be referred to here as the *go-free* concept.

The examples used here are taken from modern navigation technology, especially those involving satellite navigation systems such as GPS. (A detailed description of GPS is given in Chapter 9.) All such systems operate by measuring the transit times of the signal transmissions from each of the satellites in view to the user (usually earthbound). These transit times are then interpreted by the receiver as ranges via the velocity-of-light constant c. However, the receiver clock may not be in exact synchronism with the highly-stable satellite clocks, so there is a local clock offset (known as clock bias) that has to be estimated as well as the local position in an earth-fixed xyz coordinate frame. So, for our purposes here, after linearization as discussed in Chapter 7, we simply pick up the estmation problem as one where we need to estimate three unknown position coordinates and a clock bias based on n measurements where $n \geq 4$. Normally n is greater than 4. So, in its simplest form, we have an overdetermined linear system of noisy measurements to work with in the estimation problem.

In Sections 8.1 through 8.4, the go-free idea is discussed in detail, and complementary filtering follows in Sections 8.5 through 8.9. Then a much-used complementary filtering example in aided inertial navigation technology is presented in Sections 8.10 through 8.12.

8.1 INTRODUCTION: WHY GO FREE OF ANYTHING?

The idea of going free of certain unknowns in the suite of measurement equations is certainly not new. Differencing out the clock bias in LORAN and also in precision GPS applications are good examples of this. However, the go-free concept as used here is a generalization of the simple "differencing out" idea. As used here, go-free means that we readjust, by algebraic manipulation or otherwise, the measurement equations such that the resulting filter estimates become completely free of any assumptions regarding the statistics of the go-free variables. The key words here are "free of any assumptions." The go-free variables are not lost or discarded. They remain in the overall estimation scheme, but we are relieved of making any rash assumptions about their dynamical behavior.

In the context of Kalman filtering the go-free filter is suboptimal when compared with the hypothetical optimal filter which uses the original measurement suite; provided, of course, that we have accurate stochastic models in the full-state optimal filter. So the obvious question arises, "Why go free of anything?" The answer to this question lies mainly in the qualifying words *accurate stochastic models*. In many applied situations we simply do not have reliable random models for one or more of the variables in the measurement equations. Vehicle motion is a good example. Rather than being truly random, the motion is often better described in deterministic terms like: steady acceleration for a short period, then a coasting constant-velocity period, then more acceleration, and so forth. This kind of motion defies accurate stochastic modeling as required by the Kalman filter. Thus, it might be better, if measurement redundancy permits, to go free of making any assumptions about the dynamic behavior of the vehicle. In this way we would avoid the possibility of large estimation errors induced by mismodeling. Some loss in rms-error performance would be the price of "going-free," but this might well be better than accepting very large errors in unusual situations.

8.2 SIMPLE GPS CLOCK BIAS MODEL

Consider a simple "snapshot" GPS pseudorange example where we have six-in-view, and the problem has been linearized about some nominal position. The measurement equation then has the form:

$$\mathbf{z} = \mathbf{H}\mathbf{x} + \mathbf{v} \tag{8.2.1}$$

where

$$\begin{aligned}
\mathbf{z} &= \begin{bmatrix} z_1 & z_2 & z_3 & z_4 & z_5 & z_6 \end{bmatrix}^T \\
\mathbf{x} &= \begin{bmatrix} x_1 & x_2 & x_3 & x_4 \end{bmatrix}^T \\
\mathbf{H} &= 6 \times 4 \text{ linear connection matrix} \\
\mathbf{v} &= 6 \times 1 \text{ random measurement noise vector}
\end{aligned}$$

We will have occasion to use numerical values for comparison later, so we will let $\mathbf{H}$ be:

$$\mathbf{H} = \begin{bmatrix} h_{11} & h_{12} & h_{13} & h_{14} \\ h_{21} & h_{22} & h_{23} & h_{24} \\ \vdots & \vdots & \vdots & \vdots \\ h_{61} & h_{62} & h_{63} & h_{64} \end{bmatrix} = \begin{bmatrix} -0.7460 & 0.4689 & -0.4728 & 1.000 \\ 0.8607 & 0.3446 & -0.3747 & 1.000 \\ -0.2109 & -0.3503 & -0.9126 & 1.000 \\ 0.0619 & 0.4967 & -0.8657 & 1.000 \\ 0.7249 & -0.4760 & -0.4980 & 1.000 \\ 0.4009 & -0.1274 & -0.9072 & 1.000 \end{bmatrix} \tag{8.2.2}$$

This is a "typical" GPS geometry in that it is neither especially good nor especially poor for positioning purposes.

We have six measurements and four unknowns in this example, so we have the necessary redundancy to go free of something, if we choose to do so. Clock bias modeling is usually not very reliable, so suppose we say that we want to go completely free of clock bias (i.e., state x_4). It is obvious in this case that x_4 can be removed from the measurement equations by simple differencing. One way of doing this is to let z_1 be the reference and then subtract each of the other measurements from z_1. This then results in the reduced model:

$$\mathbf{z}' = \mathbf{H}'\mathbf{x} + \mathbf{v}' \tag{8.2.3}$$

where

$$\mathbf{z}' = \begin{bmatrix} z_1 - z_2 \\ z_1 - z_3 \\ z_1 - z_4 \\ z_1 - z_5 \\ z_1 - z_6 \end{bmatrix} \quad \mathbf{x}' = \begin{bmatrix} x_1 \\ x_2 \\ x_3 \end{bmatrix}$$

$$\mathbf{H}' = \begin{bmatrix} h_{11} - h_{21} & h_{12} - h_{22} & h_{13} - h_{23} \\ h_{11} - h_{31} & h_{12} - h_{32} & h_{13} - h_{33} \\ \vdots & \vdots & \vdots \\ h_{11} - h_{61} & h_{12} - h_{62} & h_{13} - h_{63} \end{bmatrix} \quad \mathbf{v}' = \begin{bmatrix} v_1 - v_2 \\ v_1 - v_3 \\ v_1 - v_4 \\ v_1 - v_5 \\ v_1 - v_6 \end{bmatrix} \tag{8.2.4}$$

We are now in a position to do a one-step Kalman filter solution for the reduced model. Note that we do not have to make any prior assumptions about the clock bias x_4. It does not appear anywhere in the reduced model. Also note that there will be nontrivial correlation among the elements of $\mathbf{v}'$, and this must be accounted for properly in the Kalman filter $\mathbf{R}'$ matrix. To be specific:

$$\mathbf{R}' = \begin{bmatrix} E(v_1 - v_2)^2 & E(v_1 - v_2)(v_1 - v_3) & \cdots \\ E(v_1 - v_3)(v_1 - v_2) & E(v_1 - v_3)^2 & \\ \vdots & \ddots & \vdots \\ & & E(v_1 - v_6)^2 \end{bmatrix} \tag{8.2.5}$$

For simplicity in this example we will assume that the σ 's for all of the original ε's are equal and that we have independence. Then the **R** matrix for the go-free model will be:

$$\mathbf{R}' = \sigma^2 \begin{bmatrix} 2 & 1 & 1 & 1 & 1 \\ 1 & 2 & 1 & 1 & 1 \\ 1 & 1 & 2 & 1 & 1 \\ 1 & 1 & 1 & 2 & 1 \\ 1 & 1 & 1 & 1 & 2 \end{bmatrix} \tag{8.2.6}$$

To complete our numerical example, we will let

$\sigma = 5\,m$; (pseudorange measurement sigma)

and

$\mathbf{P}_0^- = 100^2\mathbf{I}_{3x3}$; (initial rms uncertainty in position components is 100 m)

Using the specified numerical values, the go-free filter yields the following updated error covariance:

$$\mathbf{P}_0 = \begin{bmatrix} 17.0400 & 9.4083 & -13.6760 \\ 9.4083 & 33.6975 & -19.5348 \\ -13.6760 & -19.5348 & 96.2486 \end{bmatrix} \tag{8.2.7}$$

or the rms position errors are:

$\text{rms}\,(x_1) = 4.1280\,m$
$\text{rms}\,(x_2) = 5.8050\,m$
$\text{rms}\,(x_3) = 9.8106\,m$

There are no surprises in these results. The Kalman filter only goes one step, so there is no benefit of time averaging here.

Note that the filter *per se* does not provide an estimate of clock bias. However, if this is of secondary interest, an estimate of sorts can be obtained by substituting the go-free position estimates back into any one of the original six measurement equations. We will not elaborate on this further here, because it does make a difference as to which original measurement equation is used in the "back substitution." And further more, calculating the variance of the clock estimate so obtained is a bit of a hassle, so we will defer discussion of this until Section 8.4.

8.3 EULER/GOAD EXPERIMENT

In the early days of GPS, H-J Euler and C. Goad published an interesting paper on resolving the carrier phase integer ambiguities (1). This may have been the first paper to use the go-free principle in the context of Kalman filtering. Specifically, the objective of their experiment was to resolve N_1, N_2, and the wide lane ($N_1 - N_2$) integers on the basis of four measurements, code and carrier, on both L1 and L2 frequencies. Furthermore, this was to be done without precise knowledge of the satellite's orbital trajectory. This was a relatively simple single-receiver experiment. No single- or double-differencing was involved, as in the usual integer resolution scenario.

The states in the Euler/Goad experiment were:

$\rho =$ pseudorange to the satellite
$\iota/f_1^2 =$ iono code delay on $L1$ frequency ($f_1 = 1575.42\,\text{MHz}$)
$N_1 =$ integer ambiguity of $L1$ carrier phase
$N_2 =$ integer ambiguity of $L2$ carrier phase ($f_2 = 1227.6\,\text{MHz}$)

The measurement model was:

$$\underbrace{\begin{bmatrix} \rho_1 \\ \phi_1 \\ \rho_2 \\ \phi_2 \end{bmatrix}}_{\mathbf{z}} = \underbrace{\begin{bmatrix} 1 & 1 & 0 & 0 \\ 1 & -1 & \lambda_1 & 0 \\ 1 & \dfrac{f_1^2}{f_2^2} & 0 & 0 \\ 1 & -\dfrac{f_1^2}{f_2^2} & 0 & \lambda_2 \end{bmatrix}}_{\mathbf{H}} \underbrace{\begin{bmatrix} \psi \\ {}^{\iota}/f_1^2 \\ N_1 \\ N_2 \end{bmatrix}}_{\mathbf{x}} + \underbrace{\begin{bmatrix} \varepsilon_{\rho 1} \\ \varepsilon_{\phi 2} \\ \varepsilon_{\rho 2} \\ \varepsilon_{\phi 2} \end{bmatrix}}_{\boldsymbol{\varepsilon}} \tag{8.3.1}$$

We note here that the ratio f_1/f_2 is exactly 154/120. Also, the tropo and clock bias errors are common to all four measurements, so they can be lumped in with the pseudorange variable.

In this experiment Euler and Goad wanted the filter estimates of N_1 and N_2 to be completely free of modeling uncertainties in pseudorange and the iono error. There are four linear equations in the four unknowns in this measurement situation. So, with a modest pencil-and-paper effort, one can do algebraic elimination of the unwanted states and reduce the measurement model to two measurements with two unknowns of N_1 and N_2. This is a bit of a hassle, so Euler and Goad chose a different route. They reasoned that the same immunity to modeling uncertainties in States 1 and 2 could be achieved by artificially, with each recursive step, resetting the (1,1) and (2,2) terms of the $\mathbf{P}^-$ matrix to very large values, say "near infinity." This is the same thing as saying that the information content associated with prior estimates of States 1 and 2 is zero. So, for convenience (and to avoid the "near infinity" problem) Euler and Goad chose to use the alternative form of the Kalman filter (see Chapter 5, Section 5.1). Here, the $\mathbf{P}$ matrix update is done with the inverse of $\mathbf{P}$ rather than $\mathbf{P}$ itself. So, to effect the go-free constraint on States 1 and 2, they simply zeroed out the first two rows and columns of $\mathbf{P}^-$ with each recursive step. To quote Euler and Goad exactly, "This means that the previous information of ρ and ι is effectively neglected in the update of the new state vector." And it worked! It is easily verified that algebraic elimination yields the same error covariance results as the Euler/Goad method, provided that the four measurement errors are assumed to be mutually independent, and the same sigmas are used in making the comparison.

The numerical results of the experiment just described were interesting but not very encouraging. Even with measurement spans exceeding 2 hours, Euler and Goad were not able to resolve the N_1 and N_2 integers with a high degree of confidence. However, the widelane ($N_1 - N_2$) integer was resolvable, which was some consolation. Perhaps the more important and lasting contribution of the Euler/Goad paper is the method used to achieve the go-free constraint, and not the integer ambiguity results *per se*.

This whole business of artificially increasing the uncertainty of certain estimates to achieve the go-free constraint is an *ad hoc* procedure. It is without mathematical rigor and must be done carefully. If we are working directly with $\mathbf{P}^-$ rather than $(\mathbf{P}^-)^{-1}$ (i.e., the usual Kalman filter), then we must pay attention to the cross terms in $\mathbf{P}^-$ as well as the diagonal ones. For example, if we have a dynamic situation where we wish to go free of vehicle motion, we usually need to worry about velocity and acceleration states as well as position, and these three states are correlated and this must be accounted for properly.

It is best not to think in terms of increasing elements of $\mathbf{P}^-$ directly. It is better to effect the desired increase in $\mathbf{P}^-$ by increasing the appropriate terms in the $\mathbf{Q}$ matrix. Here we are concerned with the statistics of the process noise vector $\mathbf{w}_k$, and it will be obvious there if there are cross correlations among the elements that need to be accounted for. Therefore, we suggest here that we effect the desired increase in $\mathbf{P}^-$ with $\mathbf{Q}$, and not $\mathbf{P}^-$ directly. Thus, the go-free method suggested here will be called the *Boosted Q* method.

8.4 REPRISE: GPS CLOCK-BIAS MODEL REVISITED

We now return to the clock bias example in light of the go-free method introduced in the Euler/Goad experiment. In Section 8.2 we used algebraic elimination to go free of the clock bias. Now we will see if we can accomplish the same result using the Boosted-Q method. In the clock-bias example the prediction step is trivial because we only do a single update, conceptually at $t = 0$. So, in this case, boosting will be done on the (4,4) term of the initial $\mathbf{P}^-$ matrix. Therefore, leaving the initial (1,1), (2,2), and (3,3) terms at 100^2 just as before, we will set the (4,4) term at the outrageously large value of 10^{12} and see what happens. That is, let

$$\mathbf{P}^- = \begin{bmatrix} 10^4 & 0 & 0 & 0 \\ 0 & 10^4 & 0 & 0 \\ 0 & 0 & 10^4 & 0 \\ 0 & 0 & 0 & 10^{12} \end{bmatrix} m^2 \tag{8.4.1}$$

We now do the error covariance update with the Joseph update formula

$$\mathbf{P}^+ = (\mathbf{I} - \mathbf{KH})\mathbf{P}^-(\mathbf{I} - \mathbf{KH})^T + \mathbf{KRK}^T \tag{8.4.2}$$

Where $\mathbf{H}$ is the original 6×4 matrix (from Section 8.2), $\mathbf{R} = 5^2$ as before, and $\mathbf{K}$ is:

$$\mathbf{K} = \mathbf{P}^-\mathbf{H}^T\left(\mathbf{HP}^-\mathbf{H}^T + \mathbf{R}\right)^{-1} \tag{8.4.3}$$

The result is (using MATLAB):

$$\mathbf{P}^+ = \begin{bmatrix} 17.0400 & 9.4083 & -13.6760 & -12.8469 \\ 9.4083 & 33.6975 & -19.5348 & -16.8381 \\ -13.6760 & -19.5348 & 96.2486 & 68.3107 \\ -12.8469 & -16.8381 & 68.3107 & 53.9796 \end{bmatrix} \tag{8.4.4}$$

Note that the upper left 3×3 submatrix of $\mathbf{P}^+$ is the same identical error covariance that was obtained in Section 8.2 using algebraic elimination (at least within four decimal places). Furthermore, note that the (4,4) term of the Boosted-Q $\mathbf{P}^+$ matrix gives a meaningful mean-square estimation error for the go-free clock-bias state. That is, it is the best we can hope to do in a minimum-mean-square-error sense, subject to the condition that we have no prior information as to the rms value of this random variable. It is also worth noting that even though the clock bias may not be of prime interest, its error variance is of the same order as that of the position errors. Finally, note that this "extra" estimate of the clock bias comes "for free," so-to-speak, with the Boosted-Q method. And certainly, from a programming viewpoint, it is easier to just boost the **Q** matrix than to do the algebraic elimination.

8.5 THE COMPLEMENTARY FILTER

We will not look at a go-free filtering application that is best described first in terms of classical frequency-response methods. We will then show the connection to Kalman filtering later.

We begin with the simple block diagram shown in Fig. 8.1. We can now pose the following filter optimization problem: Given the spectral characteristics of the signal $y(t)$ and the additive measurement noises $n_1(t)$ and $n_2(t)$ and any cross-correlations that may exist, find the transfer functions $G_1(s)$ and $G_2(s)$ that will minimize the mean-square error in the estimate $\hat{y}(t)$. This will be recognized as an extension of the single-input Wiener problem (2). To keep things simple, we will assume that all processes are stationary and we are looking for the stationary solution. Also, if we assume Gaussian statistics throughout, there is no loss in generality in saying that the optimal solution is simply a superposition of the two noisy measurements after being passed through their respective transfer functions as shown in Fig. 8.1.

Now, in applied applications we can see an immediate difficulty in the two-input problem as just posed. Even though we can usually assume reasonable spectral models for the noises $n_1(t)$ and $n_2(t)$, more often than not the signal $y(t)$ is more of a deterministic nature, and it defies modeling as a legitimate random process. For example, in dynamic terrestrial positioning applications, the position and velocity are usually not random, but yet the filter is expected to respond faithfully to a variety of deterministic-type inputs. So, this immediately suggests a filter design where we go completely free of the dynamics of $y(t)$. With this in mind consider next the block diagram of Fig. 8.2.

Here we have the same two inputs as before, but we have imposed a constraint between the two transfer functions, namely that one be the complement of the other. With a little block diagram algebra we can easily show that the output (in the complex s domain) can be written as

$$X(s) = \underbrace{Y(s)}_{\text{Signal term}} + \underbrace{N_1(s)[1 - G(s)] + N_2(s)[G(s)]}_{\text{Error term}} \tag{8.5.1}$$

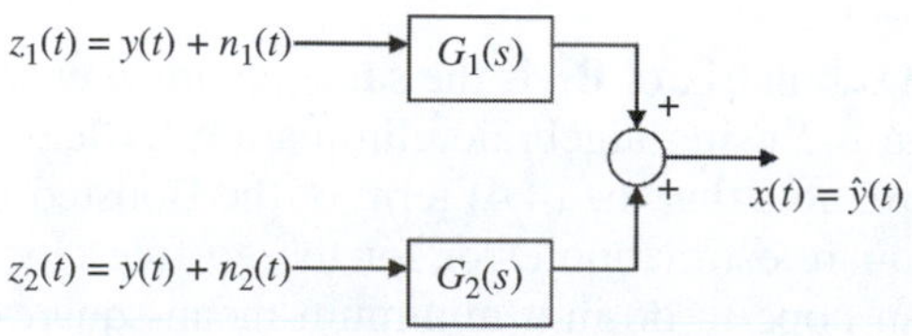

Figure 8.1 The general two-input Wiener problem.

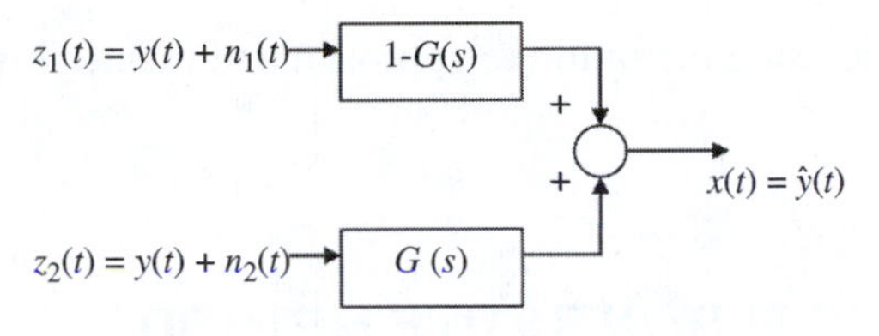

Figure 8.2 Complementary filter.

Clearly, with the complementary constraint in place the signal component goes through the system completely undistorted by the choice of $G(s)$, and the error term is completely free of the signal $y(t)$. So, the go-free condition has been achieved in the sense that the signal goes through the system undistorted, and it is estimated without making any assumptions about its spectrum or deterministic nature in any way. There is no assurance of optimality though until we consider the choice of $G(s)$ more critically.

With the choice of $G(s)$ in mind, consider the signal processing arrangement shown in Fig. 8.3.

The purpose of $G(s)$ here is to separate one noise from another and yield a good estimate of $n_1(t)$. We can then subtract $\hat{n}_1(t)$ from the raw $z_1(t)$ measurement and, hopefully, get a good estimate of $y(t)$, which is the ultimate goal. Of course, the better the estimate of $n_1(t)$, the better the estimate of $y(t)$. Mathematically, with a bit of block diagram algebra, we can write the output $x(t)$ as (in the complex s domain):

$$X(s) = Y(s) + N_1(s)[1 - G(s)] + N_2(s)[G(s)] \tag{8.5.2}$$

Lo and behold, this is the same identical expression for the signal estimate that was obtained from the "total" model shown in Fig. 8.2. Thus, we have two implementations that yield the same identical result for the two-input complementary filter. The first of these we call the total configuration because each filter operates directly on its respective total noisy measurements, whereas in the error configuration the $G(s)$ filter operates only on the subtractive combination of the measurement errors. One of the beauties of the error model shown in Fig. 8.3 is that it provides some insight into the choice of $G(s)$. Clearly, if we are to effect a meaningful separation of $n_1(t)$ from $n_2(t)$ there must be a significant difference in their respective spectral characteristics. Otherwise, the two transfer functions degenerate to positive constants that sum to unity. (This is correct, but not a very interesting optimization result.) In the nontrivial case we can either take a formal optimization approach using Kalman filtering methods, or we can use more intuitive methods where we simply guess the form of $G(s)$ for the spectral situation at hand and then, perhaps by trial-and-error, optimize with respect to parameters of $G(s)$. We will now

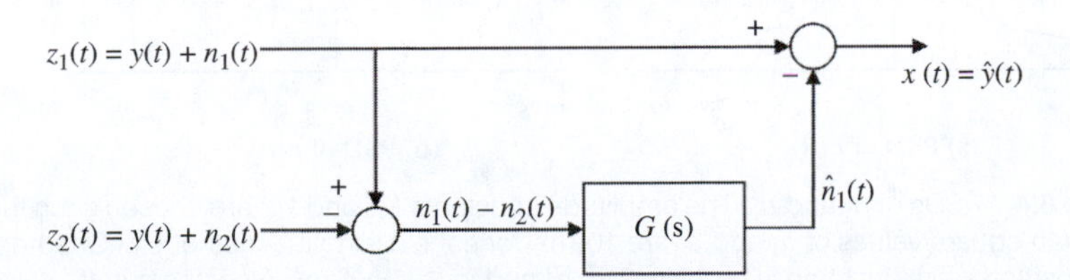

Figure 8.3 Error-state version of a complementary filter.

proceed to a simple example using the intuitive approach and then follow it with a Kalman filtering example.

8.6 SIMPLE COMPLEMENTARY FILTER: INTUITIVE METHOD

In this simple example we will assume that the power spectral densities (PSD) of the two measurement noises are as shown in Fig. 8.4. PSD_1 will be recognized as a relatively low-frequency second-order Markov process with an undamped natural frequency of ω_0 and a damping ratio of $1/\sqrt{2}$. (See Example 2.12, Chapter 2, Section 2.9.) PSD_2 is band-limited white noise where the truncation frequency is large relative to ω_0. The measurements in this example are both assumed to be distance in meters. Note that we make no assumptions about the character of the signal $y(t)$.

For the noise scenario just presented there is a significant difference in the spectral characteristics of n_1 and n_2. Therefore, just by looking at the error model in Fig. 8.3, it can be seen that $G(s)$ should be some sort of low-pass filter; that is, think of n_1 as the "signal" and n_2 as the "noise" in this sub-problem. Suppose we begin our intuitive trial-and-error analysis with the simplest of simple low-pass filters:

$$G(s) = \frac{1}{1 + Ts} = \frac{1/T}{s + 1/T} \tag{8.6.1}$$

This filter has unity gain at zero frequency and "rolls off" at 20 dB/decade at high frequency. Now, of course, having specified $G(s)$ for the z_2 channel, the filter for the z_1 channel is constrained to be:

$$1 - G(s) = 1 - \frac{1}{1 + Ts} = \frac{s}{s + 1/T} \tag{8.6.2}$$

We only have the filter time constant T left to be specified in our analysis, and we want to adjust this to yield the minimum mean-square error in the complementary filter's estimate of $y(t)$. We can now use the methods given in Chapter 3 to evaluate the mean-square values of the two error terms in Eq. (8.5.1). Call these e_1 and e_2, and

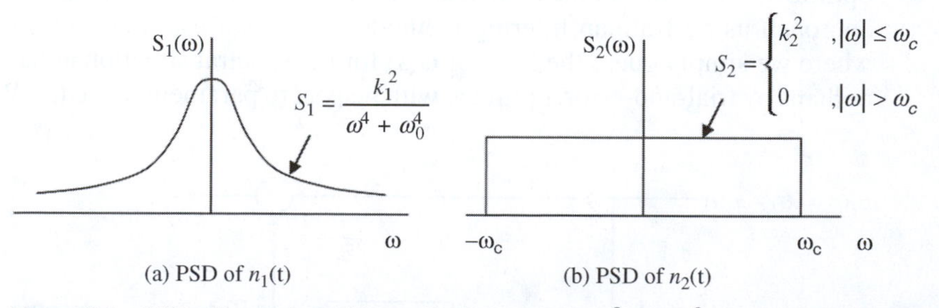

Figure 8.4 PSDs of n_1 and n_2. The amplitude constants k_1^2 and k_2^2 are chosen such that the mean-square values of n_1 and n_2 are 100 m^2 for each. Using the methods of Chapter 2 the constants work out to be $k_1^2 = 200\sqrt{2}\omega_0^3$ m^2 and $k_2^2 = \frac{100\pi}{\omega_c}$ m^2. Also, the cutoff frequency ω_c for S_2 is $20\omega_0$ where ω_0 is the bandwidth parameter associated with S_1.

their mean-square values work out to be ($k_1{}^2$ and $k_2{}^2$ are specified in the caption of Fig. 8.4):

$$E(e_1^2) = \frac{200\omega_0^2 T^2}{1 + \sqrt{2}\omega_0 T + \omega_0^2 T^2} m^2 \tag{8.6.3}$$

$$E(e_2^2) = \frac{5}{T} \tan^{-1}(20\omega_0 T) m^2 \tag{8.6.4}$$

If we now assume independence for n_1 and n_2, the total mean-square error for the complementary filter is just the sum of $E(e_1^2)$ and $E(e_2^2)$. It is this sum that is to be minimized by adjusting T. Here, we can think of T in units of $1/\omega_0$ where ω_0 is fixed, and then search for the best value of T by trial-and-error. Table 8.1 gives the results of such a search for a few values of T within the range of interest.

Clearly, the best value for the time constant T is about 0.30 on a normalized basis and the corresponding mean-square error is about 35 m^2. As a check on the reasonableness of this result, recall that the raw unfiltered z_1 and z_2 measurements were specified to have mean-square errors of 100 m^2 each, so the complementary filter does provide a considerable reduction of the error when compared with either of the measurements when considered separately with no filtering. Also, we can compare our optimized low-pass filter result with a trivial complementary filter where $G(s)$ and $[1\text{-}G(s)]$ are each set at 0.5. (This is simply averaging z_1 and z_2, which is trivial, but still legitimate.) The simple averaging filter results in a mean-square error of 50 m^2, so our low-pass filter yields a significant improvement over simple averaging, but the improvement is not overly dramatic.

We could no doubt improve on the end result presented here by considering more complex low-pass filters. However, the main point of this simple example is the methodology, not the numerical result. To summarize, the methodology is: (1) choose a filter functional form that is appropriate for the spectral characteristics of the noises in the problem at hand; then (2) adjust the filter parameters to minimize the mean-square error as determined from the error term in Eq. (8.5.1).

We will leave the intuitive approach now and proceed on to a more formal optimal approach, namely Kalman filtering. For purpose of comparison we will use the same noise scenario in the Kalman filter example as was used in the intuitive example.

Table 8.1 Total Mean-Square Error for Various T

T (in $1/\omega_0$ units)	$E(e_1^2)$ (m^2)	$E(e_2^2)$ (m^2)	Total Mean-Square Error
0.20	6.048	33.14	39.19
0.25	8.824	27.47	36.29
0.30	11.88	23.44	35.32
0.35	15.15	20.40	35.55
0.40	18.55	18.08	36.63
0.50	25.54	14.71	40.25

8.7 KALMAN FILTER APPROACH—ERROR MODEL

When Kalman filtering is viewed from the "total" configuration shown in Fig. 8.2, it is obvious where the name "complementary" comes from. The two transfer functions (i.e., filters) are complements of each other in the complex frequency domain. However, Kalman filtering has often been referred to as time-domain filtering. So, it is not at all obvious how to apply the complementary constraint shown in Fig. 8.2 to a corresponding Kalman filter. On the other hand, if we look at the equivalent error configuration in Fig. 8.3, we see no special problem in doing something similar with a Kalman filter. The purpose of $G(s)$ in the error configuration is to estimate one random variable (say n_1) in the presence of another (say n_2), and that is exactly what a Kalman filter does best! So, to begin with, we will make our complementary filter mimic the error configuration shown in Fig. 8.3. Then we will come back to the "total" configuration later.

For a Kalman filter we first need the mathematical equations for the filter states and measurements. In the error configuration n_1(t) plays the role of "signal," and n_2(t) is the corrupting measurement noise. We will look at n_2 first; it is the simpler of the two. It is bandlimited white noise. So, if we sample n_2 at the Nyquist rate, the samples will automatically be uncorrelated, and all will have the same mean-square value, namely 100 m^2 (as in the intuitive filter example). Therefore, the R_k parameter of the Kalman filter will be the scalar:

$$R_k = 100\,\text{m}^2$$

and the sampling rate will be:

$$f_{sample} = 2\left(\frac{\omega_c}{2\pi}\right) = \frac{\omega_c}{\pi} Hz$$

If we let $\omega_0 = 1$ rad/s (so we can compare with the intuitive filter), the Δt interval for the Kalman filter works out to be

$$\Delta t = \frac{\pi}{\omega_c} = \frac{\pi}{20\omega_0} = \frac{\pi}{20}\,\text{s}$$

Next, consider the signal variable n_1(t). Its spectral function is (for $\omega_0 = 1$)

$$S_1(s) = \frac{k_1^2}{s^4 + 1}; \quad k_1^2 = 200\sqrt{2}\,\text{m}^2(\text{rad/s})^3 \tag{8.7.1}$$

As discussed in Chapter 3, we can factor the spectral function into left- and right-half plane parts as follows:

$$S_1(s) = \frac{k_1}{s^2 + \sqrt{2}s + 1} \cdot \frac{k_1}{s^2 - \sqrt{2}s + 1} \tag{8.7.2}$$

We can now think conceptually of $n_1(t)$ being the result of putting unity white noise into a shaping filter as shown in Fig. 8.5.

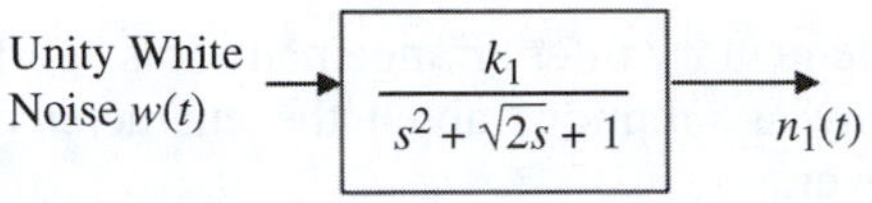

Unity White Noise $w(t)$ → $\dfrac{k_1}{s^2 + \sqrt{2}s + 1}$ → $n_1(t)$

Figure 8.5 Conceptual shaping filter to produce $n_1(t)$.

The block diagram in Fig. 8.5 then defines the differential equation for n_1(t) to be

$$\ddot{n}_1 + \sqrt{2}\dot{n}_1 + n_1 = w(t) \tag{8.7.3}$$

If we choose phase variables as our states (i.e., $x_1 = n_1$ and $x_2 = \dot{n}_1$), the state equation for n_1(t) becomes

$$\begin{bmatrix} \dot{x}_1 \\ \dot{x}_2 \end{bmatrix} = \begin{bmatrix} 0 & 1 \\ -1 & -\sqrt{2} \end{bmatrix} \begin{bmatrix} x_1 \\ x_2 \end{bmatrix} + \begin{bmatrix} 0 \\ k_1 \end{bmatrix} w(t) \tag{8.7.4}$$

We have already specified Δt to be $\pi/20$, so we can now find $\boldsymbol{\phi}_k$ and $\mathbf{Q}_k$ for our Kalman filter using the Van Loan method given in Chapter 3. Also, the $\mathbf{H}_k$ for the measurement situation in the error model is

$$\mathbf{H}_k = [1 \quad 0]$$

We now have all the key filter parameters to do covariance analysis except for the initial $\mathbf{P}_0^-$. Here the prior knowledge of x_1 and x_2 needs to be compatible with our spectral assumptions for $n_1(t)$. Using the analysis methods of Chapter 3 and the knowledge that x_2 is the derivative of x_1, we find that

$$E[x_1^2(0)] = 100\,\text{m}^2$$
$$E[x_2^2(0)] = 100\,(\text{m/s})^2$$

and that $x_1(0)$ and $x_2(0)$ are uncorrelated. Then, without benefit of any measurement information at t_0, the estimation error covariance must be

$$\mathbf{P}_0^- = \begin{bmatrix} 100 & 0 \\ 0 & 100 \end{bmatrix} \text{m}^2$$

We have now specified all the needed parameters for covariance analysis. The filter recursive equations are easily programmed in MATLAB, and the filter is found to reach a steady-state condition after about 40 steps. The resultant mean-square error for x_1 (which is n_1) is

$$\mathbf{P}(1, 1) = 21.47\,\text{m}^2$$

This compares favorably with the 35.32 mean-square error achieved with the intuitive $1/(1 + Ts)$ analog filter. Also, note from the error configuration of Fig. 8.3 that this (i.e., $21.47\,\text{m}^2$) is also the mean-square estimation error for the go-free variable $y(t)$. It is especially important to remember that we do not "throw away" the go-free variable in the complementary filter arrangement. It is "go-free" only in the sense that we go free of any assumptions about the character of $y(t)$, be it random, deterministic or whatever.

8.8 KALMAN FILTER APPROACH—TOTAL MODEL

We now return to the "total" complementary filter configuration shown in Fig. 8.2, but we wish to do the filtering in the time domain (i.e., Kalman filtering) rather than in the frequency domain. We will present this discussion as a student exercise with generous hints along the way. First note that the signal $y(t)$ is not differenced out in the total model, so it must remain as a state in the filter. Also, note that the filter has two explicit measurements here, rather than just one as in the error model. We are forced then to make some assumptions about the random process for $y(t)$. So, for simplicity, let us assume (temporarily) that $y(t)$ is a first-order Gauss-Markov process with a known σ and β, and this will be designated as State 3. The differential equation for x_3 is then (see Chapter 3)

$$\dot{x}_3 + \beta x_3 = \sqrt{2\sigma^2\beta}w_3(t) \tag{8.8.1}$$

where $w_3(t)$ is the unity white noise that is independent of the white noise driving the $n_1(t)$ process. Now append the x_3 differential equation to the second-order differential equation describing the $n_1(t)$ noise process. This leads to a pair of equations describing our system:

$$\ddot{n}_1 + \sqrt{2}\dot{n}_1 + n_1 = k_1 w_1(t) \quad (\text{As before with}\, \omega_0 = 1) \tag{8.8.2}$$

$$\dot{x}_3 + \beta x_3 = \sqrt{2\sigma^2\beta}w_3(t)$$

Or, in state-space form we have:

$$\begin{bmatrix} \dot{x}_1 \\ \dot{x}_2 \\ \dot{x}_3 \end{bmatrix} = \underbrace{\begin{bmatrix} 0 & 1 & 0 \\ -1 & -\sqrt{2} & 0 \\ 0 & 0 & -\beta \end{bmatrix}}_{\mathbf{F}} \begin{bmatrix} x_1 \\ x_2 \\ x_3 \end{bmatrix} + \underbrace{\begin{bmatrix} 0 & 0 \\ k_1 & 0 \\ 0 & \sqrt{2\sigma^2\beta} \end{bmatrix}}_{\mathbf{G}} \begin{bmatrix} w_1 \\ w_3 \end{bmatrix} \tag{8.8.3}$$

The measurement equation is obvious from Fig. 8.2, and in matrix form it is:

$$\begin{bmatrix} z_1 \\ z_2 \end{bmatrix}_k = \underbrace{\begin{bmatrix} 1 & 0 & 1 \\ 0 & 0 & 1 \end{bmatrix}}_{\mathbf{H}_k} \begin{bmatrix} x_1 \\ x_2 \\ x_3 \end{bmatrix}_k + \begin{bmatrix} 0 \\ n_2 \end{bmatrix}_k \tag{8.8.4}$$

where the subscript k indicates discrete-time samples. As before in the error model, the sampling is done at the Nyquist rate. Therefore, the filter Δt interval is

$$\Delta t = \pi/20 \quad (\text{for}\, \omega_0 = 1\, \text{and}\, \omega_c = 20\omega_0)$$

The filter $\mathbf{R}_k$ parameter is a two-tuple in the total model and is:

$$\mathbf{R}_k = \begin{bmatrix} 0^+ & 0 \\ 0 & 100 \end{bmatrix} \mathrm{m}^2$$

where 0^+ means a very small positive number introduced for numerical stability (for example, 1e-8). The $\boldsymbol{\phi}_k$ and $\mathbf{Q}_k$ parameters are determined numerically using the Van Loan method as before. But, before doing so, the σ and β for the s(t) process must be specified (temporarily, at least). Say we choose $\sigma = 10$m and $\beta = 1/(5\Delta t)$ initially, which makes $y(t)$ similar to (but not the same as) the n_1(t) and n_2(t) processes. Finally, we note that $\mathbf{W}$ in the Van Loan method is a 2×2 identity matrix because we have accounted for the noise scale factors in the $\mathbf{G}$ matrix.

We are now ready to run the three-state total filter except for specifying the initial $\mathbf{P}^-$ matrix. Using the same reasoning as before with the error-model filter, we will let the initial $\mathbf{P}^-$ be:

$$\mathbf{P}_0^- = \begin{bmatrix} 100 & 0 & 0 \\ 0 & 100 & 0 \\ 0 & 0 & \sigma^2 \end{bmatrix} \mathrm{m}^2$$

We are assuming here that we have no prior knowledge of the states initially, other than the process model assumptions.

Now run the total filter error covariance equations for about 40 steps. The filter will reach a steady-state condition, and the error covariance for the x_3 estimate works out to be (with σ set at 10)

$$\mathbf{P}(3,3) = 17.53\,\mathrm{m}^2$$

Note that this is somewhat better than the 21.47 value that was found for the error-state complementary filter. This is to be expected because the present filter is optimal with the assumption that $\sigma = 10$ m; i.e., there is no go-free constraint on the x_3 variable at this point. This filter with $\sigma = 10$ m takes advantage of this extra information and gives better results than the complementary filter with the go-free constraint. It is important to remember that the 17.53 m^2 result for the mean-square error is very much dependent on the accuracy of the assumed Markov model for $y(t)$ with σ set at 10 m.

Now to get to the total complementary filter, we return to the "Boosted Q" idea that was presented earlier in the Euler/Goad example of Section 8.3. In the present example the x_3 signal state is completely decoupled from the x_1 and x_2 states, so the value assigned to σ only affects the 3,3 term of $\mathbf{Q}_k$. Therefore, in this case all we have to do to boost the go-free $\mathbf{Q}$ term is to increase the σ parameter. To demonstrate this, try doubling σ to 20, and then rerun the error covariance program for 40 steps and observe the $\mathbf{P}(3, 3)$ term. Note that it increases. Then double σ again to 40 and repeat the run. Then keep doubling σ and making successive runs until the $P(3, 3)$ term approaches (asymptotically) a limiting value. The results of this "successive doubling" operation are shown in Table 8.2.

Note that the limiting value appears to be about 21.47 m^2, and this occurs when sigma is increased to 640 m. This limiting mean-square error (i.e., 21.47 m^2) is the same as the mean-square error that was obtained with the error-state model in Section 8.7.

Table 8.2 Results of successive increases in the signal sigma variable

Signal sigma variable (σ)	Mean Square Error of $y(t)$ Estimate (m^2)
10	17.53
20	20.32
40	21.17
80	21.39
160	21.45
320	21.46
640	21.47
1280	21.47
2560	21.47

Thus, we see that the optimal total filter morphs into the go-free (of $y(t)$) complementary filter as we increase the assumed σ associated with x_3 to an outrageously large value. The total complementary filter always involves more states than the corresponding error-states filter implementation, but boosting the Q is easy to do, and this provides an alternative way of accomplishing the desired go-free condition.

8.9 GO-FREE MONTE CARLO SIMULATION

This example is intended to illustrate the difficulty that the Kalman filter encounters in estimating a deterministic-type time function, and it also demonstrates how the go-free constraint can be used to mitigate the difficulty. Consider a simplified one-dimensional aided inertial navigation system (INS) where the primary purpose of the INS is to provide good near-continuous outputs of vehicle position, velocity, and acceleration. This information is to be outputted at the relatively high rate, say 100 Hz. It is well known though that pure inertial systems have unstable error characteristics, so they must be aided (i.e., corrected) with some external source of position information such as GPS. The aiding can be done via a Kalman filter, and the updating can be done at a much slower rate (say 1 Hz) than the internal workings of the INS. For analysis purposes, we will also say that the filter's corrections to the INS are done on a feedforward basis, so we can avoid the complications of the extended Kalman filter*.

The primary error source in the INS will be assumed to be additive accelerometer bias, and the internal integrations that produce velocity and position will be assumed to be implemented perfectly. Also, we want the accelerometer bias error to be reasonable and bounded, so we will model it as a first-order Markov process with specified σ and β parameters. Note especially, in this simulation exercise we are modeling errors as truly random processes, but the true vehicle motion will be assumed to be deterministic. The assumed truth trajectories for acceleration, velocity, and position are shown in Fig. 8.6. Note that the true initial position and velocity are

* The setting for this tutorial one-dimensional INS updated with position updates is due to J. Farrell (3). However, the solution presented here is due to the present authors.

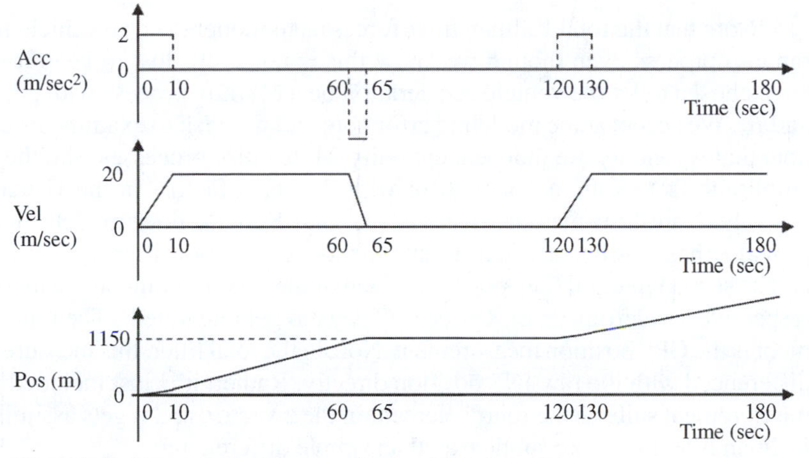

Figure 8.6 True vehicle dynamics for simulation example.

both zero and this is known to the filter. However, the accelerometer bias is a random variable, and the only thing the filter knows about it is its σ and β.

The aiding source in this exercise will be position measurements (say, from GPS), and the sampling rate will be 1 Hz. The aiding source errors will be assumed to be white.

We will consider the total Kalman filter model first and see where it leads. It is more complicated than the error-state filter, but there are lessons to be learned from it. The total filter state variables as follows:

$$\begin{aligned}
x_1 &= \text{True position} \\
x_2 &= \text{True velocity} \\
x_3 &= \text{True acceleration} \\
x_4 &= \text{Double integral of accelerometer bias error} \\
x_5 &= \text{Integral of accelerometer bias error} \\
x_6 &= \text{Accelerometer bias error (Markov process)}
\end{aligned}$$

Note that we need successive integrals of the accelerometer bias errors just as we do with the true kinematic variables. In both cases the variables are connected by differential equations, so they must be included in the state model. The state vector differential equation is then:

$$\begin{bmatrix} \dot{x}_1 \\ \dot{x}_2 \\ \dot{x}_3 \\ \dot{x}_4 \\ \dot{x}_5 \\ \dot{x}_6 \end{bmatrix} = \underbrace{\begin{bmatrix} 0 & 1 & 0 & & & \\ 0 & 0 & 1 & & \mathbf{0} & \\ 0 & 0 & -\beta_a & & & \\ & & & 0 & 1 & 0 \\ & \mathbf{0} & & 0 & 0 & 1 \\ & & & 0 & 0 & -\beta_b \end{bmatrix}}_{\mathbf{F}} \begin{bmatrix} x_1 \\ x_2 \\ x_3 \\ x_4 \\ x_5 \\ x_6 \end{bmatrix} + \underbrace{\begin{bmatrix} 0 & 0 \\ 0 & 0 \\ \sqrt{2\sigma_a^2\beta_a} & 0 \\ 0 & 0 \\ 0 & 0 \\ 0 & \sqrt{2\sigma_b^2\beta_b} \end{bmatrix}}_{\mathbf{G}} \begin{bmatrix} w_a \\ w_b \end{bmatrix} \tag{8.9.1}$$

Note that the total Kalman filter forces us to model the true vehicle motion as a random process even though we know this is not truth. Just to keep it simple, we have chosen to let the vehicle acceleration be a Markov process with parameters σ_a and β_a. We accept some modeling error here and we wish to examine its effect. Also note that w_a and w_b are independent unity white noise processes, and the respective amplitude factors are accounted for with the scale factors in the **G** matrix.

There are four measurements in the total Kalman filter model. First, the INS provides three raw uncorrected measurements: acceleration, its integral (i.e., velocity), and its second integral (i.e., position). Each of these is contaminated with the additive respective accelerometer bias effects. (They too get integrated.) The fourth measurement is the GPS position measurement. Note in the total filter, this measurement is not differenced with the raw INS position directly. Rather, it is just included in the total measurement suite as the fourth element in the **z** vector, and it gets assimilated in the Kalman filter in a more subtle way than simple differencing.

The total Kalman filter operates at a 1 Hz rate using all four measurements as a block. The intervening 99 INS measurements in between filter updates are not used directly by the filter. They are assimilated internally in the INS operating in the pure inertial mode. The total Kalman filter measurement model is then:

$$\begin{bmatrix} z_1 \\ z_2 \\ z_3 \\ z_4 \end{bmatrix} = \underbrace{\begin{bmatrix} 1 & 0 & 0 & 1 & 0 & 0 \\ 0 & 1 & 0 & 0 & 1 & 0 \\ 0 & 0 & 1 & 0 & 0 & 1 \\ 1 & 0 & 0 & 0 & 0 & 0 \end{bmatrix}}_{\mathbf{H}} \begin{bmatrix} x_1 \\ x_2 \\ x_3 \\ x_4 \\ x_5 \\ x_6 \end{bmatrix} + \underbrace{\begin{bmatrix} v_1 \\ v_2 \\ v_3 \\ v_{GPS} \end{bmatrix}}_{\mathbf{v}} \tag{8.9.2}$$

$$\mathbf{R} = \begin{bmatrix} 0^+ & 0 & 0 & 0 \\ 0 & 0^+ & 0 & 0 \\ 0 & 0 & 0^+ & 0 \\ 0 & 0 & 0 & R_{GPS} \end{bmatrix}$$

where 0^+ means a very small number (like 1E-8) inserted for numerical stability. We now wish to do a Monte Carlo simulation and inspect the system errors for various choices of parameters. To get the simulation exercise started, we will use the following set of parameters:

$$\left.\begin{aligned} \sigma_a &= 0.10\,\text{m/s}^2 \\ \beta_a &= 1/3\,\text{s}^{-1} \end{aligned}\right\}\text{Markov parameters for the modeled vehicle acceleration}$$

$$\left.\begin{aligned} \sigma_b &= 0.015\,\text{m/s}^2 \\ \beta_b &= 1/60\,\text{s}^{-1} \end{aligned}\right\}\text{Markov parameters for the accelerometer bias}^*$$

$$\mathbf{R}_{GPS} = 25\,\text{m}^2 \qquad \text{(Mean-square GPS measurement error)}$$

*More representative of a tactical grade INS than a precision navigation grade system.

We will also assume that the INS integrators are zeroed at $t = 0$, and assume that the first filter measurement occurs at $t = 1$.

The six-state total Kalman filter model is now complete. We begin by assuming that the vehicle's assumed acceleration sigma is relatively small (e.g., 0.1 m/s^2). We are saying here that most of the time the vehicle has very mild dynamics. This is a reasonable assumption for the "coast" periods in the true dynamics. However, this Markov model will be considerably in error during the short bursts of acceleration in the specified true dynamics (see Fig. 8.6). Thus, this is where we would expect to see the most pronounced mismodeling effect. Also, the GPS position updates dominate the position error, but not so much so in velocity. Thus, it will be the velocity error where we would expect to see the most mismodeling effect.

The total Kalman filter is easily programmed in MATLAB using simulated noisy measurements. The resulting velocity error for a typical run is shown in Fig. 8.7. It is clear from the plot that the velocity error becomes large every time there is an abrupt change in acceleration (which, of course, is not properly accounted for in the model). In effect, the vehicle dynamics "bleeds through" into the total filter's velocity error.

Along with the total filter's error plot in Fig. 8.7 is a plot of velocity estimation error for the corresponding complementary filter driven by the same measurement used for the total filter. Actually, the plot was obtained by boosting the **Q** associated with the assumed vehicle dynamics. This was done by increasing σ_a from 0.1 m/sec^2 to an outrageously large value of 1000 m/sec^2. As mentioned previously in Section. 8.8, this accomplishes the complementary constraint just as if we would have implemented the error-state filter initially. Note that with the complementary constraint imposed, there is no noticeable effect of the vehicle dynamics "bleeding through" into the velocity estimation error. This is the principal advantage of the complementary filter.

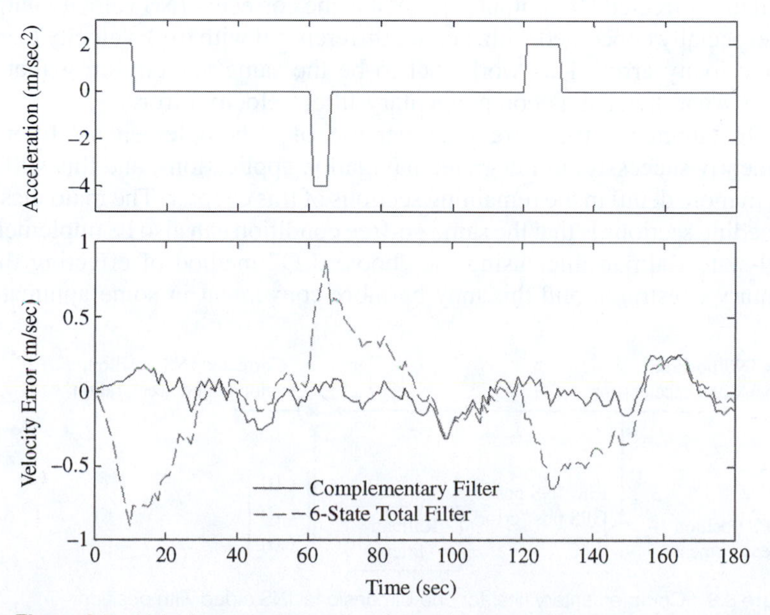

Figure 8.7 Velocity errors for total and complementary filters.

The actual three-state error-state Kalman filter can also be easily implemented using MATLAB. The model parameters are as follows:

States:

x_1 = Pure uncorrected INS position error

x_2 = Pure uncorrected INS velocity error

x_3 = Pure uncorrected INS acceleration error (Markov bias)

State Model Differential Equations:

$$\begin{bmatrix} \dot{x}_1 \\ \dot{x}_2 \\ \dot{x}_3 \end{bmatrix} = \underbrace{\begin{bmatrix} 0 & 1 & 0 \\ 0 & 0 & 1 \\ 0 & 0 & -\beta_a \end{bmatrix}}_{\mathbf{F}} \begin{bmatrix} x_1 \\ x_2 \\ x_3 \end{bmatrix} + \underbrace{\begin{bmatrix} 0 \\ 0 \\ \sqrt{2\sigma_a^2 \beta_a} \end{bmatrix}}_{\mathbf{G}} w \tag{8.9.3}$$

Measurement Equations:

$$z = \underbrace{\begin{bmatrix} 1 & 0 & 0 \end{bmatrix}}_{\mathbf{H}} \begin{bmatrix} x_1 \\ x_2 \\ x_3 \end{bmatrix} + \underbrace{v_{GPS}}_{\mathbf{v}}$$

$$\mathbf{R} = E(v_{GPS})^2 \tag{8.9.4}$$

The feedforward block diagram in Fig. 8.8 shows the implementation of a complementary filter. In this configuration the three-tuple output estimates from the filter are subtracted from the raw INS position, velocity, and acceleration outputs to yield the corrected INS outputs. Then, it is the corrected INS velocity output that we are especially concerned with, and we difference it with truth velocity to get the final INS velocity error. This works out to be the same velocity error that we see in Fig. 8.7 and labeled as complementary filter velocity error.

In summary, the error-state version of a complementary filter has been eminently successful in integrated navigation applications, and this will be continued in more detail in the remaining sections of this chapter. The main message in the preceding sections is that the same go-free condition can also be implemented with a total-state Kalman filter using the "boosted Q" method of effecting the complementary constraint, and this may be more convenient in some applications.

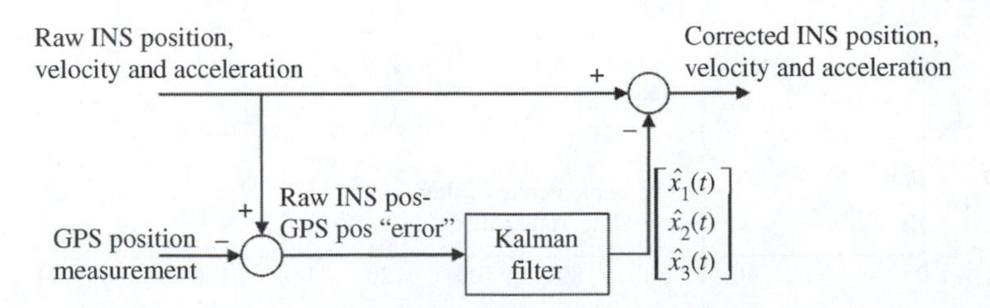

Figure 8.8 Complementary filter for one-dimensional INS aided with position updates.

8.10 INS ERROR MODELS

An INS is made up of gyroscopes (gyros, for short) and accelerometers for basic sensors. A gyro senses rotational rate (angular velocity) that mathematically integrates to give overall change in attitude over time. Similarly, an accelerometer senses linear acceleration that integrates to give velocity change, or doubly integrates to give position change over time. An INS sustains attitude, position, and velocity accuracy by accurately maintaining changes in those parameters from their initial conditions. However, due to the integration process, errors in the attitude, position, and velocity data are inherently unstable but the growth characteristics of these errors depend on the type of sensors used. The level of complexity needed for the error modeling depends on the mix of sensors in the integration and the performance expected of it.

In our previous discussion of complementary filtering, we only considered linear models for both the "total" and "error-state" filters. However, in real-life aided inertial systems we often encounter nonlinear measurements. Coping with such situations was discussed at some length in Chapter 7 where we presented the concept of an extended Kalman filter. Here we linearize about some nominal trajectory in state space, and then the filter estimates the perturbations from the reference trajectory. In aided inertial systems the INS usually provides the reference trajectory, and its deviation from truth forms the states of the Kalman filter. The filter then estimates these states (based on the aiding measurements and the INS error model). Then the estimated INS errors are subtracted from the raw INS outputs to yield the best estimates of position, velocity, and attitude. Note that this mode of operation is exactly the same as the error-state complementary filter discussed previously in Section. 8.5. Thus, this accomplishes the go-free condition that was presented earlier in the chapter.

Before proceeding to the aided INS/DME example, we need to develop some INS error models. The ones presented here are basic "no frills" models, but yet they are still reasonably accurate and are useful in a variety of terrestrial applications.

Single-Axis Inertial Error Model

We shall begin by looking at a simple model that contains the physical relationship between the gyro and the accelerometer in one axis. The following notation will be used:

$$
\begin{aligned}
\Delta x &= \text{position error} \\
\Delta \dot{x} &= \text{velocity error} \\
\Delta \ddot{x} &= \text{acceleration error} \\
\phi &= \text{platform (or attitude) error relative to level} \\
g &= \text{gravitational acceleration} \\
R_e &= \text{earth radius} \\
a &= \text{accelerometer noise} \\
\varepsilon &= \text{gyro noise in terms at angular rate}
\end{aligned}
$$

The single-axis model is instructive for the relationship it describes between the accelerometer sensor and the gyro sensor. Both sense inertial quantities, one linear

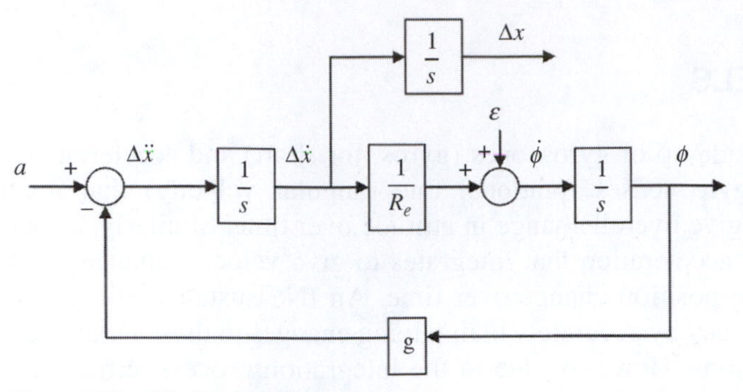

Figure 8.9 Single-axis INS error model.

acceleration and the other angular velocity. The differential equations that describe the accelerometer and the gyro errors are given as follows:

$$\Delta\ddot{x} = a - g\phi \tag{8.10.1}$$

$$\dot{\phi} = \frac{1}{R_e}\Delta\dot{x} + \varepsilon \tag{8.10.2}$$

(The block diagram for these equations is shown in Fig. 8.9.)

In Eq. (8.10.1), the error in acceleration is fundamentally due to a combination of accelerometer sensor noise and a component of gravity the accelerometer senses as a result of platform error. The platform error rate, described by Eq. (8.10.2), results from gyro sensor noise and velocity error that, when projected along the earth's surface curvature, gets translated into an angular velocity error. An accelerometer error that integrates into a velocity error gives rise to a misalignment in the perceived gravity vector due to the earth's curved surface. This misalignment results in a horizontal component that fortuitously works against the effects of the initial accelerometer error. The resulting oscillation known as the Schuler oscillation provides some stability to the horizontal errors. Note that g is assumed to be the usual earth's gravitational constant. Thus, this simple model is restricted to low dynamics.

Three-Axes Inertial Error Model

In progressing from a single-axis to a level-platform three-axes INS additional complexities arise from interaction among the three sensor pairs (4, 5). A sensor pair aligned in the north-south direction is shown in Fig. 8.10 as a transfer function block diagram denoted as the north channel. A very similar model exists for the east channel as shown in Fig. 8.11. Just as with the single-axis error model, the three-axes model is restricted to low dynamics. (See Problem 8.2 for more on this.)

The differential equations that accompany the transfer functions of Figs. 8.10 and 8.11 are given below. In the following notation, the platform rotation rate ω is an angular velocity. Bear in mind, however, that ω is not the same as the platform tilt rate error $\dot{\phi}$, which is an angular velocity *error*.

North channel:

$$\Delta\ddot{y} = a_y - g(-\phi_x) \tag{8.10.3}$$

$$-\dot{\phi}_x = \frac{1}{R_e}\Delta\dot{y} + \omega_y\phi_z - \varepsilon_x \tag{8.10.4}$$

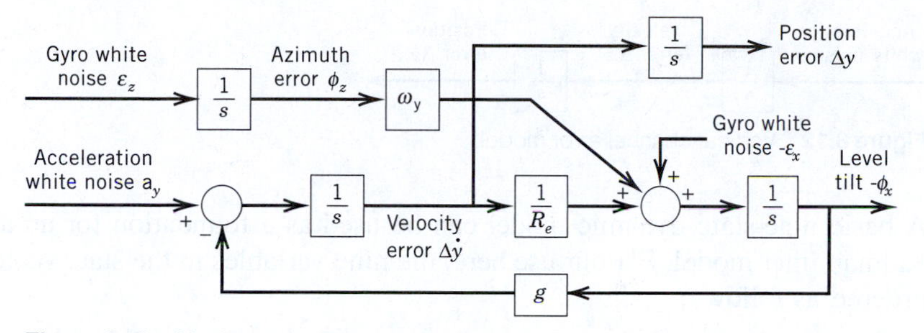

Figure 8.10 North channel error model (x is east, y is north, z is up, and ω_y= platfrom angular rate about y-axis).

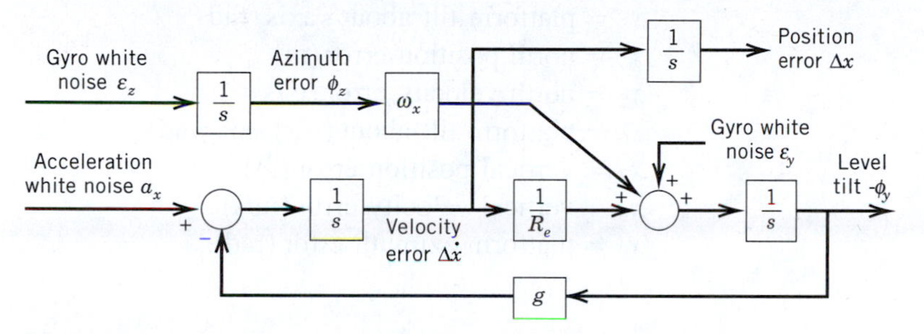

Figure 8.11 East channel error model, (x is east, y is north, z is up, and ω_x= platform angular rate about x-axis).

East channel:

$$\Delta\ddot{x} = a_x - g\phi_y \tag{8.10.5}$$

$$\dot{\phi}_y = \frac{1}{R_e}\Delta\dot{x} + \omega_x\phi_z + \varepsilon_y \tag{8.10.6}$$

Vertical channel:

$$\Delta\ddot{z} = a_z \tag{8.10.7}$$

Platform azimuth:

$$\dot{\phi}_z = \varepsilon_z \tag{8.10.8}$$

The north and east channel models take into account the previously described phenomenon that is due to the earth curvature. The vertical channel does not benefit from the Schuler phenomenon and is governed by a simpler model as shown in Fig. 8.12.*

* It can be shown that the characteristic poles for the vertical channel do not lie exactly at the origin in the s-plane. They are actually symmetrically located on the real axis, one slightly to the left of the origin and the other to the right (3). When the vertical error is observable, it is a good approximation in the Kalman filter model to assume that both poles are coincident at the origin.

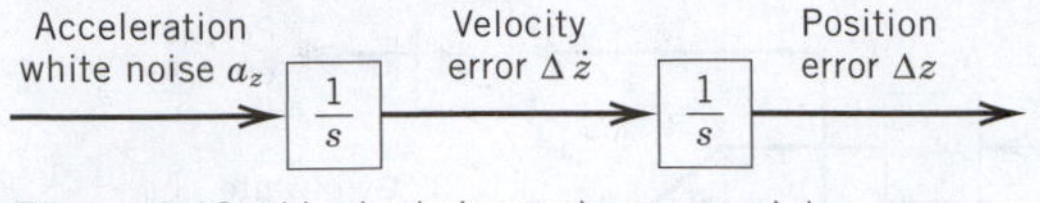

Figure 8.12 Vertical channel error model.

A basic nine-state dynamic model can be used as a foundation for an aided INS Kalman filter model. For our use here, the nine variables in the state vector will be ordered as follows:

$$
\begin{aligned}
x_1 &= \text{east position error (m)} \\
x_2 &= \text{east velocity error (m/s)} \\
x_3 &= \text{platform tilt about } y \text{ axis (rad)} \\
x_4 &= \text{north position error (m)} \\
x_5 &= \text{north velocity error (m/s)} \\
x_6 &= \text{platform tilt about } (-x) \text{ axis (rad)} \\
x_7 &= \text{vertical position error (m)} \\
x_8 &= \text{vertical velocity error (m/s)} \\
x_9 &= \text{platform azimuth error (rad)}
\end{aligned}
\tag{8.10.9}
$$

Based on Eqs. (8.10.3) through (8.10.8), we can write out the nine-dimensional vector first-order differential equation:

$$
\begin{bmatrix} \dot{x}_1 \\ \dot{x}_2 \\ \dot{x}_3 \\ \dot{x}_4 \\ \dot{x}_5 \\ \dot{x}_6 \\ \dot{x}_7 \\ \dot{x}_8 \\ \dot{x}_9 \end{bmatrix}
= \underbrace{\begin{bmatrix}
0 & 1 & 0 & 0 & 0 & 0 & 0 & 0 & 0 \\
0 & 0 & -g & 0 & 0 & 0 & 0 & 0 & 0 \\
0 & \dfrac{1}{R_e} & 0 & 0 & 0 & 0 & 0 & 0 & \omega_x \\
0 & 0 & 0 & 0 & 1 & 0 & 0 & 0 & 0 \\
0 & 0 & 0 & 0 & 0 & -g & 0 & 0 & 0 \\
0 & 0 & 0 & 0 & \dfrac{1}{R_e} & 0 & 0 & 0 & \omega_y \\
0 & 0 & 0 & 0 & 0 & 0 & 0 & 1 & 0 \\
0 & 0 & 0 & 0 & 0 & 0 & 0 & 0 & 0 \\
0 & 0 & 0 & 0 & 0 & 0 & 0 & 0 & 0
\end{bmatrix}}_{\mathbf{F}_{\text{INS}}}
\begin{bmatrix} x_1 \\ x_2 \\ x_3 \\ x_4 \\ x_5 \\ x_6 \\ x_7 \\ x_8 \\ x_9 \end{bmatrix}
+ \begin{bmatrix} 0 \\ u_{ax} \\ u_{gx} \\ 0 \\ u_{ay} \\ u_{gy} \\ 0 \\ u_{az} \\ u_{gz} \end{bmatrix}
\tag{8.10.10}
$$

From the parameters of Eq. (8.10.10), the discrete-time nine-dimensional vector first-order difference equation for the process model can be derived. Closed-form solutions for the process noise covariance matrix $\mathbf{Q}$ or the state transition matrix $\boldsymbol{\phi}_{\text{INS}}$ are not easily derived. These parameters are usually computed using the Van Loan method as discussed in Chapter 3.

The INS process model has nine states in total comprising a position, a velocity, and a platform tilt state in each of three dimensions. This is a minimal system that accounts for platform misorientation, but only allows for very little complexity in errors associated with the accelerometers and gyros. In other words, the instrument errors are grossly simplified and modeled as white noise forcing functions that drive the INS error dynamics. Also, platform angular errors are assumed to be small, and platform torquing rates are assumed to change very slowly such that they can be treated as constants. Small acceleration (relative to 1-g) is also assumed in the 9-state model given by Eq. (8.10.10). Even though the model is relatively simple, it is nonetheless workable and can be found in use in some real-life systems.

8.11 AIDING WITH POSITIONING MEASUREMENTS—INS/DME MEASUREMENT MODEL

We shall consider here an integrated navigation system that is updated with positioning measurements from a positioning system. One of the few terrestrial positioning systems still in operation today is the DME (distance-measuring equipment) system, mentioned previously in Chapter 7. The DME is a two-dimensional (horizontal) ranging system of signals traveling to and from ground-based transmitters.

Positioning systems provide position updates to an aided inertial system that suppresses position error growth to levels commensurate with the uncertainty of the position updates. Velocity error growth is naturally suppressed in the presence of position updates. The one thing that most positioning systems do not provide directly is orientation updates (of attitude and heading). In aided inertial systems, the coupling between position (and/or velocity) updates to its platform tilt states is a loose and indirect one, but nevertheless exists. More of this will be covered in Chapter 9.

INS/DME Measurement Model

The linearization of a DME measurement was discussed in Section 7.1 (Example 7.1). There, the direct slant range from the aircraft to the DME ground station was considered to be the same as horizontal range. This approximation that can be in error by a few percent for short to medium ranges is usually absorbed into the measurement noise component of the model. Once the horizontal range is obtained, it is then compared with the predicted horizontal range based on the INS position output, and the difference becomes the measurement input to the Kalman filter. The difference quantity has a linear connection to Δx and Δy as discussed in Example 7.1. Based on the same ordering of the state vector as in Eq. (8.10.9), the rows of the $\mathbf{H}_k$ matrix corresponding to the two DME stations would then be

$$\mathbf{z}_k = \begin{bmatrix} -\sin\alpha_1 & 0 & 0 & -\cos\alpha_1 & 0 & 0 & 0 & 0 & 0 \\ -\sin\alpha_2 & 0 & 0 & -\cos\alpha_2 & 0 & 0 & 0 & 0 & 0 \end{bmatrix} \mathbf{x}_k + \mathbf{v}_k \tag{8.11.1}$$

where we have written the direction cosines in terms of the bearing angle to the station α rather than θ as used in Eq. (7.1.18) (see Fig. 7.2). Note that the bearing angle is the usual clockwise angle with respect to true north, and the x-axis is assumed to be east. It is assumed, of course, that the coordinates of the DME station being interrogated are known and that the aircraft's position is known approximately from the INS position output. Thus, $\sin\alpha$ and $\cos\alpha$ are computable on-line to a first-order approximation. Range measurements from more than one DME station could be made either sequentially or simultaneously. The Kalman filter can easily accommodate to either situation by setting up appropriate rows in the $\mathbf{H}_k$ matrix corresponding to whatever measurements happen to be available.

EXAMPLE 8.1 SIMULATION OF STANDALONE AND INTEGRATED INS AND DME

For the following simulation exercise, we shall look at the performance of three different systems: (a) an integrated INS/DME system, (b) an INS-only system with initial alignment, (c) a DME-only system. The nominal aircraft motion and the DME station locations are as shown in the figure accompanying Problem 7.5 in Chapter 7. For the INS/DME system, we shall use the basic nine-state process model described in Eq. (8.10.10) and choose the following parameters for it:

Δt step size $= 1$ sec
Accelerometer white noise spectral density $= 0.0036\,(\text{m/s}^2)^2/\text{Hz}$
Gyro white noise spectral density $= 2.35\,(10^{-11})\,(\text{rad/s})^2/\text{Hz}$
$R_e = 6{,}380{,}000$ m
y-axis angular velocity $\omega_y = 0.0000727$ rad/s (earth rate at the equator)
X-axis angular velocity $\omega_x = -\dfrac{100\text{ m/s}}{R_e} = -0.0000157$ rad/sec
Initial position variance $= 10\text{ m}^2$
Initial velocity variance $= (0.001\text{ m/s})^2$
Initial attitude variance $= (0.001\text{ rad})^2$
DME white measurement error $= (15\text{ m})^2$

The INS-only system uses the same inertial sensor parameters and the same initial alignment conditions. The main difference between the INS-only system and the integrated INS/DME system is that the DME measurements are never processed, whereas the INS errors are allowed to propagate according to the natural dynamics modeled.

For the DME-only system, the aircraft motion is modeled as a random walk (in both x- and y-positions) superimposed on constant-velocity motion in the y-direction. The filter in this case is a simple two-state Kalman filter linearized about the nominal trajectory. The following parameters are used:

Δt step size $= 1$ sec
Process noise variance (in each position axis) $= 400\text{ m}^2$
Initial position variance $= 10\text{ m}^2$
DME white measurement error $= (15\text{ m})^2$

In the 200-sec run, the aircraft starts out at the location (0, −10,000) and flies north at 100 m/sec and nominally ends up at the location (0, 10,000). Fig. 8.13 shows a comparison of the standard deviations of the position error for all three systems described above, for the east component (a) and the north component (b). Although the INS position error grows without bound, the position errors are stable and smaller for the integrated INS/DME system. The cresting of the north position error near the 100-sec time mark is due to the poor DME observability in the north

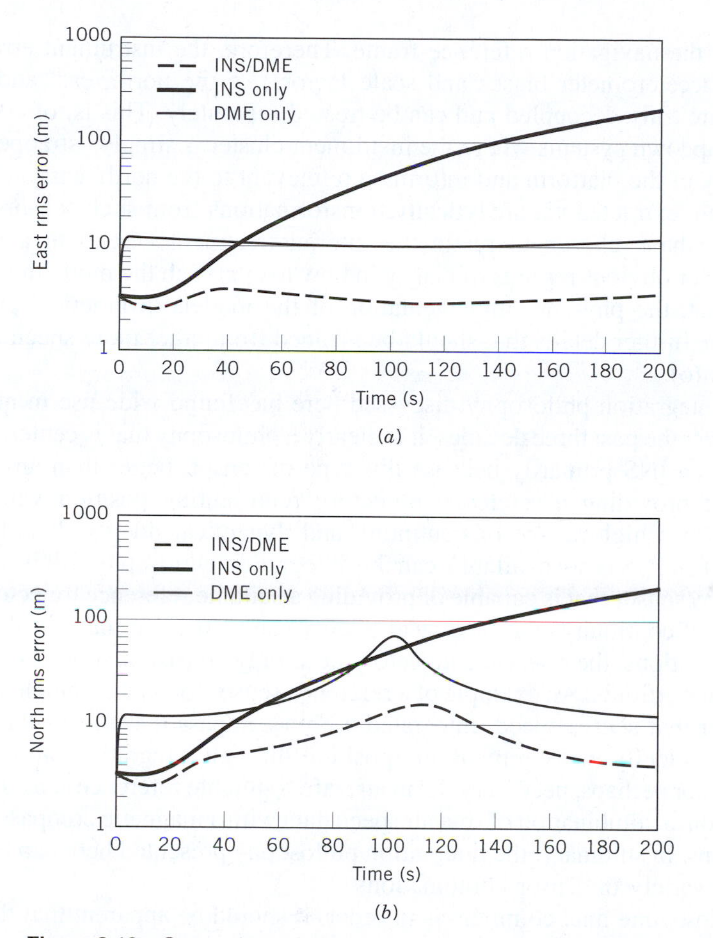

Figure 8.13 Comparison of various combinations between INS and DME sensors showing rms error time profiles for (*a*) the east component and (*b*) north component.

direction when the aircraft crosses the x-axis (see Problem 7.5). The position error for the DME-only system has characteristics similar but much larger in magnitude by comparison to those of the integrated INS/DME position error. (Note that the vertical scale is logarithmic, so the difference is larger than it appears at first glance.)

8.12 OTHER INTEGRATION CONSIDERATIONS AND CONCLUDING REMARKS

To put our treatment of the Aided Inertial Navigation system here into perspective, we should point out that gimbaled inertial systems are generally a thing of the past while most inertial systems today belong to the strapdown variety. In gimbaled systems, the three dimensions of the instrument cluster are mechanically kept fixed relative to the navigation reference frame. Therefore, the instrument errors (e.g., gyro and accelerometer biases and scale factors) of the north, east and vertical channels are truly decoupled and can be treated separately. This is, of course, not true in strapdown systems where the instrument cluster is simply "strapped down" to the body of the platform and information relevant to the north, east and vertical channels are extracted via analytically transformations from such sensors. Nevertheless, we have chosen to retain the tutorial treatment here with a gimbaled viewpoint for obvious reasons of clarity in how to approach the modeling problem. Beyond that, the physical implementation of the models in practical processing systems are further details that should be gleaned from other more specialized text references (6).

The integration philosophy discussed here has found wide use in navigation systems over the past three decades, it is clearly a philosophy that is centered around the use of an INS primarily because this type of sensor, better than any other, is capable of providing a reference trajectory representing position velocity and attitude with a high degree of continuity and dynamical fidelity. It is logical to then ask: If an INS is not available, can this integration philosophy still be useful? In general, any sensor that is capable of providing a suitable reference trajectory with a high level of continuity and dynamical fidelity can be used in place of the INS. In some applications, the reference trajectory need only consist of subsets of position, velocity, and attitude. An example of a reference sensor for a land vehicle is a wheel tachometer that senses speed, integrated with a compass that senses direction, to produce a velocity measurement and position through integration; attitude is not available, nor perhaps, necessary. In an aircraft, a suitable reference sensor might be derived from a combination of true air speed data with a magnetic compass for some applications. In summary, the integration philosophy presented here can be applied to a wide variety of sensor combinations.

To close, one final comment is in order. It should be apparent that the system integration philosophy presented in this chapter is not the only way of integrating inertial measurements with noninertial data. One only has to peruse navigation conference proceedings over the years to find countless integration schemes, each with a little different twist because of some special circumstance or choice. The scheme presented here is optimal though (within the constraint of dynamic exactness in the use of inertial measurements), and it represents the best way of managing the system integration theoretically. However, the systems engineer often does not have the luxury of designing a truly optimal system, and must at times settle for something less because of equipment constraints, cost, and so forth. Even so, the optimal methodology presented here is still valuable for analysis purposes, because it provides a lower bound on system errors for a given mix of sensors. That is, the optimal system serves as a "yardstick" that can be used for comparison purposes in evaluating the degree of suboptimality of various other candidate integration schemes.

PROBLEMS

8.1 The block diagram shown in Fig. P8.1 takes us back to the pre-digital age. It was taken from the well-known Avionics Navigation Systems book by Kayton and Fried (6), and it describes a means of blending together barometrically-derived

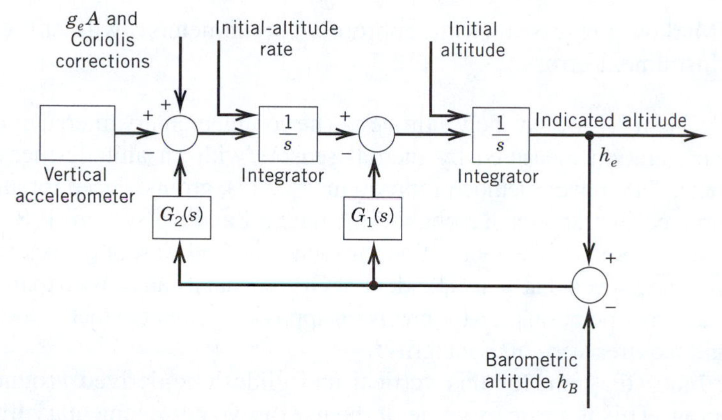

Figure P8.1

and inertially-derived altitude signals in such a way as to take advantage of the best properties of both measurements. The baro-altitude signal has fairly good accuracy in steady-state conditions, but it has sluggish response to sudden changes. The inertially-derived signal is just the opposite. Its response to sudden change is virtually instantaneous, but the steady-state error is unstable because of the double integration required in going from acceleration to position.

(a) Assume that the $G_1(s)$ and $G_2(s)$ in the diagram are simply constant gains, and that the g, A, and Coriolis corrections can be lumped together as a calculated "bias" term that makes the accelerometer output a reasonably accurate measure of vertical acceleration. Then show that the implementation shown in the figure does in fact cause the final system error to be independent of the true vertical dynamics. That is, this is truly a complementary filter implementation. Or, in terms introduced in Chapter 8, this is a "go free" implementation where the go-free variable is vertical dynamical motion. The G_1 and G_2 gain parameters can be adjusted to yield the best compromise among the contributing error sources.

[*Note:* In order to show the complementary-filter property, you must conceptually think of directly integrating the accelerometer signal twice ahead of the summer to obtain an inertially derived altitude signal contaminated with error. However, direct integration of the accelerometer output is not required in the final implementation because of cancellation of s's in numerator and denominator of the transfer function. This will be apparent alter carrying through the details of the problem. Also note that the g_e, A, and Coriolis corrections indicated in the figure are simply the gravity, accelerometer-bias, and Coriolis corrections required in order that the accelerometer output be $\ddot{h}$ (as best possible). Also, the initial conditions indicated in the figure may be ignored because the system is stable and we are interested in only the steady-state condition here.]

(b) Next, bring the analog implementation in the figure into today's digital world and develop a go-free Kalman filter model for this application. In order to keep it simple, use the boosted Q method discussed in Chapter 8. The state variables will then be true total vertical position, velocity and acceleration, and the three instrument measurement errors. First-order

Markov processes (with appropriate parameters) will suffice for the instrument errors.

8.2 In Problem 8.1, we saw how altitude can be complementary filtered by combining vertical acceleration measured by inertial sensors with an altitude measured by a baroaltimeter. In a novel method proposed in 2002 (7), ground speed for an aircraft is derived on precision approach along an Instrument Landing System (ILS) glideslope based on vertical speed. The motivation for such a method was, of course, to provide a source of ground speed that is independent of GPS so that it can be used to monitor GPS speed for a flight operation (i.e., for precision approach guidance) that is deemed safety critical and requires very high integrity.

The block diagram for this vertical and glideslope-derived ground speed is shown below. This is a rare example of the use of two complementary filters in one system. The first, designated the "Altitude Complementary Filter" block, is what was described in Problem 8.1, except that the desired output is vertical speed and not altitude. The *a priori* knowledge of the glidepath and the measured glideslope deviations combine to provide the dynamic projection vector that transforms instantaneous vertical speed to instantaneous ground speed. However, the resultant ground speed becomes very noisy because of the scale magnification of the transformation. So a second "Ground Speed Complementary Filter" block combines horizontal acceleration from inertial sensors with the noisy glideslope-derived ground speed to ultimately produce a smoothed ground speed.

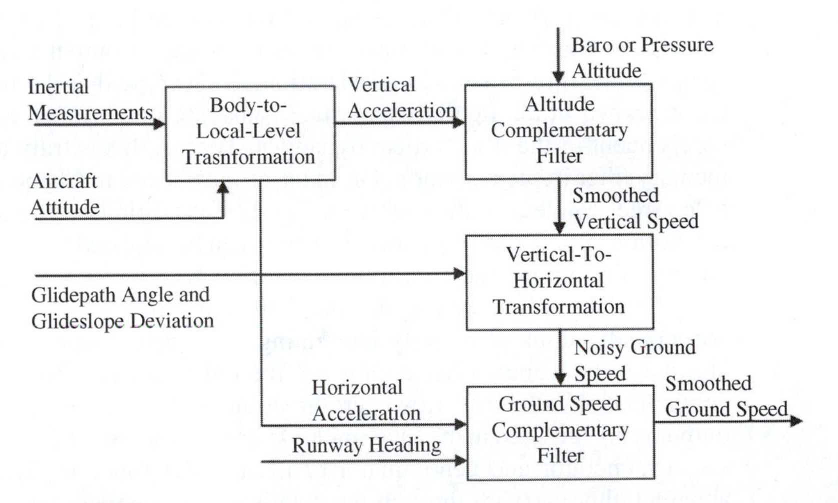

Figure P8.2

(a) Write out the transfer function for the first complementary filter that produces the smoothed vertical speed. Its inputs will be vertical acceleration and an altitude (from baro or pressure).

(b) Assuming the glidepath angle to be nominally three degrees, and that the pilot is able to steer the aircraft right down the glideslope with zero deviation, write out the transfer function for the second complementary filter that combines the noisy ground speed derived from the glidepath transformation and the horizontal acceleration.

8.3 Of the many advances made to improve the accuracy of GPS, one of the more important and yet relatively simple idea has been something called *carrier smoothing* (of the code measurements) that involves the use of a complementary filter.

As with any complementary filter, there are at least two measurements of the same quantity, in this case the signal dynamics, which we will call ψ. The code phase (or pseudorange) measurement ρ has a direct connection to ψ but it has a somewhat high level of noise η_ρ:

$$\rho = \psi_\rho + \eta_\rho \tag{P8.3.1}$$

The second measurement is a carrier phase measurement ϕ also has a connection to ψ and it has a much lower noise level η_ϕ but has an initial unknown bias N:

$$\phi = \psi_\phi + N + \eta_\phi \tag{P8.3.2}$$

(a) Assuming that $\psi_\rho \cong \psi_\phi$, draw up a block diagram for a complementary filter that will process both code phase and carrier phase measurements to produce a low-noise estimate of the code phase $\hat{\rho}$. Write out the transfer function (in terms of s) for the complementary filter.

(b) The assumption of $\psi_\rho \cong \psi_\phi$, however, is not quite true because one of its constituent components, the effect of the ionosphere on the signal, has opposite signs between the two types of measurements, i.e., $\psi_\rho = \psi + \iota$, and $\psi_\phi = \psi - \iota$. Show that the ionospheric term ι does not get eliminated so that the complementary filter is limited by the dynamics of ι.

(c) With GPS, we can also get another pair of code phase and carrier phase measurements from a given satellite at a different frequency. Suppose now that we have a set of four measurements written as follows (α is a multiplier that represents a scaling change in the size of the iono component when measured at a different frequency):

$$\begin{aligned} \rho_1 &= \psi + \iota + \eta_{\rho 1} \\ \phi_1 &= \psi - \iota + N_1 + \eta_{\phi 1} \\ \rho_2 &= \psi + \alpha\iota + \eta_{\rho 2} \\ \phi_2 &= \psi - \alpha\iota + N_2 + \eta_{\phi 2} \end{aligned} \tag{P8.3.3}$$

How can these measurements be combined for a complementary filter that fully eliminates the dynamics of ψ and ι (8).
(Hint: Note that N_1 and N_2 are biases).

8.4 The approximate INS error equations of (8.10.8) through (8.10.8) are for a slow-moving vehicle. In this model, the observability of the azimuth error ϕ_z is poor because it can only depend on earth motion (gyrocompassing). Hence, for an INS with poor gyro stability, its steady-state azimuth error can be quite large. For a faster-moving vehicle that occasionally encounters horizontal acceleration, the improved observability of ϕ_z under such conditions actually provides a substantial reduction in the error, thus momentarily stabilizing it. The INS error equations for the east and north channels (Eqs. 8.10.3 through 8.10.6.) are rewritten here with the inclusion of the lateral acceleration components A_x and A_y tie-in to the azimuth error in the horizontal acceleration error equations (additional terms are indicated with a double underscore).

East channel:

$$\begin{aligned} \Delta\ddot{x} &= a_x - g\phi_y + \underline{\underline{A_y\phi_z}} \\ \dot{\phi}_y &= \frac{1}{R_e}\Delta\dot{x} + \omega_x\phi_z + \varepsilon_y \end{aligned} \qquad \text{(P8.4.1)}$$

North channel:

$$\begin{aligned} \Delta\ddot{y} &= a_y - g(-\phi_x)\underline{\underline{-A_x\phi_z}} \\ -\dot{\phi}_x &= \frac{1}{R_e}\Delta\dot{y} + \omega_y\phi_z + \varepsilon_x \end{aligned} \qquad \text{(P8.4.2)}$$

Using the parameters for Example 8.1 in the integrated INS/DME navigation system, perform a covariance analysis to determine the time profile for the variance of the azimuth error ϕ_z for the following dynamic scenario: The nominal y-axis acceleration and velocity profiles are as shown in the accompanying figure.

The $y(t)$ profile for linearization of the **H** matrix may be approximated as constant-velocity (100 m/s) for the first 95 sec; then a constant 10-sec deceleration period; and, finally, a constant velocity of 60 m/s for the remaining 95 sec of the profile. (Note that we are not assuming this to be the *actual* flight path. This is simply the approximate reference trajectory to be used for the linearization.)

The parameter values given in Example 8.1 are to be used here, except that the gyro white noise power spectral density is to be increased to 2.35 (10^{-9}) $(\text{rad/s})^2$/Hz and the initial azimuth error is to be set at 1 degree rms. This is intended to simulate a less-stable INS as compared with the one considered in Example 8.1.

The gravity constant g and the earth radius R_e may be assumed to be 9.8 m/s^2 and 6.37 (10^6) m for this problem.

8.5 Instrument errors are often found to be simple quasibiases that wander over long time constants. These can simply be modeled' with a single-state Gauss-Markov process. Some instrument errors, however, are related in a deterministic way to the magnitude of the measured variable, the most common type being known as *scale factor* errors. We shall look at the nature of a scale factor error in combination with a bias error in this problem that involves barometric-derived altitude. Suppose that the principal relationship between the internally sensed barometric reading and the reported altitude is given by the following equations:

$$H' = b'\gamma' \qquad \text{(P8.5.1)}$$

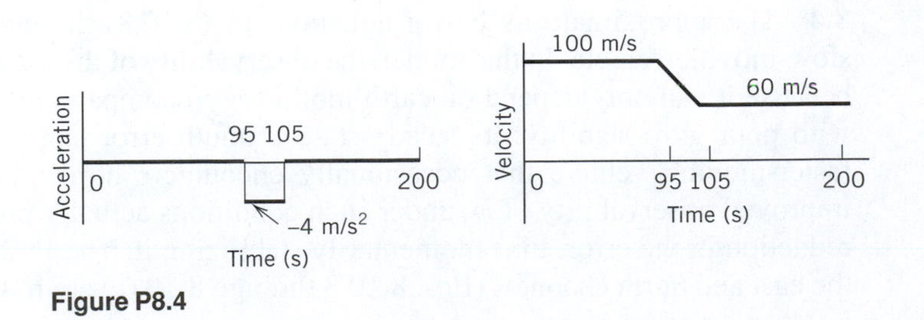

Figure P8.4

where

$H' =$ reported altitude
$b' =$ barometric reading
$\gamma' =$ barometric altitude scale factor

Consider that the barometric reading b' is made up of the correct value b plus a bias error b_ε: $b' = b + b_\varepsilon$. Consider also that the scale factor γ' is made up of the correct value γ plus an error γ_ε: $\gamma' = \gamma + \gamma_\varepsilon$.

(a) Show that we can use the 2-state measurement model shown below to account for the bias and scale factor errors (neglect second-order effects):

$$H' - H = z_k = \begin{bmatrix} 1 & H' \end{bmatrix} \begin{bmatrix} b_\varepsilon \\ \gamma_\varepsilon \end{bmatrix}_k + v_k \tag{P8.5.2}$$

(b) Suppose that the two error states are modeled as random constants:

$$\begin{bmatrix} b_\varepsilon \\ \gamma_\varepsilon \end{bmatrix}_{k+1} = \begin{bmatrix} 1 & 0 \\ 0 & 1 \end{bmatrix} \begin{bmatrix} b_\varepsilon \\ \gamma_\varepsilon \end{bmatrix}_k + \begin{bmatrix} 0 \\ 0 \end{bmatrix} \tag{P8.5.3}$$

Let $H' = 50$. Do the variances of b_ε and γ_ε go to zero in the limit? Under what condition will the variances of b_ε and γ_ε go to zero in the limit?

(c) Suppose that the two error states are modeled individually as single-state Gauss-Markov processes:

$$\begin{bmatrix} b_\varepsilon \\ \gamma_\varepsilon \end{bmatrix}_{k+1} = \begin{bmatrix} 0.9999 & 0 \\ 0 & 0.9999 \end{bmatrix} \begin{bmatrix} b_\varepsilon \\ \gamma_\varepsilon \end{bmatrix}_k + \begin{bmatrix} w_1 \\ w_2 \end{bmatrix}_k \tag{P8.5.4}$$

where

$$E(\mathbf{w}\mathbf{w}^T) = \begin{bmatrix} 2 & 0 \\ 0 & 2 \end{bmatrix} \tag{P8.5.5}$$

Let $H' = 50$. Do the variances of b_ε and γ_ε go to zero in the limit? Why?

8.6 Suppose that the integrated INS/DME situation given in Example 8.1 involves a locomotive instead of an aircraft. The locomotive is constrained to railroad tracks aligned in the north–south direction. In place of an INS, the integrated navigation system uses wheel tachometer data to provide the reference trajectory in the complementary filter arrangement. A tachometer monitors wheel revolutions to determine the relative distance traveled. In words,

Relative distance = number of wheel revolutions × wheel circumference

The model for the position error in the relative distance takes on the same form as that of a scale factor error (see Problem 8.5).

The locomotive kinematics are described by the following equations:

$$\begin{aligned} x &= 0 \\ y_{k+1} &= y_k + w_k + 10\Delta t, \quad k = 0, 1, 2, \ldots, 2000 \end{aligned} \tag{P8.6.1}$$

where w_k is an independent Gaussian sequence described by

$$w_k \sim \mathcal{N}(0, 1\,\text{m}^2)$$

The sampling interval Δt is 1 sec.

(a) Formulate a process model that includes bias and scale factor error states using random constant models for each. Also formulate a linearized measurement model using the scenario setup from Problem 7.5; use only DME station No. 2. For simplicity, a linearized Kalman filter should be used, not an extended Kalman filter. Let the initial estimation error variances for the bias and scale factor states be $(100\,\text{m})^2$ and $(0.02 \text{ per unit})^2$.

(b) Run a covariance analysis using the filter parameters worked out in (a) for $k = 0, 1, 2, \ldots, 2000$, and plot the rms estimation errors for each of the error states. Also plot the rms position error.

(c) Make another run of (b) except that, between $k = 1000$ and $k = 1800$, DMF. measurement updates are not available.

8.7 In a single-channel inertial model such as that described by the differential equation of Eq. (8.9.1), the state transition matrix computed for the discrete-time process may be simply approximated by $\boldsymbol{\phi} \approx \mathbf{I} + \mathbf{F}\Delta t$. This same first-order approximation for $\boldsymbol{\phi}$ can also be used in the integral expression for $\mathbf{Q}_k$ given by Eq. (3.9.10). When $\mathbf{F}$ is constant and $\boldsymbol{\phi}$ is first order in the step size, it is feasible to evaluate the integral analytically and obtain an expression for $\mathbf{Q}_k$ in closed form. (Each of the terms in the resulting $\mathbf{Q}_k$ are functions of Δt.)

(a) Work out the closed-form expression for $\mathbf{Q}_k$ using a first-order approximation for $\boldsymbol{\phi}$ in Eq. (3.9.10). Call this $\mathbf{Q}1$.

(b) Next, evaluate $\mathbf{Q}_k$ with MATLAB using the "Van Loan" numerical method given by Eqs. (3.9.22–3.9.25). Do this for $\Delta t = 5$ sec, 50 sec, and 500 sec. These will be referred to as $\mathbf{Q}2$ (different, of course, for each Δt).

(c) Compare the respective diagonal terms of $\mathbf{Q}1$ with those of $\mathbf{Q}2$ for $\Delta t = 5$, 50, and 500 sec.

This exercise is intended to demonstrate that one should be wary of using first-order approximations in the step size when it is an appreciable fraction of the natural period or time constant of the system.

REFERENCES CITED IN CHAPTER 8

1. H.J. Euler and C.C. Goad, "On Optimal Filtering of GPS Dual Frequency Observation Without Using Orbit Information," *Bulletin Geodesique*, 65: 130–143 (1991), Springer-Verlag.
2. R.G. Brown and P.Y.C. Hwang, *Introduction to Random Signals and Applied Kalman Filtering*, 3rd edition, New York: Wiley, 1997, Chapter 4.
3. J.A. Farrell, *Aided Navigation: GPS with High Rate Sensors*, McGraw-Hill, 2008.
4. G.R. Pitman (ed.), *Inertial Guidance*, New York: Wiley, 1962.
5. J.C. Pinson,"Inertial Guidance for Cruise Vehicles," C.T. Leondes (ed.), *Guidance and Control of Aerospace Vehicles*, New York: McGraw-Hill, 1963.
6. M. Kayton and W.R. Fried (eds.), *Avionics Navigation Systems*, 2nd Edition, New York: Wiley, 1997, p. 375.

7. R.S.Y. Young, P.Y.C. Hwang, and D.A. Anderson,"Use of Low-Cost GPS/AHRS with Head-Up Guidance Systems for CAT III Landings," *GPS Red Book Series, Vol. VII: Integrated Systems*, Institute of Navigation, 2010.
8. P.Y. Hwang, G.A. McGraw, and J.R. Bader, "Enhanced Differential GPS Carrier-Smoothed Code Processing Using Dual-Frequency Measurements," *Navigation: J. Inst. Navigation*, 46(2): 127–137 (Summer 1999).

Other General References

P.G. Savage, *Strapdown Analytics (Part 2)*, Strapdown Associates, 2000.

O. Salychev, *Applied Inertial Navigation: Problems and Solutions*, Moscow: BMSTU Press, 2004.

P.D. Groves, *Principles of GNSS, Inertial, and Multisensor Integrated Navigation Systems*, Artech House, 2008.

9

Kalman Filter Applications To The GPS And Other Navigation Systems

The Global Positioning System (GPS) has established itself over more than two decades of operation as a reliable and accurate source of positioning and timing information for navigation applications (1). Perhaps the clearest mark of its resounding success is found in the widespread use of low-cost GPS modules embedded in many modern-day utilities such as smartphones and car navigators. "GPS" is now a well-recognized acronym in our cultural lexicon. Ironically, a testament to its technological maturity actually lies in a growing emphasis of advanced research work into navigation areas beyond GPS, i.e., into so-called "GPS denied" situations. Without a doubt, the usefulness of GPS has far surpassed what had originally been envisioned by its early designers, thanks in large part to a creative and competitive community of navigation users, practitioners and developers (2). As we venture nearer a new dawn of multiple global navigation satellite systems (GNSS), that will have largely drawn from the GPS experience, we may yet see even more new and exciting concepts that take advantage of what will be a bountiful abundance of satellite navigation signals in space.

9.1 POSITION DETERMINATION WITH GPS

An observer equipped to receive and decode GPS signals must then solve the problem of position determination. In free space, there are three dimensions of position that need to be solved. Also, an autonomous user is not expected to be precisely synchronized to the satellite system time initially. In all, the standard GPS positioning problem poses four variables that can be solved from the following system of equations, representing measurements from four different satellites:

$$\begin{aligned}
\psi_1 &= \sqrt{(X_1 - x)^2 + (Y_1 - y)^2 + (Z_1 - z)^2} + c\Delta t \\
\psi_2 &= \sqrt{(X_2 - x)^2 + (Y_2 - y)^2 + (Z_2 - z)^2} + c\Delta t \\
\psi_3 &= \sqrt{(X_3 - x)^2 + (Y_3 - y)^2 + (Z_3 - z)^2} + c\Delta t \\
\psi_4 &= \sqrt{(X_4 - x)^2 + (Y_4 - y)^2 + (Z_4 - z)^2} + c\Delta t
\end{aligned} \tag{9.1.1}$$

where

$$\begin{aligned}
\psi_1, \psi_2, \psi_3, \psi_4 &= \text{noiseless pseudorange} \\
[X_i, Y_i, Z_i]^T &= \text{Cartesian position coordinates of satellite } i \\
[x, y, z]^T &= \text{Cartesian position coordinates of observer} \\
\Delta t &= \text{receiver offset from the satellite system time} \\
c &= \text{speed of light}
\end{aligned}$$

The observer position $[x, y, z]^T$ is "slaved" to the coordinate frame of reference used by the satellite system. In the case of GPS, this reference is a geodetic datum called WGS-84 (for World Geodetic System of 1984) that is earth-centered earth-fixed (3). The datum also defines the ellipsoid that crudely approximates the surface of the earth (see Fig. 9.1) Although the satellite positions are reported in WGS-84 coordinates, it is sometimes useful to deal with a locally level frame of reference, where the $x' - y'$ plane is tangential to the surface of the earth ellipsoid. As depicted in Fig. 9.1, we shall define such a locally level reference frame by having the x'-axis pointing east, the y'-axis north, and the z'-axis pointed up locally. It suffices here to say that the coordinate transformations to convert between the WGS-84

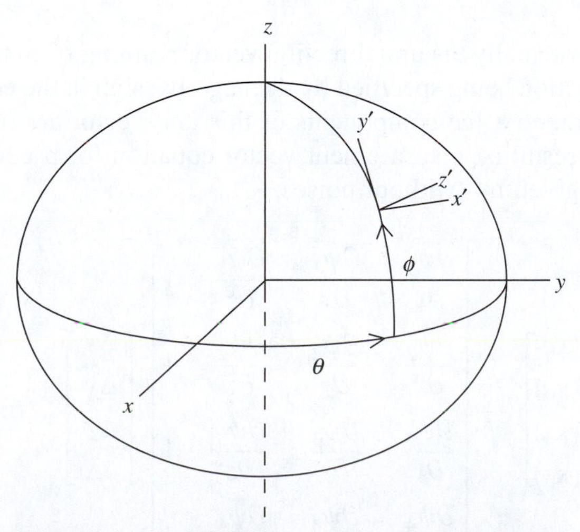

Figure 9.1 The WGS-84 coordinate reference frame (x, y, z) used by GPS and a locally level coordinate reference frame (x', y', z').

coordinates and any other derived reference frame, including the locally level one given here, are usually quite straightforward.

Measurement Linearization

The measurement situation for GPS is clearly nonlinear from Eq. (9.1.1). Linearization of a measurement of this form has already been covered in Section 7.1 and will not be reiterated here. We will simply evaluate the partial derivatives necessary to obtain the linearized equations about an approximate observer location $\mathbf{x}_0 = [x_0, y_0, z_0]^T$. This nominal point of linearization $\mathbf{x}_0$ is sometimes based on an estimate of the true observer location $\mathbf{x}$ although, in general, its choice may be arbitrary.

$$
\begin{aligned}
\frac{\partial \psi_i}{\partial x} &= -\frac{(X_i - x_0)}{\sqrt{(X_i - x_0)^2 + (Y_i - y_0)^2 + (Z_i - z_0)^2}} \\
\frac{\partial \psi_i}{\partial y} &= -\frac{(Y_i - y_0)}{\sqrt{(X_i - x_0)^2 + (Y_i - y_0)^2 + (Z_i - z_0)^2}} \\
\frac{\partial \psi_i}{\partial z} &= -\frac{(Z_i - z_0)}{\sqrt{(X_i - x_0)^2 + (Y_i - y_0)^2 + (Z_i - z_0)^2}}
\end{aligned}
\tag{9.1.2}
$$

for $i = 1, \ldots, 4$.

From a geometrical perspective, the partial derivative vector for each satellite i,

$$
\left[\frac{\partial \psi_i}{\partial x} \; \frac{\partial \psi_i}{\partial y} \; \frac{\partial \psi_i}{\partial z} \right]^T
$$

as given in Eq. (9.1.2) is actually the unit direction vector pointing from the satellite to the observer, the direction being specified by the negative sign in the equation. In classical navigation geometry, the components of this unit vector are often called *direction cosines:* The resulting measurement vector equation for pseudorange as the observable is then given by (without noise):

$$
\begin{bmatrix} \psi_1 \\ \psi_2 \\ \psi_3 \\ \psi_4 \end{bmatrix} - \begin{bmatrix} \hat{\psi}_1(\mathbf{x}_0) \\ \hat{\psi}_2(\mathbf{x}_0) \\ \hat{\psi}_3(\mathbf{x}_0) \\ \hat{\psi}_4(\mathbf{x}_0) \end{bmatrix} = \begin{bmatrix} \frac{\partial \psi_1}{\partial x} & \frac{\partial \psi_1}{\partial y} & \frac{\partial \psi_1}{\partial z} & 1 \\ \frac{\partial \psi_2}{\partial x} & \frac{\partial \psi_2}{\partial y} & \frac{\partial \psi_2}{\partial z} & 1 \\ \frac{\partial \psi_3}{\partial x} & \frac{\partial \psi_3}{\partial y} & \frac{\partial \psi_3}{\partial z} & 1 \\ \frac{\partial \psi_4}{\partial x} & \frac{\partial \psi_4}{\partial y} & \frac{\partial \psi_4}{\partial z} & 1 \end{bmatrix} \begin{bmatrix} \Delta x \\ \Delta y \\ \Delta z \\ c\Delta t \end{bmatrix}
\tag{9.1.3}
$$

where

$$\begin{aligned}
\psi_1 &= \text{noiseless pseudorange} \\
\mathbf{x}_0 &= \text{nominal point of linearization based on } [x_0, y_0, z_0]^T \\
&\quad \text{and predicted receiver time} \\
\hat{\psi}_1(\mathbf{x}_0) &= \text{predicted pseudorange based on } \mathbf{x}_0 \\
[\Delta x, \Delta y, \Delta z]^T &= \text{difference vector between true location } \mathbf{x} \text{ and } \mathbf{x}_0 \\
c\Delta t &= \text{range equivalent of the receiver timing error}
\end{aligned}$$

9.2 THE OBSERVABLES

Useful information can be derived from measurements made separately on the pseudorandom code and the carrier signal. There are many diverse signal processing schemes for the GPS signal in the commercial and military products and the block diagram shown in Fig. 9.2 is intended to represent a generic scheme. In all of these, we ultimately end up with the same types of measurements that are processed by the navigation function: pseudorange, carrier phase, and delta range (the carrier phase and delta range are related). When the measurements are formed, their noise characteristics are dependent on their respective signal-to-noise ratios after signal integration plus some pre-filtering made in the signal processing.

The observable known as *pseudorange* (sometimes also called code phase) is a timing measurement of the delay from the point of transmission to the point of receipt. This delay manifests itself in the time shift of the pseudorandom code position since the time of transmission. This measurement ideally represents the geometric range from the transmitting satellite to the receiver plus the receiver clock offset from satellite time. However, it also contains error components that affect the resulting solution.

Code (ρ^f in meters) and carrier phase (ϕ^f in cycles) measurements are represented by the following equations:

$$\rho^f = \psi + \underbrace{\xi + \tau + \iota_f}_{\beta^\rho} + \underbrace{\mu_\rho + \nu_\rho}_{\eta^\rho} \tag{9.2.1}$$

$$\lambda_f \phi^f = \psi + \underbrace{\xi + \tau - \iota_f}_{\beta^\phi} + \underbrace{\mu_\phi + \nu_\phi}_{\eta^\phi} + \lambda_f N^f \tag{9.2.2}$$

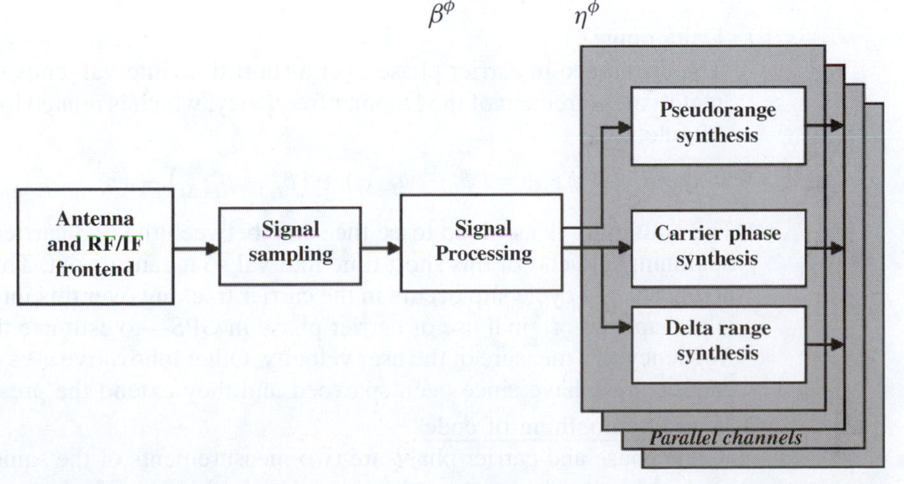

Figure 9.2 Generic GPS receiver functional block diagram.

where

ψ = ideal pseudorange consisting of geometric range and range-equivalent receiver clock error
ξ = satellite broadcast error (ephemeris and clock parameters)
τ = tropospheric refraction error
ι_f = ionospheric refraction error
μ_ρ = code multipath error
ν_ρ = pseudorange measurement noise
μ_ϕ = carrier multipath error
ν_ϕ = carrier phase measurement noise
λ_f = wavelength at frequency f
N^f = integer cycle ambiguity

Both types of measurements contain the ideal pseudorange plus various other error components. We can lump some of these error components together into at least two categories: signal-in-space (β) and receiver tracking (η).

The main differences between the two types of measurements are:

1. The carrier phase noise due to receiver tracking and multipath error are small but the carrier phase contains an integer ambiguity uncertainty that must be resolved for it to be useful. If N_f can be fully resolved, the resulting η_ϕ is very small compared to η_ρ.
2. There is a slight variation between the signal-in-space error components in that the ionospheric refraction term between the two are of opposing signs.

The processing of the code phase is straightforward and we shall see an example of this later in the chapter. Due to its associated cycle ambiguity, the carrier phase measurement is used in more specialized ways to take advantage of its high precision. Here, we list the variety of ways that the carrier phase measurement is used in different GPS applications:

1. Delta range
   The difference in carrier phase over a short time interval tends to approximate a measurement of the Doppler frequency, which is related to a measure of velocity.
$$\lambda_f \phi_t^f - \lambda_f \phi_{t-\Delta t}^f = (\psi_t - \psi_{t-\Delta t}) + \left(\beta_t^\phi - \beta_{t-\Delta t}^\phi\right) + \left(\eta_t^\phi - \eta_{t-\Delta t}^\phi\right) \quad (9.2.3)$$
   The ambiguity is assumed to be the same between the two carrier phases at beginning and end of this short time interval so it cancels out. This assumes no undetected cycle slip occurs in the carrier tracking over this interval. This is perhaps the original use of carrier phase in GPS—to estimate the doppler frequency as a measure of the user velocity. Other innovative uses of the GPS carrier phase have since been spawned and they extend the present.
2. Carrier smoothing of code
   Code phase and carrier phase are two measurements of the same "signal," i.e., ideal pseudorange, so they may be combined with a complementary

filter (see Chapter 8). In the first step of a complementary filter, we difference the two different types of measurements and the result is Kalman filtered:

$$\rho^f - \lambda_f \phi^f = -\lambda_f N^f + 2\iota_f + \underbrace{\mu_\rho + \nu_\rho}_{\eta^\rho} - \underbrace{\mu_\phi + \nu_\phi}_{\eta^\phi}$$

The filtered estimate (with estimation error ε_{CS}) would be: $-\lambda_f N^f + 2\iota_f + \varepsilon_{CS}$

Note that N^f is static but ι_f may gradually drift as a result of ionospheric refraction changes over time. The reconstituted code phase measurement corrected by the filter estimated then turns out to be:

$$\begin{aligned}\hat{\rho}^f_{CS} &= \underbrace{\psi + \xi + \tau - \iota_f + \mu_\phi + \nu_\phi + \lambda_f N^f}_{\text{Original carrier phase}} \quad \underbrace{-\lambda_f N^f + 2\iota_f + \varepsilon_{CS}}_{\text{Filter estimate}} \\ &= \psi + \xi + \tau + \iota_f + \mu_\phi + \nu_\phi + \varepsilon_{CS}\end{aligned} \tag{9.2.4}$$

When compared to the raw code phase (Eq. 9.2.1), the carrier smoothed code measurement has a smaller level of measurement noise in $(\mu_\phi + \nu_\phi + \varepsilon_{CS})$ as opposed to the original in $(\mu_\rho + \nu_\rho)$. Much of this reduction in noise, hence "smoothing," is owed to the time constant of the filter that then hinges on how much drift the ionospheric refraction is going through. This limitation is virtually eliminated by the use of dual-frequency carrier smoothing (see Problem 8.3).

3. Carrier phase differential

The differencing of measurements between two locations, one a known location and the other to be solved relative to the known location, contains relative position information between the two locations.

$$\lambda_f \phi^f_A - \lambda_f \phi^f_B = (\psi_A - \psi_B) + (\eta^\phi_A - \eta^\phi_B) + \lambda_f \left(N^f_A - N^f_B\right) \tag{9.2.5}$$

Here, the relative position information resides in $(\psi_A - \psi_B)$ while the signal-in-space error β^ϕ is eliminated but an additional integer ambiguity, $\lambda_f(N^f_A - N^f_B)$, is introduced. When this ambiguity is fully resolved, the solution has very high precision because its error is dictated by carrier phase measurement noises $(\eta^\phi_A - \eta^\phi_B)$ that are small as compared to code phase measurement noise.

4. Standalone positioning with precise satellite ephemeris/clock parameters

One of the more intriguing variations of GPS techniques to come along within the past 10 years is NASA's Global Differential GPS system (4). As an extreme form of wide-area differential GPS, the GDGPS exploits vastly advanced techniques and extensive monitoring to obtain highly-accurate determination of satellite position and clock parameters. Unlike conventional differential techniques that attempt to correct for the lumped unknown of β (Eqs. 9.2.1–9.2.2), the GDGPS solution attempts to corrects for individual components. The extensive monitoring network provides corrections for satellite orbits and clock to correct for ξ. The iono refraction error is eliminated by a proper linear combination of the carrier phases made at the two GPS frequencies:

$$
\begin{aligned}
\lambda_{f1}\phi^{f1} &= \psi + \xi + \tau - \iota_{f1} + \eta_{f1}^{\phi} + \lambda_{f1}N^{f1} \\
\lambda_{f2}\phi^{f2} &= \psi + \xi + \tau - \iota_{f2} + \eta_{f2}^{\phi} + \lambda_{f2}N^{f2} \\
\lambda_{f3}\left(\phi^{f1} - \frac{\lambda_{f1}}{\lambda_{f2}}\phi^{f2}\right) &= (\psi + \xi + \tau) + \underbrace{\lambda_{f3}\left(\eta_{f1}^{\phi} - \frac{\lambda_{f1}}{\lambda_{f2}}\eta_{f2}^{\phi}\right)}_{\eta_{f3}^{\phi}} \\
&\quad + \underbrace{\lambda_{f3}\left(N^{f1} - \frac{\lambda_{f1}}{\lambda_{f2}}N^{f2}\right)}_{\alpha}
\end{aligned} \tag{9.2.6}
$$

where the wavelength of a composite frequency f_3 is given by:

$$
\lambda_{f3} = \frac{\lambda_{f1}\lambda_{f2}^2}{\lambda_{f2}^2 - \lambda_{f1}^2}
$$

In Eq. (9.2.6), after minimizing the errors in ξ and eliminating the iono refraction, we are left with estimating out the tropo refraction error τ and the initial ambiguity α to get the best estimate of the ideal pseudorange ψ.

9.3 BASIC POSITION AND TIME PROCESS MODELS

The primary state variables given in the linearized measurement model of the ideal pseudorange in Eq. 9.1.3 are three position states (Δx, Δy, Δz) and one clock (time) bias state (Δt), which is better represented in its range-equivalent form when multiplied by the speed-of-light constant ($c\Delta t$).

When we use a Kalman filter to make the best estimate of position and time from available GPS measurements and nothing else, this problem is sometimes called *Standalone* or unaided GPS, the second term being contrasted to a system where aiding generally comes from an inertial source. The process model of such a set of three-dimensional position states is very much dependent on the dynamics encountered and, as Chapter 8 had previously made abundantly clear, such dynamics are not always easy to model because they are not all random but rather partly deterministic in nature. However, to make the best of a sometimes less-than-ideal situation, process models can be concocted to reasonably approximate the conditions at hand. But first we address the process model of a typical receiver clock.

Receiver Clock Modeling

The GPS receiver clock introduces a timing error that translates into ranging error that equally affects measurements made to all satellite (Eq. 9.1.1). This error is generally time-varying. If the satellite measurements are made simultaneously, this receiver clock error is the same on all measurements. Due to this commonality, the clock error has no effect on positioning accuracy if there are enough satellites to

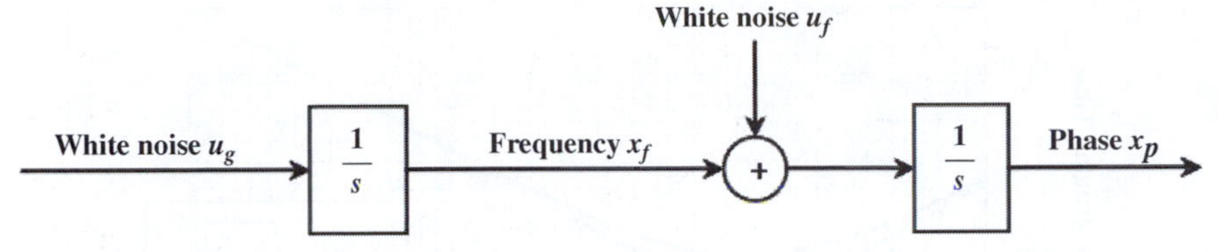

Figure 9.3 General two-state model describing clock errors. The independent white noise inputs u_f and u_g have spectral amplitudes of S_f and S_g.

solve for it and so is not included as a source of positioning error. Even so, there are advantages to properly model the receiver clock. This is illustrated in an example given at the end of the section.

A suitable clock model that makes good sense intuitively is a two-state random-process model. It simply says that we expect both the oscillator frequency and phase to random walk over reasonable spans of time. We now wish to look at how the **Q** parameters of the state model may be determined from the more conventional Allan variance parameters that are often used to describe clock drift (5). We begin by writing the expressions for the variances and covariances for the general two-state model shown in Fig. 9.3.

The clock states x_p and x_f represent the clock phase and frequency error, respectively. Let the elapsed time since initiating the white noise inputs be Δt. Then, using the methods given in Chapter 3, we have

$$
\begin{aligned}
E\left[x_p^2(\Delta t)\right] &= \int_0^{\Delta t}\int_0^{\Delta t} 1 \cdot 1 \cdot S_f \cdot \delta(u-v)\,du\,dv \\
&\quad + \int_0^{\Delta t}\int_0^{\Delta t} u \cdot v \cdot S_g \cdot \delta(u-v)\,du\,dv \\
&= S_f\Delta t + \frac{S_g\Delta t^3}{3}
\end{aligned}
\tag{9.3.1}
$$

$$
E\left[x_f^2(\Delta t)\right] = \int_0^{\Delta t}\int_0^{\Delta t} 1 \cdot 1 \cdot S_g \cdot \delta(u-v)\,du\,dv = S_g\Delta t \tag{9.3.2}
$$

$$
E\left[x_p(\Delta t)x_f(\Delta t)\right] = \int_0^{\Delta t}\int_0^{\Delta t} 1 \cdot v \cdot S_g \cdot \delta(u-v)\,du\,dv = \frac{S_g\Delta t^2}{2} \tag{9.3.3}
$$

Now let us concentrate on the variance of state x_p. In particular, let us form the rms value of x_p time-averaged over the elapsed time Δt.

$$
\begin{aligned}
\text{Avg. rms } x_p &= \frac{1}{\Delta t}\sqrt{S_f\Delta t + \frac{S_g\Delta t^3}{3}} \\
&= \sqrt{\frac{S_f}{\Delta t} + \frac{S_g\Delta t}{3}}
\end{aligned}
\tag{9.3.4}
$$

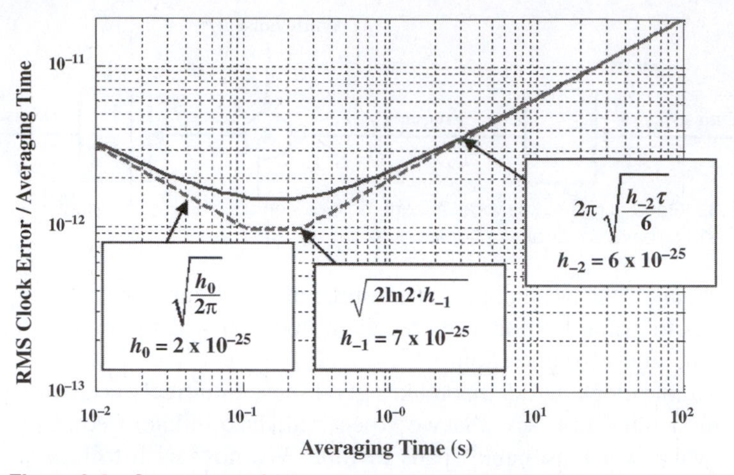

Figure 9.4 Square root of Allan variance (or Allan deviation) plot with asymptotes for a typical ovenized crystal oscillator (6).

Drift characteristics of real clocks have been studied extensively over the past few decades. Figure 9.4 shows a timing stability plot for a typical ovenized crystal clock. Such plots are known as Allan variance plots that depict the amount of rms drift that occurs over a specified period Δt, normalized by Δt. Note that over the time interval range shown in Fig. 9.4, there are three distinct asymptotic segments, the middle one of which is flat and associated with what is known as flicker noise. This segment, however, is missing from the response of the two-state model described by Eq. (9.3.4).

There was good reason for this omission in the two-state model. Flicker noise gives rise to a term in the variance expression that is of the order of Δt^2, and it is impossible to model this term exactly with a finite-order state model. To resolve this modeling dilemma, an approximate solution is to simply elevate the theoretical V of the two-state model so as to obtain a better match in the flicker floor region. This leads to a compromise model that exhibits somewhat higher drift than the true experimental values for both small and large averaging times. The amount of elevation of the V depends on the width of the particular flicker floor in question, so this calls for a certain amount of engineering judgment.

We then compare terms of similar order in Δt in 9.3.1 and the q_{11} term

$$q_{11} = \frac{h_0}{2}\Delta t + 2h_{-1}\Delta t^2 + \frac{2}{3}\pi^2 h_{-2}\Delta t^3$$

By completely ignoring the flicker (second) term, we arrive at the following correspondence:

$$\begin{aligned} S_f &\sim \frac{h_0}{2} \\ S_g &\sim 2\pi^2 h_{-2} \end{aligned} \tag{9.3.5}$$

Table 9.1 Typical Allan Variance Coefficients for Various Timing Standards (for clock error in seconds) (7)

Timing Standard	h_0	h_{-1}	h_{-2}
TCXO (low quality)	2×10^{-19}	7×10^{-21}	2×10^{-20}
TCXO (high quality)	2×10^{-21}	1×10^{-22}	3×10^{-24}
OCXO	2×10^{-25}	7×10^{-25}	6×10^{-25}
Rubidium	2×10^{-22}	4.5×10^{-26}	1×10^{-30}
Cesium	2×10^{-22}	5×10^{-27}	1.5×10^{-33}

TCXO – Temperature compensated crystal oscillator
OCXO – Ovenized crystal oscillator (temperature controlled)

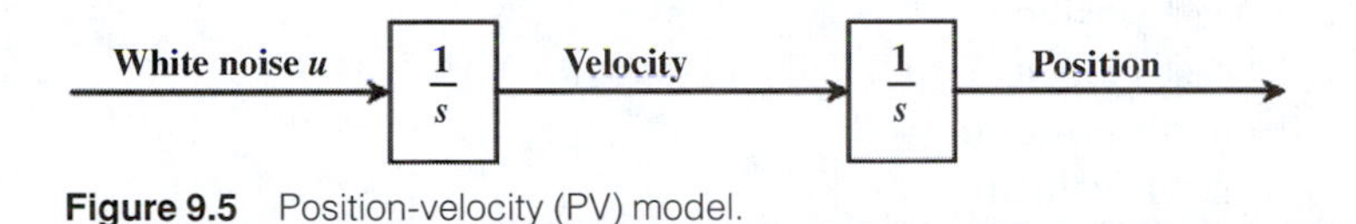

Figure 9.5 Position-velocity (PV) model.

While there are various methods to make up for ignoring the flicker term, we introduce one such approximation method that is also new, in Problem 9.4.

Precision clocks can have widely diverse Allan variance characteristics. Thus, one must treat each case individually and work out an approximate model that fits the application at hand. Table 9.1 gives typical values of h_0, h_{-1}, and h_{-2} for various types of timing standards widely used in GPS receivers. Note that the numbers given in Table 9.1 correspond to clock errors in units of seconds. When used with clock error in units of meters, the values of Table 9.1 must be multiplied by the square of the speed of light $(3 \times 10^8)^2$.

Dynamic Model for Position States

The dynamic process used in Fig. 9.3 for the clock also works reasonably well for a moving observer. Instead of phase and frequency, we replace the states with position and velocity and Fig. 9.5 depicts a Position-Velocity (PV) model for one-dimension of position. We typically represent the three dimensions of position with three such independent models.

This model simply accommodates a non-zero velocity condition whenever there is motion. The white noise u driven through the first integrator results in a random walk process for the velocity state. The position state is simply the integral of the velocity state, as shown in Fig. 9.5.

EXAMPLE 9.1

In the PV model of Fig. 9.5, each spatial dimension will have two degrees of freedom, one of position and the other of velocity. Therefore, for the GPS problem where there are three spatial dimensions and one time dimension, the state vector now becomes an eight-tuple. The PV dynamic process can be described by the following vector differential equation:

$$\begin{bmatrix} \dot{x}_1 \\ \dot{x}_2 \\ \dot{x}_3 \\ \dot{x}_4 \\ \dot{x}_5 \\ \dot{x}_6 \\ \dot{x}_7 \\ \dot{x}_8 \end{bmatrix} = \begin{bmatrix} 0&1&0&0&0&0&0&0 \\ 0&0&0&0&0&0&0&0 \\ 0&0&0&1&0&0&0&0 \\ 0&0&0&0&0&0&0&0 \\ 0&0&0&0&0&1&0&0 \\ 0&0&0&0&0&0&0&0 \\ 0&0&0&0&0&0&0&1 \\ 0&0&0&0&0&0&0&0 \end{bmatrix} \begin{bmatrix} x_1 \\ x_2 \\ x_3 \\ x_4 \\ x_5 \\ x_6 \\ x_7 \\ x_8 \end{bmatrix} + \begin{bmatrix} 0 \\ u_1 \\ 0 \\ u_2 \\ 0 \\ u_3 \\ u_f \\ u_g \end{bmatrix} \tag{9.3.6}$$

where

x_1 = east position
x_2 = east velocity
x_3 = north position
x_4 = north velocity
x_5 = altitude
x_6 = altitude rate
x_7 = range (clock) bias error
x_8 = range (clock) drift error

and the spectral densities associated with the white noise driving functions are S_p for u_1, u_2 and u_3, S_f for u_f and S_g for u_g (see Figs. 9.3 and 9.5).

From 9.3.6, the state transition matrix can be derived as

$$\mathbf{\Phi} = \left[\begin{array}{c|c|c|c} \boldsymbol{\phi} & \mathbf{0} & \mathbf{0} & \mathbf{0} \\ \hline \mathbf{0} & \boldsymbol{\phi} & \mathbf{0} & \mathbf{0} \\ \hline \mathbf{0} & \mathbf{0} & \boldsymbol{\phi} & \mathbf{0} \\ \hline \mathbf{0} & \mathbf{0} & \mathbf{0} & \boldsymbol{\phi} \end{array}\right] \quad \text{where } \boldsymbol{\phi} = \begin{bmatrix} 1 & \Delta t \\ 0 & 1 \end{bmatrix} \tag{9.3.7}$$

To obtain the process noise covariance matrix $\mathbf{Q}$, we resort to the methods given in Chapter 3. Note from Eq. 9.3.6 that we can treat the three position-velocity states variable pairs independently.

$$E\left[x_i^2(\Delta t)\right] = \int_0^{\Delta t}\int_0^{\Delta t} u \cdot v \cdot S_p \cdot \delta(u-v)\,du\,dv = \frac{S_p \Delta t^3}{3} \tag{9.3.8}$$

$$E\left[x_{i+1}^2(\Delta t)\right] = \int_0^{\Delta t}\int_0^{\Delta t} 1 \cdot 1 \cdot S_p \cdot \delta(u-v)\,du\,dv = S_p \Delta t \tag{9.3.9}$$

$$E[x_i(\Delta t)x_{i+1}(\Delta t)] = \int_0^{\Delta t}\int_0^{\Delta t} 1 \cdot v \cdot S_p \cdot \delta(u-v)\,du\,dv = \frac{S_p \Delta t^2}{2} \tag{9.3.10}$$

for $i = 1$, 3, and 5.

The equations involving the clock states x_7 and x_8 were derived from Eqs. (9.3.1) through (9.3.3). With these, the process noise covariance matrix is as follows:

$$\mathbf{Q} = \left[\begin{array}{c|c|c|c} \mathbf{Q}_{PV} & \mathbf{0} & \mathbf{0} & \mathbf{0} \\ \hline \mathbf{0} & \mathbf{Q}_{PV} & \mathbf{0} & \mathbf{0} \\ \hline \mathbf{0} & \mathbf{0} & \mathbf{Q}_{PV} & \mathbf{0} \\ \hline \mathbf{0} & \mathbf{0} & \mathbf{0} & \mathbf{Q}_{Clk} \end{array}\right]$$

where

$$\mathbf{Q}_{PV} = \begin{bmatrix} \dfrac{S_p \Delta t^3}{3} & \dfrac{S_p \Delta t^2}{2} \\ \dfrac{S_p \Delta t^2}{2} & S_p \Delta t \end{bmatrix} \quad \mathbf{Q}_{Clk} = \begin{bmatrix} S_f \Delta t + \dfrac{S_g \Delta t^3}{3} & \dfrac{S_g \Delta t^2}{2} \\ \dfrac{S_g \Delta t^2}{2} & S_g \Delta t \end{bmatrix}$$

The corresponding measurement model for pseudorange is an extension of Eq. (9.1.3) and is quite straightforward:

$$\begin{bmatrix} \rho_1 \\ \rho_2 \\ \vdots \\ \rho_n \end{bmatrix} - \begin{bmatrix} \hat{\rho}_1 \\ \hat{\rho}_2 \\ \vdots \\ \hat{\rho}_n \end{bmatrix} = \begin{bmatrix} h_x^{(1)} & 0 & h_y^{(1)} & 0 & h_z^{(1)} & 0 & 1 & 0 \\ h_x^{(2)} & 0 & h_y^{(2)} & 0 & h_z^{(2)} & 0 & 1 & 0 \\ \vdots & & \vdots & & \vdots & & \vdots & \\ h_x^{(n)} & 0 & h_y^{(n)} & 0 & h_z^{(n)} & 0 & 1 & 0 \end{bmatrix} \begin{bmatrix} x_1 \\ x_2 \\ x_3 \\ x_4 \\ x_5 \\ x_6 \\ x_7 \\ x_8 \end{bmatrix} + \begin{bmatrix} v_{\rho 1} \\ v_{\rho 2} \\ \vdots \\ v_{\rho n} \end{bmatrix} \tag{9.3.11}$$

The determination of the spectral amplitude S_p for the position random process is at best a "guesstimate" based roughly on expected vehicle dynamics. The PV model also becomes inadequate for cases where the near-constant velocity assumption does not hold, that is, in the presence of severe accelerations. To accommodate acceleration in the process model, it is appropriate to add another degree of freedom for each position state. Although we can easily add one more integrator to obtain a *Position-Velocity-Acceleration* (PVA) model, a stationary process such as the Gauss-Markov process is perhaps more appropriate than the nonstationary random walk for acceleration (Fig. 9.6). This goes in accordance with real-life physical situations where vehicular acceleration is usually brief and seldom sustained. The state vector for this PVA model then becomes 11-dimensional with the addition of three more acceleration states.

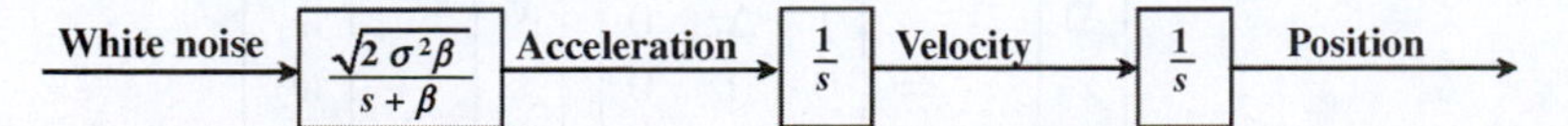

Figure 9.6 Position-velocity-acceleration model for high (acceleration) dynamics observer.

Derivation of the corresponding state transition and process noise covariance matrices will be left as part of an exercise in Problem 9.2. It should be pointed out here that although acceleration is being accounted for in some stochastic form, the exercise is an approximation at best. There is no exact way to fit deterministic accelerations into any of the random-process dynamics suitable for a Kalman filter model. With the PVA model, the Kalman filter will do a better job of estimation than the PV model, but it may still be inadequate by other measures of optimality.

9.4 MODELING OF DIFFERENT CARRIER PHASE MEASUREMENTS AND RANGING ERRORS

Delta Range Processing

Delta range (or delta pseudorange) is an approximate measurement of position change or, in the limit, of velocity (8). Since it is derived from the GPS signal carrier phase that has very low measurement noise, the quality of the information is very high and its exploitation to improve the estimation of position and velocity cannot be overemphasized. We will now look at one proper way (but not the only way) to model the delta range measurement.

We revisit the position-velocity model shown in Fig. 9.5. For an example in one spatial dimension (also ignoring any clock error states), let us define three states:

$$
\begin{aligned}
x_1(k) &= \text{position at time } t_k \\
x_2(k) &= \text{velocity at time } t_k \\
x_3(k) &= \text{position at time } t_{k-1} \; [\text{or } x_1(k-1)]
\end{aligned}
$$

The position and velocity states are related by the following differential equation:

$$
\begin{bmatrix} \dot{x}_1 \\ \dot{x}_2 \end{bmatrix} = \begin{bmatrix} 0 & 1 \\ 0 & 0 \end{bmatrix} \begin{bmatrix} x_1 \\ x_2 \end{bmatrix} + \begin{bmatrix} 0 \\ u \end{bmatrix}
$$

The discrete-time solution to the differential equation is:

$$
\begin{bmatrix} x_1 \\ x_2 \end{bmatrix}_{k+1} = \begin{bmatrix} 1 & \Delta t \\ 0 & 1 \end{bmatrix} \begin{bmatrix} x_1 \\ x_2 \end{bmatrix}_k + \begin{bmatrix} w_1 \\ w_2 \end{bmatrix}_k \tag{9.4.1}
$$

We then augment the x_3 state to the two-tuple process model of Eq. (9.4.1) to get (9):

$$
\begin{bmatrix} x_1 \\ x_2 \\ x_3 \end{bmatrix}_{k+1} = \begin{bmatrix} 1 & \Delta t & 0 \\ 0 & 1 & 0 \\ 1 & 0 & 0 \end{bmatrix} \begin{bmatrix} x_1 \\ x_2 \\ x_3 \end{bmatrix}_k + \begin{bmatrix} w_1 \\ w_2 \\ 0 \end{bmatrix}_k \tag{9.4.2}
$$

The **Q** matrix is given by:

$$\mathbf{Q} = \begin{bmatrix} \dfrac{S\Delta t^3}{3} & \dfrac{S\Delta t^2}{2} & 0 \\ \dfrac{S\Delta t^2}{2} & S\Delta t & 0 \\ 0 & 0 & 0 \end{bmatrix} \tag{9.4.3}$$

(By the very definition of $x_3(k+1) = x_1(k)$, w_3 is zero. Thus, the third row and column of **Q** are zero.)

At this time, we write out a simplified measurement model that consists of a code phase-like measurement and a delta range-like measurement.

$$\begin{bmatrix} \rho \\ \delta \end{bmatrix}_k = \begin{bmatrix} h & 0 & 0 \\ h & 0 & -h \end{bmatrix}_k \begin{bmatrix} x_1 \\ x_2 \\ x_3 \end{bmatrix}_k + \begin{bmatrix} v_\rho \\ v_\delta \end{bmatrix}_k \tag{9.4.4}$$

where $\delta_k = \lambda_f \phi_k^f - \lambda_f \phi_{k-1}^f$ is the delta range measurement as defined in Eq. (9.2.3).

This delta range model paints a more realistic picture than the simple approximation of treating it as a direct measure of velocity, as given as follows for comparison:

$$\begin{bmatrix} \rho \\ \delta \end{bmatrix}_k = \begin{bmatrix} h & 0 \\ 0 & h \end{bmatrix}_k \begin{bmatrix} x_1 \\ x_2 \end{bmatrix}_k + \begin{bmatrix} v_\rho \\ v_\delta / \Delta t \end{bmatrix}_k \tag{9.4.5}$$

whose dynamic process model is given by Eq. (9.4.1). Here the measurement noise variance

$$\mathbf{R} = \begin{bmatrix} Ev_\rho v_\rho^T & 0 \\ 0 & \dfrac{1}{\Delta t^2} Ev_\delta v_\delta^T \end{bmatrix}$$

Table 9.2 compares the steady state error (standard deviation) of the delta range model of Eqs. (9.4.2)–(9.4.4) for different values of Δt, while assuming $h = 1$,

Table 9.2 Relative comparison of velocity estimation errors as a function of sampling interval and chosen model for a given high dynamics-type environment

Delta range (and sampling) interval Δt (sec)	Steady state velocity rms estimation error (m/s)	
	Optimal delta range model (realistic)	Approx. velocity delta range as computed by suboptimal model (overoptimistic)
0.02	0.210	0.164
0.1	0.247	0.042
0.2	0.341	0.021
0.5	0.537	0.009
1.0	0.760	0.004

$S = 2\ (\text{m/s}^2)^2/\text{Hz}$. (Note that the value of S is typically much smaller if inertial aiding is available.) The table also includes the steady state error derived for the approximate delta range model of Eqs. (9.4.1) and (9.4.5). Note the gross optimism of the latter as compared to the more realistic values of the former. Although these numbers are for a simplified one-dimensional model, the comparison itself gives a qualitative feel for the differences between the models and, thus, useful insights.

The basic message here is simply that when dynamical uncertainties are relatively large, as when encountered in unaided (standalone) situations, the best we can do is with the optimal delta range model of Eqs. (9.4.2)–(9.4.4). And yet, when the delta range interval is larger than even a sizable fraction of a second, the system becomes incapable of accurately estimating instantaneous velocity. All the approximate model of Eqs. (9.4.1) and (9.4.5) is doing is fooling itself into "thinking" it has observability into the instantaneous velocity and therefore accounts for an overoptimistic velocity error covariance when in fact it can do no better than the optimal model.

Ambiguity Resolution

The usefulness of carrier phase data for high-accuracy positioning is due to the high precision measurement associated with it. The measurement model of Eq. (9.2.2) suggests that if the integer ambiguity N can be fully resolved, then the carrier phase is essentially a high precision measurement of the ideal pseudorange ψ plus a low-frequency system error β^{ϕ} but with very small high-frequency noise η^{ϕ}. β^{ϕ} is similar in size and effect to the term β^{ρ} that corrupts its code phase counterpart of Eq. (9.2.1).

This notion has been the essential foundation to an application area called Real-Time Kinematic (RTK) GPS. RTK GPS came through some wondrous innovation and evolution in the field of terrestrial surveying in the late 80s and early 90s, and has gone from its terrestrial surveying roots to dynamic positioning applications, such as automatic control of tractors in farming and precision landing of aircraft. To rid itself of the low-frequency system error β^{ϕ}, RTK GPS is based on differential principles where the solution obtained is only for the position of one observer with respect to another reference observer. The errors that corrupt the observations of two observers that are relatively close together are strong correlated and differenced out. The usual practice in RTK then is to form a "double" difference, consisting of a first and a second difference of nearly simultaneously measured carrier phases. (Fortunately, highly accurate GPS time tagging satisfies the need to do this with microsecond-level precision to avoid any dynamical compensation). The first difference of carrier phases is made for each satellite seen commonly at the two observers, and the second difference is made between first differences associated with any pair of satellites. The second difference removes the receiver clock term entirely.

We begin by re-examining Eq. (9.2.2), first leaving out its frequency argument f, then introducing indices of receiver location (A or B) and satellite ID ($SV1$ or $SV2$), and also expanding out the ideal pseudorange term ψ_A:

$$\lambda\phi_A^{(1)} = \underbrace{h^{(1)} \cdot \delta x_A + \delta t_A}_{\psi_A} + \lambda N_A^{(1)} + \beta^{(1)} + \eta_A^{(1)}$$

After forming the "double difference" combination associated with a pair of satellites, 1 and 2 say, the resulting measurement equation becomes:

$$\lambda \underbrace{\left[\left(\Delta\phi_A^{(1)} - \Delta\phi_B^{(1)}\right) - \left(\Delta\phi_A^{(2)} - \Delta\phi_B^{(2)}\right)\right]}_{\nabla\Delta\phi_{1,2}} =$$

$$\underbrace{\left(h_x^{(1)} - h_x^{(2)}\right)}_{\nabla h_x^{(1,2)}} \cdot \Delta x_{AB} + \underbrace{\left(h_y^{(1)} - h_y^{(2)}\right)}_{\nabla h_y^{(1,2)}} \cdot \Delta y_{AB} + \underbrace{\left(h_z^{(1)} - h_z^{(2)}\right)}_{\nabla h_z^{(1,2)}} \cdot \Delta z_{AB}$$

$$+\lambda \underbrace{\left[\left(N_A^{(1)} - N_B^{(1)}\right) - \left(N_A^{(2)} - N_B^{(2)}\right)\right]}_{\nabla\Delta N_{1,2}} + \underbrace{\left[\left(\eta_A^{(1)} - \eta_B^{(1)}\right) - \left(\eta_A^{(2)} - \eta_B^{(2)}\right)\right]}_{\nabla\Delta\eta_{\phi 1,2}} \tag{9.4.6}$$

Mostly notably, this equation is now devoid of a receiver clock bias term and the integer character of cycle ambiguity, given by $\nabla\Delta N_{1,2}$, is still maintained. The pairing for the double difference can be formed in many ways, but one popular method is to difference the single difference carrier phases of all satellites from one chosen reference satellite (such a scheme was illustrated in Section 8.2).

If we choose a PV model such as that adopted in Example 9.3, we would use the same first six states (position and velocity states in three spatial dimensions) but ignore the two clock states because these have been eliminated in the double difference measurement model.

$$\begin{bmatrix} \nabla\Delta\phi_{1,2} \\ \nabla\Delta\phi_{1,3} \\ \vdots \\ \nabla\Delta\phi_{1,n} \end{bmatrix} \lambda = \underbrace{\left[\begin{array}{cccccc|cccc} \nabla h_x^{(1,2)} & 0 & \nabla h_y^{(1,2)} & 0 & \nabla h_z^{(1,2)} & 0 & \lambda & 0 & \cdots & 0 \\ \nabla h_x^{(1,3)} & 0 & \nabla h_y^{(1,3)} & 0 & \nabla h_z^{(1,3)} & 0 & 0 & \lambda & & \\ \vdots & & \vdots & & \vdots & & \vdots & & \ddots & \\ \nabla h_x^{(1,n)} & 0 & \nabla h_y^{(1,n)} & 0 & \nabla h_z^{(1,n)} & 0 & 0 & & & \lambda \end{array}\right]}_{\mathbf{H}} \begin{bmatrix} x_1 \\ x_2 \\ x_3 \\ x_4 \\ x_5 \\ x_6 \\ x_7 \\ x_8 \\ \vdots \\ x_n \end{bmatrix} + \begin{bmatrix} \nabla\Delta\eta_{\phi 1,2} \\ \nabla\Delta\eta_{\phi 1,3} \\ \vdots \\ \nabla\Delta\eta_{\phi 1,n} \end{bmatrix} \tag{9.4.7}$$

States x_7 through x_n are now ambiguity states whose dynamics are trivial because these ambiguities represent some fixed initial unknown quantity, and in this particular case, even integer in nature. Therefore, these states are modeled as random biases and the portion of the **Q** matrix associated with them is simply zero.

The **R** matrix associated with the measurement noise vector term in Eq. (9.4.7) depends somewhat on the double differencing scheme that is used. In general, even if the original measurement noise covariance matrix associated with the single difference vector $\left[\left(\eta_A^{(1)} - \eta_B^{(1)}\right) \cdots \left(\eta_A^{(n)} - \eta_B^{(n)}\right)\right]^T$ is diagonal, the double difference version in Eq. (9.4.7) will not be due to non-zero cross-correlation terms.

The system of equations in Eq. (9.4.7) is not observable at a single instant in time. Very simply, there are not enough measurements to solve the given number of variables in the state vector. Rather, observability is strictly achieved over a period of time with more than one set of measurements, and actually comes from changing values of the **H** matrix over that period of time. If GPS had been made up of a constellation of geostationary satellites, we would never obtain any observability into the ambiguity states in this manner.

In practice, the ambiguities can be resolved faster by invoking an integer constraint on seeking their solution. The most widely-known and efficient search method used in practice today is the LAMBDA method (10). The Magill adaptive scheme (11) is another useful integer-constrained ambiguity estimator but does not have any kind of a search strategy that has made the LAMBDA method a far more efficient alternative. However, these methods typically do incorporate the best estimates of the integer states and their associated error covariance that may come from a Kalman filter processing model such as that described above, which is used as a starting point in the ambiguity resolution process. It should be noted that such methods can also be prone to incorrect resolution particularly if residual errors in the double difference measurements become significant, such as when processing measurements from two observers that are separated by long distances. (In such cases, their errors are no longer as strongly correlated and do not fully cancel out in the differencing.)

Tropospheric Delay Estimation

Unlike ionospheric refraction, which is frequency dependent and can largely be removed with dual-frequency measurements, Eq. (9.2.6) is an example of iono-free carrier phase, tropospheric refraction that cannot easily be observed while the measurements themselves are being used for positioning. From the early days of GPS, the tropo delay was corrected by a basic prediction model usually with Mean Sea Level altitude and satellite (local) elevation angle as inputs. Over the past decade or so, such prediction models have been tweaked with more refined data from substantial study efforts that have led to even more accurate tropo delays characteristics. Ultimately, the best prediction models require environmental measurements such as temperature, pressure and water vapor content. However, operational real-time systems would consider extra environmental sensors an unwarranted burden so the preferred solution is partly in using good prediction models that do not need such sensors and partly in estimating the residual error.

The tropospheric delay is a function of the path length through atmospheric conditions that refract the signal and the amount of delay is partially dependent on user altitude. Also, at a given user location, the tropospheric delay τ_i varies with

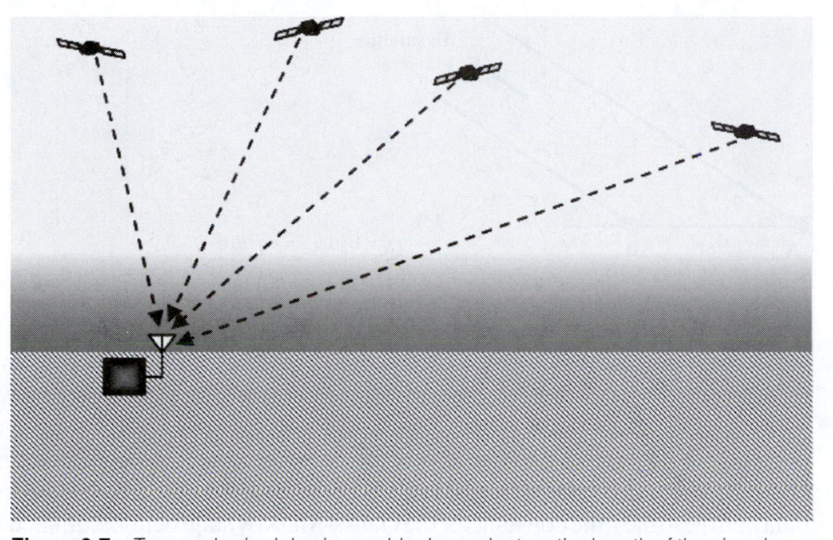

Figure 9.7 Tropospheric delay is roughly dependent on the length of the signal propagation path exposed to the atmosphere. An approximate trigonometric relationship may be used to relate the elevation-dependent tropo delays from each and every visible satellite to estimate a reference tropo delay at the local zenith.

satellite elevation θ_i roughly as a cosecant multiplier of the base tropospheric delay at zenith, τ_z (12):

$$\tau_i = f_\tau(\theta_i, h) \approx \frac{1}{\sin\theta_i}\tau_z(h) \tag{9.4.6}$$

Eq. (9.4.6) can be linearized about a predicted tropospheric delay that nearly approximates the true value. The residual difference can then be defined as $\Delta\tau_i = \tau_i - \hat{\tau}_i$, where the predicted tropospheric delay is $\hat{\tau}_i = \hat{f}_\tau(\theta_i, h)$.

Thus,

$$\begin{aligned} \Delta\tau_i &= \tau_i - \hat{f}_\tau(\theta_i, h) \\ &= \varepsilon(\theta_i)\Delta\tau_z \end{aligned} \tag{9.4.7}$$

We can then insert Eq. (9.4.7) into the pseudorange measurement model of Eq. (9.2.1) to get

$$\rho_i - \hat{f}_\tau(\theta_i, h) = \psi_i + \underbrace{\tau_i - \hat{f}_\tau(\theta_i, h)}_{\Delta\tau_i} + \xi_i + \iota_{fi} + \eta_i^\rho$$

and in terms of the zenith tropospheric delay error $\Delta\tau_z$,

$$\rho_i - \hat{f}_\tau(\theta_i, h) = \psi_i + \varepsilon(\theta_i)\Delta\tau_z + \xi_i + \iota_{fi} + \eta_i^\rho \tag{9.4.8}$$

The beauty of relating all the tropo delay errors along each line-of-sight through a map function to a term that is related to the tropo at zenith is this makes the tropo estimation observable. Where independently associating tropo delay states along each line-of-sight with each measurement would have introduced too many unrelated states. The connection through the map function to one zenith tropo error state simply introduces only one additional state.

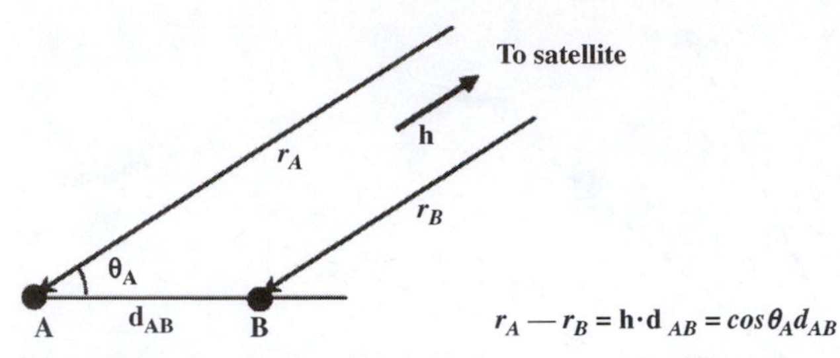

Figure 9.8 Geometric relationship between range difference and relative position between A and B.

For its process dynamics, we know that $\Delta\tau_z$ varies very slowly over time because of atmospheric dependencies, but can change more quickly when varying with altitude h. A suitable model might be a Gauss-Markov process with a time constant that depends on the rate of change of h or the vertical speed. As one might expect, any significant motion in the vertical would likely mean a change in the zenith tropospheric delay error, so the time constant for the Gauss-Markov state being estimated will need to be shortened momentarily to facilitate a quicker resettling to a new steady-state level.

Relative Positioning

There are GPS applications where the positioning solution is ultimately determined with respect to another reference location. In that regard, the solution accuracy is assessed as being relative to that reference location. Such applications include differential code phase GPS, real-time kinematic GPS, and time-relative GPS. In many cases where the relative distance between the reference location and the observer location (also called the baseline length) is very short as compared to the distances to the orbiting GPS satellites, the localized geometry becomes very simplified and linearized (Fig. 9.8).

However, if the baseline length increases to a point of significance, subtle complications with the measurement geometry will arise. It is important to understand what these subtleties are about and how the Kalman filter measurement should be properly modeled. For illustrative purposes then, we choose an example of relative positioning between two locations where the baseline length is indeed quite large. For simplicity's sake, we keep the illustration to a one-dimensional model involving one satellite that captures the essence of the problem at hand.

EXAMPLE 9.2

The measurement geometry is summarized in Fig. 9.9 (13). Unlike the simpler geometry of Fig. 9.8 the ray traces from the satellite seen at both locations A and B are no longer parallel so the connection between baseline distance d_{AB} and the measured range difference $(r_A - r_B)$ is no longer linear for this example. How then would we formulate the measurement model, when faced with angular relationships at the two locations, to the same satellite, that are different, i.e., $\boldsymbol{\theta}_A \neq \boldsymbol{\theta}_B$.

The proper way to handle this measurement situation is to first separate out the different geometries seen at A and at B, and then to linearize each one:

$$\begin{aligned} &(r_A - r_B) - (\hat{r}_A - \hat{r}_B) = (r_A - \hat{r}_A) - (r_B - \hat{r}_B) \\ &= \cos\theta_A \cdot \Delta x_A - \cos\theta_B \cdot \Delta x_B = \begin{bmatrix} \cos\theta_A & -\cos\theta_B \end{bmatrix} \begin{bmatrix} \Delta x_A \\ \Delta x_B \end{bmatrix} + \eta \end{aligned} \tag{9.4.9}$$

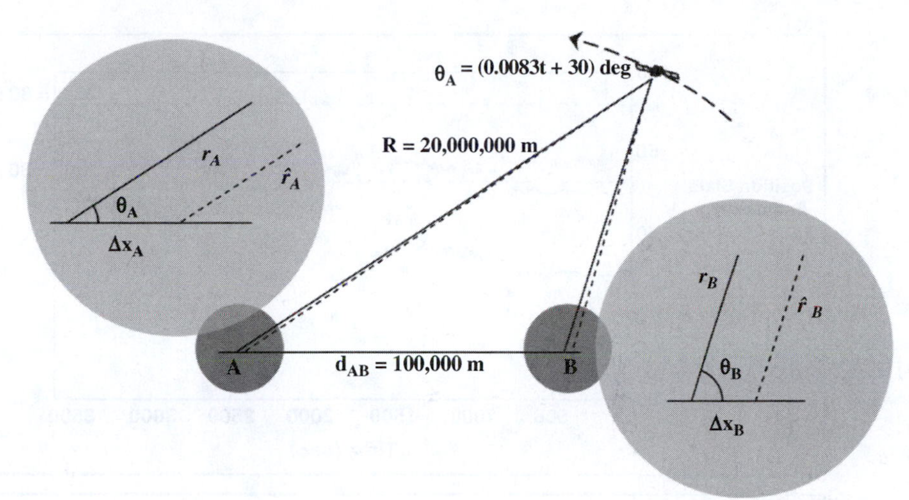

Figure 9.9 General model for relative position between A and B allowing for linearized perturbations at both locations.

In doing so, we need two states instead of one, where each state represents the linearized error at each location.

Suppose that the two locations are static and we are attempting to estimate their positions relative to one another. There will be an initial uncertainty in the state estimates given by:

$$\mathbf{P}_{\mathbf{o}}^{-} = E\left\{ \begin{bmatrix} \Delta x_A \\ \Delta x_B \end{bmatrix} \begin{bmatrix} \Delta x_A & \Delta x_B \end{bmatrix} \right\} = \begin{bmatrix} (200m)^2 & 0 \\ 0 & (200m)^2 \end{bmatrix}_k$$

But, the process model will be trivial:

$$\begin{bmatrix} \Delta \dot{x}_A \\ \Delta \dot{x}_B \end{bmatrix} = \begin{bmatrix} 0 \\ 0 \end{bmatrix} \quad \Leftrightarrow \quad \begin{bmatrix} \Delta x_A \\ \Delta x_B \end{bmatrix}_{k+1} = \begin{bmatrix} \Delta x_A \\ \Delta x_B \end{bmatrix}_k$$

In Fig. 9.10, although it takes the filter some length of time to finally converge to the true values of Δx_A and Δx_B (80 m and 30 m respectively), the relative position estimate formed from $\Delta\hat{x}_A - \Delta\hat{x}_B$ converges to the true value of 50 m almost immediately.

If we look at the Kalman filter error covariance matrix $\mathbf{P}$, clearly the square root of the two variance terms in its diagonal will not reflect the quality of the relative position estimate error, i.e., the profile in Fig. 9.10(a) is associated with Fig. 9.11(a). To obtain an error profile for Fig. 9.10(b), we need to compute it from all the terms in that $\mathbf{P}$ matrix:

$$\begin{aligned} Cov(\Delta\hat{x}_A - \Delta\hat{x}_B) &= Var(\Delta\hat{x}_A) + Var(\Delta\hat{x}_B) - 2 \cdot Cov(\Delta\hat{x}_A, \Delta\hat{x}_B) \\ &= \mathbf{P}_{11} + \mathbf{P}_{22} - 2\mathbf{P}_{12} \end{aligned}$$

When this is plotted out in Fig. 9.11(b), the convergence of the relative position estimate error can now be clearly seen to happen rather quickly after initialization.

In all relative positioning problems, the result that is of most interest is truly the estimate of $(\Delta x_A - \Delta x_B)$. However, the point of the example was to illustrate that, based on the geometry of Fig. 9.9, if we had lumped $(\Delta x_A - \Delta x_B)$ into one

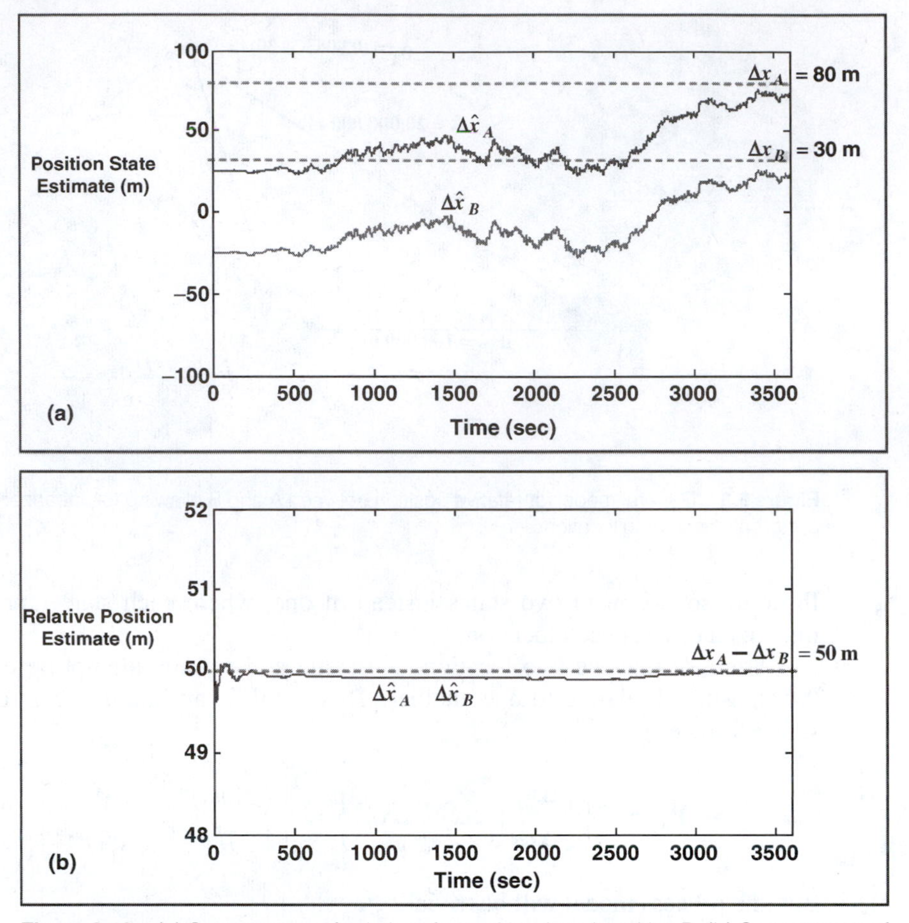

Figure 9.10 (a) Convergence of solutions for position A and position B; (b) Convergence of the difference between position A and position B.

state variable, that would have been tantamount to insisting that either location A or location B is perfectly known, i.e., $\Delta x_A = 0$ or $\Delta x_B = 0$, while there is just one unknown state to solve for at the other baseline end. This incorrect proposition would then lead to the dilemma of choosing which of the two measurement

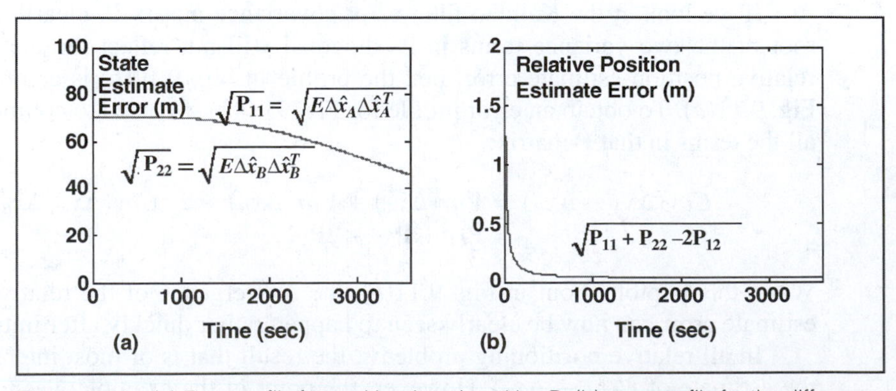

Figure 9.11 (a) Error profile for position A and position B; (b) error profile for the difference between position A and position B.

connections, $cos\ \theta_A$ or $cos\ \theta_B$ be used. In the formulation of Eq. (9.4.9), this dilemma does not exist. If the baseline is short and $cos\ \theta_A \approx cos\ \theta_B$, then this issue is moot and the lumped state of $(\Delta x_A - \Delta x_B)$ can indeed be used. This example has provided a notion of joint estimation of multiple states that have strong geometric connectivity. There are a few applications in the field of navigation that utilize this notion and we reserve some of these as exercises for the reader at the end of this chapter.

9.5 GPS-AIDED INERTIAL ERROR MODELS

GPS is inherently a positioning system that can also derive velocity and acceleration from its code and carrier phase measurements. With a single antenna, a GPS is incapable of producing any attitude or heading information without imposing special kinematic constraints. (Multi-antenna GPS systems though can be contrived to determine attitude and/or heading of the platform). An inertial navigation system, on the other hand, senses acceleration and rotational rates and can derive position, velocity, and attitude/heading from them. An INS yields very high fidelity short-term dynamical information while GPS provides noisy but very stable positioning and velocity information over the long term.

Figure 9.12 depicts a high-level block diagram that represents the mechanization of many typical GPS/INS systems. An inertial measurement unit generates acceleration (Δv) and rotation rate ($\Delta\theta$) data at high rates typically in the range of tens or hundreds of hertz. A strapdown algorithm then accumulates these incremental measurements of Δv and $\Delta\theta$ to generate a reference trajectory consisting of position, velocity and attitude/heading with high dynamic fidelity. The GPS generates data at lower rates typically one to ten hertz. These two sensors are intrinsically very complementary in nature and so the integration of these types of sensors seems very much a natural fit. In Chapter 8, we pointed out that the complementary filter structure has been used extensively for Aided Inertial systems of many kinds of navigation applications, and the GPS-Aided Inertial system is no exception.

Incorporating Inertial Error States

We can build on the core nine-state process model introduced in Section 8.10—the model is described by Eqs. (8.10.9) and (8.10.10). The reference trajectory is nine-dimensional, consisting of three states each of position, velocity, and attitude obtained from the accumulation of inertial measurements through the strapdown algorithm. In the complementary arrangement, the Kalman filter's role is to estimate the deviation of the reference trajectory from the truth trajectory. This is then the established state space. Hence, it is important to note that the state model for the

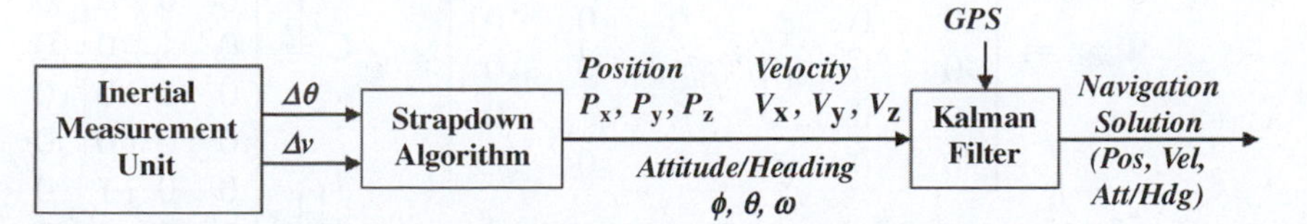

Figure 9.12 High-level block diagram of GPS/INS mechanization.

Kalman filter here consists of inertial *error* quantities, rather than "total" dynamical quantities as in the stand-alone GPS case. Correspondingly, the random-process model should reflect the random character of the errors in the inertial sensor.

For the inertial error model, the chosen navigation reference frame will follow the locally level convention established in Section 9.1, where x points east, y points north, and z is the altitude above the reference WGS-84 ellipsoid. The error models are represented by block diagrams in Figs. 8.10, 8.11, and 8.12 showing separate channels for the north, east, and altitude. These can be regarded, in part, as generic models that are representative of a typical north-oriented locally-level platform INS subject to modest dynamics.

We now consider the augmentation of additional INS error states to the process model. The acceleration and gyro error terms are simple white noise inputs in Figs. 8.10.2 through 8.10.4. To add more fidelity to the model, additional error terms with a time-correlation structure can also be included. Generally, a first-order Gauss-Markov model is sufficient for each of these types of error. To do so, an additional six states, three for accelerometer and three for gyro errors, are needed for augmenting the state vector. They are all of the form:

$$\text{Error process: } x_{k+1} = \phi x_k + w_k \tag{9.5.1}$$

where the state transition parameter $\phi = e^{-\Delta t/\tau}$, Δt being the discrete-time interval and τ the time constant of the exponential autocorrelation function that governs this process (see Example 4.3). The variance of the process noise w, if it is time-invariant, is related to the steady-state variance of the x by the following:

$$\text{Var}\,(w_k) = (1 - \phi^2)\,\text{Var}\,(x_k) \tag{9.5.2}$$

When these additional error states are augmented to the basic 9-state model, the process model can be written in the following partitioned way:

$$\begin{bmatrix} \mathbf{x}_{1-9} \\ \mathbf{x}_{10-15} \end{bmatrix}_{k+1} = \left[\begin{array}{c|c} \mathbf{\Phi}_{\mathbf{INS}} & \mathbf{C}\Delta t \\ \hline \mathbf{0} & \mathbf{\Phi}_{sens} \end{array}\right] \left[\begin{array}{c} \mathbf{x}_{1-9} \\ \hline \mathbf{x}_{10-15} \end{array}\right]_k + \left[\begin{array}{c} \mathbf{w}_{1-9} \\ \hline \mathbf{w}_{10-15} \end{array}\right] \tag{9.5.3}$$

where $\mathbf{\Phi}_{\mathbf{INS}}$ is from Eq. (8.10.10), $\mathbf{0}$ is a submatrix of zeros, $\mathbf{\Phi}_{sens}$ is a state transition submatrix for the six error states of accelerometer and gyro biases in three axes, and $\mathbf{C}$ is a submatrix that provides the appropriate additive connections of the augmented error states (x_{10} through x_{15}) to the INS dynamic equations:

$$\mathbf{\Phi}_{sens} = \begin{bmatrix} \phi_{ax} & 0 & 0 & 0 & 0 & 0 \\ 0 & \phi_{ay} & 0 & 0 & 0 & 0 \\ 0 & 0 & \phi_{az} & 0 & 0 & 0 \\ 0 & 0 & 0 & \phi_{gx} & 0 & 0 \\ 0 & 0 & 0 & 0 & \phi_{gy} & 0 \\ 0 & 0 & 0 & 0 & 0 & \phi_{gz} \end{bmatrix} \qquad \mathbf{C} = \begin{bmatrix} 0 & 0 & 0 & 0 & 0 & 0 \\ 1 & 0 & 0 & 0 & 0 & 0 \\ 0 & 0 & 0 & 1 & 0 & 0 \\ 0 & 0 & 0 & 0 & 0 & 0 \\ 0 & 1 & 0 & 0 & 0 & 0 \\ 0 & 0 & 0 & 0 & 1 & 0 \\ 0 & 0 & 0 & 0 & 0 & 0 \\ 0 & 0 & 1 & 0 & 0 & 0 \\ 0 & 0 & 0 & 0 & 0 & 1 \end{bmatrix}$$

Table 9.3 Comparison of different stability classes of inertial systems

	INS Quality			
		Tactical Grade		
	Navigation grade	High quality	Low quality	Automotive grade
Gyro bias (deg/h)	< 0.01	0.1–1.0	10.0	> 100
Gyro white noise* $(\text{deg/s}/\sqrt{\text{Hz}})$	3×10^{-5}	10^{-3}	10^{-2}	5×10^{-2}
Accelerometer bias (milli-g)	0.01–0.05	0.2–0.5	1.0–10.0	> 10
Accelerometer white noise* $(\text{milli-g}/\sqrt{\text{Hz}})$	0.003–0.01	0.05	0.1	> 0.1

The various inertial sensors categories listed in the table above are largely accepted industry definitions although the performance boundaries are less well defined. The choice of performance ranges represented in the table above result from a survey of various industry sources and paper references (14,15,16). (*Note: The units for the white noises are specified here in terms of the square root of the power spectral density.)

Table 9.3 gives a comparison of sensor error characteristics found in different inertial systems of varying qualities. The table was assembled with data taken from various sources. Here, *high-quality* refers to systems capable of standalone navigation and attitude sensing with excellent accuracy for extended durations of time (typically, hours). By comparison, *medium-quality* systems require external aiding to attain the performance offered by high-quality systems. Otherwise, medium-quality systems are capable of standalone operation over shorter durations. *Automotive grade* sensors require external aiding to provide useful performance and can only offer brief standalone operation.

Another level of sophistication that may be added to the accelerometer and gyro error models is that of accounting for scale factor error. The mathematical relationship that translates the physical quantity a sensor measures to the desired value representing that quantity generally involves a scale factor that may vary due to instrumentation uncertainties. This error may be estimated when the integrated system is in a highly observable state. Whether it is worthwhile to estimate this error depends on how significant the error is and its impact on the overall performance (see Problem 8.5).

Measurement Model And Integration Coupling

With the inclusion of GPS receiver clock states and inertial sensor bias errors (for gyro and accelerometers), we now have a 15-state dynamical process model. More sophisticated error effects, such as scale factor and misalignment, may be included that will raise the state vector dimensionality even more, but for most inertial sensor systems used today, in the medium- and low-quality categories, this model serves adequately as a foundation. From this, we now define a companion measurement model for an integrated GPS/INS system.

Of the 15 states, GPS is only able to provide information for position and velocity. We now consider two ways of integrating the GPS information into the Aided Inertial system. In one that we refer to as a *Loosely Coupled* integration, the GPS information is integrated directly in the form of position and velocity. The measurement model for this is given as follows:

Measurement Model (Loosely Coupled Integration)

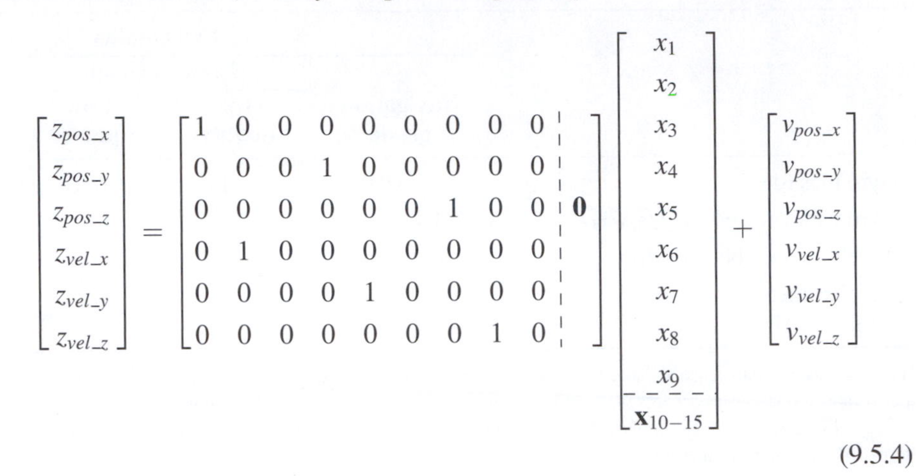

$$\begin{bmatrix} z_{pos\_x} \\ z_{pos\_y} \\ z_{pos\_z} \\ z_{vel\_x} \\ z_{vel\_y} \\ z_{vel\_z} \end{bmatrix} = \left[\begin{array}{ccccccccc|c} 1 & 0 & 0 & 0 & 0 & 0 & 0 & 0 & 0 & \\ 0 & 0 & 0 & 1 & 0 & 0 & 0 & 0 & 0 & \\ 0 & 0 & 0 & 0 & 0 & 0 & 1 & 0 & 0 & \mathbf{0} \\ 0 & 1 & 0 & 0 & 0 & 0 & 0 & 0 & 0 & \\ 0 & 0 & 0 & 0 & 1 & 0 & 0 & 0 & 0 & \\ 0 & 0 & 0 & 0 & 0 & 0 & 0 & 1 & 0 & \end{array}\right] \begin{bmatrix} x_1 \\ x_2 \\ x_3 \\ x_4 \\ x_5 \\ x_6 \\ x_7 \\ x_8 \\ x_9 \\ \hline \mathbf{x}_{10-15} \end{bmatrix} + \begin{bmatrix} v_{pos\_x} \\ v_{pos\_y} \\ v_{pos\_z} \\ v_{vel\_x} \\ v_{vel\_y} \\ v_{vel\_z} \end{bmatrix} \tag{9.5.4}$$

In this arrangement, the position and velocity may be prefiltered especially if comes from an autonomous function such as in the form of a GPS receiver. Care must be taken to account for correlation between the x-y-z components of the position and velocity due to the satellite geometry and, if prefiltered, the time correlation structure of the measurement noise that will not be a "white" sequence.

This straightforward form of integration may be contrasted with another form called *Tightly Coupled* integration where GPS pseudorange and delta range measurements are processed instead of GPS position and velocity. The obvious advantage to this approach over the Loosely Coupled one is that individual satellite measurements can still contribute valuable information to the Kalman filter even if there are less than the requisite number to form a position and velocity solution in a given sampling cycle. The added complexity of a Tightly Coupled approach comes in the form of extra dimensionality in the state vector: two clock states and if the formulation for delta range similar to Eqs. (9.4.2)–(9.4.4) is chosen, three additional delayed position states. This is a small price to pay for the improved performance of a Tightly Coupled integration, unless one is faced with a situation where the raw GPS pseudorange and delta range measurements, usually made within a GPS receiver, is not made available outside of it.

Over the past 10 years, there has been another class of aided inertial solutions known as *Ultra Tight Coupled* integration where the "raw" measurements come from one step further up the signal chain within a GPS receiver, in the form of in-phase and quadrature signal samples (commonly called I-Q samples). The Ultra Tightly Coupled integration philosophy exploits the Aided Inertial solution to also support and extend the signal tracking oftentimes beyond the regular limits of signal-to-noise ratios or interference-to-signal ratios (17,18).

EXAMPLE 9.6

Let us consider a simplified one-dimensional GPS/INS example by revisiting the single-axis INS error model described by Fig. 8.9. For the accelerometer noise a and the gyro noise ε shown in the block diagram, we specify each process to be a combination of white noise plus a first-order Gauss-Markov processes. Accelerometer and gyro sensors tend to have offset errors that are bias-like, an effect that can be captured by choosing a reasonably long time constant ($\tau = 10{,}000$ s) for the Gauss-Markov process and the appropriate standard deviation. For this example, we will choose a high-quality tactical grade IMU with representative values as given in Table 9.3.

Therefore, state process model is given as follows:

x_1 = position error
x_2 = velocity error
x_3 = platform tilt
x_4 = accelerometer bias
x_5 = gyro bias

$$\begin{bmatrix} \dot{x}_1 \\ \dot{x}_2 \\ \dot{x}_3 \\ \dot{x}_4 \\ \dot{x}_5 \end{bmatrix} = \begin{bmatrix} 0 & 1 & 0 & 0 & 0 \\ 0 & 0 & -g & 1 & 0 \\ 0 & \dfrac{1}{R_e} & 0 & 0 & 1 \\ 0 & 0 & 0 & -\dfrac{1}{\tau_a} & 0 \\ 0 & 0 & 0 & 0 & -\dfrac{1}{\tau_g} \end{bmatrix} \begin{bmatrix} x_1 \\ x_2 \\ x_3 \\ x_4 \\ x_5 \end{bmatrix} + \begin{bmatrix} 0 \\ u_a \\ u_g \\ u_{bias\_a} \\ u_{bias\_g} \end{bmatrix} \tag{9.5.7}$$

In the given process model, g is the gravitational acceleration and R_e is the Earth radius. We need to specify the white noise amplitudes of u_a, u_g, u_{bias_a}, and u_{bias_g}. From Table 9.3 for the high-quality tactical grade column, we choose the accelerometer error white noise u_a to be 0.05 milli-g/$\sqrt{\text{Hz}}$ (or 5×10^{-4} m/s^{-2}/$\sqrt{\text{Hz}}$) and the gyro error white noise u_g to be 0.001 deg/s/$\sqrt{\text{Hz}}$ (or 1.75×10^{-5} rad/s/$\sqrt{\text{Hz}}$).

Since x_4 and x_5 are first-order Gauss-Markov processes, according to Eq. (4.2.11), we must specify σ and $\beta = 1/\tau$ separately for the accelerometer and gyro bias processes.

$$u_{bias\_a} = \sqrt{2\sigma_a^2 \beta_a} = \sqrt{\frac{2\sigma_a^2}{\tau_a}}$$

and

$$u_{bias\_g} = \sqrt{2\sigma_g^2 \beta_g} = \sqrt{\frac{2\sigma_g^2}{\tau_g}}$$

From Table 9.3 for the high-quality tactical grade column, we shall choose the accelerometer bias σ_a to be 0.5 milli-g or 0.005 m/s^2 and the gyro bias σ_g to be 1.0 deg/h or 4.85×10^{-6} rad/s. We choose $\tau_a = \tau_g = 10{,}000$ s.

Let the measurement model follow the Loosely Coupled Integration form:

$$\begin{bmatrix} z_{pos} \\ z_{vel} \end{bmatrix} = \begin{bmatrix} 1 & 0 & 0 & 0 & 0 \\ 0 & 1 & 0 & 0 & 0 \end{bmatrix} \begin{bmatrix} x_1 \\ x_2 \\ x_3 \\ x_4 \\ x_5 \end{bmatrix} + \begin{bmatrix} v_{pos} \\ v_{vel} \end{bmatrix} \tag{9.5.8}$$

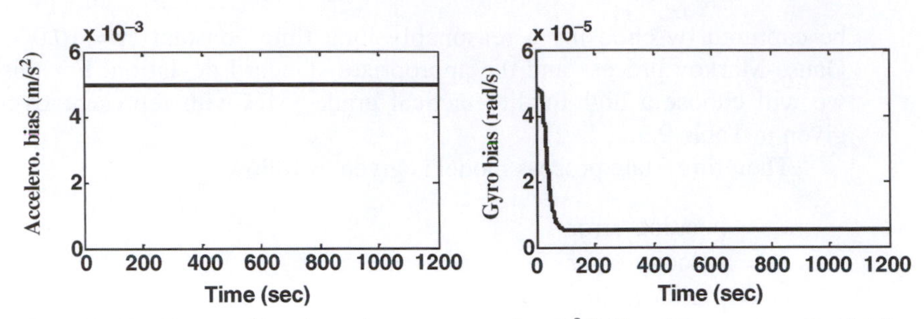

Figure 9.13 Error profiles of accelerometer error in m/s² (left) and gyro error rad/s (right).

The **R** matrix consists of specifying the variance for the position and velocity measurements. In this example, we choose $Ev_{pos}^2 = (2m)^2$ and $Ev_{vel}^2 = (0.1m/s)^2$. The time profiles for accelerometer error (square root of the 4,4 term of the **P** matrix) and for gyro error (square root of the 5,5 term of the **P** matrix) are shown in Fig. 9.13. One quickly infers that the gyro error is very observable in this dynamical system and its associated measurements, but that the accelerometer error is much less so, owing to their different rates of convergence. An error in the accelerometer results in a horizontal acceleration error but we also see a horizontal acceleration error if there is a tilt (i.e., an error in attitude) because of gravity. While at rest or under low dynamics the system, as represented by Eqs. (9.5.7)-(9.5.8), cannot easily separate these two contributions to horizontal acceleration error.

In order for the system to obtain better observability into the accelerometer bias error, it must derive a more accurate estimation of attitude from the external aiding source. However, since GPS is unable to render a direct measurement of attitude (multi-antenna systems notwithstanding), the best alternative to improve attitude estimation accuracy is to subject the platform to vertical acceleration, something only a flying platform is capable of.

We recall the accelerometer error equation from Eq. (8.10.1) for slow-moving platforms, and embellish it with a "hidden" term that represents the nominal acceleration (A_z being the vertical component of acceleration) at a given instance in time:

$$\Delta\ddot{x} = a - g\phi \underline{\underline{- A_z\phi}} \tag{9.5.9}$$

or,

$$\Delta\ddot{x} = a - (g + A_z)\phi$$

When there exists such a vertical acceleration in the platform motion and this is properly taken into account by the dynamical model, Eq. (9.5.9) in effect represents a variation from the usual gravity and its ability to sense platform tilt.

If we now consider this same example subjected to two spells of vertical acceleration that reach 1 g (9.8 m/s²) at T = 300 s and T = 600 s, the attitude error significantly improves at the first maneuver and just slightly more at the second (see Fig. 9.14).

The primary message of this example is that while accelerometer bias is less observable than gyro bias under low dynamics conditions, high dynamic maneuvers can help raise the observability to accelerometer bias errors.

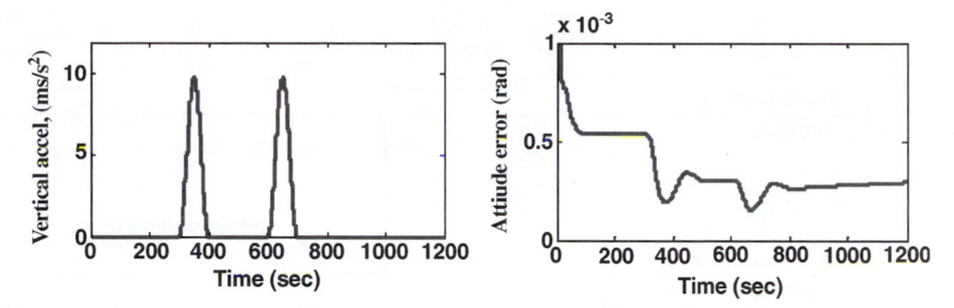

Figure 9.14 Time profile of vertical acceleration in the platform motion (left) and the resulting attitude error profile (right).

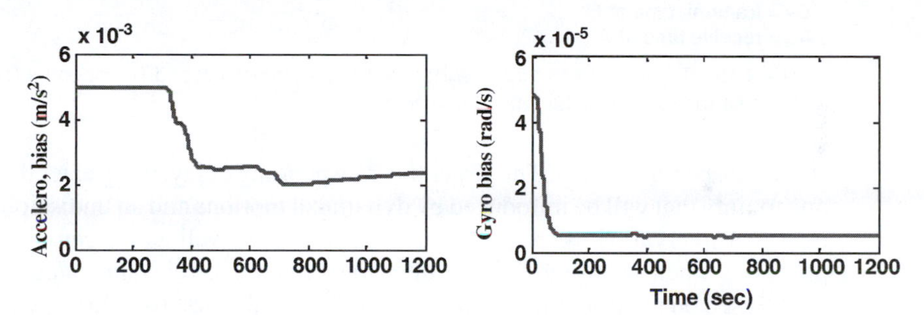

Figure 9.15 Error profile of accelerometer error in m/s^2 (left) and gyro error in rad/s (right) with the presence of vertical acceleration maneuvers as indicated by Figure 9.14.

9.6 COMMUNICATION LINK RANGING AND TIMING

There is a close relationship between communication and navigation systems. Many modern-day navigation systems involve electronic radio-frequency transmissions which are fundamental to communications systems. However, few communications systems, whose primary function is to *communicate*, are designed with the rigorous timing coherency needed to achieve meaningful performance desired by navigation systems. We shall explore the basic principles used by such systems and the use of Kalman filtering for their applications.

Ranging Determination

The fundamental measurement made between two communication radios is the time difference between the time of transmission and the time of arrival of the communication message. In general, the timing bases between the two radios are different on account of independent clocks. This timing difference then represents the propagation delay of the message taken over the physical range between the radios plus the difference between the two independent clocks. This is, in effect, a pseudorange just like the type of measurement discussed in earlier for GPS. Of course, a single measurement of pseudorange between two radios is of little use. In order to separate the range from the timing difference, a return measurement in the reverse direction between the radios is needed (Fig. 9.16). This usually happens after a

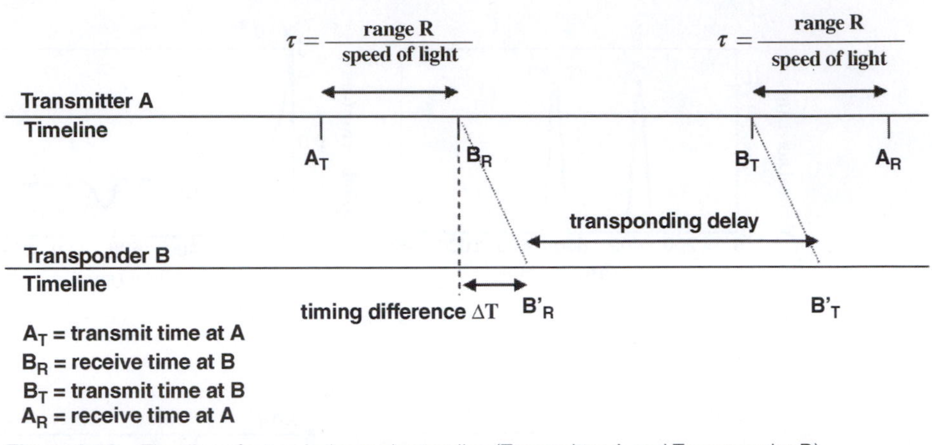

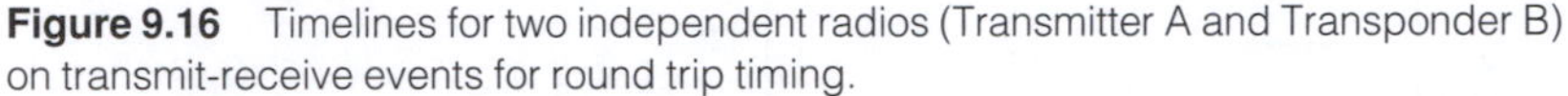

Figure 9.16 Timelines for two independent radios (Transmitter A and Transponder B) on transmit-receive events for round trip timing.

short transponding delay. The larger the transponding delay, the greater the amount of uncertainty that will be introduced by dynamical motions and an undisciplined clocks over that interval. However, we can manage this as well as we can model random processes when we use a Kalman filter to estimate the relevant states.

Based on the timelines shown in Figure 9.16, the first pseudorange is formed as the range equivalent of the receive time at B minus the transmit time at A, i.e., $\rho_{AB} = c\,(B'_R - A_T)$, where c is the speed of light. Similarly, the second pseudorange is $\rho_{BA} = c\,(A_R - B'_T)$. Both pseudoranges are connected differently to the relevant parameters of interest, R and ΔT. Thus, we can form the following two-tuple measurement model:

$$\begin{bmatrix} \rho_{AB} \\ \rho_{BA} \end{bmatrix} = \begin{bmatrix} 1 & 1 \\ 1 & -1 \end{bmatrix} \begin{bmatrix} R \\ \Delta T \end{bmatrix} \tag{9.6.1}$$

This method is sometimes called round trip timing or two-way time transfer. If the communication ranging and timing is being made between two mobile platforms, then the range R is dynamic as is the timing difference ΔT due to the real nature of clocks. And oftentimes, the two pseudorange measurements cannot be made simultaneously one after the other. All of this can be elegantly handled by a Kalman filter as we shall see in the next example.

EXAMPLE 9.7

For this example, suppose that we have a specific need to synchronize the timing between an aircraft and a ground station so the method used is a two-way time transfer as described above. In our example, the clock that is timing the radio in the aircraft is a temperature compensated crystal (TCXO) while the clock in the ground station unit is a highly-stable rubidium. We are therefore entitled to break up the ΔT state found in Eq. (9.6.1) into separate representations of two distinct clocks. Also for R, we can capture the dynamics of the two nodes but in our present example, the dynamics of the ground node is trivial so we only need to represent the dynamics of the aircraft.

Therefore, we choose for simplicity, a one-dimensional inertial aided aircraft system that is equipped with a TCXO-type clock for a communication ranging radio. On the ground, the radio is timed by a rubidium standard and its dynamics are modeled to be stationary.

The process model can be formulated then to be:

$$\begin{bmatrix} \dot{x}_1 \\ \dot{x}_2 \\ \dot{x}_3 \\ \dot{x}_4 \\ \dot{x}_5 \\ \dot{x}_6 \end{bmatrix} = \left[\begin{array}{cc|cc|cc} 0 & 1 & \multicolumn{2}{c|}{\mathbf{0}} & \multicolumn{2}{c}{\mathbf{0}} \\ 0 & 0 & & & & \\ \hline & & 0 & 1 & \multicolumn{2}{c}{\mathbf{0}} \\ \multicolumn{2}{c|}{\mathbf{0}} & 0 & 0 & & \\ \hline & & & & 0 & 1 \\ \multicolumn{2}{c|}{\mathbf{0}} & \multicolumn{2}{c|}{\mathbf{0}} & 0 & 0 \end{array}\right] \begin{bmatrix} x_1 \\ x_2 \\ x_3 \\ x_4 \\ x_5 \\ x_6 \end{bmatrix} + \begin{bmatrix} 0 \\ u \\ u_{f1} \\ u_{g1} \\ u_{f2} \\ u_{g2} \end{bmatrix} \tag{9.6.2}$$

where

x_1 = inter-node range
x_2 = inter-node range rate
x_3 = aircraft clock bias error
x_4 = aircraft clock drift error
x_5 = ground clock bias error
x_6 = ground clock drift error

The corresponding measurement model then becomes:

$$\begin{bmatrix} \rho_{AB} \\ \rho_{BA} \end{bmatrix} = \begin{bmatrix} 1 & 0 & 1 & 0 & -1 & 0 \\ 1 & 0 & -1 & 0 & 1 & 0 \end{bmatrix} \begin{bmatrix} x_1 \\ x_2 \\ x_3 \\ x_4 \\ x_5 \\ x_6 \end{bmatrix} + \begin{bmatrix} v_{AB} \\ v_{BA} \end{bmatrix} \tag{9.6.3}$$

Oftentimes, the two measurements are not made simultaneously but rather sequentially. The beauty of processing communication link ranging and timing measurements in a Kalman filter model is that the sequential nature of these paired measurements are easily handled by the filter that also takes into account the time variation of the states according to the assumed dynamical model.

In a simulation, we choose a model with the following parameters:

$$\begin{aligned} S_{\mathrm{u}} &= 0.01\,(\mathrm{m/s^2})^2/\mathrm{Hz} \\ S_{\mathrm{f1}} &= 1.348 \times 10^{-4}\,(\mathrm{m/s})^2/\mathrm{Hz} \\ S_{\mathrm{g1}} &= 3.548 \times 10^{-6}\,(\mathrm{m/s^2})^2/\mathrm{Hz} \\ S_{\mathrm{f2}} &= 9 \times 10^{-6}\,(\mathrm{m/s})^2/\mathrm{Hz} \\ S_{\mathrm{g2}} &= 1.8 \times 10^{-12}\,(\mathrm{m/s^2})^2/\mathrm{Hz} \end{aligned}$$

The measurement noise variance for each pseudorange is $(2\,\mathrm{m})^2$. The initial error covariance was set to be at 10^6 for each of the six states. Measurements are made every 10 seconds and there is a 5-second interval between each pair of measurements.

The rms error from the Kalman filter covariance analysis is shown over a period of 120 seconds in Fig. 9.17. The ranging rms error is obtained from the $\mathbf{P}_{11}$ term of the $\mathbf{P}$ matrix. However, the relative timing rms error needs to be obtained from: $\mathbf{P}_{33} + \mathbf{P}_{55} - 2\mathbf{P}_{35}$ (see Example 9.5 for a similar evaluation of the relative

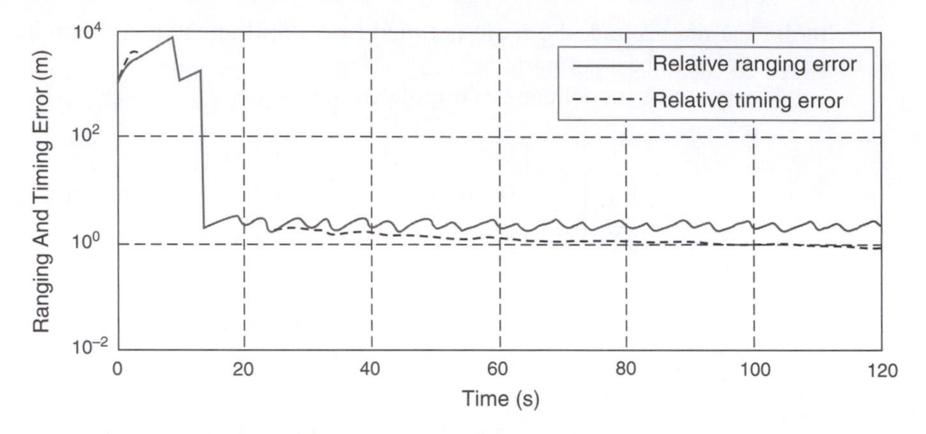

Figure 9.17 Time profile of ranging (solid) and timing (dashed) error for two-way communication link measurements with 30-second updates, with 5-second interval between paired measurements.

difference between two states, which is what the estimation error of relative timing is all about). The ranging and timing errors are at their lowest at each 30-second update except for the very first pair of measurements at $t = 0$ and $t = 4$. It requires a second pair of measurements to fully resolve all the states, including the rate states.

9.7 SIMULTANEOUS LOCALIZATION AND MAPPING (SLAM)

Robotic vehicles of many forms are beginning to permeate our world in a variety of situations. Unmanned airborne, ground, and underwater vehicles have developed capabilities to autonomously roam, explore and monitor their environments. One of the many enabling technologies that have given these vehicles their autonomous capabilities is robotic navigation. Over the past two decades, this burgeoning field has developed some innovative concepts, most of which are based on a rather simple notion called Simultaneous Localization And Mapping, or SLAM (19). At its most fundamental level, autonomous robots are expected to operate in unknown environments. With SLAM, robots learn about their environment through *mapping* while navigating in them. The mapping of its own immediate environment requires the robot to determine the locations of definable objects or landmarks in their vicinity based on knowledge of its position and orientation. Then, as the robot moves within this environment, the locations of those very same objects and landmarks are used to determine the robot's position in a process called *localization*. The ego-sensor sees an *egocentric* point of view so the determination of the robot's position is sometimes known as an estimation of ego-motion (20).

Visual SLAM

While the concept of SLAM can encompass the use of any kind of sensor with the appropriate capability to determine the location of landmarks from the ego-sensor platform, it is the visual sensors that have dominated recent developments in this field. In particular, optical image sensors in the form of videocameras have become exceedingly affordable while delivering progressively increasing quality. It is the result of such high-quality imagery that allows the ego-sensor to identify landmarks and track each and every one without ambiguity over time from sequential images and obtain precise angular measurements when doing so. The sensor used must be able to derive range and bearing to the various landmarks in the local scenery. For example, stereo videocameras are generally needed to obtain range to the landmarks using binocular principles of distance measurement. Scanning lidar systems that can sweep a fine laser beam to capture the returns in order to measure the direction and range to all objects that are nearby in the scenery, have also been considered. More recently within the past 10 years, a single videocamera solution has been proposed (21) although this solution, by itself, is unobservable and has stability issues but it can be made to work reasonably well with an external piece of aiding measurement, such as a dead-reckoning solution that is readily available from the wheeled sensors of a robotic vehicle.

EXAMPLE 9.8: A ONE-DIMENSIONAL SLAM

To get a better understanding of SLAM behavior in terms of how its accuracy changes over time, we simplify the model to a bare bones one-dimensional system where the ego-motion is constrained to somewhere along a linear axis as are the position of the landmarks. The measurement between a landmark and the ego-sensor is that of range (see Fig. 9.18).

1D SLAM Process Model:

$$\begin{bmatrix} \Delta x \\ \Delta \dot{x} \\ \Delta y \end{bmatrix} = \begin{bmatrix} 1 & \Delta t & 0 \\ 0 & 1 & 0 \\ 0 & 0 & 1 \end{bmatrix} \begin{bmatrix} \Delta x \\ \Delta \dot{x} \\ \Delta y \end{bmatrix} + \begin{bmatrix} w_1 \\ w_2 \\ 0 \end{bmatrix} \tag{9.7.1}$$

1D SLAM Measurement Model:

$$\underbrace{r - \hat{r}}_{z} = \begin{bmatrix} 1 & 0 & -1 \end{bmatrix} \begin{bmatrix} \Delta x \\ \Delta \dot{x} \\ \Delta y \end{bmatrix} + v \tag{9.7.2}$$

We specify the remaining Kalman filter parameters to be R = 1 and

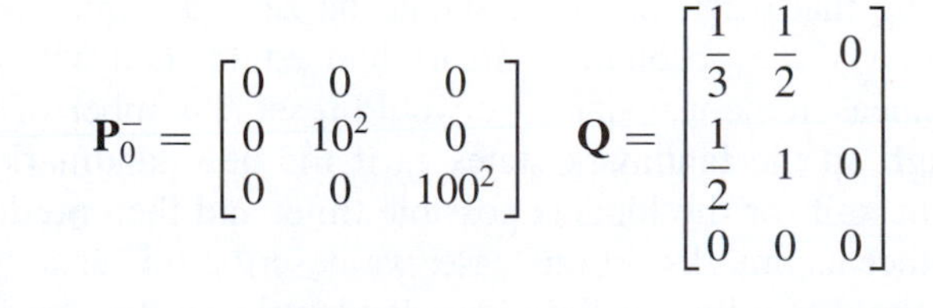

$$\mathbf{P}_0^- = \begin{bmatrix} 0 & 0 & 0 \\ 0 & 10^2 & 0 \\ 0 & 0 & 100^2 \end{bmatrix} \quad \mathbf{Q} = \begin{bmatrix} \frac{1}{3} & \frac{1}{2} & 0 \\ \frac{1}{2} & 1 & 0 \\ 0 & 0 & 0 \end{bmatrix}$$

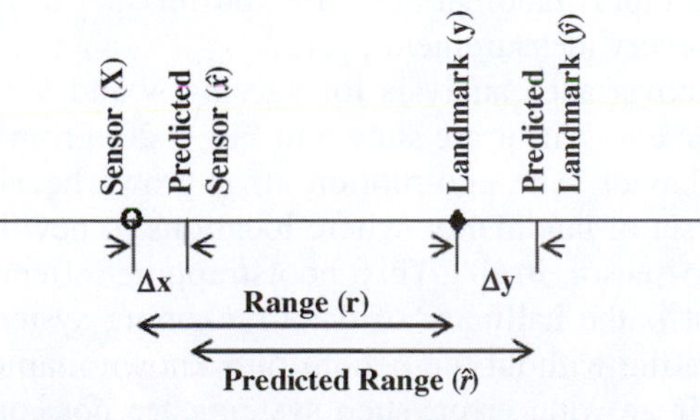

Figure 9.18 Measurement model for one-dimensional SLAM.

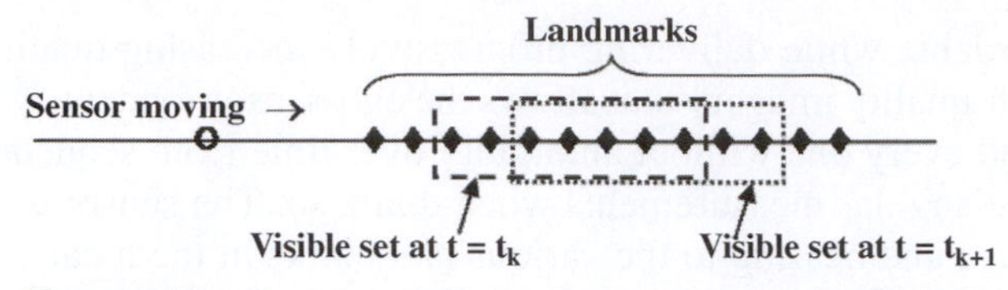

Figure 9.19 Conceptual picture shows seven landmarks being available to range from at each measurement sampling time and two landmarks are "turned over" between every cycle; hence, this example represents $N=7$, $b=2$.

Note that the measurement model has similarities to the relative positioning case of Example 9.5. In both cases, two states become strongly correlated from what is essentially a relative measurement that connects them together. While the relative distance between the two states is observable, the two states are individually not so, such that we did not start the problem out by having some known or perfect information about the initial position.

As the ego-sensor moves along its one-dimensional degree of freedom, the availability of that landmark for a measurement has a limited window. As the local environment changes, such as when a robotic platform wanders about an unfamiliar building, new landmarks continually appear to replace visible landmarks after a brief sequence of measurements. To simulate a similar condition with our one-dimensional model formulated above, we introduce a "buffer" of N discrete but limited landmarks, each one represented by one element of the state vector (Fig. 9.19). With the original two states for our PV model for the ego-motion, our state vector now becomes $N+2$ dimensional. Now, we can conduct an error covariance analysis to see how it behaves over time.

If our scenario is contrived to replace a fixed number of older, existing landmarks, say b, with the same number of new ones at every measurement update, we can simulate this by simply resetting b number of state of the error covariance matrix $\mathbf{P}$ at every cycle. The reset of each state involves initializing the diagonal term associated with that state to be a large value, signifying a lack of *a priori* knowledge of the landmark location. At the same time, the row and column criss-crossing that particular state should be zeroed to nullify any built-up correlation the state of the obsolete landmark had established with all the other states. With each measurement update, we would reset b number of states, uniformly cycling through all the landmark states as if the new landmarks enter the "visible set," remain in it for the longest possible time, and then predictably exit it just like all the other landmarks. In this case, we describe this as a "turnover" of b landmarks (i.e., the discarding of the oldest landmarks in the "buffer" set in favor of being replaced by a new one) at every measurement cycle.

When we perform a covariance analysis for varying N and b, the estimation error for the position of the ego-sensor are shown in Fig. 9.20. From this, we see a not-entirely-surprising behavior. The ego-motion error grows because it gets its position from a changing set of landmarks, where locations of new landmarks are determined from the ego-sensor itself. This bootstrapping effect leads to an accumulation of error that is the hallmark of dead-reckoning systems, a class to which SLAM, when operating without the benefit of a known map of landmarks, truly belongs. Due to their growing error, such systems are not considered fully observable in the rigorous sense but rather are considered partially observable (22).

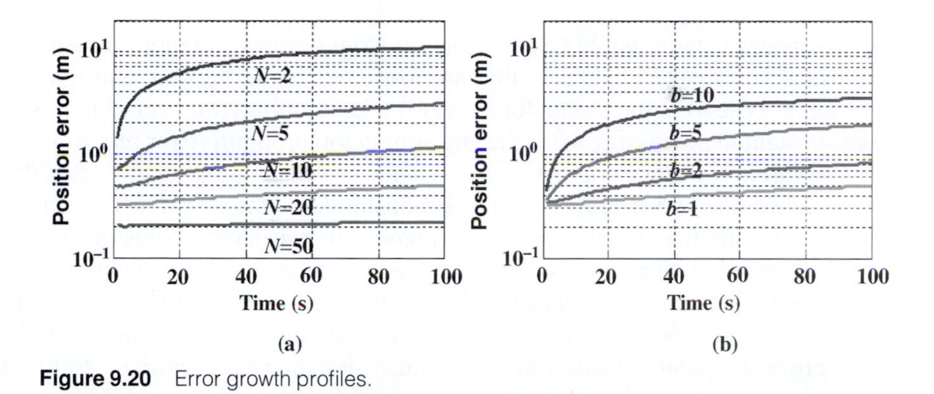

Figure 9.20 Error growth profiles.

Yet, a partially observable system can still be very useful in its own right if its error growth only gradually increases over time from its initial value.

Figure 9.20(a) show the case of replacing landmarks but where the "buffer" of persisting landmarks was of different sizes (N), even as we keep the number of landmark turnover to be one for every measurement cycle, i.e., $b = 1$. Then in Fig. 9.20(b), we fixed the number of landmarks to be 20 ($N = 20$) and varied the number of landmarks that turn over every cycle. These two plots simply affirm the advantage of having numerous landmarks with as small a number of turnovers as possible. The position error growth is minimized, when doing so.

The constant error growth is a result of the introduction of new landmarks with unknown locations as older landmarks are lost. If motion stops and the scenery remains constant, the landmarks will stop turning over so the position error will no longer grow. If previously-established landmarks are revisited, the position error may actually be lowered because the landmark position error determined at an earlier time would be lower than if established for the first time at the time of the revisit. All these different conditions can be easily analyzed using the Kalman filter.

Directional Uncertainty and Other Complexities

Our simplistic one-dimensional SLAM model example introduced before only considered the range relationship between the ego-sensor and the landmarks lined up along the constrained course of travel. Of course, the complexity of the problem increases considerably when we also account for orientation. In SLAM, the directional location of the landmarks with respect to the ego-sensor depends on the ego-sensor's own correct sense of orientation. Any misorientation of this ego-sensor would directly add to the directional and leveling error of the landmarks from the ego-sensor. We leave this next layer of complexity as a problem of 2D SLAM at the end of this chapter.

In the real 3D world, the mis-orientation can have up to three degrees of freedom requiring as many states to be added to three position states. However, it should be apparent that the bulk of the states are made up to represent landmark positions, three states each. If we extend the 2D model into the 3D world, the error state vector we need would consist of 3 position, 3 velocity, 3 orientation, and 3 position per landmark, or $(9 + 3N)$ where N is the maximum number of landmarks we can choose to track at once. If we are to accommodate tracking 20 landmarks, the number of states would be 69 in total. It would grow even more if there is a need to track "dormant" landmarks that are lost, after being viewed, and may be viewed again. Clearly, the brute force way of accounting for everything will lead to substantial processing loads that are simply too prohibitive for real-time operations. A variety of other computational schemes that are simplifications of the optimal models have been proposed and demonstrated in the research literature.

In addition to the navigation algorithms we have addressed here using the Kalman filter, there are considerable challenges beyond these navigation algorithms. One of the foremost challenges is the problem of data association where the landmarks must be identified and correctly and re-associated every processing cycle. With vision sensors for example, this may involve intensive image processing algorithms. For the navigation processing, the extensive body of research work on SLAM produced over the years has produced innovative algorithms that provide improved computational efficiency and robustness to nonlinear conditions, including the growing use of Particle filter methods, an introductory treatment of which was given in Chapter 7. With the foundation laid by our tutorial treatment here, we leave the reader to pursue the finer details of such works in the published literature (23,24,25).

9.8 CLOSING REMARKS

There can be no doubt that the Kalman filter has had a long and fruitful history in enhancing navigation performance in real-world applications. GPS, and the subsequent current pursuit to find non-GPS alternatives to deliver similar levels of GPS performance, have continued this reliance on this useful tool and, especially, the systems methodology that comes with it.

The Kalman filter remains an important tool whether used for first-order analysis in defining a solution concept or for real-time computational processing in navigation implementations. In this chapter, we have kept the treatment of the various issues to a tutorial pace. Readers interested in delving further and deeper into other applications of Kalman filtering can find numerous references in other text books, trade journals, and on the Internet.

PROBLEMS

9.1 Oftentimes, the analysis of a positioning problem with GPS can be simplified considerably by reducing it to a one-dimensional problem to capture just its essence. In this problem, we will consider such a simplified scenario. See the figure showing where the coordinates of two satellites are given as a function of time T. Also, the coordinates of "nominal position" of the GPS receiver are fixed at [0 , 6,378,000] m.

(a) Formulate the **H** matrix for a two-state Kalman filter (consisting of position Δx and clock error Δt). Assume that both states are independent random walk processes, with

$$\mathbf{Q} = \begin{bmatrix} (2.0\,m)^2 & 0 \\ 0 & (0.1\,m)^2 \end{bmatrix}$$

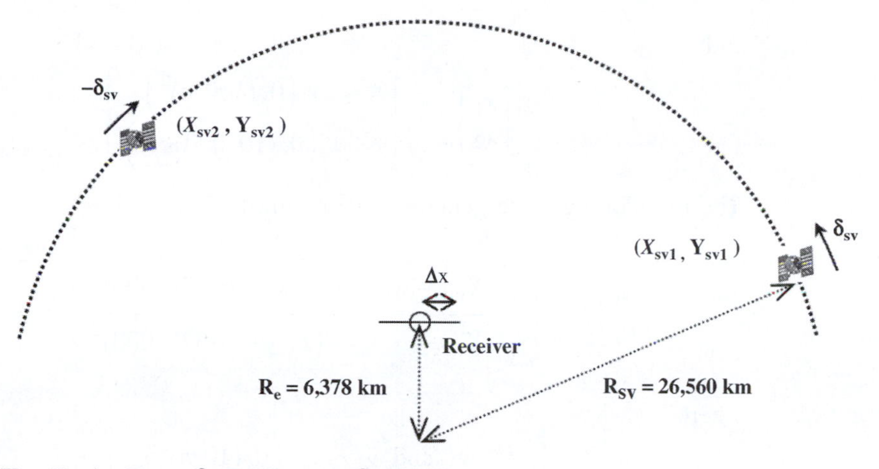

$(X_{sv1}, Y_{sv1}) = [R_{sv}\cos(\delta_{sv}T+30^o), R_{sv}\sin(\delta_{sv}T+30^o)]$

$(X_{sv2}, Y_{sv2}) = [R_{sv}\cos(-\delta_{sv}T+135^o), R_{sv}\sin(-\delta_{sv}T+135^o)]$

$\delta_{sv} = 180/43200\ ^o/s$

Figure P9.1

and a measurement noise covariance matrix for the uncorrelated pseudorange measurements from both satellites to be

$$\mathbf{R} = \begin{bmatrix} (10.0\,m)^2 & 0 \\ 0 & (10.0\,m)^2 \end{bmatrix}$$

Then, run an error covariance analysis with the Kalman filter for 10,000 seconds starting with the following initial **P** matrix:

$$\mathbf{P}_0^- = \begin{bmatrix} (100\,m)^2 & 0 \\ 0 & (10,000\,m)^2 \end{bmatrix}$$

Plot the standard deviation of the position state (square root of the 1,1-term of the updated error covariance matrix) as a function of time.

(b) To perform a Monte Carlo simulation of the same scenario, we need to first generate a random process of the changing position and clock error over time and then to next generate noisy pseudorange measurements based on this random "truth" states and the moving but known satellite locations. Generate samples for a time window from T = 0 to T = 10,000 s. Use a step size of 1 second.

Generate the position and clock error random walk processes according to:

$$\begin{bmatrix} x_1 \\ x_2 \end{bmatrix}_{k+1} = \begin{bmatrix} 1 & 0 \\ 0 & 1 \end{bmatrix} \begin{bmatrix} x_1 \\ x_2 \end{bmatrix}_k + \begin{bmatrix} w_1 \\ w_2 \end{bmatrix}_k$$

where

$$w_1 \sim \mathcal{N}\left(0, (2\,m)^2\right) \text{ and } w_2 \sim \mathcal{N}\left(0, (0.1\,m)^2\right)$$

and

$$\begin{bmatrix} x_1 \\ x_2 \end{bmatrix}_0 = \begin{bmatrix} \sim \mathcal{N}\left(0, (100\,m)^2\right) \\ \sim \mathcal{N}\left(0, (10,000\,m)^2\right) \end{bmatrix}$$

The pseudoranges are generated according to:

$$\begin{bmatrix} \rho_1 \\ \rho_2 \end{bmatrix}_k = \begin{bmatrix} \sqrt{(X_{sv1} - x_1)^2 + (Y_{sv1} - 6,378,000)^2} + x_2 \\ \sqrt{(X_{sv2} - x_1)^2 + (Y_{sv2} - 6,378,000)^2} + x_2 \end{bmatrix} + \begin{bmatrix} v_1 \\ v_2 \end{bmatrix}_k$$

where

$$v_1 \text{ and } v_2 \sim \mathcal{N}\left(0, (10\,m)^2\right)$$

Process the measurements with a two-state Kalman filter and plot the estimate for the position state along with the truth position over the 10,000-second window. Then, plot the position estimation error, i.e., difference between the position estimate and the truth position. Plot the standard deviation obtained from the position variance term of the error covariance matrix to compare the reasonableness of the one-sigma bound to the estimation error samples.

9.2 The geometric dilution of precision (commonly called GDOP) is a well known measure used in the satellite navigation community to specify the solution accuracy based on the geometry of a given set of visible satellites. In essence, it is the "snapshot" rms error of the combined three-dimensional position and time solution that results from a one-meter rms error in the measurements. The term *snapshot* implies that the solution is entirely dependent on the satellite measurements made at one specific instance of time. In other words, the solution is unfiltered.

If an **H** matrix is set up as the linear connection between an n-tuple measurement vector and a four-tuple state vector that consists of three position error states and one range (clock) error state, then GDOP is the square root of the trace of a covariance matrix formed from $(\mathbf{H}^T\mathbf{H})^{-1}$.

We can similarly compute GDOP from a Kalman filter set up with the same four-tuple state vector and **H** matrix, and choose the other filter parameters appropriately. Show that the Kalman filter's updated **P** covariance matrix, after the first measurement update step, is equivalent to $(\mathbf{H}^T\mathbf{H})^{-1}$.

(*Hint*: Review the use of the Information Filter measurement update for this.)

9.3 The usual two-state clock model is given by the transfer function block diagram of Fig. 9.3. To generate a random sequence of samples that would be representative of the clock model, we must first write out a discrete-time process model that specifies the state transition matrix ϕ and the process noise covariance matrix **Q**, from which a pair of random number sequences can be generated to drive the process model.

(a) Generate a random sequence of two-tuple vectors (representing clock phase and frequency errors) by choosing $S_f = 1.0 \times 10^{-21}$ (s/s)^2/Hz and $S_g = 1.2 \times 10^{-22}$ (s/s^2)^2/Hz, the relevant spectral amplitudes given by Eqs. 9.3.1 to 9.3.3. (Be careful to note that the units of the clock random process in this

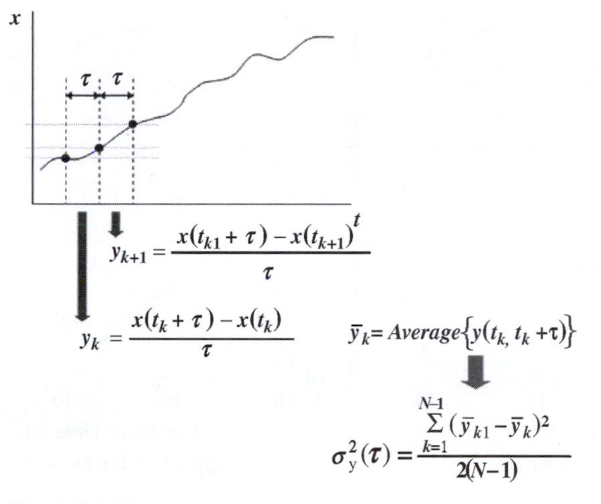

Figure P9.3

case are seconds, and not in the range equivalent form of meters for the clock description in Section 9.3. They are, of course, related through the square of the speed of light.) Use a sampling time $\Delta t = 0.1$ s, and set the initial values of the clock phase to 0 s and the clock frequency to 10^{-8} s/s.

(b) The empirical Allan variance plot is generated by computing a certain type of statistics of the same data record for different averaging intervals τ.

Intuitively, the figure depicts the Allan variance to be a characterization of the statistical variation in the fractional frequency approximated by successive estimates of y. Also, the original Allan variance formula (called the two-sample variance in reference to the paired successive estimates) was somewhat wasteful in not taking advantage of all the data available especially in the calculations involving large τ intervals. A later version known as the overlapping Allan variance remedied that shortcoming.

When working with the clock phase x, however, the direct equation to compute the overlapping Allan variance is given by (26):

$$\sigma_y^2(\tau) = \frac{1}{2(N-2m)\tau^2} \sum_{i=1}^{N-2m} [x_{i+2m} - 2x_{i+m} + x_i]^2$$

where N is the total number of samples, and m is the integer number of samples making up the averaging interval τ. If τ_0 is the intrinsic sampling interval, then $\tau = m\tau_0$.

The square root of the Allan variance (or Allan deviation) is usually plotted on a log-log scale. Compute the Allan deviation of the random process generated in Part (a) for the different values of τ from 0.1 to 100 s and plot it out.

(c) Compute and plot the theoretical Allan deviation from asymptotic parameters related to the noise spectral amplitude given for the clock in Part (a). Does your plot here compare well against the empirical plot for Part (b)?

9.4 In the "Receiver Clock Modeling" subsection in Section 9.3, it was pointed out that a finite-order state model cannot adequately account for the flicker noise

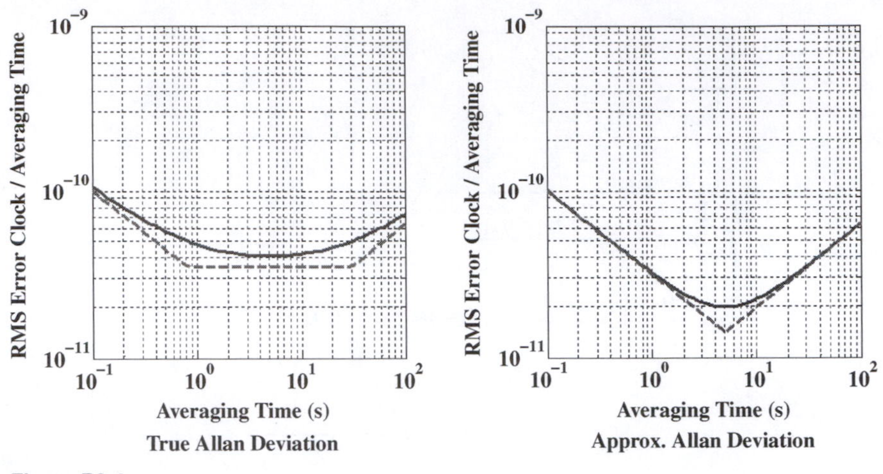

Figure P9.4

component. The relationships given by Eq. (9.3.5) were derived by simply ignoring the flicker noise term. When comparing the figures below for a particular clock model ($h_0 = 2 \times 10^{-21}, h_{-1} = 9 \times 10^{-22}, h_{-2} = 6 \times 10^{-24}$), the Allan deviation curve droops more than it should in the approximate model that ignores the flicker noise component. There have been various suggestions to compensate for the missing term (27) but they are poor approximations at best unless one is willing to invest in one or more additional states to represent the clock.

In this problem, we will explore the use of an additive component to the usual clock model that will approximately capture the effect of bumping up the curve near the valley where the flicker noise component resides.

(a) To determine the most suitable additive component to use, choose one of the following models by generating a random sequence with each and then evaluating its Allan deviation with the processing algorithm written in Problem 9.3:
   (i) First-order Gauss-Markov model with $\sigma = 1$ and $\beta = 1/3$
   (ii) Second-order Gauss-Markov model with $\sigma = 1$ and $\omega_0 = 1/3$
   (iii) Integrated Gauss-Markov model with $\sigma = 1$ and $\beta = 1/3$

   Which of these will work best to "boost" the V-shaped curve of the approximate Allan deviation curve to emulate the flicker noise component?

(b) Form a new clock model that additively combines the two-state approximate model with the chosen "boost" model from Part (a) with appropriately tuned parameters to achieve the desired results, i.e., as close as possible to the true Allan deviation.

(c) What is the minimum number of states needed to represent this combined model?

9.5 Unlike most real-life positioning problems, the positioning of a train is unique in being essentially a one-dimensional measurement situation—its one degree of freedom is along the track to which it is constrained. To take advantage of this reduction in the model dimensionality, the exact trajectory of the railroad track must be known and linearized for the approximate vicinity of the train's location.

(a) Begin by assuming the standard four-variable GPS measurement model:

$$\bar{z} = \rho - \hat{\rho} = [h_x(t) \quad h_y(t) \quad h_z(t) \quad 1] \begin{bmatrix} \Delta x \\ \Delta y \\ \Delta z \\ c\Delta t \end{bmatrix}$$

Let the rail track (see the figure) be described by the set of parametric equations:

$$x = A\cos\psi$$
$$y = B\sin\psi\cos\beta$$
$$z = B\sin\psi\sin\beta$$

where ψ is the parametric variable and β is a fixed angle of inclination of the elliptical track whose semimajor and semiminor axes are A and B. Using a linearized approximation of the above set of parametric equations, rewrite the measurement model such that the state varibles comprise just ψ and $c\Delta t$. How is the new linear measurement connection vector $\mathbf{h}$ written in terms of $h_x(t), h_y(t), h_z(t)$?

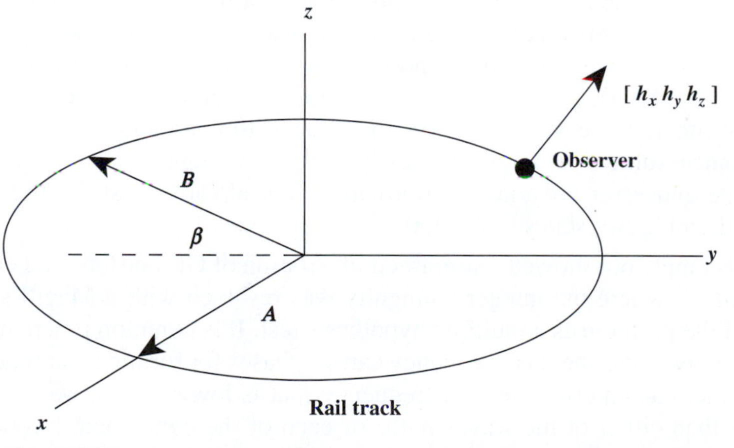

Figure P9.5

(b) What is the minimum number of satellites required to solve the positioning problem of the train with given track trajectory information, as is the case here?

(c) Formulate a similar measurement model that allows for dynamical motion (start out with the position-velocity model of Eqs. 9.3.6 and 9.3.11) by also assuming the same track described above.

(d) What is the random-process model (specify $\boldsymbol{\phi}$ and $\mathbf{Q}$ matrices) for the modified state vector that corresponds to the position-velocity model.

9.6 In differential carrier phase methods, there is a set of unknown cycle ambiguities that must be resolved in order to extract the high precision positioning that the carrier phase measurement is ultimately capable of providing (see Section 9.2). When these so-called double differenced measurements are formed (differencing across receivers followed by differencing across satellites), the ambiguities embedded in these measurements are therefore associated, not with individual satellites, but rather with pairs of satellites. For example, a double differenced carrier phase might be written out as

$$\phi_{AB}^{(pq)} = \left(\phi_A^{(p)} - \phi_B^{(p)}\right) - \left(\phi_A^{(q)} - \phi_B^{(q)}\right) \tag{P9.6.1}$$

where the carrier phase measurements made at Receivers A and B for satellites p and q are combined as shown.

When processing a particular double differenced measurement with a Kalman filter, the cycle ambiguity estimated using this measurement is therefore associated with a pair of satellites. Over time, given the dynamic nature of satellite tracking, we must be able to accommodate a seamless transition when the pairing of satellites is changed in the event a satellite is lost from the tracking list.

Consider the following scenario. From $t=0$ to 300 seconds, the tracking list of satellites consist of: SV1, SV2, SV3, SV4, and SV5. During this time, we chose the following pairing of satellites for the differencing scheme: (SV1-SV2), (SV1-SV3), (SV1-SV4), and (SV1-SV5). For this, the associated ambiguity states are N_{12}, N_{13}, N_{14}, and N_{15}. The Kalman filter would provide estimates of these states and a 4×4 submatrix of the error covariance **P** matrix. Then, at $t=300$, tracking of SV1 was discontinued so we would then choose next to rearrange the pairings to be (SV2-SV3), (SV2-SV4), and (SV2-SV5), and form new ambiguity states N_{23}, N_{24}, and N_{25}. Write out the steps to form the state estimates and the associated error covariance submatrix for these newly-defined ambiguity states, from the state estimate and error covariance information that already exist for the previously-defined ambiguity states at $t=300$.

9.7 Example 6.4 showed a simplified illustration of GPS differential carrier phase positioning where the integer ambiguity was resolved with a Magill scheme that treated the problem as a multiple hypothesis test. It is common practice in the GPS community to use the dual frequency carrier phases for forming a composite carrier phase that has an effective beat frequency that is lower, i.e., a wavelength that is longer than either of the wavelengths of each of the component frequencies. The widelane wavelength combined with a coarse bound provided by the pseudoranges permitted the eventual resolution of the widelane integer ambiguity.

In this problem, we take a different tack at solving the problem by leaving the two frequencies as separate measurements and, therefore, as separate integer ambiguities to resolve. Formulate a Magill scheme where the hypothesis space is two-dimensional.

(a) Write out the three-tuple measurement equation that includes a pseudorange measurement, a carrier phase measurement at the L1 frequency, and another carrier phase measurement at the L2 frequency.
(b) Complete a Kalman filter description by specifying $\boldsymbol{\phi}$, **Q**, and **R**.
(c) Generate a measurement sequence using random numbers for the Gaussian noise corrupting the measurements according to the geometry given in Example 6.4. Design a Magill Kalman filter to process the measurements. Demonstrate the successful convergence of the hypothesis that represents the correct pair of integer ambiguities.

(d) Letting the initial uncertainty of the position and ambiguity states to be large, say, 10^6, run a 10-second solution of the Magill hypothesis tester. Plot the convergence of the *a posteriori* probability of the correct hypothesis and the profile of the weighted state estimates associated with all hypotheses.

9.8 In this problem, we will first re-create the two-way comm ranging and timing model of Example 9.7 but with a Monte Carlo simulation where noisy measurements are generated and processed by the Kalman filter and the states estimated accordingly.

(a) First generate a simple two-dimensional aircraft dynamic profile as given in the accompanying figure. The scenario starts with the aircraft being at (−10,000 m, 0), flying for 200 seconds at 100 meters per second due north (i.e., positive y direction). Generate a 200-second sequence of line-of-sight range from the aircraft to the ground station. Also, generate a 200-second sequence of clock bias errors for both independent clocks using two-state clock models and parameters given in Example 9.7. Form a one-way pseudorange measurement from the aircraft to the ground station once every 30 seconds, and then form a return one-way pseudorange measurement from the ground station to the aircraft 5 seconds after each aircraft-to-ground transmission.

(b) Formulate a Kalman filter that cycles through once every second (i.e., $\Delta t = 1$ s) for 200 seconds of the flight profile and for processing the measurements whenever they are available. Compute the difference of the clock bias error estimates made between the aircraft and the ground and subtract this from the truth values of the clock bias error difference that had been simulated in Part (a). This represents the error in the relative timing between the aircraft and ground receivers. Plot this relative timing error together with the standard deviation bounds derived from the error covariance **P** matrix of the Kalman filter.

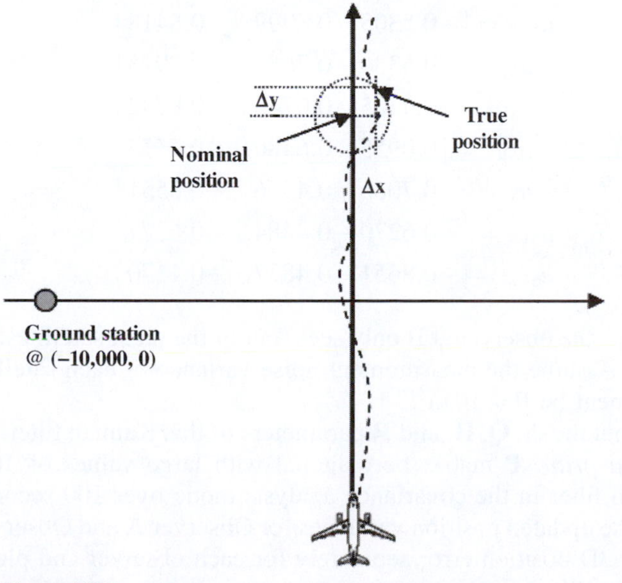

Figure P9.8

9.9 One of the main features of GPS that has made it so versatile and useful is its allowance for the receiver clock to run independently of system timing, which makes GPS a pseudoranging-type system. In 1996, Misra presented a profound finding about a certain weakness of pseudoranging-type systems in terms of yielding slightly poorer measurement geometries (28). Essentially, the key point is that this issue goes away if the receiver clock is known perfectly. Later on, Hwang, et al. (29) extended this finding to differential GPS systems and the use of two-way time transfer via a communication link to synchronize a receiver to its reference base station. It should be pointed out while this degradation of the measurement geometries associated with pseudoranging systems is small in nominal conditions, it can also quickly deteriorate with marginal conditions.

In this problem, we explore the use of accurate relative timing between two receivers and show how one viewing good satellite geometry can assist the other viewing poor satellite geometry. Consider an observer at Location A solving for his GPS position and time with a five-state model consisting of three position states and two clock states. Define the first two position states to be related to the horizontal dimensions and the third state to the vertical dimension. Each position state is a random walk process with a process noise variance of $10\,\text{m}^2$ per step, each time step interval being one second in duration. Let the clock take on the model represented by Fig. 9.3 and whose process noise covariance is defined by Eqs. (9.3.1)–(9.3.3). Let $S_f = 1.348 \times 10^{-4}\,(\text{m/s})^2/\text{Hz}$ and $S_g = 3.548 \times 10^{-6}\,(\text{m/s}^2)^2/\text{Hz}$. Set the initial uncertainties of all states to be equally large, say 10^4.

Now formulate a Kalman filter model with a state vector that is made up of two five-state subvectors that would represent two observers, one a Location A and the other at Location B. The state vector now has 10 states. Suppose that the observers at the two locations have different local environments such that the observer at A sees eight satellites while the observer at B sees only four of the eight satellites seen at A. The unit direction vectors from the satellites to the observer at A are given by:

$$
\begin{aligned}
u_{SV1} &= [-0.2031 \quad 0.6498 \quad -0.7325] \\
u_{SV2} &= [-0.5305 \quad 0.0999 \quad -0.8418] \\
u_{SV3} &= [0.6336 \quad -0.7683 \quad -0.0913] \\
u_{SV4} &= [0.4705 \quad 0.0745 \quad -0.8792] \\
u_{SV5} &= [0.0955 \quad 0.6480 \quad -0.7557] \\
u_{SV6} &= [0.7086 \quad -0.4356 \quad -0.5551] \\
u_{SV7} &= [-0.6270 \quad -0.4484 \quad -0.6371] \\
u_{SV8} &= [-0.8651 \quad -0.4877 \quad -0.1170]
\end{aligned}
$$

At the same time, the observer at B only sees four of the eight satellites: SV1, SV2, SV4, and SV5. Assume the measurement noise variance of each satellite pseudorange measurement be $\text{R} = 10\,\text{m}^2$.

(a) Write out the $\boldsymbol{\phi}$, **Q**, **H**, and **R** parameters of this Kalman filter. Letting the initial *a priori* **P** matrix be diagonal with large values of 10^4, run this Kalman filter in the covariance analysis mode over 100 seconds, storing away the updated position variances for Observer A and Observer B. Form an rms 3D position error separately for each observer and plot the result over the 100-second duration.

(b) Suppose that a two-way ranging and timing set of measurements is made over a communication link between Observer A and Observer B once a second, only after the initial 100-second period prescribed in Part (a). Assume that the two-way ranging and timing process outlined in Eq. (9.6.1) is computed separately, but that the outcome of completing one pair of two way measurements is an estimate of the relative timing error between the clocks biases maintained at Observer A and Observer B's radios. To process this scalar relative timing measurement in the 10-state Kalman filter formulated in Part (a), what is the **H** matrix (which is truly a 1×10 vector in this case) for this? Assuming that the measurement noise variance **R** for this scalar measurement is $4\ \text{m}^2$, rerun the Kalman filter of Part (a) for 100 seconds followed by another 20 seconds where the relative timing measurement is processed once a second as well. Plot the rms 3D position error for each observer noting the improvement achieved when the relative timing measurement started contributing to the estimation solution. What is the percentage improvement in the rms 3D position error of Observer B after 120 seconds as compared to the error after 100 seconds, just prior to the contribution of the relative timing measurement?

9.10 If we expand the 1D SLAM model of Example 9.8 into a two-dimensional solution space, we will use a measurement of range and direction between the ego-sensor and the available landmarks. Assume that our state vector now consists of the following elements:

$$\begin{aligned}
x_1 &= \text{Ego-sensor } x\text{-position error} \\
x_2 &= \text{Ego-sensor } x\text{-velocity error} \\
x_3 &= \text{Ego-sensor } y\text{-position error} \\
x_4 &= \text{Ego-sensor } y\text{-velocity error} \\
x_5 &= \text{Bearing error} \\
x_6 &= \text{Landmark 1 } x\text{-position error} \\
x_7 &= \text{Landmark 1 } y\text{-position error} \\
&\vdots \\
x_{4+2N} &= \text{Landmark } N\ x\text{-position error} \\
x_{5+2N} &= \text{Landmark } N\ y\text{-position error}
\end{aligned}$$

With the ego-sensor position and the landmark positions being two-dimensional, the measurements to each landmark would be range and bearing. Since the bearing measurement is generally made by the ego-sensor with respect to the orientation of the body of the platform that may not perfectly know its orientation to the outside world, we assume a nontrivial bearing error term $\Delta\theta$ that directly relates to the bearing measurement.

The general nonlinear measurement model is:

$$\begin{bmatrix} r \\ \theta \end{bmatrix} = \begin{bmatrix} \sqrt{(x_L - x)^2 + (y_L - y)^2} + v_r \\ \tan^{-1}\left(\dfrac{y_L - y}{x_L - x}\right) - \Delta\theta + v_\theta \end{bmatrix} \tag{P9.10.1}$$

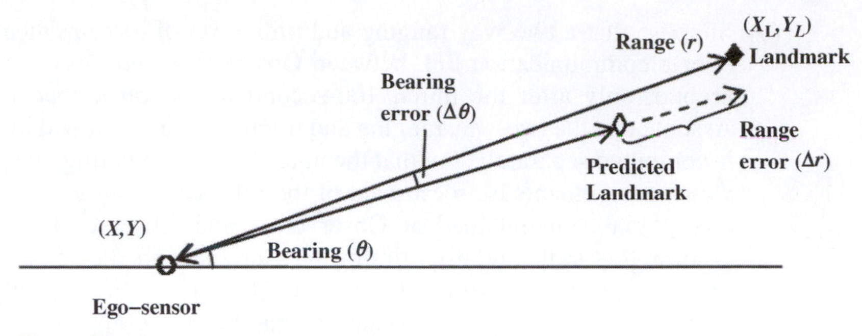

Figure P9.10

We can now fit this linearized form into the full measurement model that connects to measurements to the state vector. Here, we show a measurement model for two landmarks.

(a) Write out the process model for the given state vector, assuming that the ego-sensor have two independent sets of position-velocity models for the x and y axes each with a driving white noise function of spectral amplitude $1\ \mathrm{m/s^2/\sqrt{Hz}}$ and the states having zero initial uncertainties. Assume that the bearing error is a random walk with a driving white noise function of spectral amplitude $0.0001\ \mathrm{rad/s/\sqrt{Hz}}$ and with an initial uncertainty of 0.01 radians. Also, assume each two-state landmark sub-vector to be random constants with a suitably large initial uncertainty.

(b) Write out the measurement model after linearizing Eq. (P9.10.1). Assume the measurement noise to have a sigma of 0.001 radians.

REFERENCES CITED IN CHAPTER 9

1. *Global Positioning System*, Vol. I–VII, Red Book Series, The Institute of Navigation, Washington, DC, 1980–2010.
2. G. Gibbons, "GNSS Interoperability," *Inside GNSS*, January/February 2011, pp. 28–31.
3. *Department of Defense World Geodetic System 1984—Its Definition and Relationships with Local Geodetic Systems*, Defense Mapping Agency tech. rept. No. 8350.2, Sept. 1987.
4. R.J. Muellerschoen, Y.E. Bar-Sever, W.I. Bertiger, D.A. Stowers, "NASA's Global DGPS for high precision users," *GPS World*, 12(1): 14–20 (2001).
5. D.W. Allan, "Statistics Of Atomic Frequency Standards," *Proceedings of the IEEE*, 54(2): 221–230 (Feb. 1966).
6. I. Abramzon, S. Baranushkin, A. Gubarev, O. Rotova, V. Tapkov, "High-Stability Miniature OCXOs Based On Advanced IHR Technology," white paper available from Magic Xtal Ltd., Omsk, Russia (http://magicxtal.com).
7. J.R. Vig, *Quartz Resonator & Oscillator Tutorial*, March 2007.
8. P. Misra, P. Enge, *Global Positioning System: Signals, Measurements, and Performance*, 2nd Edition, Ganga-Jamuna Press: Massachusetts, 2006, p. 153.
9. L. Levy, notes from "Applied Kalman Filtering," Course 447, Navtech Seminars, 2007.
10. P.J.G. Teunissen, "A New Method for Fast Carrier Phase Ambiguity Estimation," *Proceedings of the IEEE Position, Location and Navigation Symposium PLANS' 94*, Las Vegas, NV, April 11–15, pp. 562–573.
11. P.Y. Hwang and R.G. Brown, "GPS Geodesy: Experimental Results Using the Kalman Filter Approach," *Proceedings of the IEEE Electronic and Aerospace Systems Conference (EASCON '85)*, Washington, DC, Oct. 28–30, 1985, pp. 9–18.

12. S.M. Lichten, "Precise Estimation of Tropospheric Path Delays with GPS Techniques," TDA Progress Report 42-100, JPL, NASA, 1990.
13. P. Y. Hwang, notes from "Applied Kalman Filtering," Course 447, Navtech Seminars, 2008.
14. M.G. Petovello, M.E. Cannon, G. LaChapelle, "Benefits of Using a Tactical-Grade IMU for High-Accuracy Positioning," *Navigation: J. Inst. of Navigation*, 51(1): 1–12 (Spring 2004).
15. E. Foxlin,"Motion Tracking Requirements and Technologies," in *Handbook of Virtual Environments: Design, Implementation, and Applications*, K. Stanney (ed.), Lawrence Erlbaum Associates, New Jersey, 2002, p. 183.
16. S. Nasiri, "A Critical Review of MEMS Gyroscopes Technology and Commercialization Status, InvenSense white paper.
17. A.S. Abbott, W.E. Lillo, "Global Positioning Systems and Inertial Measuring Unit Ultratight Coupling Method," U.S. Patent 6,516,021, The Aerospace Corporation, CA, 2003.
18. D. Gustafson, J. Dowdle, "Deeply Integrated Code Tracking: Comparative Performance Analysis," *Proceedings of Institute of Navigation GPS/GNSS Conference*," Portland, OR, 2003.
19. J.J. Leonard, H.F. Durrant-Whyte, "Simultaneous Map Building and Localization For An Autonomous Mobile Robot," *Proceedings of the IEEE Int. Workshop on Intelligent Robots and Systems, pages* 1442–1447, Osaka, Japan, November 1991.
20. H.F. Durrant-Whyte, T. Bailey, "Simultaneous Localization and Mapping (SLAM): Part I The Essential Algorithms," *Robotics and Automation Magazine*, 13: 99–110, (2006).
21. A.J. Davison, I.D. Reid, N.D. Molton, O. Stasse, "MonoSLAM: Real-Time Single Camera SLAM," *IEEE Transactions on Pattern Analysis and Machine Intelligence*, 29(6): 1052–1067 (June 2007).
22. T.A. Vidal Calleja, *Visual Navigation in Unknown Environments*, Ph.D Dissertation, Universitat Politecnica de Catalunya, Barcelona, 2007.
23. L.M. Paz, P. Jensfelt, J.D. Tardos, J. Neira, "EKF SLAM updates in O(n) with Divide and Conquer SLAM," *Proceedings of the IEEE Int. Conf. Robotics and Automation*, Rome, Italy, April 2007.
24. M. Montemerlo and S. Thrun, "Simultaneous localization and mapping with unknown data association using FastSLAM," *Proceeding of IEEE International Conferences on Robotics and Automation*, 2003, pp. 1985–1991.
25. A. Brooks and T. Bailey,"HybridSLAM: Combining FastSLAM and EKF-SLAM for Reliable Mapping," *Workshop on the Algorithmic Fundamentals of Robotics*, Guanajuato, Mexico, Dec. 2008.
26. W.J. Riley, *Handbook of Frequency Stability Analysis*, National Institute of Standards and Technology (NIST) Special Publication 1065, U.S. Government Printing Office, Washington, DC, 2008, p. 16.
27. R.G. Brown, P.Y. Hwang, *Introduction to Random Signals and Applied Kalman Filtering*, 3rd edition, New York: Wiley, 1997, pp. 431, 450.
28. P. Misra, "The Role of the Clock in a GPS Receiver," *GPS World*, 7(4): 60–6 (1996).
29. P.Y. Hwang, G.A. McGraw, B.A. Schnaufer, D.A. Anderson, "Improving DGPS Accuracy with Clock Aiding Over Communication Links," *Proceedings of the ION GNSS-2005*, pp, 1961–1970.

APPENDIX A

Laplace and Fourier Transforms

Both Laplace and Fourier transforms are used frequently in random signal analysis. Laplace transforms are especially useful in analysis of systems that are governed by linear differential equations with constant coefficients. In this application the transformation makes it possible to convert the original problem from the world of differential equations to simpler algebraic equations. Fourier transforms, on the other hand, simply furnish equivalent descriptions of signals in the time and frequency domains. For example, the autocorrelation function (time) and power spectral density function (frequency) are Fourier transform pairs.

Short tables of common Laplace and Fourier transforms will now be presented for reference purposes.

A.1 THE LAPLACE TRANSFORM

Electrical engineers usually first encounter Laplace transforms in circuit analysis, and then again in linear control theory. In both cases the central problem is one of finding the system response to an input initiated at $t = 0$. Since the time history of the system prior to $t = 0$ is summarized in the form of the initial conditions, the ordinary one-sided Laplace transform serves us quite well. Recall that it is defined as

$$F(s) = \int_{0+}^{\infty} f(t)e^{-st}\,dt \tag{A.1}$$

The defining integral is, of course, insensitive to $f(t)$ for negative t; but, for reasons that will become apparent shortly, we *arbitrarily* set $f(t) = 0$ for $t < 0$ in one-sided transform theory. The integral of Eq. (A.1) has powerful convergence properties because of the e^{-st} term. We know it will always converge somewhere in the

Table A.1 Common One-Sided Laplace Transform Pairs[a]

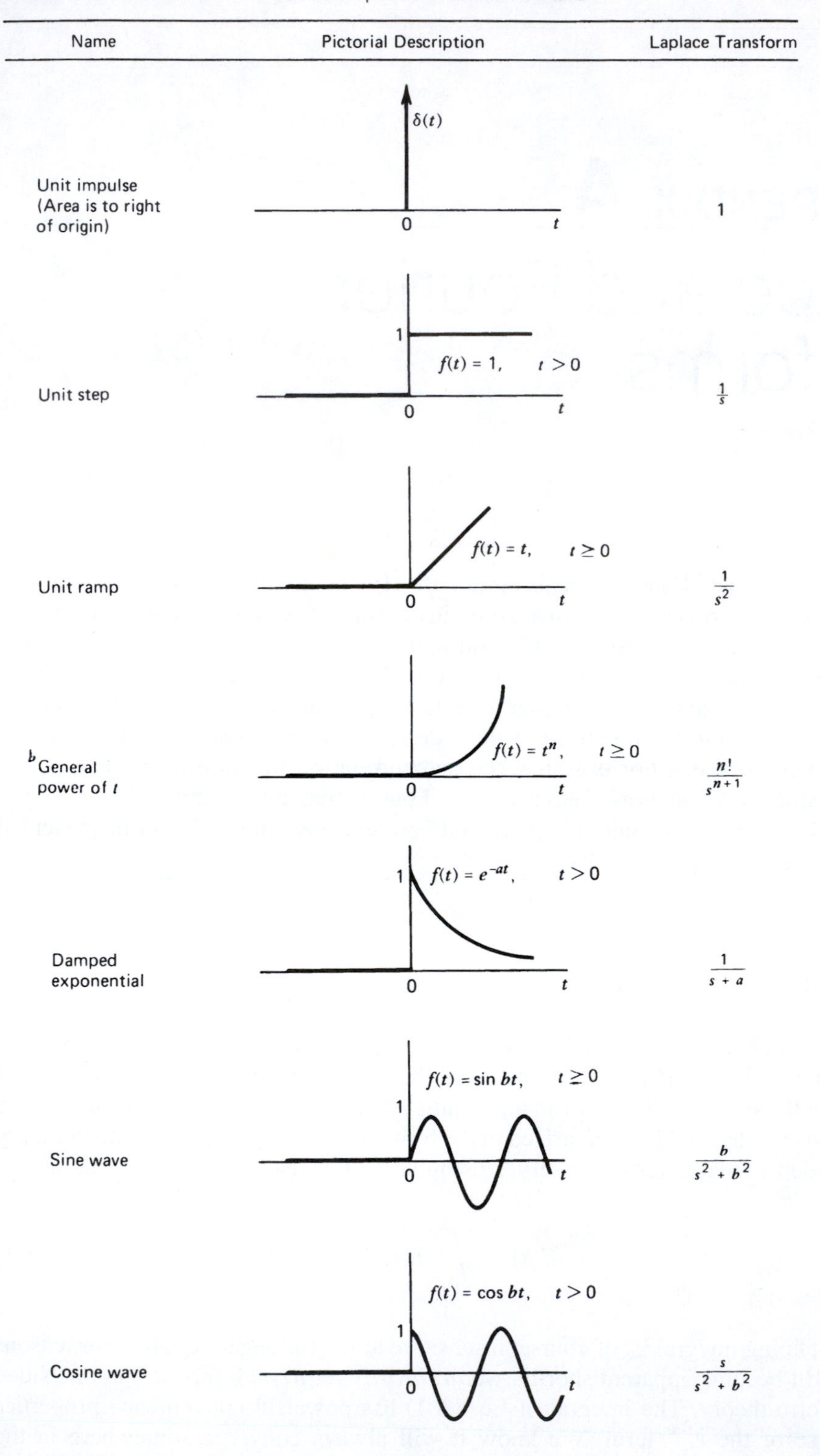

Name	Pictorial Description	Laplace Transform
Unit impulse (Area is to right of origin)	$\delta(t)$	1
Unit step	$f(t) = 1, \quad t > 0$	$\frac{1}{s}$
Unit ramp	$f(t) = t, \quad t \geq 0$	$\frac{1}{s^2}$
[b]General power of t	$f(t) = t^n, \quad t \geq 0$	$\frac{n!}{s^{n+1}}$
Damped exponential	$f(t) = e^{-at}, \quad t > 0$	$\frac{1}{s+a}$
Sine wave	$f(t) = \sin bt, \quad t \geq 0$	$\frac{b}{s^2+b^2}$
Cosine wave	$f(t) = \cos bt, \quad t > 0$	$\frac{s}{s^2+b^2}$

Table A.1 (*Continued*)

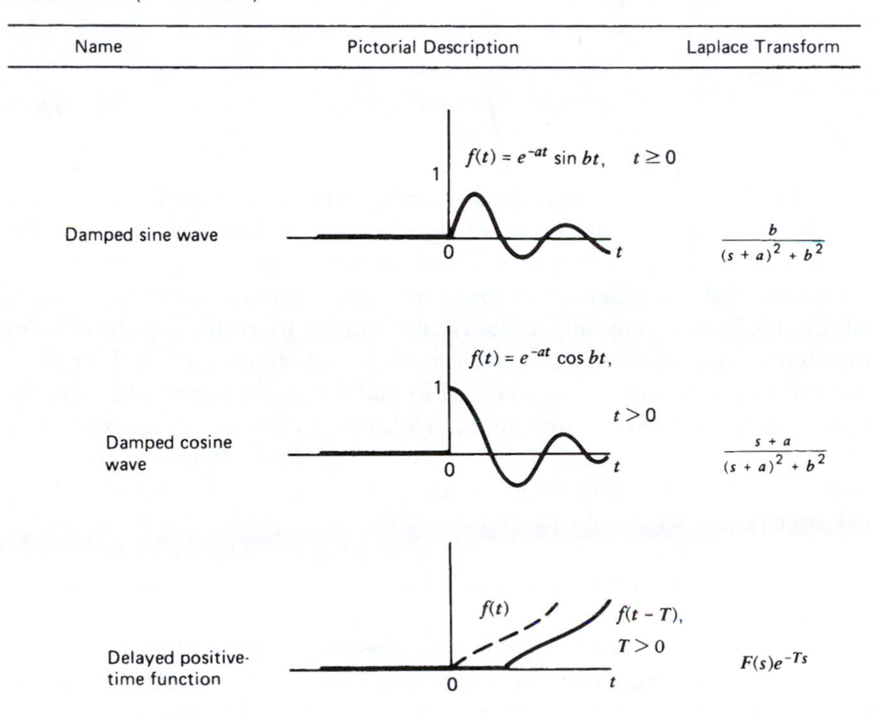

Name	Pictorial Description	Laplace Transform
Damped sine wave	$f(t) = e^{-at} \sin bt, \quad t \geq 0$	$\dfrac{b}{(s+a)^2 + b^2}$
Damped cosine wave	$f(t) = e^{-at} \cos bt, \quad t > 0$	$\dfrac{s+a}{(s+a)^2 + b^2}$
Delayed positive-time function	$f(t)$; $f(t-T), \; T > 0$	$F(s)e^{-Ts}$

[a] Time functions having a discontinuity at $t = 0$ are intentionally left undefined at the origin in this table. The missing value does not affect the direct transform.
[b] When n is not an integer, $n!$ must be interpreted as the gamma function; $\Gamma(n+1)$.

right-half s-plane, provided that we consider only inputs (and responses) that increase no faster than at some fixed exponential rate. This is usually the case in circuits and control problems, and hence the actual region of convergence is of little concern. A common region of convergence is tacitly assumed to exist somewhere, and we simply adopt a table look-up viewpoint for getting back and forth between the time and complex s domains. For reference purposes a brief list of common transform pairs is given in Table A.1. Note again that we have intentionally defined all time functions in the table to be zero for $t < 0$. We will have occasion later to refer to such functions as *positive-time* type functions. It is also worth mentioning that the impulse function of one-sided transform theory is considered to have all its area to the right of the origin in the limiting process. Thus it is a positive-time function. (The word *function* is abused a bit in describing an impulse, but this is common usage, so it will be continued here.)

A.2 THE FOURIER TRANSFORM

The Fourier transform is used widely in communications theory where we often wish to consider signals that are nontrivial for both positive and negative time. Thus, a two-sided transform is appropriate. Recall that the Fourier transform of $f(t)$ is defined as

$$F(j\omega) = \int_{-\infty}^{\infty} f(t)e^{-j\omega t}\,dt \tag{A.2}$$

We know, through the evolution of the Fourier transform from the Fourier series, that $F(j\omega)$ has the physical significance of signal spectrum. The parameter ω in Eq. (A.2) is $(2\pi) \times$ (frequency in hertz), and in elementary signal analysis we usually consider ω to be real. This leads to obvious convergence problems with the defining integral, Eq. (A.2), and is usually circumvented simply by restricting the class of time functions being considered to those for which convergence exists for real ω. The two exceptions to this are constant (*d-c*) and harmonic (sinusoidal) signals. These are usually admitted by going through a limiting process that leads to Dirac delta functions in the ω domain. Even though the class of time functions allowed is somewhat restrictive, the Fourier transform is still very useful because many physical signals just happen to fit into this class (e.g., pulses and finite-energy signals). If we take convergence for granted, we can form a table of transform pairs, just as we did with Laplace transforms, and Table A.2 gives a brief list of common Fourier transform pairs.

For those who are more accustomed to one-sided Laplace transforms than Fourier transforms, there are formulas for getting from one to the other. These are especially useful when the time functions have either even or odd symmetry. Let $f(t)$ be a time function for which the Fourier transform exists, and let

$$\begin{aligned} \mathfrak{F}[f(t)] &= \text{Fourier transform of } f(t) \\ F(s) &= \text{one-sided Laplace transform of } f(t) \end{aligned}$$

Then, if $f(t)$ is even,

$$\mathfrak{F}[f(t)] = F(s)|_{s=j\omega} + F(s)|_{s=-j\omega} \tag{A.3}$$

or if $f(t)$ is odd,

$$\mathfrak{F}[f(t)] = F(s)|_{s=j\omega} - F(s)|_{s=-j\omega} \tag{A.4}$$

These formulas follow directly from the defining integrals of the two transforms.

Table A.2 Common Fourier Transform Pairs

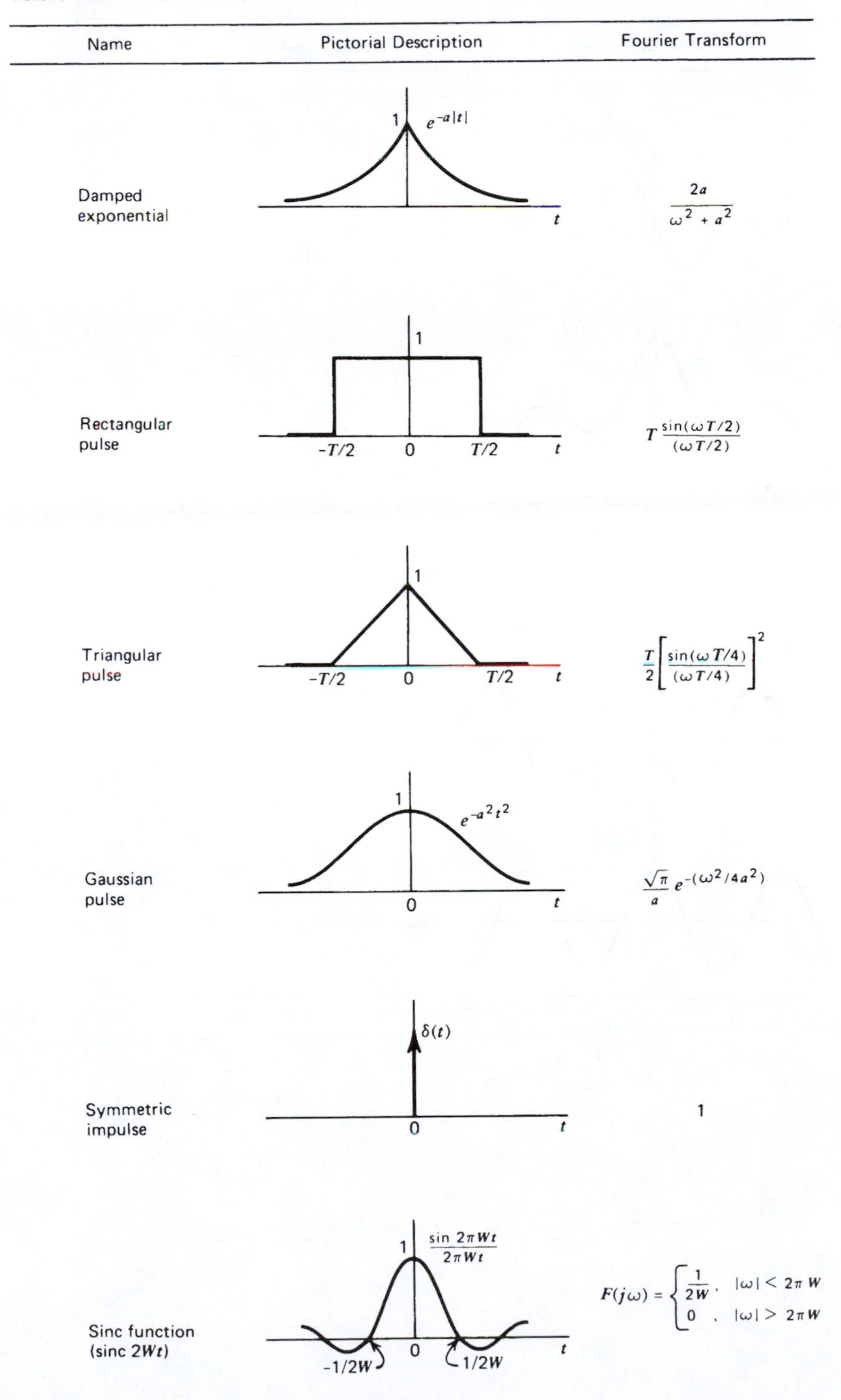

Name	Pictorial Description	Fourier Transform
Damped exponential	$e^{-a\|t\|}$; peak 1; axis t	$\frac{2a}{\omega^2 + a^2}$
Rectangular pulse	height 1; $-T/2$, 0, $T/2$; axis t	$T\frac{\sin(\omega T/2)}{(\omega T/2)}$
Triangular pulse	peak 1; $-T/2$, 0, $T/2$; axis t	$\frac{T}{2}\left[\frac{\sin(\omega T/4)}{(\omega T/4)}\right]^2$
Gaussian pulse	$e^{-a^2t^2}$; peak 1; 0; axis t	$\frac{\sqrt{\pi}}{a}e^{-(\omega^2/4a^2)}$
Symmetric impulse	$\delta(t)$; 0; axis t	1
Sinc function (sinc $2Wt$)	$\frac{\sin 2\pi Wt}{2\pi Wt}$; peak 1; $-1/2W$, 0, $1/2W$; axis t	$F(j\omega) = \begin{cases} \frac{1}{2W}, & \|\omega\| < 2\pi W \\ 0, & \|\omega\| > 2\pi W \end{cases}$

Table A.2 (*Continued*)

Name	Pictorial Description	Fourier Transform
Damped sine wave	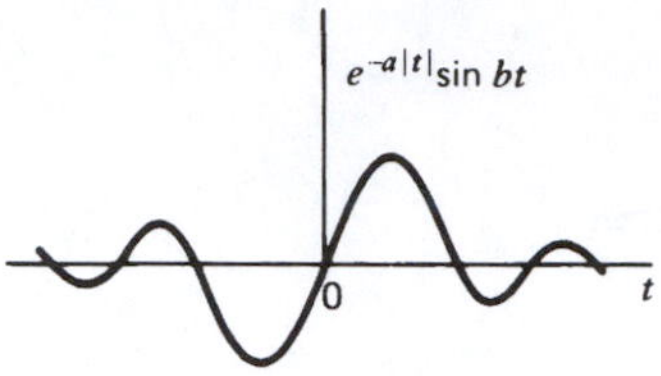	$\frac{ja}{a^2+(\omega+b)^2} - \frac{ja}{a^2+(\omega-b)^2}$
Damped cosine wave	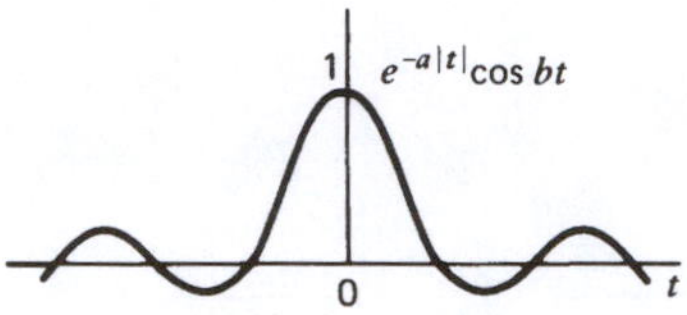	$\frac{a}{a^2+(\omega+b)^2} + \frac{a}{a^2+(\omega-b)^2}$
Constant	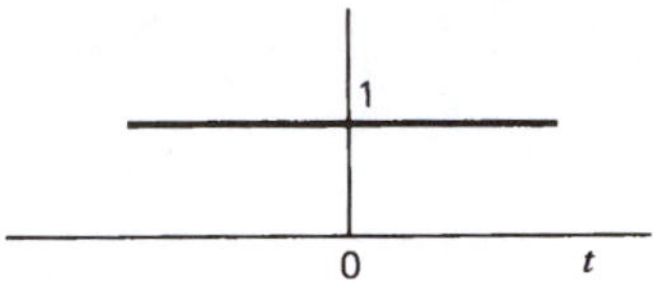	$2\pi\,\delta(\omega)$
Sine wave	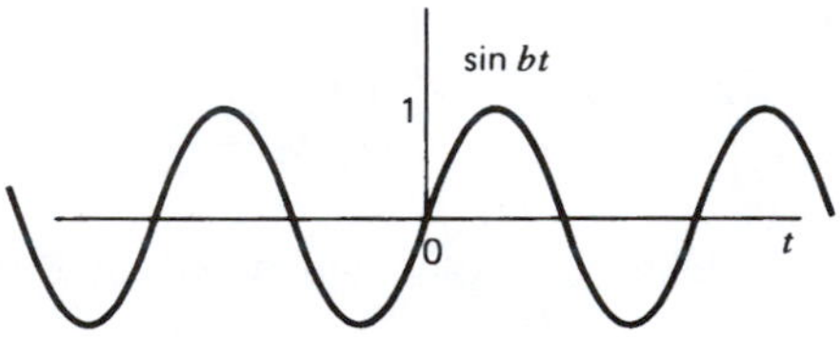	$j\pi\,\delta(\omega+b) - j\pi\,\delta(\omega-b)$
Cosine wave	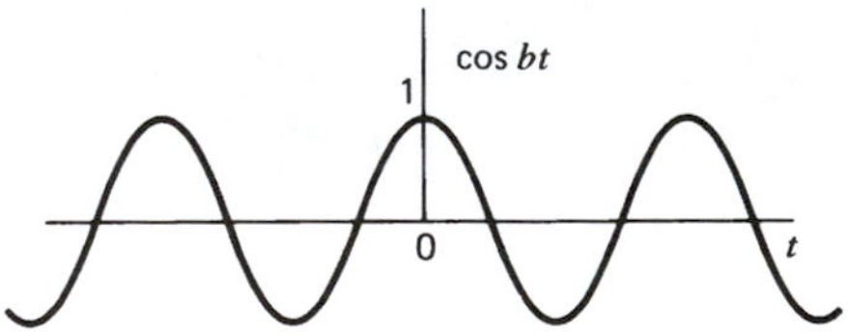	$\pi\,\delta(\omega+b) + \pi\,\delta(\omega-b)$

APPENDIX B

The Continuous Kalman Filter

About a year after his paper on discrete-data filtering, R. E. Kalman coauthored a second paper with R. S. Bucy on continuous filtering (1). This paper also proved to be a milestone in the area of optimal filtering. Our approach here will be somewhat different from theirs, in that we will derive the continuous filter equations as a limiting case of the discrete equations as the step size becomes small.* Philosophically, it is of interest to note that we begin with the discrete equations and then go to the continuous equations. So often in numerical procedures, we begin with the continuous dynamical equations; these are then discretized and the discrete equations become approximations of the continuous dynamics. Not so with the Kalman filter! The discrete equations are exact and stand in their own right, provided, of course, that the difference equation model of the process is exact and not an approximation.

The continuous Kalman filter is probably not as important in applications as the discrete filter, especially in real-time systems. However, the continuous filter is important for both conceptual and theoretical reasons, so this appendix will be devoted to the basics of continuous filtering.

B.1 TRANSITION FROM THE DISCRETE TO CONTINUOUS FILTER EQUATIONS

First, we assume the process and measurement models to be of the form:

$$\text{Process model:} \quad \dot{\mathbf{x}} = \mathbf{F}\mathbf{x} + \mathbf{G}\mathbf{u} \tag{B.1.1}$$

$$\text{Measurement model:} \quad \mathbf{z} = \mathbf{H}\mathbf{x} + \mathbf{v} \tag{B.1.2}$$

* One has to be careful in applying the methods of ordinary differential calculus to stochastic differential equations. Such methods are legitimate here only because we are dealing exclusively with linear dynamical systems [see, e.g., Jazwinski (2)]. It is worth noting that it is only the estimate equation that is stochastic. The error covariance equation (which is nonlinear) is deterministic. It depends only on the model parameters, which are not random and are assumed to be known.

where

$$E\left[\mathbf{u}(t)\mathbf{u}^T(\tau)\right] = \mathbf{Q}\delta(t-\tau) \tag{B.1.3}$$

$$E\left[\mathbf{v}(t)\mathbf{v}^T(\tau)\right] = \mathbf{R}\delta(t-\tau) \tag{B.1.4}$$

$$E\left[\mathbf{u}(t)\mathbf{v}^T(\tau)\right] = \mathbf{0} \tag{B.1.5}$$

We note that in Eqs. (B.1.1) and (B.1.2), $\mathbf{F}$, $\mathbf{G}$, and $\mathbf{H}$ may be time-varying. Also, by analogy with the discrete model, we assume that $\mathbf{u}(t)$ and $\mathbf{v}(t)$ are vector white noise processes with zero crosscorrelation. The covariance parameters $\mathbf{Q}$ and $\mathbf{R}$ play roles similar to $\mathbf{Q}_k$ and $\mathbf{R}_k$ in the discrete filter, but they do not have the same numerical values. The relationships between the corresponding discrete and continuous filter parameters will be derived presently.

Recall that for the discrete filter,

$$\mathbf{Q}_k = E\left[\mathbf{w}_k\mathbf{w}_k^T\right] \tag{B.1.6}$$

$$\mathbf{R}_k = E\left[\mathbf{v}_k\mathbf{v}_k^T\right] \tag{B.1.7}$$

To make the transition from the discrete to continuous case, we need the relations between $\mathbf{Q}_k$ and $\mathbf{R}_k$ and the corresponding $\mathbf{Q}$ and $\mathbf{R}$ for a small step size Δt. Looking at $\mathbf{Q}_k$ first and referring to Eq. (3.9.10), we note that $\boldsymbol{\phi} \approx \mathbf{I}$ for small Δt and thus

$$\mathbf{Q}_k \approx \underset{\text{small } \Delta t}{\int\int} \mathbf{G}(\xi)E\left[\mathbf{u}(\xi)\mathbf{u}^T(\eta)\right]\mathbf{G}^T(\eta)\,d\xi\,d\eta \tag{B.1.8}$$

Next, substituting Eq. (B.1.3) into (B.1.8) and integrating over the small Δt interval yield

$$\mathbf{Q}_k = \mathbf{GQG}^T\,\Delta t \tag{B.1.9}$$

The derivation of the equation relating $\mathbf{R}_k$ and $\mathbf{R}$ is more subtle. In the continuous model $\mathbf{v}(t)$ is white, so simple sampling of $\mathbf{z}(t)$ leads to measurement noise with infinite variance. Hence, in the sampling process, we have to imagine averaging the continuous measurement over the Δt interval to get an equivalent discrete sample. This is justified because $\mathbf{x}$ is not white and may be approximated as a constant over the interval. Thus, we have

$$\begin{aligned}\mathbf{z}_k &= \frac{1}{\Delta t}\int_{t_{k-1}}^{t_k} \mathbf{z}(t)\,dt = \frac{1}{\Delta t}\int_{t_{k-1}}^{t_k} [\mathbf{Hx}(t) + \mathbf{v}(t)]\,dt \\ &\approx \mathbf{Hx}_k + \frac{1}{\Delta t}\int_{t_{k-1}}^{t_k} \mathbf{v}(t)\,dt\end{aligned} \tag{B.1.10}$$

The discrete-to-continuous equivalence is then

$$\mathbf{v}_k = \frac{1}{\Delta t} \int_{\text{small } \Delta t} \mathbf{v}(t)\, dt \tag{B.1.11}$$

From Eq. (B.1.7) we have

$$E\left[\mathbf{v}_k \mathbf{v}_k^T\right] = \mathbf{R}_k = \frac{1}{\Delta t^2} \int\int_{\text{small } \Delta t} E\left[\mathbf{v}(u)\mathbf{v}^T(v)\right] du\, dv \tag{B.1.12}$$

Substituting Eq. (B.1.4) into (B.1.12) and integrating yield the desired relationship

$$\mathbf{R}_k = \frac{\mathbf{R}}{\Delta t} \tag{B.1.13}$$

At first glance, it may seem strange to have the discrete measurement error approach ∞ as $\Delta t \to 0$. However, this is offset by the sampling rate becoming infinite at the same time.

In making the transition from the discrete to continuous case, we first note from the error covariance projection equation (i.e., $\mathbf{P}_{k+1}^- = \boldsymbol{\phi}_k \mathbf{P}_k \boldsymbol{\phi}_k^T + \mathbf{Q}_k$) that $\mathbf{P}_{k+1}^- \to \mathbf{P}_k$ as $\Delta t \to 0$. Thus, we do not need to distinguish between a priori and a posteriori $\mathbf{P}$ matrices in the continuous filter. We proceed with the derivation of the continuous gain expression. Recall that the discrete Kalman gain is given by (see Fig. 4.1)

$$\mathbf{K}_k = \mathbf{P}_k^- \mathbf{H}_k^T \left(\mathbf{H}_k \mathbf{P}_k^- \mathbf{H}_k^T + \mathbf{R}_k\right)^{-1} \tag{B.1.14}$$

Using Eq. (B.1.3) and noting that $\mathbf{R}/\Delta t \gg \mathbf{H}_k \mathbf{P}_k^- \mathbf{H}_k^T$ lead to

$$\mathbf{K}_k = \mathbf{P}_k^- \mathbf{H}_k^T \left(\mathbf{H}_k \mathbf{P}_k^- \mathbf{H}_k + \mathbf{R}/\Delta t\right)^{-1} \approx \mathbf{P}_k^- \mathbf{H}_k^T \mathbf{R}^{-1} \Delta t$$

We can now drop the subscripts and the super minus on the right side and we obtain

$$\mathbf{K}_k = \left(\mathbf{P}\mathbf{H}^T \mathbf{R}^{-1}\right) \Delta t \tag{B.1.15}$$

We *define* the continuous Kalman gain as the coefficient of Δt in Eq. (B.1.15), that is,

$$\mathbf{K} \triangleq \mathbf{P}\mathbf{H}^T \mathbf{R}^{-1} \tag{B.1.16}$$

Next, we look at the error covariance equation. From the projection and update equations (Fig. 5.9), we have

$$\begin{aligned}
\mathbf{P}_{k+1}^- &= \boldsymbol{\phi}_k \mathbf{P}_k \boldsymbol{\phi}_k^T + \mathbf{Q}_k \\
&= \boldsymbol{\phi}_k (\mathbf{I} - \mathbf{K}_k \mathbf{H}_k) \mathbf{P}_k^- \boldsymbol{\phi}_k^T + \mathbf{Q}_k \\
&= \boldsymbol{\phi}_k \mathbf{P}_k^- \boldsymbol{\phi}_k^T - \boldsymbol{\phi}_k \mathbf{K}_k \mathbf{H}_k \mathbf{P}_k^- \boldsymbol{\phi}_k^T + \mathbf{Q}_k
\end{aligned} \tag{B.1.17}$$

We now approximate $\boldsymbol{\phi}_k$ as $\mathbf{I} + \mathbf{F}\Delta t$ and note from Eq. (B.1.15) that $\mathbf{K}_k$ is of the order of Δt. After we neglect higher-order terms in Δt, Eq. (B.1.17) becomes

$$\mathbf{P}_{k+1}^- = \mathbf{P}_k^- + \mathbf{F}\mathbf{P}_k^-\Delta t + \mathbf{P}_k^-\mathbf{F}^T\Delta t - \mathbf{K}_k\mathbf{H}_k\mathbf{P}_k^- + \mathbf{Q}_k \tag{B.1.18}$$

We next substitute the expressions for $\mathbf{K}_k$ and $\mathbf{Q}_k$, Eqs. (B.1.15) and (B.1.9), and form the finite difference expression

$$\frac{\mathbf{P}_{k+1}^- - \mathbf{P}_k^-}{\Delta t} = \mathbf{F}\mathbf{P}_k^- + \mathbf{P}_k^-\mathbf{F}^T - \mathbf{P}_k^-\mathbf{H}^T\mathbf{R}^{-1}\mathbf{H}_k\mathbf{P}_k^- + \mathbf{G}\mathbf{Q}\mathbf{G}^T \tag{B.1.19}$$

Finally, passing to the limit as $\Delta t \to 0$ and dropping the subscripts and super minus lead to the matrix differential equation

$$\begin{aligned} \dot{\mathbf{P}} &= \mathbf{F}\mathbf{P} + \mathbf{P}\mathbf{F}^T - \mathbf{P}\mathbf{H}^T\mathbf{R}^{-1}\mathbf{H}\mathbf{P} + \mathbf{G}\mathbf{Q}\mathbf{G}^T \\ \mathbf{P}(0) &= \mathbf{P}_0 \end{aligned} \tag{B.1.20}$$

Next, consider the state estimation equation. Recall the discrete equation is

$$\hat{\mathbf{x}}_k = \hat{\mathbf{x}}_k^- + \mathbf{K}_k\left(\mathbf{z}_k - \mathbf{H}_k\hat{\mathbf{x}}_k^-\right) \tag{B.1.21}$$

We now note that $\hat{\mathbf{x}}_k^- = \boldsymbol{\phi}_{k-1}\hat{\mathbf{x}}_{k-1}$. Thus, Eq. (B.1.21) can be written as

$$\hat{\mathbf{x}}_k = \boldsymbol{\phi}_{k-1}\hat{\mathbf{x}}_{k-1} + \mathbf{K}_k(\mathbf{z}_k - \mathbf{H}_k\boldsymbol{\phi}_{k-1}\hat{\mathbf{x}}_{k-1}) \tag{B.1.22}$$

Again, we approximate $\boldsymbol{\phi}$ as $\mathbf{I} + \mathbf{F}\Delta t$. Then, neglecting higher-order terms in Δt and noting that $\mathbf{K}_k = \mathbf{K}\Delta t$ lead to

$$\hat{\mathbf{x}}_k - \hat{\mathbf{x}}_{k-1} = \mathbf{F}\hat{\mathbf{x}}_{k-1}\Delta t + \mathbf{K}\Delta t(\mathbf{z}_k - \mathbf{H}_k\hat{\mathbf{x}}_{k-1}) \tag{B.1.23}$$

Finally, dividing by Δt, passing to the limit, and dropping the subscripts yield the differential equation

$$\hat{\mathbf{x}} = \mathbf{F}\hat{\mathbf{x}} + \mathbf{K}(\mathbf{z} - \mathbf{H}\hat{\mathbf{x}}) \tag{B.1.24}$$

Equations (B.1.16), (B.1.20) and (B.1.24) comprise the continuous Kalman filter equations and these are summarized in Fig. B.1. If the filter were to be implemented on-line, note that certain equations would have to be solved in real time as indicated in Fig. B.1. Theoretically, the differential equation for $\mathbf{P}$ could be solved off-line, and the gain profile could be stored for later use on-line. However, the main $\hat{\mathbf{x}}$ equation must be solved on-line, because $\mathbf{z}(t)$, that is, the noisy measurement, is the input to the differential equation.

The continuous filter equations as summarized in Fig. B.1 are innocent looking because they are written in matrix form. They should be treated with respect, though. It does not take much imagination to see the degree of complexity that results when they are written out in scalar form. If the dimensionality is high, an analog implementation is completely unwieldy.

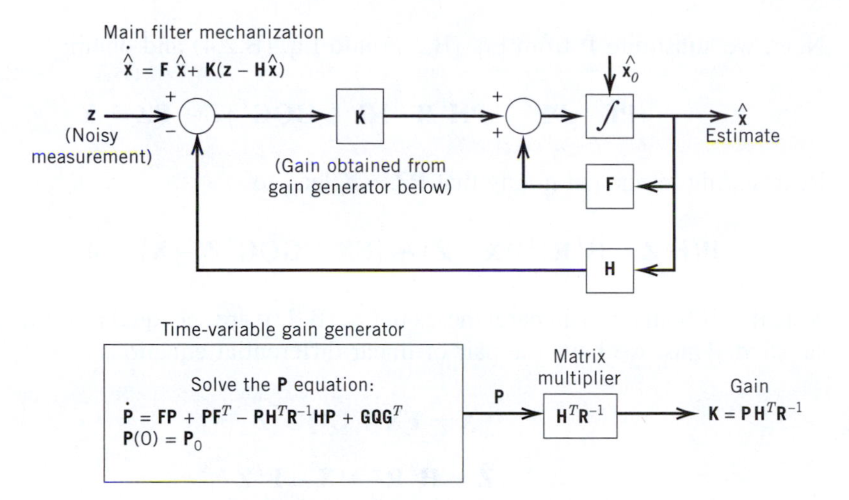

Figure B.1 On-line block diagram for the continuous Kalman filter.

Note that the error covariance equation must be solved in order to find the gain, just as in the discrete case. In the continuous case, though, a differential rather than difference equation must be solved. Furthermore, the differential equation is nonlinear because of the $\mathbf{PH}^T\mathbf{R}^{-1}\mathbf{HP}$ term, which complicates matters. This will be explored further in the next section.

B.2 SOLUTION OF THE MATRIX RICCATI EQUATION

The error covariance equation

$$\begin{aligned} \dot{\mathbf{P}} &= \mathbf{FP} + \mathbf{PF}^T - \mathbf{PH}^T\mathbf{R}^{-1}\mathbf{HP} + \mathbf{GQG}^T \\ \mathbf{P}(0) &= \mathbf{P}_0 \end{aligned} \tag{B.2.1}$$

is a special form of nonlinear differential equation known as the matrix Riccati equation. This equation has been studied extensively, and an analytical solution exists for the constant-parameter case. The general procedure is to transform the single nonlinear equation into a system of two simultaneous linear equations; of course, analytical solutions exist for linear differential equations with constant coefficients. Toward this end we assume that $\mathbf{P}$ can be written in product form as

$$\mathbf{P} = \mathbf{XZ}^{-1}, \quad \mathbf{Z}(0) = \mathbf{I} \tag{B.2.2}$$

or

$$\mathbf{PZ} = \mathbf{X} \tag{B.2.3}$$

Differentiating both sides of Eq. (B.2.3) leads to

$$\dot{\mathbf{P}}\mathbf{Z} + \mathbf{P}\dot{\mathbf{Z}} = \dot{\mathbf{X}} \tag{B.2.4}$$

Next, we substitute $\dot{\mathbf{P}}$ from Eq. (B.2.1) into Eq. (B.2.4) and obtain

$$(\mathbf{PF} + \mathbf{PF}^T - \mathbf{PH}^T\mathbf{R}^{-1}\mathbf{HP} + \mathbf{GQG}^T)\mathbf{Z} + \mathbf{P}\dot{\mathbf{Z}} = \dot{\mathbf{X}} \tag{B.2.5}$$

Rearranging terms and noting that $\mathbf{PZ} = \mathbf{X}$ lead to

$$\mathbf{P}(\mathbf{F}^T\mathbf{Z} - \mathbf{H}^T\mathbf{R}^{-1}\mathbf{HX} + \dot{\mathbf{Z}}) + (\mathbf{FX} + \mathbf{GQG}^T\mathbf{Z} - \dot{\mathbf{X}}) = \mathbf{0} \tag{B.2.6}$$

Note that if both terms in parentheses in Eq. (B.2.6) are set equal to zero, equality is satisfied. Thus, we have the pair of linear differential equations

$$\dot{\mathbf{X}} = \mathbf{FX} + \mathbf{GQG}^T\mathbf{Z} \tag{B.2.7}$$

$$\dot{\mathbf{Z}} = \mathbf{H}^T\mathbf{R}^{-1}\mathbf{HX} - \mathbf{F}^T\mathbf{Z} \tag{B.2.8}$$

with initial conditions

$$\begin{aligned} \mathbf{X}(0) &= \mathbf{P}_0 \\ \mathbf{Z}(0) &= \mathbf{I} \end{aligned} \tag{B.2.9}$$

These can now be solved by a variety of methods, including Laplace transforms. Once $\mathbf{P}$ is found, the gain $\mathbf{K}$ is obtained as $\mathbf{PH}^T\mathbf{R}^{-1}$, and the filter parameters are determined. An example illustrates the procedure.

EXAMPLE B.1

Consider a continuous filter problem where the signal and noise are independent and their autocorrelation functions are

$$R_s(\tau) = e^{-|\tau|} \quad \left[\text{or } S_s(s) = \frac{2}{-s^2 + 1}\right] \tag{B.2.10}$$

$$R_n(\tau) = \delta(\tau) \quad (\text{or } S_n = 1) \tag{B.2.11}$$

Since this is a one-state system, x is a scalar. Let x equal the signal. The additive measurement noise is white and thus no augmentation of the state vector is required. The process and measurement models are then

$$\dot{x} = -x + \sqrt{2}u, \quad u = \text{unity white noise} \tag{B.2.12}$$

$$z = x + v \qquad v = \text{unity white noise} \tag{B.2.13}$$

Thus, the system parameters are

$$F = -1, \quad G = \sqrt{2}, \quad Q = 1, \quad R = 1 \quad H = 1$$

The differential equations for X and Z are then

$$\begin{aligned}\dot{X} &= -X + 2Z, \quad X(0) = P_0 \\ \dot{Z} &= X + Z, \quad Z(0) = 1\end{aligned} \tag{B.2.14}$$

Equations (B.2.14) may be solved readily using Laplace-transform techniques. The result is

$$\begin{aligned}X(t) &= P_0 \cosh\sqrt{3}\,t + \frac{(2-P_0)}{\sqrt{3}}\sinh\sqrt{3}\,t \\ Z(t) &= \cosh\sqrt{3}\,t + \frac{(P_0+1)}{\sqrt{3}}\sinh\sqrt{3}\,t\end{aligned} \tag{B.2.15}$$

The solution for P may now be formed as $P = XZ^{-1}$:

$$P = \frac{P_0 \cosh\sqrt{3}\,t + \dfrac{(2-P_0)}{\sqrt{3}}\sinh\sqrt{3}\,t}{\cosh\sqrt{3}\,t + \dfrac{(P_0+1)}{\sqrt{3}}\sinh\sqrt{3}\,t} \tag{B.2.16}$$

Once P is found, the gain K is given by

$$K = PH^T R^{-1}$$

and the filter yielding $\hat{x}$ is determined.

It is of special interest to look at the stationary (i.e., steady-state) filter solution. This is obtained by letting t be large in Eq. (B.2.16). The expression for P then reduces to (noting that $P_0 = 1$)

$$P \to \frac{1 \cdot e^{\sqrt{3}t} + \dfrac{2-1}{\sqrt{3}}e^{\sqrt{3}t}}{e^{\sqrt{3}t} + \dfrac{1+1}{\sqrt{3}}e^{\sqrt{3}t}} = \sqrt{3} - 1 \tag{B.2.17}$$

The Kalman filter block diagram for this example is then as shown in Fig. B.2. This can be systematically reduced to yield the following overall transfer function

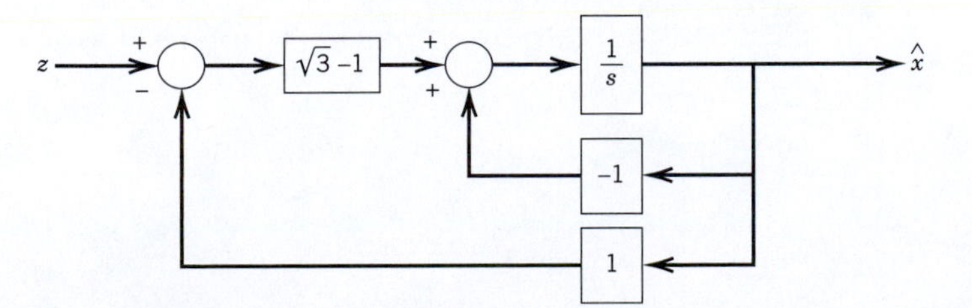

Figure B.2 Stationary Kalman filter for Example B.1.

relating $\hat{x}$ to z:

$$G(s) = \frac{\text{Laplace transform of } \hat{x}}{\text{Laplace transform of } z} = \frac{\sqrt{3} - 1}{s + \sqrt{3}} \quad \text{(B.2.18)}$$

This is the same result obtained using Wiener methods [See Ref. (3)].

REFERENCES CITED IN APPENDIX B

1. R. E. Kalman and R. S. Bucy, "New Results in Linear Filtering and Prediction," *Trans. ASME, J. Basic Engr.*, 83: 95–108 (1961).
2. A. H. Jazwinski, *Stochastic Processes and Filtering Theory*, New York: Academic Press, 1970.
3. R. G. Brown and P. Y. C. Hwang, *Introduction to Random Signals and Applied Kalman Filtering*, 3rd edition. New York: Wiley, 1997.

Index